VOLUME	EDITOR-IN-CHIEF	PAGES	
33	CHARLES C. PRICE	115	*Out of print*
34	WILLIAM S. JOHNSON	121	*Out of print*
35	T. L. CAIRNS	122	*Out of print*
36	N. J. LEONARD	120	
37	JAMES CASON	109	*Out of print*
38	JOHN C. SHEEHAN	120	
39	MAX TISHLER	114	

Collective Vol. IV   A revised edition of Annual Volumes 30–39
        NORMAN RABJOHN, *Editor-in-Chief*   1036

40	MELVIN S. NEWMAN	114	
41	JOHN D. ROBERTS	118	*Out of print*
42	VIRGIL BOEKELHEIDE	118	*Out of print*
43	B. C. McKUSICK	124	*Out of print*
44	WILLIAM E. PARHAM	131	
45	WILLIAM G. DAUBEN	118	
46	E. J. COREY	146	
47	WILLIAM D. EMMONS	140	
48	PETER YATES	164	
49	KENNETH B. WIBERG	124	*Out of print*

Collective Vol. V   A revised edition of Annual Volumes 40–49
        HENRY E. BAUMGARTEN, *Editor-in-Chief*   1234

50	RONALD BRESLOW	136
51	RICHARD E. BENSON	209
52	HERBERT O. HOUSE	192
53	ARNOLD BROSSI	193
54	ROBERT E. IRELAND	155
55	SATORU MASAMUNE	134

Cumulative Indices to Collective Volumes I, II, III, IV, V

RALPH L. AND RACHEL H. SHRINER, *Editors*

ORGANIC SYNTHESES
CUMULATIVE INDICES

ORGANIC SYNTHESES

COLLECTIVE VOLUMES I, II, III, IV, V

CUMULATIVE INDICES

Edited by

RALPH L. SHRINER AND RACHEL H. SHRINER

ADVISORY BOARD

Former Editors of Volumes of Organic Syntheses

1976

JOHN WILEY AND SONS

NEW YORK · LONDON · SYDNEY · TORONTO

Library of Congress Catalog Card Number: 21-17747

ISBN 0-471-78885-6

Printed in the United States of America

10 9 8 7 6 5 4 3 2

PREFACE

This volume constitutes a single reference source to the classified indices in all five *Collective Volumes* of *Organic Syntheses*. Thus, users of *Organic Syntheses* may now consult this one set of *Cumulative Indices* instead of those in five separate volumes. In addition to the time-saving factor, it is hoped that the usefulness of the *Collective Volumes* will be enhanced.

Annual Volumes of *Organic Syntheses* have been published since 1921. As these volumes were published and distributed, chemists from all parts of the world reported improvements and helpful modifications and precautions to the secretaries of the Editorial Boards. These comments were carefully collected and the editor of each Collective Volume incorporated them into the *Collective Volumes* at ten-year intervals. Additional references to methods of preparation were also inserted. Thus the *Collective Volumes* are more than mere reprints of the annual volumes. The publication of the *Collective Volumes* extended over a 41-year span:

Collective Volume	Editor-in-Chief	Revision of Annual Volumes	Pages
I (1932)	Henry Gilman	1–9	564
I (1941, Revised)	Henry Gilman and A. Harold Blatt	1–9	580
II (1943)	A. Harold Blatt	10–19	654
III (1955)	Evan C. Horning	20–29	890
IV (1963)	Norman Rabjohn	30–39	1036
V (1973)	Henry E. Baumgarten	40–49	1234

The preface to each collective volume, written by the Editor-in-Chief of that volume, contains valuable information and should be consulted. Naturally over the years there have been changes in editorial policy, selection of compounds for checking, illustrations of useful reactions, purity of products, and changes in nomenclature. In the early *Annual Volumes,* common names were used for the products of the syntheses. In later volumes, some titles were based on the international systems of nomenclature, such as, the Geneva System (1892), International Union of Chemistry (IUC) rules (1940), and names adopted by the International Union of Pure and Applied Chemistry (1949). After about 1940 some titles of the preparations were those in use in indices of *Chemical Abstracts* and these names were also added beneath the common title name. The *Collective Volumes* retain the titles from the *Annual Volumes* plus the CA index name, but no uniform system of nomenclature is used in all five volumes.

To help solve nomenclature problems, to facilitate locating compounds in the *Organic Syntheses Collective Volumes,* and to assist in later literature surveys, Index No. 1, Cumulative Contents Index, has an alphabetical index of the preparations according to the main title names followed by the latest Chemical Abstracts Index name. Index No. 2 has the compounds alphabetized by the CA Index Name followed by the common name. The CA Registry Numbers are given in both indices. All citations are to *Collective Volume* numbers and pages in that *Collective Volume.* (See pages 1 and 87.)

The sequence of the classified indices in this volume corresponds to frequency of usage. Thus, specific compounds (Indices Nos. 1 and 2) are the most sought after, followed by Type of Reaction (No. 3), and Type of Compound (No. 4). Users of the Cumulative Index should note that each of the indices has an introductory paragraph describing the material in that particular index. The coordination and consolidation of the indices has posed some problems. Most of the preparations in the first twenty Annual Volumes were single-step procedures to produce a specific compound. Later, multistep syntheses were selected which involved one, two, or more intermediates. Beginning in 1961 with *Annual Volume* 41, sections on the merits of the preparation were introduced. Also, emphasis was placed on unique model procedures illustrating important types of reactions rather than a specific compound. The accompanying discussion sections are difficult to index, but by consulting the Name Index, the Reaction Type Index or the Type of Compound Index, information may be found concerning the scope of the reactions. Duplicate entries have been consolidated and minor reagents used in the work-up of the products are not indexed. The purification of solvents and special reagents, not in the original *Collective Volumes* I, and II have been indexed, combined with III, IV and V, and inserted in Index No. 6. Author Indices for *Collective Volumes* I, II, and III have been compiled and integrated with those of *Collective Volumes* IV and V to give a complete Index (No. 8).

Especial attention of all users of *Organic Syntheses* is called to the numerous precautions, warnings, and hazards cited in the procedures and notes. Special warning notices have been added in each volume (annual and collective) on the basis of information supplied by chemists using the procedures. *Organic Syntheses* is written for competently trained chemists, but the occasional appearance of hazard notices suggests that all procedures and chemicals should be treated with due caution and carried out with all modern safety practices. Neither the Editorial Board nor the publishers assume any liability with respect to the use of the preparations. The procedures have been carried out by the submitting authors many times and checked by a senior editor and his co-workers in order to assure reproducibility.

For information concerning hazards of individual compounds chemists should consult, *The Toxic Substances List 1974 Edition,* published by the National Institute for Occupational Safety and Health (USPHS), Rockville, Md. 20852. The first part of this volume has explanations for the selection of toxicity basis and reprints of *Rules and Regulations* from the Federal Register (1972–1974). Then follow 817 pages with 13,000 names of inorganic and organic chemicals and 29,000 synonyms and codes. Arrangement is alphabetical with TSL Compound Number. Toxicity data for various species of animals, routes of administration and information concerning possible hazardous compounds are given. The *"Toxic Substance Prime Name"* is the Chemical Abstracts Index Name with the Chemical Abstracts Service (CAS) Registry Numbers (See p. 2) and the Wiswesser Line Notation (WLN); In the book, a short but comprehensive description of WLN is given and an alphanumeric list of WLN names with citation of 6,000 compound numbers in TSL. This volume is updated annually.

Since the editing of a *Collective Volume* takes 3 or 4 years of the "spare time" of the editor-in-chief, there is a time lag in publication. To aid users, Appendix A lists the alphabetized contents indices of the most recent *Annual Volumes,* 50 through 54 (1970–1974).

Appendix B outlines the development and operations of Organic Syntheses, Inc. Special attention is called to the article in Annual Volume 50, "Fifty Years of Organic Syntheses" by Roger Adams (1889–1971), one of the founders of *Organic Syntheses* and its leader during the years 1920–1971.

The editors of the current *Annual Volumes* of *Organic Syntheses* invite the submission of interesting syntheses of compounds of research utility and also procedures illustrating new unique reactions. Suggestions for time-saving modifications and real improvements in published procedures are desired for inclusion in the next collective volume. All correspondence should be sent to the current secretary of the Board of Editors, Wayland E. Noland, School of Chemistry, University of Minnesota, Minneapolis, Minn. 55455.

Each recent *Annual Volume* has information concerning the "Submission of Preparations", and the secretary will provide a style guide for preparing the written procedures. Authors should consult these *Cumulative Indices* to avoid duplicating preparations of compounds or types of reactions already published.

This volume is to recognize and express appreciation to the organic chemists who have edited the *Collective Volumes of Organic Syntheses* and served as secretaries to the Board of Editors. Their altruistic labors have benefitted thousands of chemists.

Henry Gilman, of Iowa State University, edited *Annual Volume* 6 in 1926 and *Collective Volume* I in 1932.

Charles F. H. Allen, of Eastman Kodak Co., served as the first Secretary to the Board of Editors, 1927–1937, and edited *Annual Volume* 20 in 1940.

A. Harold Blatt, of Queens College was the second Secretary to the Board of Editors, 1938–1943. He collaborated with Henry Gilman to edit a revised edition of *Collective Volume I* in 1941 and then edited *Collective Volume* II in 1943.

Evan C. Horning, now at Baylor College of Medicine, was Secretary to the Editorial Board, 1944–1949, and edited *Collective Volume* III in 1953.

Norman Rabjohn, at the University of Missouri, was Secretary to the Editorial Board, 1950–1958, and edited *Collective Volume* IV in 1963.

Henry Baumgarten, at the University of Nebraska, was Secretary to the Editorial Board, 1959–1968, and edited *Collective Volume* V in 1973.

All these received great help from the editors of *Annual Volumes* I–49; their names are listed on the inside of the front cover and they are members of the Advisory Board shown on the title page.

RALPH L. SHRINER
RACHEL H. SHRINER

Dallas, Texas
June 1975

Henry Gilman

Charles F. H. Allen

A. Harold Blatt

Evan C. Horning

Norman Rabjohn

Henry Baumgarten

CONTENTS

INDEX 1 CUMULATIVE CONTENTS INDEX 1
(*According to Title Names of Preparations*)

INDEX 2 CUMULATIVE CONTENTS INDEX 87
(*According to Chemical Abstracts Index Names*)

INDEX 3 TYPE OF REACTION INDEX 169

INDEX 4 TYPE OF COMPOUND INDEX 221

INDEX 5 FORMULA INDEX 273

INDEX 6 SOLVENTS AND REAGENTS INDEX 307

INDEX 7 APPARATUS INDEX 313

INDEX 8 CUMULATIVE AUTHOR INDEX 317

INDEX 9 CUMULATIVE GENERAL INDEX 339

INDEX 10 APPENDIX A. CONTENTS INDEX OF ANNUAL VOL-
UMES, 50–54 419

INDEX 11 APPENDIX B. ORGANIC SYNTHESES 423
(*Origin, Development, Organization, Operations*)

ORGANIC SYNTHESES
CUMULATIVE INDICES

CUMULATIVE CONTENTS INDEX

According to Title Names of Preparations

In the front of each *Collective Volume of Organic Syntheses* are contents pages that list the names of the compounds whose syntheses are given. Each name is printed in boldfaced capital letters as the title heading for each preparation in that volume. Intermediates and synonyms are not listed here; they may be found in the General Index. The names were those that were in common use at the time of checking and editing. Beneath the title name of each preparation appears the Chemical Abstracts (CA) Index Name in parentheses, when it differs from the main name. These entries are the systematic CA Index names preferred in the year of publication of that specific volume. Also, these Chemical Abstracts names were not incorporated in the general indices of *Collective Volumes* I, II, III, IV, but were included in that of *Collective Volume* V.

The tremendous expansion of organic chemistry, both in numbers of compounds and complexity of structure, plus the inauguration of the computer retrieval system by Chemical Abstracts Service (CAS) in the 1960s, necessitated changes in the systematic Chemical Abstracts Index Names. To expedite the search for compounds, via either common names or Chemical Abstracts Index Names, this Cumulative Contents Index No. 1 shows alphabetically the common title name in capital letters with the *Collective Volume* number and page. Beneath is the Chemical Abstracts Index Name for the Ninth Collective Index Period (1972–1976), plus the Chemical Abstracts Service Registry Number. The next Contents Index (No. 2) is arranged alphabetically by the Chemical Abstracts Index Name plus Registry Number, followed by the common title name and *Collective Volume* number and page (See p. oo).

The **CAS Registry Number** is a computer-assigned number which uniquely identifies each chemical compound. Within the CAS Registry

1

System, the Registry Number links the current CA Index Name, the Molecular Formula, and synonyms under which each compound has appeared in the original literature. To use the CAS chemical information tools most effectively, research chemists, searching for literature references other than those cited in *Organic Syntheses,* should consult the introductory pages to *Chemical Abstracts* and the CA Volume Indices. CA Index Name selection policies are described in detail in the CA Volume 76 Index Guide. Information regarding the CAS Registry System may be found in the CAS Registry Handbook-Number Section. It should be noted that these CA Index names frequently differ from IUPAC nomenclature. Also, the CAS Registry Numbers are entirely different from Beilstein's "System Nummer".

The *Toxic Substance List 1974 Edition,* mentioned in the preface, also lists compounds by their Chemical Abstracts Index Name and CAS Registry Numbers. However, the latter have nine digits. The last numbers are identical to those given here in Indices 1 and 2 but sufficient zeros are inserted to make nine digits. For example:-

Methane, diazo- [334-88-3]

is listed in TSL as CAS 000334883

Acetamide, 2-chloro- [79-07-2]

is listed in TSL as CAS 000079072

These Chemical Abstracts Index Names and CAS Registry Numbers were supplied by Chemical Abstracts Services. The editors wish to thank the CAS staff for their generosity and cooperation. Hopefully, these multiple listings will help to solve the nomenclature problem for the previous *Collective Volumes* and also provide a sound basis for future Collective Volume Indices.

A

ABIETIC ACID, IV, 1;
   1-Phenanthrenecarboxylic acid, 1,2,3,4,4a,4b,5,6,10,10a-decahydro-1,4a–dimethyl-7-(1-methylethyl)-, [1*R*-(1α,4aβ,4bα,10aα)]-[514-10-3]

ACENAPHTHENEQUINONE, III, 1;
   1,2-Acenaphthalenedione [82-86-0]

ACENAPHTHENOL-7, III, 3;
   1-Acenaphthylenol,1,2-dihydro-[6306-07-6]

β-(3-ACENAPHTHOYL)PROPIONIC ACID, III, 6;
   5-Acenaphthylenebutanoic acid, 1,2-dihydro-γ-oxo- [16294-60-3]

ACETAL, I, 1;
  Ethane, 1,1-diethoxy [105-57-7]

ACETAMIDE, I, 3;
  Acetamide [60-35-5]

ACETAMIDINE HYDROCHLORIDE, I, 5;
  Ethanimidamide, monohydrochloride [124-42-5]

3-ACETAMIDO-2-BUTANONE, IV, 5;
  Acetamide,N-(1-methyl-2-oxopropyl)- [6628-81-5]

2-ACETAMIDO-3,4,6-TRI-O-ACETYL-2-DEOXY-α-D-GLUCOPYRANOSYL CHLORIDE,
    V, 1;
  α-D-Glucopyranosyl chloride, 2-(acetylamino)-2-,deoxy-,3,4,6-triacetate [3068-34-6]

p-ACETAMINOBENZENESULFINIC ACID, I, 7;
  Benzenesulfinic acid, 4-(acetylamino)- [710-24-7]

p-ACETAMINOBENZENESULFONYL CHLORIDE, I, 8;
  Benzenesulfonyl chloride, 4-(acetylamino)- [121-60-8]

α-ACETAMINOCINNAMIC ACID, II, 1;
  2-Propenoic acid, 2-(acetylamino)-3-phenyl- [5469-45-4]

ACETOACETANILIDE, III, 10;
  Butanamide, 3-oxo-N-phenyl- [102-01-2]
ACETOBROMOGLUCOSE, III, 11;
  α-D-Glucopyranosyl bromide, tetraacetate [572-09-8]

ACETO p-CYMENE, II, 3;
  Ethanone, 1-[2-methyl-5-(1-methylethyl)phenyl]- [1202-08-0]

ACETOL, II, 5;
  2-Propanone, 1-hydroxy- [116-09-6]

ACETONE CYANOHYDRIN, II, 7;
  Propanenitrile, 2-hydroxy-2-methyl- [75-86-5]

ACETONE DIBUTYL ACETAL, V, 5;
  Butane, 1,1'-[(1-methylethylidene)bis(oxy)]bis- [141-72-0]

ACETONEDICARBOXYLIC ACID, I, 10;
  Pentanedioic acid, 3-oxo- [542-05-2]

2-ACETOTHIENONE, II, 8; III, 14;
  Ethanone,1-(2-thienyl)- [88-15-3]

ACETOXIME, I, 318;
  2-Propenone, oxime [127-06-0]

3β-ACETOXYETIENIC ACID, V, 8;
  Androst-5-ene-17-carboxylic acid, 3-(acetyloxy)-,(3β,17β)- [7150-18-7]

ACETYLACETONE, III, 16;
  2,4-Pentanedione [123-54-6]

N-(*p*-ACETYLAMINOPHENYL)RHODANINE, IV, 6;
Acetamide,*N*-[4-(4-oxo-2-thioxo-3-thiazolidinyl)phenyl]- [53663-36-8]

9-ACETYLANTHRACENE, IV, 8;
Ethanone, 1-(9-anthracenyl)- [784-04-3]

ACETYLBENZOYL, III, 20;
1,2-Propanedione, 1-phenyl- [579-07-7]

α-ACETYL-δ-CHLORO-γ-VALEROLACTONE, IV, 10;
2(3*H*)-Furanone, 3-acetyl-5-(chloromethyl)dihydro- [3154-75-4]

1-ACETYLCYCLOHEXANOL, IV, 13;
Ethanone, 1-(1-hydroxycyclohexyl)- [1123-27-9]

1-ACETYLCYCLOHEXENE, III, 22;
Ethanone, 1-(1-cyclohexen-1-yl)- [932-66-1]

ACETYLENEDICARBOXYLIC ACID, II, 10;
2-Butynedioic acid [142-45-0]

2-ACETYLFLUORENE, III, 23;
Ethanone, 1-(9*H*-fluoren-2-yl)- [781-73-7]

ACETYLGLYCINE, II, 11;
Glycine, *N*-acetyl- [543-24-8]

ACETYLMANDELIC ACID, I, 12;
Benzeneacetic acid, α-(acetyloxy)- [5438-68-6]

ACETYLMANDELYL CHLORIDE, I, 12;
Benzeneacetyl chloride, α-(acetyloxy)- [1638-63-7]

3-ACETYLOXINDOLE, V, 12;
2*H*-Indol-2-one, 3-acetyl-1,3-dihydro- [17266-70-5]

9-ACETYLPHENANTHRENE, III, 26;
Ethanone, 1-(9-phenanthrenyl)- [2039-77-2]

2-*p*-ACETYLPHENYLHYDROQUINONE, IV, 15;
Ethanone,1-(2′,5′-dihydroxy[1,1′-biphenyl]-4-yl)- [3948-13-8]

δ-ACETYL-*n*-VALERIC ACID, IV, 19;
Heptanoic acid, 6-oxo [3128-07-2]

ACID AMMONIUM *o*-SULFOBENZOATE, I, 14;
Benzoic acid, 2-sulfo-, monoammonium salt [6939-89-5]

ACID ANHYDRIDES, III, 28

ACONITIC ACID, II, 12;
1-Propene-1,2,3-tricarboxylic acid [499-12-7]

ACRIDONE, II, 15;
9(10*H*)-Acridinone [578-95-0]

ACROLEIN, I, 15;
2-Propenal [107-02-8]

ACROLEIN ACETAL, II, 17; IV, 21;
1-Propene,3,3-diethoxy- [3054-95-3]

ACRYLIC ACID, III, 30;
2-Propenoic acid [79-10-7]

ADAMANTANE, V, 16;
Tricyclo[3.3.1.1^{3,7}]decane [281-23-2]

1-ADAMANTANECARBOXYLIC ACID, V, 20;
Tricyclo[3.3.1.1^{3,7}]decane-1-carboxylic acid [828-51-3]

ADIPIC ACID, I, 18;
Hexanedioic acid [124-04-9]

β-ALANINE, II, 19; III, 34;
β-Alanine [107-95-9]

dl-ALANINE, I, 21;
DL-Alanine [302-72-7]

ALKYL AND ALKYLENE BROMIDES, I, 25;

ALLANTOIN, II, 21;
Urea,(2,5-dioxo-4-imidazolidinyl)- [97-59-6]

ALLENE, V, 22;
1,2-Propadiene [463-49-0]

ALLOXAN MONOHYDRATE, III, 37; IV, 23;
2,4,6(1H,3H,5H)-Pyrimidinetrione,5,5-dihydroxy- [3237-50-1]

ALLOXANTIN DIHYDRATE, III, 42; IV, 25;
[5,5'-Bipyrimidine]-2,2',4,4',6,6'(1H,1'H,3H,3'H,5H,5'H)-hexone, dihydrate
[6011-27-4]

ALLYL ALCOHOL, I, 42;
2-Propen-1-ol [107-18-6]

ALLYLAMINE, II, 24;
2-Propen-1-amine [107-11-9]

ALLYL BROMIDE, I, 27;
1-Propene, 3-bromo- [106-95-6]

ALLYL CYANIDE, I, 46;
3-Butenenitrile [109-75-1]

2-ALLYLCYCLOHEXANONE, III, 44; V, 25;
Cyclohexanone, 2-(2-propenyl)- [94-66-6]

ALLYL LACTATE, III, 46;
Propanoic acid, 2-hydroxy-,2-propenyl ester [5349-55-3]

ALUMINUM tert-BUTOXIDE, III, 48;
2-Propanol, 2-methyl, aluminum salt [556-91-2]

AMINOACETAL, III, 50;
 Ethanamine, 2,2-diethoxy- [645-36-3]

AMINOACETONE SEMICARBAZONE HYDROCHLORIDE, V, 27;
 Hydrazinecarboxamide,2-(2-amino-1-methylethylidene)-,monohydrochloride [10469-70-2]

9-AMINOACRIDINE, III, 53;
 9-Acridinamine [90-45-9]

2-AMINO-4-ANILINO-6-(CHLOROMETHYL)-s-TRIAZINE, IV, 29;
 1,3,5-Triazine-2,4-diamine,6-(chloromethyl-N-phenyl [30355-60-3]

o-AMINOBENZALDEHYDE, III, 56;
 Benzaldehyde, 2-amino [529-23-7]

p-AMINOBENZALDEHYDE, IV, 31;
 Benzaldehyde, 4-amino- [556-18-3]

m-AMINOBENZALDEHYDE DIMETHYLACETAL, III, 59;
 Benzenamine,3-(dimethoxymethyl)- [53663-37-9]

2-AMINOBENZOPHENONE, IV, 34;
 Methanone, (2-aminophenyl)phenyl- [2835-77-0]

o-AMINOBENZYL ALCOHOL, III, 60;
 Benzenemethanol, 2-amino- [5344-90-1]

γ-AMINOBUTYRIC ACID, II, 25;
 Butanoic acid, 4-amino- [56-12-2]

α-AMINO-n-CAPROIC ACID, I, 48;
 Norleucine [5157-09-5]

ε-AMINOCAPROIC ACID, II, 28; IV, 39;
 Hexanoic acid, 6-amino- [60-32-2]

2-AMINO-p-CYMENE, III, 63;
 Benzenamine,2-methyl-5-(methylethyl)- [2051-53-8]

α-AMINODIETHYLACETIC ACID, III, 66;
 Butanoic acid, 2-amino-2-ethyl- [2566-29-2]

5-AMINO-2,3-DIHYDRO-1,4-PHTHALAZINEDIONE, III, 69;
 1,4-Phthalazinedione, 5-amino-2,3-dihydro- [521-31-3]

4-AMINO-2,6-DIMETHYLPYRIMIDINE, III, 71;
 4-Pyrimidinamine, 2,6-dimethyl- [461-98-3]

2-AMINOFLUORENE, V, 30;
 9H-Fluoren-2-amine [153-78-6]

AMINOGUANIDINE BICARBONATE, III, 73;
 Carbonic acid, comp'd, with hydrazinecarboximidamide(1:1) [2582-30-1]

α-AMINOISOBUTYRIC ACID, II, 29;
 Alanine, 2-methyl- [62-57-7]

AMINOMALONONITRILE *p*-TOLUENESULFONATE, V, 32;
Propanedinitrile,amino-,mono(4-methylbenzenesulfonate) [5098-14-6]

2-AMINO-6-METHYLBENZOTHIAZOLE, III, 76;
2-Benzothiazolamine, 6-methyl- [2536-91-6]

1-AMINO-1-METHYLCYCLOHEXANE, V, 35;
Cyclohexanamine,1-methyl- [6526-78-9]

2-AMINO-4-METHYLTHIAZOLE, II, 31;
2-Thiazolamine, 4-methyl- [1603-91-4]

3-AMINO-2-NAPHTHOIC ACID, III, 78;
2-Naphthalenecarboxylic acid, 3-amino- [5959-52-4]

1,2-AMINONAPHTHOL HYDROCHLORIDE, II, 33;
2-Naphthalenol,1-amino-, hydrochloride [1198-27-2]

1,4-AMINONAPHTHOL HYDROCHLORIDE, I, 49; II, 39;
1-Naphthalenol,4-amino-, hydrochloride [5959-56-8]

1-AMINO-2-NAPHTHOL-4-SULFONIC ACID, II, 42;
1-Naphthalenesulfonic acid, 4-amino-3-hydroxy- [116-63-2]

2-AMINO-4-NITROPHENOL, III, 82;
Phenol, 2-amino-4-nitro- [99-57-0]

2-AMINO-3-NITROTOLUENE, IV, 42;
Benzenamine,2-methyl-6-nitro- [570-24-1]

dl-α-AMINOPHENYLACETIC ACID, III, 84;
Benzeneacetic acid, α-amino-(±)- [2835-06-5]

p-AMINOPHENYLACETIC ACID, I, 52;
Benzeneacetic acid, 4-amino- [1197-55-3]

p-AMINOPHENYL DISULFIDE, III, 86;
Benzenamine, 4,4′-dithiobis- [722-27-0]

DL-α-AMINO-α-PHENYLPROPIONIC ACID, III, 88;
Benzeneacetic acid, α-amino-α-methyl-(±) [6945-32-0]

DL-β-AMINO-β-PHENYLPROPIONIC ACID, III, 91;
Benzenepropanoic acid, β-amino-, (±)- [3646-50-2]

β-AMINOPROPIONITRILE, III, 93;
Propanenitrile, 3-amino- [151-18-8]

3(5)-AMINOPYRAZOLE, V, 39;
1*H*-Pyrazol-3-amine [1820-80-0]

3-AMINOPYRIDINE, IV, 45;
3-Pyridinamine [462-08-8]

1-AMINOPYRIDINIUM IODIDE, V, 43;
Pyridinium,1-amino-, iodide [6295-87-0]

p-AMINOTETRAPHENYLMETHANE, IV, 47;
  Benzenamine, 4-(triphenylmethyl)- [22948-06-7]

3-AMINO-III-1,2,4-TRIAZOLE, III, 95;
  1*H*-1,2,4-Triazol-3-amine [61-82-5]

4-AMINO-4H-1,2,4-TRIAZOLE, III, 96;
  4*H*-1,2,4-Triazol-4-amine [584-13-4]

4-AMINOVERATROLE, II, 44;
  Benenamine, 3,4-dimethoxy- [6315-89-5]

AMMONIUM SALT OF AURIN TRICARBOXYLIC ACID, I, 54;
  Benzoic acid, 5-[(3-carboxy-4-hydroxyphenyl)(3-carboxy-4-oxo-2,5-cyclohexadien-1-yli-
    dene)methyl]-2-hydroxy-,triammonium salt [569-58-4]

n-AMYLBENZENE, II, 47;
  Benzene, pentyl [538-68-1]

iso-AMYL BROMIDE, I, 27;
  Butane, 1-bromo-3-methyl- [107-82-4]

ANHYDRO-2-HYDROXIMERCURI-3-NITROBENZOIC ACID, I, 56;
  3*H*-2,1-Benzoxamercurol-3-one, 7-nitro- [53663-14-2]

o-ANISALDEHYDE, V, 46;
  Benzaldehyde, 2-methoxy- [135-02-4]

ANISOLE, I, 58;
  Benzene, methoxy- [100-66-3]

9-ANTHRALDEHYDE, III, 98;
  9-Anthracenecarboxaldehyde [642-31-9]

ANTHRONE, I, 60;
  9(10*H*)-Anthracenone [90-44-8]

APPARATUS FOR CATALYTIC REDUCTION, I, 61;

D-ARABINOSE, III, 101;
  D-Arabinose [10323-20-3]

l-ARABINOSE, I, 67;
  L-Arabinose [5328-37-0]

d-ARGININE HYDROCHLORIDE, II, 49;
  L-Arginine, monohydrochloride [1119-34-2]

AROMATIC ALDEHYDES, MESITALDEHYDE, V, 49;
  Benzaldehyde, 2,4,6-trimethyl- [487-68-3]

ARSANILIC ACID, I, 70;
  Arsonic acid, (4-aminophenyl)- [98-50-0]

ARSENOACETIC ACID, I, 73;
  Acetic acid, 2,2′-diarsenediyl)bis- [544-27-4]

ARSONOACETIC ACID, I, 73;
Acetic acid, arsono- [107-38-0]

p-ARSONOPHENOXYACETIC ACID, I, 75;
Acetic acid, (4-arsonophenoxy)- [53663-15-3]

ARYLBENZENES:3,4-DICHLOROBIPHENYL, V, 51;
1,1'-Biphenyl, 3,4-dichloro [2974-92-7]

ARYLUREAS, IV, 49;
p-BROMOPHENYLUREA, IV, 49;
Urea, (4-bromophenyl)- [1967-25-5]

p-ETHOXYPHENYLUREA, IV, 52;
Urea, (4-ethoxyphenyl)- [150-69-6]

DL-ASPARTIC ACID, IV, 55;
DL-Aspartic acid [617-45-8]

ATROLACTIC ACID, IV, 58;
Benzeneacetic acid, α-hydroxy-α-methyl- [515-30-0]

AZELAIC ACID, II, 53;
Nonanedioic acid [123-99-9]

AZELANITRILE, IV, 62;
Nonanedinitrile [1675-69-0]

AZLACTONE OF α-BENZOYLAMINO-β-(3,4-DIMETHOXYPHENYL)-ACRYLIC ACID,
II, 55;
5(4H)-Oxazolone, 4-[(3,4-dimethoxyphenyl)methylene]-2-phenyl- [5415-56-5]

AZOBENZENE, III, 103;
Diazene, diphenyl- [103-33-3]

1,1'-AZO-bis-1-CYCLOHEXANENITRILE, IV, 66;
Cyclohexanecarbonitrile, 1,1'-azobis- [2094-98-6]

AZOXYBENZENE, II, 57;
Diazene, diphenyl-, 1-oxide [495-48-7]

B

BARBITURIC ACID, II, 60;
2,4,6(1H,3H,5H)-Pyrimidinetrione [67-52-7]

BENZALACETONE, I, 77;
3-Buten-2-one, 4-phenyl- [122-57-6]

BENZALACETONE DIBROMIDE, III, 105;
2-Butanone,3,4-dibromo-4-phenyl- [6310-44-7]

BENZALACETOPHENONE, I, 78;
2-Propen-1-one, 1,3-diphenyl- [94-41-7]

BENZALANILINE, I, 80;
Benzenamine, N-(phenylmethylene)- [538-51-2]

BENZALPHTHALIDE, II, 61;
1(3*H*)-Isobenzofuranone, 3-(phenylmethylene)- [575-61-1]

BENZALPINACOLONE, I, 81;
1-Penten-3-one, 4,4-dimethyl-1-phenyl- [538-44-3]

BENZANILIDE, I, 82;
Benzamide, *N*-phenyl- [93-98-1]

BENZANTHRONE, II, 62;
7*H*-Benz[*de*]anthracen-7-one [82-05-3]

BENZENEBORONIC ANHYDRIDE, IV, 68;
Boroxin, triphenyl- [3262-89-3]

BENZENEDIAZONIUM-2-CARBOXYLATE, V, 54;
Benzenediazonium, 2-carboxy-, hydroxide, inner salt [1608-42-0]

BENZENESULFONYL CHLORIDE, I, 84;
Benzenesulfonyl chloride [98-09-9]

BENZHYDRYL β-CHLOROETHYL ETHER, IV, 72;
Benzene, 1,1'-[(2-chloroethoxy)methylene]bis- [32669-06-0]

BENZIL, I, 87;
Ethanedione, diphenyl- [134-81-6]

BENZILIC ACID, I, 89;
Benzeneacetic acid, α-hydroxy-α-phenyl- [76-93-7]

BENZIMIDAZOLE, II, 65;
1*H*-Benzimidazole [51-17-2]

BENZOFURAZAN OXIDE, IV, 74;
Benzofurazan, 1-oxide [480-96-6]

BENZOGUANAMINE, IV, 78;
1,3,5-Triazine-2,4-diamine, 6-phenyl- [91-76-9]

BENZOHYDROL, I, 90;
Benzenemethanol, α-phenyl- [91-01-0]

BENZOHYDROXAMIC ACID, II, 67;
Benzamide, *N*-hydroxy- [495-18-1]

BENZOIC ANHYDRIDE, I, 91;
Benzoic acid, anhydride [93-97-0]

BENZOIN, I, 94;
Ethanone, 2-hydroxy-1,2-diphenyl- [119-53-9]

BENZOIN ACETATE, II, 69;
Ethanone, 2-(acetyloxy)-1,2-diphenyl- [574-06-1]

BENZOPHENONE, I, 95;
Methanone, diphenyl- [119-61-9]

BENZOPHENONE OXIME, II, 70;
Methanone, diphenyl-, oxime [574-66-3]

BENZOPINACOL, II, 71;
1,2-Ethandiol, 1,1,2,2-tetraphenyl- [464-72-2]

β-BENZOPINACOLONE, II, 73;
Ethanone, tetraphenyl- [466-37-5]

1,2,3-BENZOTHIADIAZOLE 1,1-DIOXIDE, V, 60;
1,2,3-Benzothiadiazole, 1,1-dioxide [37150-27-9]

1,2,3-BENZOTRIAZOLE, III, 106;
1H-Benzotriazole [95-14-7]

BENZOYLACETANILIDE, III, 108; IV, 80;
Benzenepropanamide, β-oxo-N-phenyl- [959-66-0]

β-BENZOYLACRYLIC ACID, III,109;
2-Butenoic acid, 4-oxo-4-phenyl- [583-06-2]

ε-BENZOYLAMINO-α-BROMOCAPROIC ACID, II, 74;
Hexanoic acid, 6-(benzoylamino)-2-bromo- [1700-05-6]

ε-BENZOYLAMINOCAPROIC ACID, II, 76;
Hexanoic acid, 6-(benzoylamino)- [956-09-2]

BENZOYLCHOLINE CHLORIDE, IV, 84;
Ethanaminium, 2-(benzoyloxy)-N,N,N-trimethyl-, chloride [2964-09-2]

BENZOYLCHOLINE IODIDE, IV, 84;
Ethanaminium, 2-(benzoyloxy)-N,N,N-trimethyl-, iodide [17518-43-3]

BENZOYL CYANIDE, III, 112;
Benzeneacetonitrile, α-oxo- [613-90-1]

BENZOYL DISULFIDE, III, 116;
Disulfide, dibenzoyl [644-32-6]

BENZOYLENE UREA, II, 79;
2,4(1H,3H)-Quinazolinedione [86-96-4]

BENZOYL FLUORIDE, V, 66;
Benzoyl fluoride [455-32-3]

BENZOYLFORMIC ACID, III, 114;
Benzeneacetic acid, α-oxo- [611-73-4]

3-BENZOYLOXYCYCLOHEXENE, V, 70;
2-Cyclohexen-1-ol, benzoate [3352-93-0]

BENZOYL PIPERIDINE, I, 99;
Piperidine, 1-benzoyl- [776-75-0]

β-BENZOYLPROPIONIC ACID, II, 81;
Benzenebutanoic acid, γ-oxo- [2051-95-8]

3-BENZOYLPYRIDINE, IV, 88;
Methanone, phenyl-3-pyridinyl- [5424-19-1]

BENZYLACETOPHENONE, I, 101;
1-Propanone, 1,3-diphenyl- [1083-30-3]

N-BENZYLACRYLAMIDE, V, 73;
2-Propenamide, N-(phenylmethyl)- [13304-62-6]

2-BENZYLAMINOPYRIDINE, IV, 91;
2-Pyridinamine, N-(phenylmethyl)- [6935-27-9]

BENZYLANILINE, I, 102;
Benzenemethanamine, N-phenyl- [103-32-2]

BENZYL BENZOATE, I, 104;
Benzoic acid, phenylmethyl ester [120-51-4]

BENZYL CHLOROMETHYL KETONE, III, 119;
2-Propanone, 1-chloro-3-phenyl- [937-38-2]

BENZYL CYANIDE, I, 107;
Benzeneacetonitrile [140-29-4]

2-BENZYLCYCLOPENTANONE, V, 76;
Cyclopentanone, 2-(phenylmethyl)- [2867-63-2]

α-BENZYLIDENE-γ-PHENYL-Δ^{β,γ}-BUTENOLIDE, V, 80;
2(3H)-Furanone, 5-phenyl-3-(phenylmethylene)- [4361-96-0]

3-BENZYL-3-METHYLPENTANOIC ACID, IV, 93;
Benzenebutanoic acid, β-ethyl-β-methyl- [53663-16-4]

cis-2-BENZYL-3-PHENYLAZIRIDINE, V, 83;
Aziridine, 2-phenyl-3-(phenylmethyl)-, cis- [1605-08-9]

BENZYL PHTHALIMIDE, II, 83;
1H-Isoindole-1,3(2H)-dione, 2-(phenylmethyl)- [2142-01-0]

1-BENZYLPIPERAZINE, V, 88;
Piperazine, 1-(phenylmethyl)- [2759-28-6]

BENZYLTRIMETHYLAMMONIUM ETHOXIDE, IV, 98;
Benzenemethanaminium, N,N,N-trimethyl-, ethoxide [27292-06-4]

BETAINE HYDRAZIDE HYDROCHLORIDE, II, 85;
Ethanaminium, 2-hydrazino-N,N,N-trimethyl-2-oxo-,chloride [123-46-6]

BIALLYL, III, 121;
1,5-Hexadiene, [592-42-7]

BICYCLO[2.2.1]HEPTEN-7-ONE, V, 91;
Bicyclo[2.2.1]hept-2-en-7-one, [694-71-3]

exo-cis-BICYCLO[3.3.0]OCTANE-2-CARBOXYLIC ACID, V, 93;
1-Pentalenecarboxylic acid, octahydro-,(1α,3aα,6aα)- [18209-43-3]

BICYCLO[2.1.0]PENTANE, V, 96;
  Bicyclo[2.1.0]pentane [185-94-4]

BIPHENYLENE, V, 54;
  Biphenylene, [259-79-0]

2,2'-BIPYRIDINE, V, 102;
  2,2'-Bipyridine, [366-18-7]

1,2-BIS(n-BUTYLTHIO)BENZENE, V, 107;
  Benzene,1,2-bis(butylthio)-, [53663-38-0]

BISCHLOROMETHYL ETHER, IV, 101;
  Methane, oxybis[chloro- [542-88-1]

BIS-(β-CYANOETHYL)AMINE, III, 93;
  Propanenitrile,3,3'-iminobis- [111-94-4]

4,4'BIS(DIMETHYLAMINO)BENZIL, V, 111;
  Ethanedione, bis[4-(dimethylamino)phenyl]- [17078-27-2]

BIS(1,3-DIPHENYLIMIDAZOLIDINYLIDENE-2), V, 115;
  Imidazolidene,2-(1,3-diphenyl-2-imidazolidinylidene)-1,3-diphenyl- [2179-89-7]

BROMAL, II, 87;
  Acetaldehyde, tribromo- [115-17-3]

BROMOACETAL, III, 123;
  Ethane,2-bromo-1,1-diethoxy- [2032-35-1]

N-BROMOACETAMIDE, IV, 104;
  Acetamide,N-bromo- [79-15-2]

BROMOACETONE, II, 88;
  2-Propanone, 1-bromo- [598-31-2]

3-BROMOACETOPHENONE, V, 117;
  Ethanone,1-(3-bromophenyl)- [2142-63-4]

p-BROMOACETOPHENONE, I, 109;
  Ethanone,1-(4-bromophenyl)- [99-90-1]

2-BROMOALLYLAMINE, V,121;
  2-Propen-1-amine, 2-bromo- [6943-51-7]

N-(2-BROMOALLYL)ETHYLAMINE, V, 124;
  2-Propen-1-amine, 2-bromo-N-ethyl- [871-23-8]

3-BROMO-4-AMINOTOLUENE, I, 111;
  Benzenamine, 2-bromo-4-methyl- [583-68-6]

α-BROMOBENZALACETONE, III, 125;
  3-Buten-2-one, 3-bromo-4-phenyl- [31207-17-7]

p-BROMOBENZALDEHYDE, II, 89;
  Benzaldehyde,4-bromo- [1122-91-4]

p-BROMOBIPHENYL, I, 113;
1,1'-Biphenyl,4-bromo- [92-66-0]

α-BROMO-*n*-CAPROIC ACID, I, 115;
Hexanoic acid, 2-bromo- [616-05-7]

BROMOCYCLOPROPANE, V, 126;
Cyclopropane, bromo- [4333-56-6]

p-BROMODIPHENYLMETHANE, V, 130;
Benzene, 1-bromo-4-(phenylmethyl)- [2116-36-1]

2-BROMOETHANOL, I, 117;
Ethanol, 2-bromo- [540-51-2]

β-BROMOETHYLAMINE HYDROBROMIDE, II, 91;
Ethanamine, 2-bromo-,hydrobromide [2576-47-8]

β-BROMOETHYLPHTHALIMIDE, I, 119; IV, 106;
1*H*-Isoindole-1,3(2*H*)-dione,2-(2-bromoethyl)- [574-98-1]

1-BROMO-2-FLUOROBENZENE, V, 133;
Benzene,1-bromo-2-fluoro- [1072-85-1]

1-BROMO-2-FLUOROHEPTANE, V, 136;
Heptane, 1-bromo-2-fluoro- [1786-32-9]

α-BROMOHEPTALDEHYDE, III, 127;
Heptanal,2-bromo- [16486-84-3]

4-BROMO-2-HEPTENE, IV, 108;
2-Heptene,4-bromo- [22118-57-6]

3-BROMO-4-HYDROXYTOLUENE, III, 130;
Phenol,2-bromo-4-methyl- [6627-55-0]

α-BROMOISOVALERIC ACID, II, 93;
Butanoic acid,2-bromo-3-methyl- [565-74-2]

p-BROMOMANDELIC ACID, IV, 110;
Benzeneacetic acid, 4-bromo-α-hydroxy- [6940-50-7]

BROMOMESITYLENE, II, 95;
Benzene,2-bromo-1,3,5-trimethyl- [576-83-0]

2-BROMO-4-METHYLBENZALDEHYDE, V, 139;
Benzaldehyde, 2-bromo-4-methyl- [824-54-4]

2-BROMO-3-METHYLBENZOIC ACID, IV, 114;
Benzoic acid, 2-bromo-3-methyl- [53663-39-1]

α-BROMONAPHTHALENE, I, 121;
Naphthalene, 1-bromo- [90-11-9]

2-BROMONAPHTHALENE, V, 142;
Naphthalene, 2-bromo- [580-13-2]

6-BROMO-2-NAPHTHOL, III, 132;
  2-Naphthalenol,6-bromo- [15231-91-1]

m-BROMONITROBENZENE, I, 123;
  Benzene,1-bromo-3-nitro- [585-79-5]

2-BROMO-3-NITROBENZOIC ACID, I, 125;
  Benzoic acid, 2-bromo-3-nitro- [573-54-6]

p-BROMOPHENACYL BROMIDE, I, 127;
  Ethanone,2-bromo-1-(4-bromophenyl)- [99-73-0]

9-BROMOPHENANTHRENE, III, 134;
  Phenanthrene, 9-bromo- [573-17-1]

o-BROMOPHENOL, II, 97;
  Phenol, 2-bromo- [95-56-7]

p-BROMOPHENOL, I, 128;
  Phenol, 4-bromo- [106-41-2]

3-BROMOPHTHALIDE, V, 145;
  1(3*H*)-Isobenzofuranone,3-bromo- [6940-49-4]

β-BROMOPROPIONIC ACID, I, 131;
  Propanoic acid, 3-bromo- [590-92-1]

3-BROMOPYRENE, V, 147;
  Pyrene,1-bromo- [1714-29-0]

2-BROMOPYRIDINE, III, 136;
  Pyridine, 2-bromo- [109-04-6]

4-BROMORESORCINOL, II, 100;
  1,3-Benzenediol, 4-bromo- [6626-15-9]

3-BROMOTHIOPHENE, V, 149;
  Thiophene, 3-bromo- [872-31-1]

m-BROMOTOLUENE, I, 133;
  Benzene,1-bromo-3-methyl- [591-17-3]

o-BROMOTOLUENE, I, 135;
  Benzene,1-bromo-2-methyl- [95-46-5]

p-BROMOTOLUENE, I, 136;
  Benzene,1-bromo-4-methyl- [106-38-7]

4-BROMO-o-XYLENE, III, 138;
  Benzene,4-bromo-1,2-dimethyl- [583-71-1]

1,3-BUTADIENE, II, 102;
  1,3-Butadiene [106-99-0]

o-n-BUTOXYNITROBENZENE, III, 140;
  Benzene, 1-butoxy-2-nitro- [7252-51-9]

7-*t*-BUTOXYNORBORNADIENE, V, 151;
  Bicyclo[2.2.1]hepta-2,5-diene,7-(1,1-dimethylethoxy)- [877-06-5]

t-BUTYL ACETATE, III, 141;
  Acetic acid,1,1-dimethylethyl ester [540-88-5]

t-BUTYL ACETOACETATE, V, 155;
  Butanoic acid, 3-oxo-,1,1-dimethylethyl ester [1694-31-1]

n-BUTYLACETYLENE, IV, 117;
  1-Hexyne [693-02-7]

n-BUTYL ACRYLATE, III, 146;
  2-Propenoic acid, butyl ester [141-32-2]

t-BUTYLAMINE, III, 148;
  2-Propanamine,2-methyl- [75-64-9]

t-BUTYLAMINE HYDROCHLORIDE, III, 153;
  2-Propanamine,2-methyl-,hydrochloride [10017-37-5]

t-BUTYL AZIDOFORMATE, V, 157;
  Carbonazidic acid,1,1-dimethylethyl ester [1070-19-5]

t-BUTYL AZODIFORMATE, V, 160;
  Diazenedicarboxylic acid, bis(1,1-dimethylethyl) ester [870-50-8]

n-BUTYLBENZENE, III, 157;
  Benzene, butyl- [104-51-8]

n-BUTYL BORATE, II, 106;
  Boric acid(H_3BO_3), tributyl ester [688-74-4]

n-BUTYL BROMIDE, I, 28, 37;
  Butane, 1-bromo- [109-65-9]

sec-BUTYL BROMIDE, I, 38;
  Butane,2-bromo- [78-76-2]

t-BUTYL BROMIDE, I, 38;
  Propane,2-bromo-2-methyl- [507-19-7]

n-BUTYL,*n*-BUTYRATE, I, 138;
  Butanoic acid, butyl ester [109-21-7]

sec-BUTYL-α-*n*-CAPROYLPROPIONATE, IV, 120;
  Octanoic acid,2-methyl-3-oxo-, 1-methylpropyl ester [53663-40-4]

n-BUTYL CARBAMATE, I, 140;
  Carbamic acid, butyl ester [592-35-8]

t-BUTYL CARBAMATE, V, 162;
  Carbamic acid, 1,1-dimethylethyl ester [4248-19-5]

t-BUTYL CARBAZATE, V, 166;
  Hydrazinecarboxylic acid, 1,1-dimethylethyl ester [870-46-2]

n-BUTYL CHLORIDE, I, 142;
  Butane, 1-chloro- [109-69-3]

t-BUTYL CHLORIDE, I, 144;
  Propane, 2-chloro-2-methyl- [507-20-0]

t-BUTYL CYANOACETATE, V, 171;
  Acetic acid, cyano-, 1,1-dimethylethyl ester [1116-98-9]

trans-4-*t*-BUTYLCYCLOHEXANOL, V, 175;
  Cyclohexanol, 4-(1,1-dimethylethyl)-,trans- [21862-63-5]

t-BUTYL DIAZOACETATE, V, 179;
  Acetic acid, diazo-, 1,1-dimethylethyl ester [35059-50-8]

n-BUTYL GLYOXYLATE, IV, 124;
  Acetic acid, oxo, butyl ester [6295-06-3]

t-BUTYL HYPOCHLORIDE, IV, 125; V, 184; *Warning*, V, 183;
  Hypochlorous acid, 1,1-dimethylethyl ester [507-40-4]

2-*n*-BUTYL-2-METHYLCYCLOHEXANONE, V, 187;
  Cyclohexanone, 2-butyl-2-methyl- [1197-78-0]

n-BUTYL NITRITE, II, 108;
  Nitrous acid, butyl ester [544-16-1]

2-*t*-BUTYL-3-PHENYLOXAZIRANE, V, 191;
  Oxaziridine, 2-(1,1-dimethylethyl)-3-phenyl- [7731-34-2]

n-BUTYL PHOSPHATE, II, 109;
  Phosphoric acid, tributyl ester [126-73-8]

1-*n*-BUTYLPYRROLIDINE, III, 159;
  Pyrrolidine,1-butyl- [767-10-2]

n-BUTYL SULFATE, II, 111;
  Sulfuric acid, dibutyl ester [625-22-9]

n-BUTYL SULFITE, II, 112;
  Sulfurous acid, dibutyl ester [626-85-7]

n-BUTYL *p*-TOLUENESULFONATE, I, 145;
  Benzenesulfonic acid, 4-methyl-, butyl ester [778-28-9]

2-BUTYN-1-OL, IV, 128;
  2-Butyn-1-ol [764-01-2]

BUTYRCHLORAL, IV, 130;
  Butanal, 2,2,3-trichloro- [76-36-8]

BUTYROIN, II, 114;
  4-Octanone, 5-hydroxy- [496-77-5]

n-BUTYRYL CHLORIDE, I, 147;
  Butanoyl chloride [141-75-3]

C

D,L-10-CAMPHORSULFONIC ACID(REYCHLER'S ACID), V, 194;
  Bicyclo[2.2.1]heptane-1-methanesulfonic acid, 7,7-dimethyl-2-oxo-, (±)- [5872-08-2]

D,L-10-CAMPHORSULFONYL CHLORIDE, V, 196;
  Bicyclo[2.2.1]heptane-1-methanesulfonyl chloride, 7,7-dimethyl-2-oxo-, (±)- [6994-93-0]

n-CAPROIC ANHYDRIDE, III, 164;
  Hexanoic acid, anhydride [2051-49-2]

3-CARBETHOXYCOUMARIN, III, 165;
  2H-1-Benzopyran-3-carboxylic acid, 2-oxo-, ethyl ester [1846-76-0]

2-CARBETHOXYCYCLOOCTANONE, V, 198;
  Cyclooctanecarboxylic acid, 2-oxo-, ethyl ester [4017-56-5]

2-CARBETHOXYCYCLOPENTANONE, II, 116;
  Cyclopentanecarboxylic acid, 2-oxo-, ethyl ester [611-10-9]

β-CARBETHOXY-γ,γ-DIPHENYLVINYLACETIC ACID, IV, 132;
  Butanedioic acid, (diphenylmethylene)-, 1-ethyl ester [5438-22-2]

CARBOBENZOXY CHLORIDE, AND DERIVATIVES, III, 167;
  Carbonochloridic acid, phenylmethyl ester [501-53-1]

β-CARBOMETHOXYPROPIONYL CHLORIDE, III, 169;
  Butanoic acid, 4-chloro-4-oxo-, methyl ester [1490-25-1]

1,1'-CARBONYLDIIMIDAZOLE, V, 201;
  1H-Imidazole, 1,1'-carbonylbis- [530-62-1]

CARBOXYMETHOXYLAMINE HEMIHYDROCHLORIDE, III, 172;
  Acetic acid, (aminooxy)-, hydrochloride (2:1) [6921-14-4]

o-CARBOXYPHENYLACETONITRILE, III, 174;
  Benzoic acid, 2-(cyanomethyl)- [6627-91-4]

β-(o-CARBOXYPHENYL) PROPIONIC ACID, IV, 136;
  Benzenepropanoic acid, 2-carboxy- [776-79-4]

CASEIN, II, 120;
  Casein [9000-71-9]

CATALYST, RANEY NICKEL, W-2, III, 181;
CATALYST, RANEY NICKEL, W-6, III, 176;
  Nickel [7440-02-0]

CATECHOL, I, 149;
  1,2-Benzenediol [120-80-9]

CELLOBIOSE, II, 122;
  D-Glucose, 4-O-β-D-glucopyranosyl- [528-50-7]

α-CELLOBIOSE OCTAACETATE, II, 124;
  D-Glucose, 4-O-α-D-glucopyranosyl-, octaacetate [5346-90-7]

CETYLMALONIC ESTER, IV, 141;
  Propanedioic acid, hexadecyl-, diethyl ester [41433-81-2]

CHELIDONIC ACID, II, 126;
  4*H*-Pyran-2,6-dicarboxylic acid, 4-oxo- [99-32-1]

CHLOROACETAMIDE, I, 153;
  Acetamide, 2-chloro- [79-07-2]

CHLOROACETONITRILE, IV, 144;
  Acetonitrile, chloro- [107-14-2]

p-CHLOROACETYLACETANILIDE, III, 183;
  Acetamide, *N*-[4-(chloroacetyl)phenyl]- [140-49-8]

α-CHLOROACETYL ISOCYANATE, V, 204;
  Acetyl isocyanate, chloro- [4461-30-7]

3-(*o*-CHLOROANILINO)PROPIONITRILE, IV, 146;
  Propanenitrile, 3-[(2-chlorophenyl)amino]- [94-89-3]

9-CHLOROANTHRACENE, V, 206;
  Anthracene, 9-chloro- [716-53-0]

α-CHLOROANTHRAQUINONE, II, 128;
  9,10-Anthracenedione, 1-chloro- [82-44-0]

m-CHLOROBENZALDEHYDE, II, 130;
  Benzaldehyde, 3-chloro- [587-04-2]

p-CHLOROBENZALDEHYDE, II, 133;
  Benzaldehyde, 4-chloro- [104-88-1]

o-CHLOROBENZOIC ACID, II, 135;
  Benzoic acid, 2-chloro- [118-91-2]

CHLORO-*p*-BENZOQUINONE, IV, 148;
  2,5-Cyclohexadiene-1,4-dione, 2-chloro- [695-99-8]

o-CHLOROBENZOYL CHLORIDE, I, 155;
  Benzoyl chloride, 2-chloro- [609-65-4]

N-CHLOROBETAINYL CHLORIDE, IV, 154;
  Ethanaminium, 2-chloro-*N,N,N*-trimethyl-2-oxo-, chloride [53684-57-4]

o-CHLOROBROMOBENZENE, III, 185;
  Benzene, 1-bromo-2-chloro- [694-80-4]

4-CHLOROBUTYL BENZOATE, III, 187;
  1-Butanol, 4-chloro-, benzoate [946-02-1]

γ-CHLOROBUTYRONITRILE, I, 156;
  Butanenitrile, 4-chloro- [628-20-6]

2-CHLOROCYCLOHEXANOL, I, 158;
  Cyclohexanol, 2-chloro- [1561-86-0]

2-CHLOROCYCLOHEXANONE, III, 188;
Cyclohexanone, 2-chloro- [822-87-7]

N-CHLOROCYCLOHEXYLIDENEIMINE, V, 208;
Cyclohexanimine, N-chloro- [6681-70-5]

trans-2-CHLOROCYCLOPENTANOL, IV, 157;
Cyclopentanol, 2-chloro-, trans- [20377-80-4]

CHLORODIISOPROPYLPHOSPHINE, V, 211;
Phosphinous chloride, bis(1-methylethyl)- [40244-90-4]

1-CHLORO-2,6-DINITROBENZENE, IV, 160;
Benzene, 2-chloro-1,3-dinitro- [606-21-3]

β-CHLOROETHYL METHYL SULFIDE, II, 136;
Ethane, 1-chloro-2-(methylthio)- [542-81-4]

2-CHLORO-1-FORMYL-1-CYCLOHEXENE, V, 215;
1-Cyclohexene-1-carboxaldehyde, 2-chloro- [1680-73-5]

ω-CHLOROISONITROSOACETOPHENONE, III, 191;
Benzeneethanimidoyl chloride, N-hydroxy-α-oxo- [4937-87-5]

2-CHLOROLEPIDINE, III, 194;
Quinoline, 2-chloro-4-methyl- [634-47-9]

p-CHLOROMERCURIBENZOIC ACID, I, 159;
Mercury, (4-carboxyphenyl)chloro- [59-85-8]

O-CHLOROMERCURIPHENOL, I, 161;
Mercury, chloro(2-hydroxyphenyl)- [90-03-9]

2-CHLORO-2-METHYLCYCLOHEXANONE AND 2-METHYL-2-CYCLOHEXANONE,
IV, 162;
Cyclohexanone, 2-chloro-2-methyl- [10409-46-8]
2-Cyclohexen-1-one, 2-methyl- [1121-18-2]

bis(CHLOROMETHYL)ETHER (Hazard Note), V, 218;
Methane, oxybis[chloro- [542-88-1]

1-CHLOROMETHYLNAPHTHALENE, III, 195;
Naphthalene, 1-(chloromethyl)- [86-52-2]

CHLOROMETHYLPHOSPHONOTHIOIC DICHLORIDE, V, 218;
Phosphonothioic dichloride, (chloromethyl)- [1983-27-3]

2-CHLOROMETHYLTHIOPHENE, III, 197;
Thiophene, 2-(chloromethyl)- [765-50-4]

2-CHLORONICOTINONITRILE, IV, 166;
3-Pyridinecarbonitrile, 2-chloro- [6602-54-6]

m-CHLORONITROBENZENE, I, 162;
Benzene, 1-chloro-3-nitro- [121-73-3]

p-CHLOROPHENOXYMETHYL CHLORIDE, V, 221;
  Benzene, 1-chloro-4-(chloromethoxy)- [21151-56-4]

α-CHLOROPHENYLACETIC ACID, IV, 169;
  Benzeneacetic acid, α-chloro- [4755-72-0]

o-CHLOROPHENYLCYANAMIDE, IV, 172;
  Cyanamide, (2-chlorophenyl)- [45765-25-1]

p-CHLOROPHENYL ISOTHIOCYANATE, I, 165; V, 223;
  Benzene, 1-chloro-4-isothiocyanato- [2131-55-7]

m-CHLOROPHENYLMETHYLCARBINOL, III, 200;
  Benzenemethanol, 3-chloro-α-methyl [6939-95-3]

α-(4-CHLOROPHENYL)-γ-PHENYLACETOACETONITRILE, IV, 174;
  Benzenebutanenitrile, α-(4-chlorophenyl)-β-oxo- [35741-47-0]

1-(*p*-CHLOROPHENYL)-3-PHENYL-2-PROPANONE, IV, 176;
  2-Propanone, 1-(4-chlorophenyl)-3-phenyl- [35730-03-1]

p-CHLOROPHENYL SALICYLATE, IV, 178;
  Benzoic acid, 2-hydroxy-, 4-chlorophenyl ester [2944-58-3]

o-CHLOROPHENYLTHIOUREA, IV, 180;
  Thiourea, (2-chlorophenyl)- [5344-82-1]

β-CHLOROPROPIONALDEHYDE ACETAL, II, 137;
  Propane, 3-chloro-1,1-diethoxy- [35573-93-4]

β-CHLOROPROPIONIC ACID, I, 166;
  Propanoic acid, 3-chloro- [107-94-8]

γ-CHLOROPROPYL ACETATE, III, 203;
  1-Propanol, 3-chloro-, acetate [628-09-1]

2-CHLOROPYRIMIDINE, IV, 182;
  Pyrimidine, 2-chloro- [1722-12-9]

m-CHLOROSTYRENE, III, 204;
  Benzene, 1-chloro-3-ethenyl- [2039-85-2]

CHLOROSULFONYL ISOCYANATE, V, 226;
  Sulfuryl chloride isocyanate [1189-71-5]

2-CHLOROTHIIRANE 1,1-DIOXIDE, V, 231;
  Thiirane, chloro-, 1,1-dioxide [10038-13-8]

o-CHLOROTOLUENE, I, 170;
  Benzene, 1-chloro-2-methyl- [95-49-8]

p-CHLOROTOLUENE, I, 170;
  Benzene, 1-chloro-4-methyl- [106-43-4]

1-CHLORO-1,4,4-TRIFLUOROBUTADIENE, V, 235;
  1,4-Butadiene, 1-chloro-1,4,4-trifluoro- [764-14-7]

2-CHLORO-1,1,2-TRIFLUOROETHYL ETHYL ETHER, IV, 184;
Ethane, 2-chloro-1-ethoxy-1,1,2-trifluoro- [310-71-4]

3-CHLORO-2,2,3-TRIFLUOROPROPIONIC ACID, V, 239;
Propanoic acid, 3-chloro-2,2,3-trifluoro- [425-97-8]

β-CHLOROVINYL ISOAMYL KETONE, IV, 186;
1-Hepten-3-one, 1-chloro-6-methyl- [18378-90-0]

CHOLANE-24-AL, V, 242;
Cholan-24-al [26606-02-0]

CHOLESTANONE, II, 139;
Cholestan-3-one, (5a)- [566-88-1]

CHOLESTANYL METHYL ETHER, V, 245;
Cholestane, 3-methoxy-, (3β,5α)- [1981-90-4]

Δ⁴-CHOLESTEN-3,6-DIONE, IV, 189;
Cholest-4-ene-3,6-dione [984-84-9]

CHOLESTENONE, III, 207;
Cholest-4-en-3-one [601-57-0]

Δ⁴-CHOLESTEN-3-ONE, IV, 192; 195;
Cholest-4-en-3-one [601-57-0]

CHOLESTEROL, IV, 195;
Cholest-5-en-3-ol(3β)- [57-88-5]

Δ⁵-CHOLESTEN-3-ONE, IV, 195;
Cholest-5-en-3-one [601-54-7]

CINNAMYL BROMIDE, V, 249;
Benzene (3-bromo-1-propenyl)- [4392-24-9]

CITRACONIC ANHYDRIDE AND CITRACONIC ACID, II, 140;
2,5-Furandione, 3-methyl- [616-02-4]
2-Butenedioic acid, 2-methyl, (Z)- [498-23-7]

COPPER CHROMITE CATALYST, II, 142;
Chromium copper oxide, (CuCr₂O₄) [12018-10-9]

COUMALIC ACID, IV, 201;
2H-Pyran-5-carboxylic acid, 2-oxo- [500-05-0]

COUMARILIC ACID, III, 209;
2-Benzofurancarboxylic acid [496-41-3]

COUMARONE, V, 251;
Benzofuran [271-89-6]

COUPLING OF o-TOLUIDINE AND CHICAGO ACID, II, 145;
1,3-Naphthalenedisulfonic acid, 6,6'-[(3,3'-dimethyl[1,1'-biphenyl]-4,4'-
diyl)bis(azo)]bis[4-amino-5-hydroxy-, tetrasodium salt [314-13-6]

CREATININE, I, 172;
4*H*-Imidazol-4-one, 2-amino-1,5-dihydro-1-methyl- [60-27-5]

CREOSOL, IV, 203;
Phenol, 2-methoxy-4-methyl- [93-51-6]

p-CRESOL, I, 175;
Phenol, 4-methyl- [106-44-5]

γ-CROTONOLACTONE, V, 255;
2(5*H*)-Furanone [497-23-4]

CROTYL DIAZOACETATE, V, 258;
Acetic acid, diazo-, 2-butenyl ester, (*E*)- [14746-03-3]

CUPFERRON, I, 177;
Benzenamine, *N*-hydroxy-*N*-nitroso-, ammonium salt [135-20-6]

CYANOACETAMIDE, I, 179;
Acetamide, 2-cyano- [107-91-5]

1-CYANOBENZOCYCLOBUTENE, V, 263;
Bicyclo[4.2.0]octa-1,3,5-triene-7-carbonitrile [6809-91-2]

N-2-CYANOETHYLANILINE, IV, 205;
Propanenitrile, 3-(phenylamino)- [1075-76-9]

CYANOGEN BROMIDE, II, 150;
Cyanogen bromide, [506-68-3]

CYANOGEN IODIDE, IV, 207;
Iodine cyanide, I(CN) [506-78-5]

7-CYANOHEPTANAL, V, 266;
Octanenitrile, 8-oxo- [13050-09-4]

2-CYANO-6-METHYLPYRIDINE, V, 269;
2-Pyridinecarbonitrile, 6-methyl- [1620-75-3]

3-CYANO-6-METHYL-2(1)-PYRIDONE, IV, 210;
3-Pyridinecarbonitrile, 1,2-dihydro-6-methyl-2-oxo- [4241-27-4]

ω-CYANOPELARGONIC ACID, III, 768;
Nonanoic acid, 9-cyano- [5810-19-5]

9-CYANOPHENANTHRENE, III, 212;
9-Phenanthrenecarbonitrile [2510-55-6]

α-CYANO-β-PHENYLACRYLIC ACID, I, 181;
2-Propenoic acid, 2-cyano-3-phenyl- [1011-92-3]

1-CYANO-3-PHENYLUREA, IV, 213;
Urea,*N*-cyano-*N'*-phenyl- [41834-91-7]

CYCLOBUTANECARBOXYLIC ACID, III, 213;
Cyclobutanecarboxylic acid [3721-95-7]

1,1-CYCLOBUTANEDICARBOXYLIC ACID, III, 213;
  1,1-Cyclobutanedicarboxylic acid [5445-51-2]

CYCLOBUTYLAMINE, V, 273;
  Cyclobutanamine [2516-34-9]

1,2-CYCLODECANEDIOL, IV, 216;
  1,2-Cyclodecanediol [21014-77-7]

CYCLODECANONE, IV, 218; **V**, 277;
  Cyclodecanone [1502-06-3]

cis-**CYCLODODECENE, V**, 281;
  Cyclododecene, (Z)- [1129-89-1]

CYCLOHEPTANONE, IV, 221;
  Cycloheptanone [502-42-1]

1,3-CYCLOHEXADIENE, V, 285;
  1,3-Cyclohexadiene [592-57-4]

trans-**1,2-CYCLOHEXANEDIOL, III**, 217;
  1,2-Cyclohexanediol, *trans*- [1460-57-7]

1,4-CYCLOHEXANEDIONE, V, 288;
  1,4-Cyclohexanedione [637-88-7]

1,2-CYCLOHEXANEDIONE DIOXIME, IV, 229;
  1,2-Cyclohexanedione, dioxime [492-99-9]

CYCLOHEXANONE DIALLYL ACETAL, V, 292;
  Cyclohexane, 1,1-bis(2-propenyloxy)- [53608-84-7]

CYCLOHEXENE, I, 183;
  Cyclohexene [110-83-8]

CYCLOHEXENE OXIDE, I, 185;
  7-Oxabicyclo[4.1.0]heptane [286-20-4]

CYCLOHEXENE SULFIDE, IV, 232;
  7-Thiabicyclo[4.1.0]heptane [286-28-2]

2-CYCLOHEXENONE, V, 294;
  2-Cyclohexen-1-one [930-68-7]

CYCLOHEXYLBENZENE, II, 151;
  Benzene, cyclohexyl- [827-52-1]

3-CYCLOHEXYL-2-BROMOPROPENE, I, 186;
  Cyclohexane, (2-bromo-2-propenyl)- [53608-85-8]

CYCLOHEXYLCARBINOL, I, 188;
  Cyclohexanemethanol [100-49-2]

**CYCLOHEXYLIDENECYANOACETIC ACID AND 1-CYCLOHEXENYLACETONI-
TRILE, IV**, 234;

Acetic acid, cyanocyclohexylidene- [37107-50-9]
1-Cyclohexene-1-acetonitrile [6975-71-9]

CYCLOHEXYLIDENECYCLOHEXANE, V, 297;
Cyclohexane, cyclohexylidene- [4233-18-5]

CYCLOHEXYL ISOCYANIDE, V, 300;
Cyclohexane, isocyano- [931-53-3]

CYCLOHEXYL METHYL KETONE, V, 775;
Ethanone, 1-cyclohexyl [823-76-7]

2-CYCLOHEXYLOXYETHANOL, V, 303;
Ethanol, 2-(cyclohexyloxy)- [1817-88-5]

α-CYCLOHEXYLPHENYLACETONITRILE, III, 219;
Benzeneacetonitrile, α-cyclohexyl- [3893-23-0]

3-CYCLOHEXYLPROPINE, I, 191;
Cyclohexane, 2-propynyl- [17715-00-3]

1,2-CYCLONONADIENE, V, 306;
1,2-Cyclononadiene [1123-11-1]

CYCLOÖCTANONE, V, 310;
Cyclooctanone [502-49-8]

trans-CYCLOOCTENE, V, 315;
Cyclooctene, (E)- [931-89-5]

CYCLOPENTADIENE AND 3-CHLOROCYCLOPENTENE, IV, 238;
1,3-Cyclopentadiene [542-92-7]
Cyclopentene, 3-chloro- [96-40-2]

CYCLOPENTANECARBOXALDEHYDE, V, 320;
Cyclopentanecarboxaldehyde [872-53-7]

CYCLOPENTANONE, I, 192;
Cyclopentanone [120-92-3]

2-CYCLOPENTENE-1,4-DIONE, V, 324;
4-Cyclopentene-1,3-dione [930-60-9]

2-CYCLOPENTENONE, V, 326;
2-Cyclopenten-1-one [930-30-3]

CYCLOPROPANECARBOXYLIC ACID, III, 221;
Cyclopropanecarboxylic acid [1759-53-1]

CYCLOPROPYLBENZENE, V, 328; Same as PHENYLCYCLOPROPANE, V, 929;
Benzene, cyclopropyl- [873-49-4]

CYCLOPROPYL CYANIDE: III, 223;
Cyclopropanecarbonitrile [5500-21-0]

m-CYMENE, V, 332;
Benzene, 1-methyl-3-(1-methylethyl)- [535-77-3]

CYSTEIC ACID MONOHYDRATE, III, 226;
L-Alanine, 3-sulfo-, hemihydrate [53643-49-5]

l-CYSTINE, I, 194;
L-Cystine [56-89-3]

D

DECAMETHYLENE BROMIDE, III, 227;
Decane, 1,10-dibromo- [4101-68-2]

DECAMETHYLENEDIAMINE, III, 229;
1,10-Decanediamine [646-25-3]

DECAMETHYLENE GLYCOL, II, 154;
1,10-Decanediol [112-47-0]

DEHYDROACETIC ACID, III, 231;
2*H*-Pyran-2,4(3*H*)-dione, 3-acetyl-6-methyl- [520-45-6]

DEOXYANISOIN, V, 339;
Ethanone, 1,2-bis(4-methoxyphenyl)- [120-44-5]

DESOXYBENZOIN, II, 156;
Ethanone, 1,2-diphenyl- [451-40-1]

nor-DESOXYCHOLIC ACID, III, 234;
24-Norcholan-23-oic acid, 3,12-dihydroxy-, (3α,5β,12α)- [53608-86-9]

DESYL CHLORIDE, II, 159;
Ethanone, 2-chloro-1,2-diphenyl- [447-31-4]

DIACETONAMINE HYDROGEN OXALATE, I, 196;
2-Pentanone,4-amino-4-methyl-, ethanedioate (2:1) [53608-87-0]

DIACETONE ALCOHOL, I, 199;
2-Pentanone, 4-hydroxy-4-methyl- [123-42-2]

3,12-DIACETOXY-*bisnor*-CHOLANYLDIPHENYLETHYLENE, III, 237;
Chol-23-ene-3,12-diol, 24,24-diphenyl-, diacetate, (3α,5β,12α)- [53608-88-1]

DIACETYL *d*-TARTARIC ANHYDRIDE, IV, 242;
2,5-Furandione, 3,4-bis(acetyloxy)dihydro-, (3*S*-trans)- [19523-83-2]

DIALLYLAMINE, I, 201;
2-Propen-1-amino, *N*-2-propenyl- [124-02-7]

DIALLYLCYANAMIDE, I, 203;
Cyanamide, di-2-propenyl- [538-08-9]

4,4'-DIAMINOAZOBENZENE, V, 341;
Benzenamine, 4,4'-azobis- [538-41-0]

DIAMINOBIURET, III, 404;
Imidodicarbonic dihydrazide [4375-11-5]

2,5-DIAMINO-3,4-DICYANOTHIOPHENE, IV, 243;
3,4-Thiophenedicarbonitrile, 2,5-diamino [17989-89-8]

4,4'-DIAMINODIPHENYLSULFONE, III, 239;
Benzenamine, 4,4'-sulfonylbis- [80-08-0]

2,4-DIAMINO-6-HYDROXYPYRIMIDINE, IV, 245;
4(1H)-Pyrimidinone, 2,6-diamino- [56-06-4]

DIAMINOMALEONITRILE(HYDROGEN CYANIDE TETRAMER), V, 344;
2-Butenedinitrile, 2,3-diamino, (Z)- [1187-42-4]

1,2-DIAMINO-4-NITROBENZENE, III, 242;
1,2-Benzenediamine, 4-nitro- [99-56-9]

2,3-DIAMINOPYRIDINE, V, 346;
2,3-Pyridinediamine [452-58-4]

2,4-DIAMINOTOLUENE, II, 160;
1,3-Benzenediamine, 4-methyl- [95-80-7]

DIAMINOURACIL HYDROCHLORIDE, IV, 247;
2,4(1H,3H)-Pyrimidinedione, 4,5-diamino-, monohydrochloride [53608-89-2]

DIAZOAMINOBENZENE, II, 163;
1-Triazene,1,3-diphenyl [136-35-6]

DIAZOMETHANE, II, 165; **III,** 244; **IV,** 250; **V,** 351;
Methane,diazo- [334-88-3]

DIBENZALACETONE, II, 167;
1,4-Pentadien-3-one, 1,5-diphenyl- [538-58-9]

1,4-DIBENZOYLBUTANE, II, 169;
1,6-Hexanedione, 1,6-diphenyl- [3375-38-0]

***trans*-DIBENZOYLETHYLENE, III,** 248;
2-Butene-1,4-dione, 1,4-diphenyl-, (E)- [959-28-4]

DIBENZOYLMETHANE, I, 205; **III,** 251;
1,3-Propanedione,1,3-diphenyl- [120-46-7]

DIBROMOACETONITRILE, IV, 254;
Acetonitrile, dibromo [3252-43-5]

2,6-DIBROMOANILINE, III, 262;
Benzenamine, 2,6-dibromo- [608-30-0]

9,10-DIBROMOANTHRACENE, I, 207;
Anthracene, 9,10-dibromo- [523-27-3]

4,4'-DIBROMOBIPHENYL, IV, 256;
1,1'-Biphenyl,4,4'-dibromo- [92-86-4]

1,2-DIBROMOCYCLOHEXANE, II, 171;
Cyclohexane,1,2-dibromo- [5401-62-7]

2,6-DIBROMO-4-NITROPHENOL, II, 173;
Phenol, 2,6-dibromo-4-nitro- [99-28-5]

2,3-DIBROMOPROPENE, I, 209;
1-Propene, 2,3-dibromo- [513-31-5]

2,6-DIBROMOQUINONE-4-CHLOROIMIDE, II, 175;
2,5-Cyclohexadien-1-one, 2,6-dibromo-4-(chloroimino)- [537-45-1]

α,β-DIBROMOSUCCINIC ACID, II, 177;
Butanedioic acid, 2,3-dibromo- [526-78-3]

β-DI-n-BUTYLAMINOETHYLAMINE, III, 254;
1,2-Ethanediamine,N,N-dibutyl- [3529-09-7]

γ-DI-n-BUTYLAMINOPROPYLAMINE, III, 256;
1,3-Propanediamine,N,N-dibutyl- [102-83-0]

DI-n-BUTYLCARBINOL, II, 179;
5-Nonanol [623-93-8]

DI-n-BUTYLDIVINYLTIN, IV, 258;
Stannane, dibutyldiethenyl- [7330-43-0]

DI-t-BUTYL MALONATE, IV, 261;
Propanedioic acid, bis(1,1-dimethylethyl) ester [541-16-2]

DI-t-BUTYL NITROXIDE, V, 355;
Nitroxide, bis(1,1-dimethylethyl) [2406-25-9]

DI-β-CARBETHOXYETHYLMETHYLAMINE, III, 258;
β-Alanine,N-(3-ethoxy-3-oxopropyl)-N-methyl-, ethyl ester [6315-60-2]

α,α-DICHLOROACETAMIDE, III, 260;
Acetamide,2,2-dichloro- [683-72-7]

DICHLOROACETIC ACID, II, 181;
Acetic acid, dichloro- [79-43-6]

α,γ-DICHLOROACETONE, I, 211;
2-Propanone, 1,3-dichloro- [534-07-6]

2,6-DICHLOROANILINE, III, 262;
Benzenamine, 2,6-dichloro- [608-31-1]

4,4'-DICHLORODIBUTYL ETHER, IV, 266;
Butane, 1,1'-oxybis[4-chloro- [6334-96-9]

1,1-DICHLORO-2,2-DIFLUOROETHYLENE, IV, 268;
Ethene, 1,1-dichloro-2,2-difluoro- [79-35-6]

β,β-DICHLORO-p-DIMETHYLAMINOSTYRENE, V, 361;
Benzenamine,4-(2,2-dichloroethenyl)-N,N-dimethyl- [6798-58-9]

2,2'-DICHLORO-α,α'-EPOXYBIBENZYL, V, 358;
Oxirane,2.3-bis(2-chlorophenyl)- [53608-92-7]

2,2-DICHLOROETHANOL, IV, 271;
Ethanol, 2,2-dichloro- [598-38-9]

DICHLOROMETHYLENETRIPHENYLPHOSPHORANE, V, 361;
Phosphorane,(dichloromethylene)triphenyl- [6779-08-4]

DICHLOROMETHYL METHYL ETHER, V, 365;
Methane, dichloromethoxy- [4885-02-3]

2,6-DICHLORONITROBENZENE, V, 367;
Benzene, 1,3-dichloro-2-nitro- [601-88-7]

2,6-DICHLOROPHENOL, III, 267;
Phenol, 2,6-dichloro, [87-65-0]

DI-(p-CHLOROPHENYL)ACETIC ACID, III, 270; Correction, V, 370;
Benzeneacetic acid, 4-chloro-α-(4-chlorophenyl)- [83-05-6]

4,7-DICHLOROQUINOLINE, III, 272;
Quinoline, 4,7-dichloro- [86-98-6]

3,4-DICHLORO-1,2,3,4-TETRAMETHYLCYCLOBUTENE, V, 370;
Cyclobutene,3,4-dichloro-1,2,3,4-tetramethyl- [1194-30-5]

1,1'-DICYANO-1,1'-BICYCLOHEXYL, IV, 273;
[1,1'-Bicyclohexyl]-1,1'dicarbonitrile, [18341-40-7]

1,2-DI-1(1-CYANO)CYCLOHEXYLHYDRAZINE, IV, 274;
Cyclohexanecarbonitrile,1,1'-hydrazobis- [17643-01-5]

DICYANOKETENE ETHYLENE ACETAL, IV, 276;
Propanedinitrile, 1,3-dioxolan-2-ylidene- [5694-65-5]

DICYCLOPROPYL KETONE, IV, 278;
Methanone, dicyclopropyl- [1121-37-5]

DIETHYL ACETAMIDOMALONATE, V, 373;
Propanedioic acid, (acetylamino)-, diethyl ester [1068-90-2]

DIETHYLAMINOACETONITRILE, III, 275;
Acetonitrile,(diethylamino)- [3010-02-4]

1-DIETHYLAMINO-3-BUTANONE, IV, 281;
2-Butanone, 4-(diethylamino)- [3299-38-5]

β-DIETHYLAMINOETHYL ALCOHOL, II, 183;
Ethanol, 2-(diethylamino)- [100-37-8]

DIETHYL AMINOMALONATE HYDROCHLORIDE, V, 376;
Propanedioic acid, amino-, diethyl ester, hydrochloride [13433-00-6]

N,N'-DIETHYLBENZIDINE, IV, 283;
[1,1'-Biphenyl]-4,4'-diamine,N,N'-diethyl- [6290-86-4]

DIETHYL[O-BENZOYL]ETHYLTARTRONATE, V, 379;
Propanedioic acid,(benzoyloxy)ethyl-, diethyl ester [6259-78-5]

DIETHYL BENZOYLMALONATE, IV, 285;
Propanedioic acid, benzoyl-, diethyl ester [1087-97-4]

DIETHYL BIS(HYDROXYMETHYL)MALONATE, V, 381;
Propanedioic acid, bis(hydroxymethyl)-, diethyl ester [20605-01-0]

DIETHYL 1,1-CYCLOBUTANEDICARBOXYLATE, IV, 288;
1,1-Cyclobutanedicarboxylic acid, diethyl ester [3779-29-1]

DIETHYL Δ²-CYCLOPENTENYLMALONATE, IV, 291;
Propanedioic acid, 2-cyclopenten-1-yl-, diethyl ester [53608-93-8]

DIETHYL ETHYLIDENEMALONATE, IV, 293;
Propanedioic acid, ethylidene-, diethyl ester [1462-12-0]

DIETHYL cis-HEXAHYDROPHTHALATE, IV, 304;
1,2-Cyclohexanedicarboxylic acid, diethyl ester,cis- [17351-07-4]

DIETHYL β-KETOPIMELATE, V, 384;
Heptanedioic acid, 3-oxo, diethyl ester [40420-22-2]

DIETHYL MERCAPTOACETAL, IV, 295;
Ethanethiol, 2,2-diethoxy- [53608-94-9]

DIETHYL METHYLENEMALONATE, IV, 298;
Propanedioic acid, methylene-, diethyl ester [3377-20-6]

DIETHYL γ-OXOPIMELATE, IV, 302;
Heptanedioic acid, 4-oxo-, diethyl ester [6317-49-3]

DIETHYL SUCCINATE, V, 993;
Butanedioic acid, diethyl ester [123-25-1]

DIETHYL cis-Δ⁴-TETRAHYDROPHTHALATE, IV, 304;
4-Cyclohexene-1,2-dicarboxylic acid, diethyl ester,cis- [4841-85-4]

DIETHYLTHIOCARBAMYL CHLORIDE, IV, 307;
Carbamothioic chloride, diethyl- [88-11-9]

N,N-DIETHYL-1,2,2-TRICHLOROVINYLAMINE, V, 387;
Ethenamine, 1,2,2-trichloro-N,N-diethyl- [686-10-2]

DIETHYL ZINC, II, 184;
Zinc, diethyl- [557-20-0]

4,4'-DIFLUOROBIPHENYL, II, 188;
1,1'-Biphenyl, 4,4'-difluoro- [398-23-2]

β,β-DIFLUOROSTYRENE, V, 390;
Benzene, (2,2-difluoroethenyl)- [405-42-5]

2,2-DIFLUOROSUCCINIC ACID, V, 393;
Butanedioic acid, 2,2-difluoro- [665-31-6]

α,α-DIFLUOROTOLUENE, V, 396;
Benzene,(difluoromethyl)- [455-31-2]

9,10-DIHYDROANTHRACENE, V, 398;
Anthracene, 9,10-dihydro- [613-31-0]

1,4-DIHYDROBENZOIC ACID, V, 400;
2,5-Cyclohexadiene-1-carboxylic acid [4794-04-1]

DIHYDROCHOLESTEROL, II, 191;
Cholestan-3-ol (3β,5α) [80-97-7]

2,5-DIHYDRO-2,5-DIMETHOXYFURAN, V, 403;
Furan, 2,5-dihydro-2,5-dimethoxy- [332-77-4]

1,3-DIHYDROISOINDOLE, V, 406;
1H-Isoindole, 2,3-dihydro- [496-12-8]

3,4-DIHYDRO-2-METHOXY-4-METHYL-2H-PYRAN, IV, 311;
2H-Pyran, 3,4-dihydro-2-methoxy-4-methyl- [53608-95-0]

3,4-DIHYDRO-1,2-NAPHTHALIC ANHYDRIDE, II, 194;
Naphtho[1,2-c]furan-1,3-dione, 4,5-dihydro- [37845-14-0]

9,10-DIHYDROPHENANTHRENE, IV, 313;
Phenanthrene, 9,10-dihydro- [776-35-2]

2,3-DIHYDROPYRAN, III, 276;
2H-Pyran, 3,4-dihydro- [110-87-2]

DIHYDRORESORCINOL, III, 278;
1,3-Cyclohexanedione [504-02-9]

1,3-DIHYDRO-3,5,7-TRIMETHYL-2H-AZEPIN-2-ONE, V, 408;
2H-Azepin-2-one, 1,3-dihydro-3,5,7-trimethyl- [936-85-6]

2,5-DIHYDROXYACETOPHENONE, III, 280;
Ethanone, 1-(2,5-dihydroxyphenyl)- [490-78-8]

2,6-DIHYDROXYACETOPHENONE, III, 281;
Ethanone, 1-(2,6-dihydroxyphenyl)- [699-83-2]

2,5-DIHYDROXY-p-BENZENEDIACETIC ACID, III, 286;
1,4-Benzenediacetic acid, 2,5-dihydroxy- [5488-16-4]

3,5-DIHYDROXYBENZOIC ACID, III, 288;
Benzoic acid, 3,5-dihydroxy- [99-10-5]

3,3'-DIHYDROXYBIPHENYL, V, 412;
[1,1'-Biphenyl]-3,3'-diol [612-76-0]

DIHYDROXYCYCLOPENTENE, V, 414;
3-Cyclopentene-1,2-diol,cis- [694-29-1]

3,5-DIHYDROXYCYCLOPENTENE, V, 414;
2-Cyclopentene-1,4-diol [4157-01-1]

9,10-DIHYDROXYSTEARIC ACID, IV, 317;
Octadecanoic acid, 9,10-dihydroxy- [120-87-6]

1,4-DIIODOBUTANE, IV, 321;
Butane, 1,4-diiodo- [628-21-7]

1,6-DIIODOHEXANE, IV, 323;
Hexane, 1,6-diiodo- [629-09-4]

2,6-DIIODO-p-NITROANILINE, II, 196;
Benzenamine, 2,6-diiodo-4-nitro- [5398-27-6]

DIISOPROPYL METHYLPHOSPHONATE, IV, 325;
Phosphonic acid, methyl, bis(1-methylethyl) ester [1445-75-6]

DIISOVALERYLMETHANE, III, 291;
4,6-Nonanedione, 2,8-dimethyl- [7307-08-6]

1,2-DIMERCAPTOBENZENE, V, 419;
1,2-Benzenedithiol [17534-155]

DIMESITYLMETHANE, V, 422;
Benzene, 1,1'-methylenebis[2,4,6-trimethyl- [733-07-3]

2,6-DIMETHOXYBENZONITRILE, III, 293;
Benzonitrile, 2,6-dimethoxy- [16932-49-3]

7,7-DIMETHOXYBICYCLO[2.2.1]HEPTENE, V, 424;
Bicyclo[2.2.1]hept-2-ene, 7,7-dimethoxy- [875-04-7]

3,3'-DIMETHOXYBIPHENYL, III, 295;
1,1'-Biphenyl, 3,3'-dimethoxy- [6161-50-8]

2,3-DIMETHOXYCINNAMIC ACID, IV, 327;
2-Propenoic acid, 3-(2,3-dimethoxyphenyl)- [7461-60-1]

6,7-DIMETHOXY-3,4-DIHYDRO-2-NAPHTHOIC ACID, III, 300;
2-Naphthalenecarboxylic acid, 3,4-dihydro-6,7-dimethoxy- [53684-50-7]

trans-4,4'-DIMETHOXYSTILBENE, V, 428;
Benzene, 1,1'-(1,2-ethenediyl)bis[4-methoxy-, (E)- [15638-14-9]

DIMETHYL ACETYLENECARBOXYLATE, IV, 329;
2-Butynedioic acid, dimethyl ester [762-42-5]

β,β-DIMETHYLACRYLIC ACID, III, 302;
2-Butenoic acid, 3-methyl- [541-47-9]

p-DIMETHYLAMINOBENZALDEHYDE, I, 214; IV, 331;
Benzaldehyde, 4-(dimethylamino)- [100-10-7]

p-DIMETHYLAMINOBENZOPHENONE, I, 217;
Methanone, [4-(dimethylamino)phenyl]phenyl- [530-44-9]

β-DIMETHYLAMINOETHYL CHLORIDE HYDROCHLORIDE, IV, 333;
Ethanamine, 2-chloro-N,N-dimethyl-, hydrochloride [4584-46-7]

6-(DIMETHYLAMINO)FULVENE, V, 431;
Methanamine, 1-(2,4-cyclopentadien-1-ylidene)-N,N-dimethyl- [696-68-4]

N,N-DIMETHYLAMINOMETHYLFERROCENE METHIODIDE, V, 434;
Methanaminium, 1-ferrocenyl-N,N,N-trimethyl-, iodide [12086-40-7]

α-*N*,*N*-DIMETHYLAMINOPHENYLACETONITRILE, V, 437;
Benzeneacetonitrile, α-(dimethylamino)- [827-36-1]

β-DIMETHYLAMINOPROPIOPHENONE HYDROCHLORIDE, III, 305;
1-Propanone, 3-(dimethylamino)-1-phenyl-, hydrochloride [879-72-1]

2-(DIMETHYLAMINO)PYRIMIDINE, IV, 336;
2-Pyrimidinamine, *N*,*N*-dimethyl- [5621-02-3]

3,4-DIMETHYLANILINE, III, 307;
Benzenamine, 3,4-dimethyl- [95-64-7]

2,3-DIMETHYLANTHRAQUINONE, III, 310;
9,10-Anthracenedione, 2,3-dimethyl- [6531-35-7]

3,3'-DIMETHYLBIPHENYL, III, 295;
1,1'-Biphenyl, 3,3'-dimethyl- [612-75-9]

2,3-DIMETHYL-1,3-BUTADIENE, III, 312;
1,3-Butadiene, 2,3-dimethyl- [513-81-5]

3,5-DIMETHYL-4-CARBETHOXY-2-CYCLOHEXEN-1-ONE, III, 317;
2-Cyclohexene-1-carboxylic acid, 2,6-dimethyl-4-oxo-, ethyl ester [6102-15-4]

2,4-DIMETHYL-5-CARBETHOXYPYRROLE, II, 198;
1*H*-Pyrrole-2-carboxylic acid, 3,5-dimethyl-, ethyl ester [2199-44-2]

4,6-DIMETHYLCOUMALIN, IV, 337;
2*H*-Pyran-2-one, 4,6-dimethyl- [675-09-2]

5,5-DIMETHYL-1,3-CYCLOHEXANEDIONE, II, 200;
1,3-Cyclohexanedione, 5,5-dimethyl- [126-81-8]

DIMETHYL CYCLOHEXANONE-2,6-DICARBOXYLATE, V, 439;
1,3-Cyclohexanedicarboxylic acid, 2-oxo-, dimethyl ester [25928-05-6]

3,5-DIMETHYL-2-CYCLOHEXEN-1-ONE, III, 317;
2-Cyclohexen-1-one, 3,5-dimethyl- [1123-09-7]

N,*N*-DIMETHYLCYCLOHEXYLMETHYLAMINE, IV, 339;
Cyclohexanemethanamine, *N*,*N*-dimethyl- [16607-80-0]

2,4-DIMETHYL-3,5-DICARBETHOXYPYRROLE, II, 202;
1*H*-Pyrrole-2,4-dicarboxylic acid, 3,5-dimethyl-, diethyl ester [2436-79-5]

2,7-DIMETHYL-2,7-DINITROÖCTANE, V, 445;
Octane, 2,7-dimethyl-2,7-dinitro- [53684-51-8]

2,6-DIMETHYL-3,5-DIPHENYL-4*H*-PYRAN-4-ONE, V, 450;
4*H*-Pyran-4-one, 2,6-dimethyl-3,5-diphenyl- [33731-54-3]

DIMETHYLETHYNYLCARBINOL, III, 320;
3-Butyn-2-ol, 2-methyl- [115-19-5]

DIMETHYLFURAZAN, IV, 342;
Furazan, dimethyl- [4975-21-7]

β,β-DIMETHYLGLUTARIC ACID, IV, 345;
Pentanedioic acid, 3,3-dimethyl- [4839-46-7]

DIMETHYLGLYOXIME, II, 204;
2,3-Butanedione, dioxime [95-45-4]

4,6-DIMETHYL-1-HEPTEN-4-OL, V, 452;
1-Hepten-4-ol, 4,6-dimethyl- [32189-75-6]

5,5-DIMETHYLHYDANTOIN, III, 323;
2,4-Imidazolidinedione, 5,5-dimethyl- [77-71-4]

sym-DIMETHYLHYDRAZINE DIHYDROCHLORIDE, II, 208;
Hydrazine, 1,2-dimethyl-, dihydrochloride [306-37-6]

unsym-DIMETHYLHYDRAZINE HYDROCHLORIDE, II, 211;
Hydrazine, 1,1-dimethyl-, monohydrochloride [593-82-8]

DIMETHYLKETENE, IV, 348;
1-Propen-1-one, 2-methyl- [598-26-5]

DIMETHYLKETENE β-LACTONE DIMER, V, 456;
2-Oxetanone, 3,3-dimethyl-4-(1-methylethylidene)- [3173-79-3]

2,5-DIMETHYLMANDELIC ACID, III, 326;
Benzeneacetic acid, α-hydroxy-2,5-dimethyl- [5766-40-5]

DIMETHYL 3-METHYLENECYCLOBUTANE-1,2-DICARBOXYLATE, V, 459;
1,2-Cyclobutanedicarboxylic acid, 3-methylene-, dimethyl ester [53684-52-9]

DIMETHYL OCTADECANEDIOATE, V, 463;
Octadecanedioic acid, dimethyl ester [1472-93-1]

2,7-DIMETHYLOXEPIN, V, 467;
Oxepin, 2,7-dimethyl- [1487-99-6]

5,5-DIMETHYL-2-n-PENTYLTETRAHYDROFURAN, IV, 350;
Furan, tetrahydro-2,2-dimethyl-5-pentyl- [53684-53-0]

α,α-DIMETHYL-β-PHENETHYLAMINE, V, 471;
Benzeneethanamine, α,α-dimethyl- [122-09-8]

3,5-DIMETHYLPYRAZOLE, IV, 351;
1H-Pyrazole, 3,5-dimethyl- [67-51-6]

2,6-DIMETHYLPYRIDINE, II, 214;
Pyridine, 2,6-dimethyl- [108-48-5]

2,4-DIMETHYLPYRROLE, II, 217;
1H-Pyrrole, 2,4-dimethyl- [625-82-1]

2,5-DIMETHYLPYRROLE, II, 219;
1H-Pyrrole, 2,5-dimethyl- [625-84-3]

2,2-DIMETHYLPYRROLIDINE, IV, 354;
Pyrrolidine, 2,2-dimethyl- [35018-15-6]

1,5-DIMETHYL-2-PYRROLIDONE, III, 328;
2-Pyrrolidinone, 1,5-dimethyl- [5075-92-3]

5,5-DIMETHYL-2-PYRROLIDONE, IV, 357;
2-Pyrrolidinone, 5,5-dimethyl- [5165-28-6]

2,4-DIMETHYLQUINOLINE, III, 329;
Quinoline, 2,4-dimethyl- [1198-37-4]

N,N-DIMETHYLSELENOUREA, IV, 359;
Selenourea, N,N-dimethyl- [5117-16-8]

2,4-DIMETHYLTHIAZOLE, III, 332;
Thiazole, 2,4-dimethyl- [541-58-2]

asym-DIMETHYLUREA, IV, 361;
Urea, N,N-dimethyl- [598-94-7]

2,4-DINITROANILINE, II, 221;
Benzenamine, 2,4-dinitro- [97-02-9]

2,6-DINITROANILINE, IV, 364;
Benzenamine, 2,6-dinitro- [606-22-4]

3,5-DINITROANISOLE, I, 219;
Benzene, 1-methoxy-3,5-dinitro- [5327-44-6]

2,4-DINITROBENZALDEHYDE, II, 223;
Benzaldehyde, 2,4-dinitro- [528-75-6]

p-DINITROBENZENE, II, 225;
Benzene, 1,4-dinitro- [100-25-4]

2,4-DINITROBENZENESULFENYL CHLORIDE, V, 474;
Benzenesulfenyl chloride, 2,4-dinitro- [528-76-7]

2,5-DINITROBENZOIC ACID, III, 334;
Benzoic acid, 2,5-dinitro- [610-28-6]

3,5-DINITROBENZOIC ACID, III, 337;
Benzoic acid, 3,5-dinitro- [99-34-3]

p,p'-DINITROBIBENZYL, IV, 367;
Benzene, 1,1'-(1,2-ethanediyl)bis[4-nitro- [736-30-1]

2,2'-DINITROBIPHENYL, III, 339;
1,1'-Biphenyl, 2,2'-dinitro- [2436-96-6]

1,4-DINITROBUTANE, IV, 368;
Butane, 1,4-dinitro- [4286-49-1]

3,4-DINITRO-3-HEXENE, IV, 372;
3-Hexene, 3,4-dinitro- [53684-54-1]

2,4-DINITROIODOBENZENE, V, 478;
Benzene, 1-iodio-2,4-dinitro- [709-49-9]

1,4-DINITRONAPHTHALENE, III, 341;
Naphthalene, 1,4-dinitro- [6921-26-2]

Di-*o*-NITROPHENYL DISULFIDE, I, 220;
Disulfide, bis(2-nitrophenyl) [1155-00-6]

2,4-DINITROPHENYLHYDRAZINE, II, 228;
Hydrazine, (2,4-dinitrophenyl)- [119-26-6]

3,5-DINITRO-*o*-TOLUNITRILE, V, 480;
Benzonitrile, 2-methyl-3,5-dinitro- [948-31-2]

1,6-DIOXO-8*a*-METHYL-1,2,3,4,6,7,8,8*a*-OCTAHYDRONAPHTHALENE, V, 486;
1,6(2*H*,7*H*)-Naphthalenedione, 3,4,8,8*a*-tetrahydro-8*a*-methyl- [20007-72-1]

DIPHENALDEHYDE, V, 489;
[1,1'-Biphenyl]-2,2'-dicarboxaldehyde [1210-05-5]

DIPHENALDEHYDIC ACID, V, 493;
[1,1'-Biphenyl]-2-carboxylic acid, 2'-formyl- [6720-26-9]

DIPHENIC ACID, I, 222;
[1,1'-Biphenyl]-2,2'-dicarboxylic acid [482-05-3]

DIPHENYLACETALDEHYDE, IV, 375;
Benzeneacetaldehyde, α-phenyl- [947-91-1]

DIPHENYLACETIC ACID, I, 224;
Benzeneacetic acid, α-phenyl- [117-34-0]

α,α-DIPHENYLACETONE, III, 343;
2-Propanone, 1,1-diphenyl- [781-35-1]

DIPHENYLACETONITRILE, III, 347;
Benzeneacetonitrile, α-phenyl- [86-29-3]

DIPHENYLACETYLENE, III, 350; IV, 377;
Benzene, 1,1'-(1,2-ethynediyl)*bis*- [501-65-5]

1,4-DIPHENYL-5-AMINO-1,2,3-TRIAZOLE, AND 4-PHENYL-5-ANILINO-1,2,3-TRI-
AZOLE, IV, 380;
1*H*-1,2,3-Triazol-5-amine, 1,4-diphenyl- [29704-63-0]
1*H*-1,2,3-Triazol-4-amine, *N*,5-diphenyl- [53684-55-2]

N,N'-DIPHENYLBENZAMIDINE, IV, 383;
Benzenecarboximidamide, *N*,*N*'-diphenyl- [2556-46-9]

DIPHENYL-*p*-BROMOPHENYLPHOSPHINE, V, 496;
Phosphine, (4-bromophenyl)diphenyl- [734-59-8]

1,4-DIPHENYL-1,3-BUTADIENE, II, 229; V, 499;
Benzene, 1,1'-(1,3-butadiene-1,4-diyl)*bis*- [886-65-7]

DIPHENYLCARBODIIMIDE (METHODS I AND II), V, 501; 504;
Benzenamine, *N*,*N*'-methanetetraylbis- [622-16-2]

α,β-DIPHENYLCINNAMONITRILE, IV, 387;
Benzeneacetonitrile, α-(diphenylmethylene)- [6304-33-2]

1,1-DIPHENYLCYCLOPROPANE, V, 509;
Benzene, 1,1'-cyclopropylidenebis- [3282-18-6]

DIPHENYLCYCLOPROPENONE, V, 514;
2-Cyclopropen-1-one, 2,3-diphenyl- [886-38-4]

DIPHENYLDIACETYLENE, V, 517;
Benzene, 1,1'-(1,3-butadiyne-1,4-diyl)bis- [886-66-8]

DIPHENYLDIAZOMETHANE, III, 351;
Benzene, 1,1'-(diazomethylene)bis- [883-40-9]

1,1-DIPHENYLETHYLENE, I, 226;
Benzene, 1,1'-ethenylidenebis- [530-48-3]

4,5-DIPHENYLGLYOXALONE, II, 231;
2H-Imidazol-2-One, 1,5-dihydro-4,5-diphenyl- [53684-56-3]

2,3–DIPHENYLINDONE (2,3-DIPHENYL-1-INDENONE), III, 353;
1H-Inden-1-one, 2,3-diphenyl- [1801-42-9]

DIPHENYLIODONIUM IODIDE, III, 355;
Iodonium, diphenyl, iodide [2217-79-0]

DIPHENYLKETENE, III, 356;
Ethenone, diphenyl [525-06-4]

DIPHENYL KETIMINE, V, 520;
Benzenemethanimine, α-phenyl- [1013-88-3]

DIPHENYLMERCURY, I, 228;
Mercury, diphenyl- [587-85-9]

DIPHENYLMETHANE, II, 232;
Benzene, 1,1'-methylenebis- [101-81-5]

DIPHENYLMETHANE IMINE HYDROCHLORIDE, II, 234;
Benzenemethanimine, α-phenyl-, hydrochloride [5319-67-5]

1,1-DIPHENYLPENTANE, V, 523;
Benzene, 1,1'-pentylidenebis- [1726-12-1]

α,β-DIPHENYLPROPIONIC ACID, V, 526;
Benzenepropanoic acid, α-phenyl- [3333-15-1]

β,β-DIPHENYLPROPIOPHENONE, II, 236;
1-Propanone, 1,3,3-triphenyl- [606-86-0]

2,4-DIPHENYLPYRROLE, III, 358;
1H-Pyrrole, 2,4-diphenyl- [3274-56-4]

DIPHENYL SELENIDE, II, 238;
Benzene, 1,1'-selenobis- [1132-39-4]

DIPHENYLSELENIUM DICHLORIDE, II, 240;
Selenium, dichlorodiphenyl- [2217-81-4]

DIPHENYL SUCCINATE, IV, 390;
Butanedioic acid, diphenyl ester [621-14-7]

2,3-DIPHENYLSUCCINONITRILE, IV, 392;
Butanedinitrile, 2,3-diphenyl- [5424-86-2]

DIPHENYL SULFIDE, II, 242;
Benzene, 1,1'-thiobis- [139-66-2]

DIPHENYL TRIKETONE, II, 244;
Propanetrione, diphenyl- [643-75-4]

sym-DIPHENYLUREA, I, 453;
Urea, N,N'-diphenyl- [102-07-8]

p-DITHIANE, IV, 396;
1,4-Dithiane [505-29-3]

DITHIZONE, III, 360;
Diazenecarbothioic acid, phenyl-,2-phenylhydrazide [60-10-6]

1,1-DI-p-TOLYLETHANE, I, 229;
Benzene, 1,1'-ethylidenebis [4-methyl-][530-45-0]

DI-p-TOLYLMERCURY, I, 231;
Mercury, bis(4-methylphenyl)- [537-64-4]

trans-1,2-DIVINYLCYCLOBUTANE, V, 528;
Cyclobutane, 1,2-diethenyl-, trans- [6553-48-6]

cis-1,2-DIVINYLCYCLOBUTANE, V, 528;
Cyclobutane, 1,2-diethenyl-, cis- [16177-46-1]

DOCOSANEDIOIC ACID, V, 533;
Docosanedioic acid [505-56-6]

trans-2-DODECENOIC ACID, IV, 398;
2-Dodecenoic acid, (E)- [32466-54-9]

n-DODECYL BROMIDE, I, 29; II, 246;
Dodecane, 1-bromo- [143-15-7]

n-DODECYL MERCAPTAN, III, 363;
1-Dodecanethiol [112-55-0]

n-DODECYL p-TOLUENESULFONATE, III, 366;
Benzenesulfonic acid, 4-methyl-, dodecyl ester [10157-76-3]

DURENE, II, 248;
Benzene, 1,2,4,5-tetramethyl- [95-93-2]

DUROQUINONE, II, 254;
2,5-Cyclohexadiene-1,4-dione, 2,3,5,6-tetramethyl- [527-17-3]

DYPNONE, III, 367;
2-Buten-1-one, 1,3-diphenyl- [495-45-4]

E

EPIBROMOHYDRIN, II, 256;
Oxirane, (bromomethyl)- [3132-64-7]

EPICHLOROHYDRIN, I, 233; II, 256;
Oxirane, (chloromethyl)- [106-89-8]

ERUCIC ACID, II, 258;
13-Docosenoic acid, (Z)- [112-86-7]

ETHANEDITHIOL, IV, 401;
1,2-Ethanedithiol [540-63-6]

ETHOXYACETIC ACID, II, 260;
Acetic acid, ethoxy- [627-03-2]

ETHOXYACETYLENE, IV, 404;
Ethyne, ethoxy- [927-80-0]

3-ETHOXY-2-CYCLOHEXENONE, V, 539;
2-Cyclohexen-1-one, 3-ethoxy- [5323-87-5]

β-ETHOXYETHYL BROMIDE, III, 370;
Ethane, 1-bromo-2-ethoxy- [592-55-2]

2-ETHOXY-1-NAPHTHALDEHYDE, III, 98;
1-Naphthalenecarboxaldehyde, 2-ethoxy- [19523-57-0]

β-ETHOXYPROPIONALDEHYDE ACETAL, III, 371;
Propane, 1,1,3-triethoxy- [7789-92-6]

β-ETHOXYPROPIONITRILE, III, 372;
Propanenitrile, 3-ethoxy- [2141-62-0]

ETHYL ACETOACETATE, I, 235;
Butanoic acid, 3-oxo-, ethyl ester [141-97-9]

ETHYL ACETONEDICARBOXYLATE, I, 237;
Pentanedioic acid, 3-oxo-, diethyl ester [105-50-0]

ETHYL ACETOPYRUVATE, I, 238;
Pentanoic acid, 2,4-dioxo-, ethyl ester [615-79-2]

ETHYL ACETOSUCCINATE, II, 262;
Propanedioic acid, acetyl-, diethyl ester [570-08-1]

ETHYL α-ACETYL-β-(2,3-DIMETHOXYPHENYL)PROPIONATE, IV, 408;
Benzenepropanoic acid, α-acetyl-2,3-dimethoxy-, ethyl ester [53608-80-3]

ETHYL ADIPATE, II, 264;
Hexanedioic acid, diethyl ester [141-28-6]

N-ETHYLALLENIMINE, V, 541;
Aziridine, 1-ethyl-2-methylene- [872-39-9]

ETHYL p-AMINOBENZOATE, I, 240;
Benzoic acid, 4-amino-, ethyl ester [94-09-7]

ETHYL β-ANILINOCROTONATE, III, 374;
2-Butenoic acid, 3-(phenylamino)-, ethyl ester [31407-07-5]

ETHYL AZODICARBOXYLATE, III, 375; IV, 411; Warning, V, 544;
Diazenedicarboxylic acid, diethyl ester [1972-28-7]

ETHYL BENZALMALONATE, III, 377;
Propanedioic acid, (phenylmethylene)-, diethyl ester [5292-53-5]

ETHYL BENZOYLACETATE, II, 266; III, 379; IV, 415;
Benzenepropanoic acid, β-oxo, ethyl ester [94-02-0]

ETHYL BENZOYLDIMETHYLACETATE, II, 268;
Benzenepropanoic acid, α,α-dimethyl-β-oxo-, ethyl ester [25491-42-3]

ETHYL BENZOYLFORMATE, I, 241;
Benzeneacetic acid, α-oxo-, ethyl ester [1603-79-8]

ETHYL BROMIDE, I, 29, 36;
Ethane, bromo- [74-96-4]

ETHYL BROMOACETATE, III, 381;
Acetic acid, bromo, ethyl ester [105-36-2]

ETHYL γ-BROMOBUTYRATE, V, 545;
Butanoic acid, 4-bromo, ethyl ester [2969-81-5]

ETHYL BROMOMALONATE, I, 245;
Propanedioic acid, bromo-, diethyl ester [685-870]

ETHYL β-BROMOPROPIONATE, I, 246;
Propanoic acid, 3-bromo-, ethyl ester [539-74-2]

ETHYL n-BUTYLACETOACETATE, I, 248;
Hexanoic acid, 2-acetyl-, ethyl ester [1540-29-0]

ETHYL n-BUTYLCYANOACETATE, III, 385;
Hexanoic acid, 2-cyano-, ethyl ester [7391-39-1]

ETHYL n-BUTYLMALONATE, I, 250;
Propanedioic acid, butyl, diethyl ester [133-08-4]

ETHYL t-BUTYL MALONATE, IV, 417;
Propanedioic acid, 1,1-dimethylethyl ethyl ester [32864-38-3]

N-ETHYL-p-CHLOROANILINE, IV, 420;
Benzenamine, 4-chloro-N-ethyl- [13519-75-0]

ETHYL CHLOROFLUOROACETATE, IV, 423;
Acetic acid, chlorofluoro-, ethyl ester [401-56-9]

2-ETHYLCHROMONE, III, 387;
4H-1-Benzopyran-4-one, 2-ethyl- [14736-30-2]

ETHYL CINNAMATE, I, 252;
2-Propenoic acid, 3-phenyl-, ethyl ester [103-36-6]

ETHYL CYANOACETATE, I, 254;
Acetic acid, cyano-, ethyl ester [105-56-6]

ETHYL CYCLOHEXYLIDENEACETATE, V, 547;
Acetic acid, cyclohexylidene, ethyl ester [1552-92-7]

ETHYL DIACETYLACETATE, III, 390;
Butanoic acid, 2-acetyl-3-oxo-, ethyl ester [603-69-0]

ETHYL DIAZOACETATE, III, 392; IV, 424;
Acetic acid, diazo, ethyl ester [623-73-4]

ETHYL α,β-DIBROMO-β-PHENYLPROPIONATE, II, 270;
Benzenepropenoic acid, α,β-dibromo-, ethyl ester [5464-70-0]

ETHYL DIETHOXYACETATE, IV, 427;
Acetic acid, diethoxy, ethyl ester [6065-82-3]

ETHYL 6,7-DIMETHOXY-3-METHYLINDENE-2-CARBOXYLATE, V, 550;
1H-Indene-2-carboxylic acid, 6,7-dimethoxy-3-methyl-, ethyl ester [5415-54-3]

ETHYL p-DIMETHYLAMINOPHENYLACETATE, V, 552;
Benzeneacetic acid, 4-(dimethylamino)-, ethyl ester [17078-29-4]

1-ETHYL-3-(3-DIMETHYLAMINO)PROPYLCARBODIIMIDE HYDROCHLORIDE, V, 555;
1,3-Propanediamine,N'-(ethylcarbonimidoyl)-N,N-dimethyl-,monohydrochloride [25952-53-8]

1-ETHYL-3-(3-DIMETHYLAMINO)PROPYLCARBODIIMIDE METHIODIDE, V, 555;
1-Propanaminium,3-[(ethylcarbonimidoyl)amino]-N,N,N-trimethyl-, iodide [22572-40-3]

ETHYL 2,4-DIPHENYLBUTANOATE, V, 559;
Benzenebutanoic acid, α-phenyl-, ethyl ester [53608-81-4]

ETHYL ENANTHYLSUCCINATE, IV, 430;
Butanedioic acid, (1-oxoheptyl)-, diethyl ester [41117-78-6]

ETHYLENE CYANOHYDRIN, I, 256;
Propanenitrile, 3-hydroxy- [109-78-4]

ETHYLENE SULFIDE, V, 562;
Thiirane [420-12-2]

ETHYLENE THIOUREA, III, 394;
2-Imidazolidinethione [96-45-7]

ETHYLENIMINE, IV, 433;
Aziridine [151-56-4]

ETHYL ETHOXALYLPROPIONATE, II, 272;
Pentanedioic acid, 2-oxo-, diethyl ester [5965-53-7]

ETHYL ETHOXYACETATE, II, 260;
Acetic acid, ethoxy, ethyl ester [817-95-8]

ETHYL ETHOXYMETHYLENEMALONATE, III, 395;
Propanedioic acid, (ethoxymethylene)-, diethyl ether [87-13-8]

ETHYL ETHYLENETETRACARBOXYLATE, II, 273;
Ethenetetracarboxylic acid, tetraethyl ester [6174-95-4]

ETHYL (1-ETHYLPROPENYL)METHYLCYANOACETATE, III, 397;
3-Pentenoic acid, 2-cyano-3-ethyl-2-methyl-, ethyl ester [53608-83-6]

ETHYL (1-ETHYLPROPYLIDENE)CYANOACETATE, III, 399;
2-Pentenoic acid, 2-cyano-3-ethyl-, ethyl ester [868-04-2]

ETHYL 1,16-HEXADECANEDICARBOXYLATE, III, 401;
Octadecanedioic acid, diethyl ester [1472-90-8]

2-ETHYLHEXANONITRILE, IV, 436;
Hexanenitrile, 2-ethyl- [4528-39-6]

ETHYL HYDRAZINECARBOXYLATE, III, 404;
Hydrazinecarboxylic acid, ethyl ester [4114-31-2]

ETHYL HYDROGEN SEBACATE, II, 276;
Decanedioic acid, monoethyl ester [693-55-0]

ETHYL β-HYDROXY-β,β-DIPHENYLPROPIONATE, V, 564;
Benzenepropanoic acid, β-hydroxy-β-phenyl-, ethyl ester [894-18-8]

ETHYL INDOLE-2-CARBOXYLATE, V, 567;
1H-Indole-2-carboxylic acid, ethyl ester [3770-50-1]

ETHYL ISOCYANIDE, IV, 438;
Ethane, isocyano- [624-79-3]

ETHYL α-ISOPROPYLACETOACETATE, III, 405;
Butanoic acid, 2-acetyl-3-methyl-, ethyl ester [1522-46-9]

ETHYL N-METHYLCARBAMATE, II, 278;
Carbamic acid, methyl-, ethyl ester [105-40-8]

β-ETHYL-β-METHYLGLUTARIC ACID, IV, 441;
Pentanedioic acid, 3-ethyl-3-methyl- [5345-01-7]

ETHYL METHYLMALONATE, II, 279;
Propanedioic acid, methyl-, diethyl ester [609-08-5]

·4-ETHYL-2-METHYL-2-OCTENOIC ACID, IV, 444;
2-Octenoic acid, 4-ethyl-2-methyl- [6975-97-9]

5-ETHYL-2-METHYLPYRIDINE, IV, 451;
Pyridine, 5-ethyl-2-methyl- [104-90-5]

α-ETHYL α-METHYLSUCCINIC ACID; V, 572;
Butanedioic acid, 2-ethyl-2-methyl- [631-31-2]

ETHYL α-NAPHTHOATE, II, 282;
1-Naphthalenecarboxylic acid, ethyl ester [3007-97-4]

ETHYL α-NITROBUTYRATE, IV, 454;

Butanoic acid, 2-nitro-, ethyl ester [2531-81-9]

ETHYL ORTHOCARBONATE, IV, 457;

Ethane,1,1′,1″,1‴-[methanetetrayltetrakis(oxy)]tetrakis- [78-09-1]

ETHYL ORTHOFORMATE, I, 258;

Ethane,1,1′,1″-[methylidynetris(oxy)]tris- [122-51-0]

ETHYL OXALATE, I, 261;

Ethanedioic acid, diethyl ester [95-92-1]

ETHYL OXOMALONATE, I, 266;

Propanedioic acid, oxo-, monoethyl ester [27728-17-2]

ETHYL β,β-PENTAMETHYLENEGLYCIDATE, IV, 459;

1-Oxaspiro[2.5]octane-2-carboxylic acid, ethyl ester [6975-17-3]

ETHYL PHENYLACETATE, I, 270;

Benzeneacetic acid, ethyl ester [101-97-3]

ETHYL α-PHENYLACETOACETATE, II, 284;

Benzeneacetic acid, α-acetyl-, ethyl ester [5413-05-8]

ETHYL PHENYLCYANOACETATE, IV, 461;

Benzeneacetic acid, α-cyano-, ethyl ester [4553-07-5]

ETHYL PHENYLCYANOPYRUVATE, II, 287;

Benzenepropanoic acid, β-cyano-α-oxo-, ethyl ester [6362-63-6]

ETHYL(1-PHENYLETHYLIDENE)CYANOACETATE, IV, 463;

2-Butenoic acid, 2-cyano-3-phenyl-, ethyl ester [18300-89-5]

ETHYL N-PHENYLFORMIMIDATE, IV, 464;

Methanimidic acid, N-phenyl-, ethyl ester [6780-49-0]

ETHYL β-PHENYL-β-HYDROXYPROPIONATE, III, 408;

Benzenepropanoic acid, β-hydroxy-, ethyl ester [5764-85-2]

ETHYL PHENYLMALONATE, II, 288;

Propanedioic acid, phenyl-, diethyl ester [83-13-6]

ETHYL PHTHALIMIDOMALONATE, I, 271;

Propanedioic acid, (1,3-dihydro-1,3-dioxo-2H-isoindol-2-yl)-, diethyl ester [5680-61-5]

N-ETHYLPIPERIDINE, V, 575;

Piperidine, 1-ethyl- [766-09-6]

ETHYL PROPANE-1,1,2,3-TETRACARBOXYLATE, I, 272;

1,1,2,3-Propanetetracarboxylic acid, tetraethyl ester [635-03-0]

4-ETHYLPYRIDINE, III, 410;

Pyridine, 4-ethyl- [536-75-4]

ETHYL 2-PYRIDYLACETATE, III, 413;

2-Pyridineacetic acid, ethyl ester [2739-98-2]

ETHYL α-(1-PYRROLIDYL)PROPIONATE, IV, 466;
1-Pyrrolidineacetic acid, α-methyl-, ethyl ester [26846-86-6]

ETHYL PYRUVATE, IV, 467;
Propanoic acid, 2-oxo-, ethyl ester [617-35-6]

N-ETHYL-m-TOLUIDINE, II, 290;
Benzenamine, N-ethyl-3-methyl- [102-27-2]

ETHYL N-TRICARBOXYLATE, III, 415;
Nitridotricarbonic acid, triethyl ester [3206-31-3]

ETHYL n-TRIDECYLATE, II, 292;
Tridecanoic acid, ethyl ester [28267-29-0]

1-ETHYNYLCYCLOHEXANOL, III, 416;
Cyclohexanol, 1-ethynyl- [78-27-3]

1,1'-ETHYNYLENE-bis-CYCLOHEXANOL, IV, 471;
Cyclohexanol, 1,1'-(1,2-ethynediyl)bis- [78-54-6]

o-EUGENOL, III, 418;
Phenol, 2-methoxy-6-(2-propenyl)- [579-60-2]

F

FERROCENE, IV, 473;
Ferrocene [102-54-5]

FERROCENYLACETONITRILE, V, 578;
Ferrocene, (cyanomethyl)- [1316-91-2]

FLAVONE, IV, 478;
4H-1-Benzopyran-4-one, 2-phenyl- [525-82-6]

9-FLUORENECARBOXYLIC ACID, IV, 482;
9H-Fluorene-9-carboxylic acid [1989-33-9]

FLUORENONE-2-CARBOXYLIC ACID, III, 420;
9H-Fluorene-2-carboxylic acid, 9-oxo- [784-50-9]

FLUOROBENZENE, II, 295;
Benzene, fluoro- [462-06-6]

p-FLUOROBENZOIC ACID, II, 299;
Benzoic acid, 4-fluoro- [456-22-4]

2-FLUOROHEPTANOIC ACID, V, 580;
Heptanoic acid, 2-fluoro- [1578-58-1]

FORMAMIDINE ACETATE, V, 582;
Methanimidamide, monoacetate [3473-63-0]

5-FORMYL-4-PHENANTHROIC ACID, IV, 484;
4-Phenanthrenecarboxylic acid, 5-formyl- [5684-15-1]

FUMARIC ACID, II, 302;
2-Butenedioic acid (E)- [110-17-8]

FUMARONITRILE, IV, 486;
2-Butenedinitrile, (E)- [764-42-1]

FUMARYL CHLORIDE, III, 422;
2-Butenedioyl dichloride, (E)- [627-63-4]

FURAN, I, 274;
Furan [110-00-9]

2-FURANCARBOXYLIC ACID, I, 276;
2-Furoic acid [88-14-2]

FURFURAL, I, 280;
2-Furancarboxaldehyde [98-01-1]

2-FURFURALACETONE, I, 283;
3-Buten-2-one, 4-(2-furanyl)- [623-15-4]

FURFURAL DIACETATE, IV, 489;
Methanediol, 2-furanyl, diacetate [613-75-2]

2-FURFURYL MERCAPTAN, IV, 491;
2-Furanmethanethiol, [98-02-2]

2-FUROIC ACID, IV, 493;
2-Furancarboxylic acid [88-14-2]

FURYLACRYLIC ACID, III, 425;
2-Propenoic acid, 3-(2-furanyl)- [539-47-9]

3-(2-FURYL)ACRYLONITRILE, V, 585;
2-Propenenitrile, 3-(2-furanyl)- [7187-01-1]

2-FURYLCARBINOL, I, 276;
2-Furanmethanol [98-00-0]

2-FURYLMETHYL ACETATE, I, 285;
2-Furanmethanol, acetate [623-17-6]

G

GALLACETOPHENONE, II, 304;
Ethanone,1-(2,3,4-trihydroxyphenyl)- [528-21-2]

β-GENTIOBIOSE OCTAACETATE, III, 428;
β-D-Glucopyranose,6-O-(2,3,4,6-tetra-O-acetyl-β-D-glucopyranosyl)-, tetraacetate [4613-78-9]

d-GLUCOSAMINE HYDROCHLORIDE, III, 430;
D-Glucose, 2-amino-2-deoxy-, hydrochloride [66-84-2]

β-d-GLUCOSE-1,2,3,4-TETRAACETATE, III, 432;
β-D-Glucopyranose, 1,2,3,4-tetraacetate [13100-46-4]

β-d-GLUCOSE-2,3,4,6-TETRAACETATE, III, 434;
β-D-Glucopyranose, 2,3,4,6-tetraacetate [3947-62-4]

d-GLUTAMIC ACID, I, 286;
D-Glutamic acid [6893-26-1]

GLUTARIC ACID, I, 289; IV, 496;
Pentanedioic acid [110-94-1]

GLUTARIMIDE, IV, 496;
2,6-Piperidinedione [1121-89-7]

dl-GLYCERALDEHYDE, II, 305;
Propanal, 2,3-dihydroxy, (±)- [56-82-6]

dl-GLYCERALDEHYDE DIETHYL ACETAL, II, 307;
1,2-Propanediol,3,3-diethoxy-, (±)- [10487-05-5]

GLYCEROL α,γ-DIBROMOHYDRIN, II, 308;
2-Propanol, 1,3-dibromo- [96-21-9]

GLYCEROL α,γ-DICHLOROHYDRIN, I, 292;
2-Propanol, 1,3-dichloro- [96-23-1]

GLYCEROL α-MONOCHLOROHYDRIN, I, 294;
1,2-Propanediol, 3-chloro- [96-24-2]

α-GLYCERYL PHENYL ETHER, I, 296;
1,2-Propanediol, 3-phenoxy- [538-43-2]

GLYCINE, I, 298;
Glycine [56-40-6]

GLYCINE *t*-BUTYL ESTER, V, 586;
Glycine,1,1-dimethylethyl ester [6456-74-2]

GLYCINE ETHYL ESTER HYDROCHLORIDE, II, 310;
Glycine, ethyl ester, hydrochloride [623-33-6]

GLYCOLONITRILE, III, 436;
Acetonitrile, hydroxy- [107-16-4]

GLYOXAL BISULFITE, III, 438;
1,2-Ethanedisulfonic acid, 1,2-dihydroxy- [18381-20-9]

GUANIDINE NITRATE, I, 302; *Warning*, V, 589;
Guanidine, mononitrate [506-93-4]

GUANIDOACETIC ACID, III, 440;
Glycine, *N*-(aminoiminomethyl)- [352-97-6]

GUANYLTHIOUREA, IV, 502;
Thiourea, (aminoiminomethyl)- [2114-02-5]

D-GULONIC-γ-LACTONE, IV, 506;
D-Gulonic acid, γ-lactone [6322-07-2]

H

HEMIMELLITENE, IV, 508;
Benzene, 1,2,3-trimethyl- [526-73-8]

HEMIN, III, 442;
Iron, chloro[7,12-diethenyl-3,8,13,17-tetramethyl-21H,23H-porphine-2,18-dipropa-
noato(2-)-N^{21}, N^{22},N^{23},N^{24}-, (*SP*-5-13)- [15489-47-1]

HENDECANEDIOIC ACID, IV, 510;
Undecanedioic acid [1852-04-6]

unsym-HEPTACHLOROPROPANE, II, 312;
Propane, 1,1,1,2,2,3,3-heptachloro- [594-89-8]

HEPTALDOXIME, II, 313;
Heptanal, oxime [629-31-2]

n-HEPTAMIDE, IV, 513;
Heptanamide [628-62-6]

n-HEPTANOIC ACID, II, 315;
Heptanoic acid [111-14-8]

2-HEPTANOL, II, 317;
2-Heptanol [543-49-7]

4-HEPTANONE, V, 589;
4-Heptanone [123-19-3]

n-HEPTYL ALCOHOL, I, 304;
1-Heptanol [111-70-6]

n-HEPTYLAMINE, II, 318;
1-Heptanamine [111-68-2]

3-*n*-HEPTYL-5-CYANOCYTOSINE, IV, 515;
5-Pyrimidinecarbonitrile, 1-heptyl-1,2,3,6-tetrahydro-6-imino-2-oxo- [53608-90-5]

o-*n*-HEPTYLPHENOL, III, 444;
Phenol, 2-heptyl- [5284-22-0]

n-HEXADECANE, II, 320;
Hexadecane [544-76-3]

n-HEXADECYL IODIDE, II, 322;
Hexadecane, 1-iodo- [544-77-4]

HEXAHYDROGALLIC ACID AND HEXAHYDROGALLIC ACID TRIACETATE, V, 591;
Cyclohexanecarboxylic acid, 3,4,5-trihydroxy-, (1α,3α,4α,5α)- [53796-39-7]
Cyclohexanecarboxylic acid, 3,4,5-tris(acetyloxy)-, 1α,3α,4α,5α)- [53796-40-0]

HEXAHYDRO-1,3,5-TRIPROPIONYL-*s*-TRIAZINE, IV, 518;
1,3,5-Triazine, hexahydro-1,3,5-tris(1-oxopropyl)- [30805-19-7]

HEXAHYDROXYBENZENE, V, 595;
Benzenehexol [608-80-0]

n-HEXALDEHYDE, II, 323;
Hexanal [66-25-1]

HEXAMETHYLBENZENE, IV, 520;
Benzene, hexamethyl- [87-85-4]

2,3,4,5,6,6-HEXAMETHYL-2,4-CYCLOHEXADIEN-1-ONE, V, 598;
2,4-Cyclohexadien-1-one, 2,3,4,5,6,6-hexamethyl- [3854-96-4]

HEXAMETHYLENE CHLOROHYDRIN, III, 446;
1-Hexanol, 6-chloro- [2009-83-8]

HEXAMETHYLENE DIISOCYANATE, IV, 521;
Hexane, 1,6-diisocyanato- [822-06-0]

HEXAMETHYLENE GLYCOL, II, 325;
1,6-Hexanediol [629-11-8]

HEXAMETHYLPHOSPHOROUS TRIAMIDE, V, 602;
Phosphorous triamide, hexamethyl [1608-26-0]

HEXAPHENYLBENZENE, V, 604;
1,1':2',1''-Terphenyl, 3',4',5',6'-tetraphenyl- [992-04-1]

1,3,5-HEXATRIENE, V, 608;
1,3,5-Hexatriene [2235-12-3]

n-HEXYL ALCOHOL, I, 306;
1-Hexanol [111-27-3]

n-HEXYL FLUORIDE, IV, 525;
Hexane, 1-fluoro- [373-14-8]

HIPPURIC ACID, II, 328;
Glycine, N-benzoyl- [495-69-2]

l-HISTIDINE MONOHYDROCHLORIDE, II, 330;
L-Histidine, monohydrochloride [645-35-2]

HOMOPHTHALIC ACID, V, 612;
Benzeneacetic acid, 2-carboxy- [89-51-0]

HOMOPHTHALIC ACID AND ANHYDRIDE, III, 449;
1H-2-Benzopyran-1,3(4H)-dione [703-59-3]

HOMOVERATRIC ACID, II, 333;
Benzeneacetic acid, 3,4-dimethoxy- [93-40-3]

HYDRAZINE SULFATE, I, 309;
Hydrazine, sulfate (1:1) [10034-93-2]

α-HYDRINDONE, II, 336;
1H-Inden-1-one, 2,3-dihydro- [83-33-0]

HYDROCINNAMIC ACID, I, 311;
Benzenepropanoic acid [501-52-0]

HYDROGEN BROMIDE (ANHYDROUS), II, 338;
Hydrobromic acid [10035-10-6]

HYDROGEN CYANIDE (ANHYDROUS), I, 314;
Hydrocyanic acid [74-90-8]

HYDROQUINONE DIACETATE, III, 452;
1,4-Benzenediol, diacetate [1205-91-0]

m-HYDROXYBENZALDEHYDE, III, 453;
Benzaldehyde, 3-hydroxy- [100-83-4]

p-HYDROXYBENZOIC ACID, II, 341;
Benzoic acid, 4-hydroxy- [99-96-7]

4-HYDROXY-1-BUTANESULFONIC ACID SULTONE, IV, 529;
1,2-Oxathiane, 2,2-dioxide [1633-83-6]

2-HYDROXYCINCHONINIC ACID, III, 456;
4-Quinolinecarboxylic acid, 1,2-dihydro-2-oxo- [15733-89-8]

2-HYDROXY-3,5-DIIODOBENZOIC ACID, II, 343;
Benzoic acid, 2-hydroxy-3,5-diiodo- [133-91-5]

β-(2-HYDROXYETHYLMERCAPTO)PROPIONITRILE, III, 458;
Propanenitrile, 3-[(2-hydroxyethyl)thio]- [15771-37-6]

β-HYDROXYETHYL METHYL SULFIDE, II, 345;
Ethanol, 2-(methylthio)- [5271-38-5]

3-HYDROXYGLUTARONITRILE, V, 614;
Pentanedinitrile, 3-hydroxy- [13880-89-2]

HYDROXYHYDROQUINONE TRIACETATE, I, 317;
1,2,4-Benzenetriol, triacetate [613-03-6]

2-HYDROXYISOPHTHALIC ACID, V, 617;
1,3-Benzenedicarboxylic acid, 2-hydroxy- [606-19-9]

HYDROXYLAMINE HYDROCHLORIDE, I, 318;
Hydroxylamine, hydrochloride [5470-11-1]

HYDROXYMETHYLFERROCENE, V, 621;
Ferrocene, (hydroxymethyl)- [1273-86-5]

4(5)-HYDROXYMETHYLIMIDAZOLE HYDROCHLORIDE, III, 460;
1*H*-Imidazole-4-methanol, monohydrochloride [32673-41-9]

2-HYDROXY-3-METHYLISOCARBOSTYRYL, V, 623;
1(2*H*)-Isoquinolinone, 2-hydroxy-3-methyl- [7114-79-6]

2-HYDROXY-1-NAPHTHALDEHYDE, III, 463;
1-Naphthalenecarboxaldehyde, 2-hydroxy- [708-06-5]

2-HYDROXY-1,4-NAPHTHOQUINONE, III, 465;
1,4-Naphthalenedione, 2-hydroxy- [83-72-7]

6-HYDROXYNICOTINIC ACID, IV, 532;
3-Pyridinecarboxylic acid, 1,6-dihydro-6-oxo- [5006-66-6]

2-HYDROXY-5-NITROBENZYL CHLORIDE, III, 468;
Phenol, 2-(chloromethyl)-4-nitro- [2973-19-5]

5-HYDROXYPENTANAL, III, 470;
Pentanal, 5-hydroxy- [4221-03-8]

p-HYDROXYPHENYLPYRUVIC ACID, V, 627;
Benzenepropanoic acid, 4-hydroxy-α-oxo- [156-39-8]

β-HYDROXYPROPIONIC ACID, I, 321;
Propanoic acid, 3-hydroxy- [503-66-2]

3-HYDROXYPYRENE, V, 632;
1-Pyrenol [5315-79-7]

3-HYDROXYQUINOLINE, V, 635;
3-Quinolinol [580-18-7]

3-HYDROXYTETRAHYDROFURAN, IV, 534;
3-Furanol, tetrahydro- [453-20-3]

2-HYDROXYTHIOPHENE, V, 642;
Thiophene-2-ol [17236-58-7]

HYDROXYUREA, V, 645;
Urea, hydroxy- [127-07-1]

I

IMIDAZOLE, III, 471;
1*H*-Imidazole [288-32-4]

2-INDANONE, V, 647;
2*H*-Inden-2-one, 1,3-dihydro- [615-13-4]

INDAZOLE, III, 475; IV, 536; V, 650;
1*H*-Indazole [271-44-3]

INDOLE, III, 479;
1*H*-Indole [120-72-9]

INDOLE-3-ACETIC ACID, V, 654;
1*H*-Indole-3-acetic acid [87-51-4]

INDOLE-3-ALDEHYDE, IV, 539;
1*H*-Indole-3-carboxaldehyde [487-89-8]

INDOLE-3-CARBONITRILE, V, 656;
1*H*-Indole-3-carbonitrile [5457-28-3]

p-IODOANILINE, II, 347;
Benzenamine, 4-iodo- [540-37-4]

5-IODOANTHRANILIC ACID, II, 349;
Benzoic acid, 2-amino-5-iodo- [5326-47-6]

IODOBENZENE, I, 323; II, 351;
Benzene, iodo- [591-50-4]

IODOBENZENE DICHLORIDE, III, 482;
Benzene, (dichloroiodo)- [932-72-9]

m-IODOBENZOIC ACID, II, 353;
Benzoic acid, 3-iodo- [618-51-9]

p-IODOBENZOIC ACID, I, 325;
Benzoic acid, 4-iodo- [619-58-9]

IODOCYCLOHEXANE, IV, 543;
Cyclohexane, iodo- [626-62-0]

o-IODOPHENOL, I, 326;
Phenol, 2-iodo- [533-58-4]

p-IODOPHENOL, II, 355;
Phenol, 4-iodo- [540-38-5]

IODOSOBENZENE, III, 483; V, 658;
Benzene, iodosyl- [536-80-1]

IODOSOBENZENE DIACETATE, V, 660;
Iodine, bis(acetato-*O*)phenyl- [3240-34-4]

N-IODOSUCCINIMIDE, V, 663;
2,5-Pyrrolidinedione, 1-iodo- [516-12-1]

2-IODOTHIOPHENE, II, 357; IV, 545;
Thiophene, 2-iodo- [3437-95-4]

4-IODOVERATROLE, IV, 547;
Benzene, 4-iodo-1,2-dimethoxy- [5460-32-2]

IODOXYBENZENE, III, 485; V, 665;
Benzene, iodoxy- [696-33-3]

ISATIN, I, 327;
1*H*-Indole-2,3-dione [91-56-5]

ISATOIC ANHYDRIDE, III, 488;
2*H*-3,1-Benzoxazine-2,4(1*H*)-dione [118-48-9]

ISOBUTYL BROMIDE, II, 358;
Propane, 1-bromo-2-methyl- [78-77-3]

ISOBUTYRAMIDE, III, 490;
Propanamide, 2-methyl- [563-83-7]

ISOBUTYRONITRILE, III, 493;
Propanenitrile, 2-methyl- [78-82-0]

ISODEHYDROACETIC ACID AND ETHYL ISODEHYDROACETATE, IV, 549;
2*H*-Pyran-3-carboxylic acid, 4,6-dimethyl-2-oxo- [33953-26-3]
2*H*-Pyran-5-carboxylic acid, 4,6-dimethyl-2-oxo-, ethyl ester [3385-34-0]

ISODURENE, II, 360;
Benzene, 1,2,3,5-tetramethyl- [527-53-7]

dl-ISOLEUCENE, III, 495;
DL-Isoleucine [443-79-8]

ISONITROSOPROPIOPHENONONE, II, 363;
1,2-Propanedione, 1-phenyl-, 2-oxime [119-51-7]

ISOPHORONE OXIDE, IV, 552;
  7-Oxabicyclo[4.1.0]heptan-2-one, 4,4,6-trimethyl- [10276-21-8]

ISOPHTHALALDEHYDE, V, 668;
  1,3-Benzenedicarboxaldehyde [626-19-7]

ISOPRENE CYCLIC SULFONE, III, 499;
  Thiophene, 2,5-dihydro-3-methyl-, 1,1-dioxide [1193-10-8]

2-ISOPROPYLAMINOETHANOL, III, 501;
  Ethanol, 2-[(1-methylethyl)amino]- [109-56-8]

dl-ISOPROPYLIDENEGLYCEROL, III, 502;
  1,3-Dioxolane-4-methanol, 2,2-dimethyl-, (±)- [22323-83-7]

ISOPROPYL LACTATE, II, 365;
  Propanoic acid, 2-hydroxy-, 1-methylethyl ester [617-51-6]

ISOPROPYL THIOCYANATE, II, 366;
  Thiocyanic acid 1-methylethyl ester [625-59-2]

3-ISOQUINUCLIDONE, V, 670;
  2-Azobicyclo[2.2.2]octan-3-one [3306-69-2]

β-ISOVALEROLACTAM-N-SULFONYL CHLORIDE AND β-ISOVALEROLACTAM, V, 673;
  Azetidinesulfonyl chloride, 2,2-dimethyl-4-oxo [17174-96-8]
  2-Azetidinone, 4,4-dimethyl- [4879-95-2]

ITACONIC ANHYDRIDE, AND, ITACONIC ACID, II, 368;
  2,5-Furandione, dihydro-3-methylene- [2170-03-8]
  Butanedioic acid, methylene- [97-65-4]

ITACONYL CHLORIDE, IV, 554;
  Butanedioyl dichloride, methylene- [1931-60-8]

J

JULOLIDINE, III, 504;
  1H,5H-Benzo[ij]quinolizine, 2,3,6,7-tetrahydro- [479-59-4]

K

KETENE, I, 330; V, 679;
  Ethenone [463-51-4]

KETENE DIETHYLACETAL, III, 506;
  Ethene, 1,1-diethoxy- [2678-54-8]

KETENE DIMER, III, 508;
  2-Oxetanone, 4-methylene- [674-82-8]

KETENE DI(2-METHOXYETHYL) ACETAL, V, 684;
  2,5,7,10-Tetraoxaundecane, 6-methylene- [5130-02-9]

α-KETOGLUTARIC ACID, III, 510; V, 687;
  Pentanedioic acid, 2-oxo- [328-50-7]

6-KETOHENDECANEDIOIC ACID, IV, 555;
Undecanedioic acid, 6-oxo- [3242-53-3]

2-KETOHEXAMETHYLENIMINE, II, 371;
2H-Azepin-2-one, hexahydro- [105-60-2]

β-KETOISOOCTALDEHYDE DIMETHYL ACETAL, IV, 558;
2-Hexanone, 1,1-dimethoxy-5-methyl- [53684-58-5]

D,L,-KETOPINIC ACID, V, 689;
Bicyclo[2.2.1]heptane-1-carboxylic acid, 7,7-dimethyl-, (±)- [53684-59-6]

KRYPTOPYRROLE, III, 513;
1H-pyrrole, 3-ethyl-2,4-dimethyl- [517-22-6]

L

LACTAMIDE, III, 516;
Propanamide, 2-hydroxy- [2043-43-8]

18,20-LACTONE OF 3β-ACETOXY-20β-HYDROXY-5-PREGNENE-18-OIC ACID, V, 692;
Pregn-5-en- 18-oic acid, 3-(acetyloxy)-20-hydroxy-, γ-lactone, (3β,20R)- [3020-10-8]

LAURONE, IV, 560;
12-Tricosanone [540-09-0]

LAURYL ALCOHOL, II, 372;
1-Dodecanol [112-53-8]

LAURYLMETHYLAMINE, IV, 564;
1-Dodecanamine, N-methyl- [7311-30-0]

LEPIDINE, III, 519;
Quinoline, 4-methyl- [491-35-0]

dl-LEUCINE, III, 523;
DL-Leucine [328-39-2]

LEVOPIMARIC ACID, V, 699;
1-Phenanthrenecarboxylic acid,1,2,3,4,4a,4b,5,9,10,10a-decahydro-1,4a-dimethyl-7-(1-methylethyl)-,[1R-(1α,4aβ,4bα,10aα)]- [79-54-9]

LEVULINIC ACID, I, 335;
Pentanoic acid, 4-oxo- [123-76-2]

LINOLEIC ACID, III, 526;
9,12-Octadecadienoic acid (Z,Z)- [60-33-3]

LINOLENIC ACID, III, 531;
9,12,15-Octadecatrienoic acid, (Z,Z,Z)- [463-40-1]

dl-LYSINE HYDROCHLORIDES, MONO, II, 374;
DL-Lysine, monohydrochloride [70-53-1]

dl-LYSINE HYDROCHLORIDES, DI-, II, 374;
DL-Lysine, dihydrochloride [617-68-5]

M

MALONIC ACID, II, 376;
   Propanedioic acid [141-82-2]

MALONONITRILE, II, 379; III, 535;
   Propanedinitrile [109-77-3]

MANDELAMIDE, III, 536;
   Benzeneacetamide, α-hydroxy- [4410-31-5]

MANDELIC ACID, I, 336; III, 538;
   Benzeneacetic acid, α-hydroxy- [90-64-2]

D-MANNOSE, III, 541;
   D-Mannose [3458-28-4]

l-MENTHONE, I, 340;
   Cyclohexanone, 5-methyl-2-(1-methylethyl)-,(2S-trans)- [14073-97-3]

l-MENTHOXYACETIC ACID, III, 544;
   Acetic acid, [[5-methyl-2-(1-methylethyl)cyclohexyl]oxy]-, [1R-(1a,2p,5a)]- [40248-63-3]

l-MENTHOXYACETYL CHLORIDE, III, 547;
   Acetyl chloride, [[5-methyl-2-(1-methylethyl)cyclohexyl]oxy]-, [1R-(1a,2p,5a)]- [15356-62-4]

2-MERCAPTO-4-AMINO-5-CARBETHOXYPYRIMIDINE, IV, 566;
   5-Pyrimidinecarboxylic acid, 4-amino-1,2-dihydro-2-thioxo-, ethyl ester [774-07-2]

2-MERCAPTOBENZIMIDAZOLE, IV, 569;
   2H-Benzimidazole-2-thione, 1,3-dihydro- [583-39-1]

2-MERCAPTO-4-HYDROXY-5-CYANOPYRIMIDINE, IV, 566;
   5-Pyrimidinecarbonitrile, 1,2,3,4-tetrahydro-4-oxo-2-thioxo- [23945-49-5]

2-MERCAPTOPYRIMIDINE, V, 703;
   2(1H)-Pyrimidinethione [1450-85-7]

MERCURY DI-β-NAPHTHYL, II, 381;
   Mercury, di-2-naphthalenyl- [19510-26-0]

MESACONIC ACID, II, 382;
   2-Butenedioic acid, 2-methyl-, (E)- [498-24-8]

MESITALDEHYDE, III, 549;
   Benzaldehyde, 2,4,6-trimethyl- [487-68-3]

MESITOIC ACID, III, 553; V, 706;
   Benzoic acid, 2,4,6-trimethyl- [480-63-7]

MESITOIC ACID AND MESITOYL CHLORIDE, III, 555;
   Benzoyl chloride, 2,4,6-trimethyl- [938-18-1]

MESITYLACETIC ACID, III, 557;
   Benzeneacetic acid, 2,4,6-trimethyl- [4408-60-0]

MESITYLENE, I, 341;
   Benzene, 1,3,5-trimethyl- [108-67-8]

MESITYL OXIDE, I, 345;
3-Penten-2-one, 4-methyl- [141-79-7]

METHACRYLAMIDE, III, 560;
2-Propenamide, 2-methyl- [79-39-0]

METHANESULFINYL CHLORIDE, V, 709;
Methanesulfinyl chloride [676-85-7]

METHANESULFONYL CHLORIDE, IV, 571;
Methanesulfonyl chloride [124-63-0]

dl-METHIONINE, II, 384;
DL-Methionine [59-51-8]

METHOXYACETONITRILE, II, 387;
Acetonitrile, methoxy- [1738-36-9]

ω-METHOXYACETOPHENONE, III, 562;
Ethanone, 2-methoxy-1-phenyl- [4079-52-1]

m-METHOXYBENZALDEHYDE, III, 564;
Benzaldehyde, 3-methoxy- [591-31-1]

1-(2-METHOXYCARBONYLPHENYL)PYRROLE, V, 716;
Benzoic acid, 2-(1H-pyrrol-1-yl)-, methyl ester [10333-67-2]

2-METHOXYDIPHENYL ETHER, III, 566;
Benzene, 1-methoxy-2-phenoxy- [1695-04-1]

6-METHOXY-8-NITROQUINOLINE, III, 568;
Quinoline, 6-methoxy-8-nitro- [85-81-4]

o-METHOXYPHENYLACETONE, IV, 573;
2-Propanone, 1-(2-methoxyphenyl)- [5211-62-1]

p-METHOXYPHENYLACETONITRILE, IV, 576;
Benzeneacetonitrile, 4-methoxy- [104-47-2]

1-(p-METHOXYPHENYL)-5-PHENYL-1,3,5-PENTANETRIONE, V, 718;
1,3,5-Pentanetrione, 1-(4-methoxyphenyl)-5-phenyl- [1678-17-7]

2-(p-METHOXYPHENYL)-6-PHENYL-4-PYRONE, V, 721;
4H-Pyran-4-one, 2-(4-methoxyphenyl)-6-phenyl- [14116-43-9]

METHYL p-ACETYLBENZOATE, IV, 579;
Benzoic acid, 4-acetyl-, methyl ester [3609-53-8]

METHYLAMINE HYDROCHLORIDE, I, 347;
Methanamine, hydrochloride [593-51-1]

1-METHYLAMINOANTHRAQUINONE, III, 573;
9,10-Anthracenedione, 1-(methylamino)- [82-38-2]

1-METHYLAMINO-4-BROMOANTHRAQUINONE, III, 575;
9,10-Anthracenedione, 1-bromo-4-(methylamino)- [128-93-8]

METHYL n-AMYL KETONE, I, 351;
2-Heptanone [110-43-0]

β-METHYLANTHRAQUINONE, I, 353;
9,10-Anthracenedione, 2-methyl- [84-54-8]

METHYL BENZENESULFINATE, V, 723;
Benzenesulfinic acid, methyl ester [670-98-4]

o-METHYLBENZYL ALCOHOL, IV, 582;
Benzenemethanol, 2-methyl- [89-95-2]

2-METHYLBENZYLDIMETHYLAMINE, IV, 585;
Benzenemethanamine,N,N,2-trimethyl- [4525-48-8]

METHYL BENZYL KETONE, II, 389;
2-Propanone, 1-phenyl- [103-79-7]

10-METHYL-10,9-BORAZAROPHENANTHRENE, V, 727;
Dibenz[c,e][1,2]azaborine, 5,6-dihydro-6-methyl- [15813-13-5]

METHYL β-BROMOPROPIONATE, III, 576;
Propanoic acid, 3-bromo-, methyl ester [3395-91-3]

METHYL 5-BROMOVALERATE, III, 578;
Pentanoic acid, 5-bromo-, methyl ester [5454-83-1]

METHYL BUTADIENOATE, V, 734;
2,3-Butadienoic acid, methyl ester [18913-35-4]

N-METHYLBUTYLAMINE, V, 736;
1-Butanamine N-methyl- [110-68-9]

O-METHYLCAPROLACTIM, IV, 588;
2H-Azepine, 3,4,5,6-tetrahydro-7-methoxy- [2525-16-8]

4-METHYLCARBOSTYRIL, III, 580;
2(1H)-Quinolinone, 4-methyl- [607-66-9]

4-METHYLCOUMARIN, III, 581;
2H-1-Benzopyran-2-one, 4-methyl- [607-71-6]

3-METHYLCOUMARONE, IV, 590;
Benzofuran, 3-methyl- [21535-97-7]

METHYL ω-CYANOPELARGONATE, III, 584;
Nonanoic acid, 9-cyano-, methyl ester [53663-26-6]

1-METHYLCYCLOHEXANECARBOXYLIC ACID, V, 739;
Cyclohexanecarboxylic acid, 1-methyl- [1123-25-7]

2-METHYL-1,3-CYCLOHEXANEDIONE, V, 743;
1,3-Cyclohexanedione, 2-methyl- [1193-55-1]

METHYL CYCLOPENTANECARBOXYLATE, IV, 594;
Cyclopentanecarboxylic acid, methyl ester [4630-80-2]

2-METHYLCYCLOPENTANE-1,3-DIONE, V, 747;
1,3-Cyclopentanedione, 2-methyl- [765-69-5]

METHYL CYCLOPROPYL KETONE, IV, 597;
Ethanone, 1-cyclopropyl- [765-43-5]

2-METHYL-2,5-DECANEDIOL, IV, 601;
2,5-Decanediol, 2-methyl- [53731-34-3]

N-METHYL-3,4-DIHYDROXYPHENYLALANINE, III, 586;
DL-Tyrosine, 3-hydroxy-N-methyl- [53663-27-7]

N-METHYL-2,3-DIMETHOXYBENZYLAMINE, IV, 603;
Benzenemethanamine, 2,3-dimethoxy-N-methyl- [53663-28-8]

N-METHYL-1,2-DIPHENYLETHYLAMINE AND HYDROCHLORIDE, IV, 605;
Benzeneethanamine, N-methyl-α-phenyl- [53663-25-5]
Benzeneethanamine, N-methyl-α-phenyl-, hydrochloride [7400-77-3]

trans-2-METHYL-2-DODECENOIC ACID, IV, 608;
2-Dodecenoic acid, 2-methyl-, (E)- [53663-29-9]

METHYLENEAMINOACETONITRILE, I, 355;
Acetonitrile, (methyleneamino)- [109-82-0]

METHYLENE BROMIDE, I, 357;
Methane, dibromo- [74-95-3]

METHYLENECYCLOHEXANE, V, 751;
Cyclohexane, methylene- [1192-37-6]

METHYLENECYCLOHEXANE AND N,N-DIMETHYLHYDROXYLAMINE HYDRO-
CHLORIDE, IV, 612;
Cyclohexane, methylene- [1192-37-6]
Methanamine, N-hydroxy-N-methyl-, hydrochloride [16645-06-0]

METHYLENECYCLOHEXANE OXIDE, V, 755;
1-Oxaspiro[2.5]octane [185-70-6]

2-METHYLENEDODECANOIC ACID, IV, 616;
Dodecanoic acid, 2-methylene- [52756-21-5]

METHYLENE IODIDE, I, 358;
Methane, diiodo- [75-11-6]

4-METHYLESCULETIN, I, 360;
2H-1-Benzopyran-2-one, 6,7-dihydroxy-4-methyl- [529-84-0]

dl-METHYLETHYLACETIC ACID, I, 361;
Butanoic acid, 2-methyl-, (±)- [600-07-7]

N-METHYLETHYLAMINE, V, 758;
Ethanamine, N-Methyl- [624-78-2]

1-METHYL-3-ETHYLOXINDOLE, IV, 620;
2H-Indol-2-one, 3-ethyl-1,3-dihydro-1-methyl- [2525-35-1]
9-METHYLFLUORENE, IV, 623;
9H-Fluorene, 9-methyl- [2523-37-7]
N-METHYLFORMANILIDE, III, 590;
Formamide, N-methyl-N-phenyl- [93-61-8]

5-METHYLFURFURAL, II, 393;
2-Furancarboxaldehyde, 5-methyl- [620-02-0]

5-METHYLFURFURYLDIMETHYLAMINE, IV, 626;
2-Furanmethanamine, N,N,5-trimethyl- [14496-35-6]

3-METHYL-2-FUROIC ACID AND 3-METHYLFURAN, IV, 628;
2-Furancarboxylic acid, 3-methyl [4412-96-8]
Furan, 3-methyl- [930-27-8]

α-METHYL d-GLUCOSIDE, I, 364;
α-D-Glucopyranoside, methyl [97-30-3]

β-METHYLGLUTARIC ACID, III, 591;
Pentanedioic acid, 3-methyl- [626-51-7]

β-METHYLGLUTARIC ANHYDRIDE, IV, 630;
2H-Pyran-2,6(3H)-dione, dihydro-4-methyl- [4166-53-4]

METHYLGLYOXAL-ω-PHENYLHYDRAZONE, IV, 633;
Propanal, 2-oxo-, 1-(phenylhydrazone) [5391-74-2]

3-METHYLHEPTANOIC ACID, V, 762;
Heptanoic acid, 3-methyl- [53663-30-2]

5-METHYL-5-HEXEN-2-ONE, V, 767;
5-Hexen-2-one, 5-methyl- [3240-09-3]

METHYL-n-HEXYLCARBINOL, I, 366;
2-Octanol [123-96-6]

METHYLHYDRAZINE SULFATE, II, 395;
Hydrazine, methyl-, sulfate (1:1) [302-15-8]

METHYL HYDROGEN HENDECANEDIOATE, IV, 635;
Undecanedioic acid, monomethyl ester [3927-60-4]

4-METHYL-6-HYDROXYPYRIMIDINE, IV, 638;
4(1H)-Pyrimidinone, 6-methyl- [3524-87-6]

2-METHYL-4-HYDROXYQUINOLINE, III, 593;
4-Quinolinol, 2-methyl- [607-67-0]

METHYLIMINODIACETIC ACID, II, 397;
Glycine, N-(carboxymethyl)-N-methyl- [4408-64-4]

1-METHYL-2-IMINO-β-NAPHTHOTHIAZOLINE, III, 595;
Naphtho[1,2-d]thiazol-2(1H)-imine, 1-methyl- [53663-31-3]

1-METHYLINDOLE, V, 769;
1H-Indole, 1-methyl- [603-76-9]

2-METHYLINDOLE, III, 597;
1H-Indole, 2-methyl- [95-20-5]

METHYL IODIDE, II, 399;
Methane, iodo- [74-88-4]

METHYL ISOCYANIDE, V, 772;
Methane, isocyano- [593-75-9]

METHYL ISOPROPYL CARBINOL, II, 406;
2-Butanol, 3-methyl- [598-75-4]

METHYL ISOPROPYL KETONE, II, 408;
2-Butanone, 3-methyl- [563-80-4]

1-METHYLISOQUINOLINE, IV, 641;
Isoquinoline, 1-methyl- [1721-93-3]

METHYL ISOTHIOCYANATE, III, 599;
Methane, isothiocyanato- [556-61-6]

S-METHYL ISOTHIOUREA SULFATE, II, 411;
Carbamimidothioic acid, methyl ester, sulfate (2:1) [867-44-7]

METHYLISOUREA HYDROCHLORIDE, IV, 645;
Carbamimidic acid, methyl ester, monohydrochloride [5329-33-9]

METHYL 4-KETO-7-METHYLOCTANOATE, III, 601;
Octanoic acid, 7-methyl-4-oxo-, methyl ester [53663-32-4]

α-METHYL MANNOSIDE, I, 371;
α-D-Mannoside, methyl [27939-30-6]

2-METHYLMERCAPTO-N-METHYL-Δ^2-PYRROLINE, V, 780;
1H-Pyrrole, 2,3-dihydro-1-methyl-5-(methylthio)- [25355-52-6]

METHYL 3-METHYL-2-FUROATE, IV, 649;
2-Furancarboxylic acid, 3-methyl-, methyl ester [6141-57-7]

METHYL γ-METHYL-γ-NITROVALERATE, IV, 652;
Pentanoic acid, 4-methyl-4-nitro-, methyl ester [16507-02-1]

METHYL MYRISTATE, III, 605;
Tetradecanoic acid, methyl ester [124-10-7]

N-METHYL-1-NAPHTHYLCYANAMIDE, III, 608;
Cyanamide, methyl-1-naphthalenyl- [53663-33-5]

1-METHYL-1-(1-NAPHTHYL)-2-THIOUREA, III, 609;
Thiourea, N-methyl-N-1-naphthalenyl- [53663-34-6]

METHYL NITRATE, II, 412;
Nitric acid, methyl ester [598-58-3]

METHYL m-NITROBENZOATE, I, 372;
Benzoic acid, 3-nitro-, methyl ester [618-95-1]

3-METHYL-4-NITROPYRIDINE-1-OXIDE, IV, 654;
Pyridine, 3-methyl-4-nitro, 1-oxide [1074-98-2]

METHYL OXALATE, II, 414;
Ethanedioic acid, dimethyl ester [553-90-2]

3-METHYLOXINDOLE, IV, 657;
2H-Indol-2-One, 1,3-dihydro-3-methyl- [1504-06-9]

METHYL PALMITATE, III, 605;
Hexadecanoic acid, methyl ester [112-39-0]

3-METHYL-1,5-PENTANEDIOL, IV, 660;
1,5-Pentanediol, 3-methyl- [4457-71-0]

3-METHYLPENTANE-2,4-DIONE, V, 785;
2,4-Pentanedione, 3-methyl- [815-57-6]

3-METHYLPENTANOIC ACID, II, 416;
Pentanoic acid, 3-methyl [105-43-1]

β-METHYL-β-PHENYL-α,α'-DICYANOGLUTARIMIDE, IV, 662;
3,5-Piperidinedicarbonitrile, 4-methyl-2,6-dioxo-4-phenyl- [6936-95-4]

β-METHYL-β-PHENYLGLUTARIC ACID, IV, 664;
Pentanedioic acid, 3-methyl-3-phenyl- [4160-92-3]

α-METHYL-α-PHENYLHYDRAZINE, II, 418;
Hydrazine, 1-methyl-1-phenyl- [618-40-6]

1-METHYL-3-PHENYLINDANE, IV, 665;
1H-Indene, 2,3-dihydro-1-methyl-3-phenyl- [6416-39-3]

3-METHYL-1-PHENYLPHOSPHOLENE OXIDE, V, 787;
1H-Phosphole, 2,3-dihydro-4-methyl-1-phenyl-,1-oxide [707-61-9]

METHYL PHENYL SULFOXIDE, V, 791;
Benzene, (methylsulfinyl)- [1193-82-4]

1-METHYL-2-PYRIDONE, II, 419;
2(1H)-Pyridinone, 1-methyl- [694-85-9]

4-METHYLPYRIMIDINE, V, 794;
Pyrimidine, 4-methyl- [3438-46-8]

METHYL PYRUVATE, III, 610;
Propanoic acid, 2-oxo-, methyl ester [600-22-6]

METHYL RED, I, 374;
Benzoic acid, 2-[[4-(dimethylamino)phenyl]azo]- [493-52-7]

METHYL SEBACAMATE, III, 613;
Decanoic acid, 10-amino-10-oxo-, methyl ester [53663-35-7]

METHYLSUCCINIC ACID, III, 615;
Butanedioic acid, methyl- [498-21-5]

METHYL 2-THIENYL SULFIDE, IV, 667;
Thiophene, 2-(methylthio)- [5780-36-9]

METHYL β-THIODIPROPIONATE, IV, 669;
Propanoic acid, 3,3'-thiobis-, dimethyl ester [4131-74-2]

3-METHYLTHIOPHENE, IV, 671;
Thiophene, 3-methyl- [616-44-4]

METHYLTHIOUREA, III, 617;
Thiourea, methyl- [598-52-7]

METHYL *p*-TOLYL SULFONE, IV, 674;
Benzene, 1-methyl-4-(methylsulfonyl)- [3185-99-7]

1-METHYL-3-*p*-TOLYLTRIAZENE, V, 797;
1-Triazene, 1-methyl-3-(4-methylphenyl)- [21124-13-0]

6-METHYLURACIL, II, 422;
2,4(1*H*,3*H*)-Pyrimidinedione, 6-methyl- [626-48-2]

β-METHYL-δ-VALEROLACTONE, IV, 677;
2*H*-Pyran-2-one, tetrahydro-4-methyl- [1121-84-2]

MONOBENZALPENTAERYTHRITOL, IV, 679;
1,3-Dioxane-5,5-dimethanol, 2-phenyl- [2425-41-4]

MONOBROMOPENTAERYTHRITOL, IV, 681;
1,3-Propanediol, 2-(bromomethyl)-2-(hydroxymethyl)- [19184-65-7]

MONOCHLOROMETHYL ETHER, I, 377;
Methane, chloromethoxy- [107-30-2]

N-MONO- AND *N,N*-DISUBSTITUTED UREAS AND THIOUREAS, V, 801;

MONOPERPHTHALIC ACID, III, 619; V, 805;
Benzenecarboperoxoic acid, 2-carboxy- [2311-91-3]

MONOVINYLACETYLENE, IV, 683;
1-Buten-3-yne [689-97-4]

1-MORPHOLINO-1-CYCLOHEXENE, V, 808;
Morpholine, 4-(1-cyclohexen-1-yl)- [670-80-4]

MUCOBROMIC ACID, III, 621; IV, 688;
2-Butenoic acid, 2,3-dibromo-4-oxo-, (Z)- [488-11-9]

MUCONIC ACID, III, 623;
2,4-Hexadienedioic acid [505-70-4]

MYRISTIC ACID, I, 379; III, 605;
Tetradecanoic acid [544-63-8]

N

1-NAPHTHALDEHYDE, IV, 690;
1-Naphthalenecarboxaldehyde [66-77-3]

β-NAPHTHALDEHYDE, III, 626;
2-Naphthalenecarboxaldehyde [66-99-9]

2,3-NAPHTHALENEDICARBOXYLIC ACID, V, 810;
2,3-Naphthalenedicarboxylic acid [2169-87-1]

2,6-NAPHTHALENEDICARBOXYLIC ACID, V, 813;
2,6-Naphthalenedicarboxylic acid [1141-38-4]

NAPHTHALENE-1,5-DISULFONYL CHLORIDE, IV, 693;
1,5-Naphthalenedisulfonyl dichloride [1928-01-4]

1,5-NAPHTHALENEDITHIOL, IV, 695;
1,5-Naphthalenedithiol [5325-88-2]

1,2-NAPHTHALIC ANHYDRIDE, II, 423;
Naphtho[1,2-c]furan-1,3-dione [5343-99-7]

α-NAPHTHOIC ACID, II, 425;
1-Naphthalenecarboxylic acid [86-55-5]

β-NAPHTHOIC ACID, II, 428;
2-Naphthalenecarboxylic acid [93-09-4]

β-NAPHTHOL PHENYLAMINOMETHANE, I, 381;
2-Naphthalenol, 1-(aminophenylmethyl)- [481-82-3]

α-NAPHTHONITRILE, III, 631;
1-Naphthalenecarbonitrile [86-53-3]

1,2-NAPHTHOQUINONE, II, 430;
1,2-Naphthalenedione [524-42-5]

1,4-NAPHTHOQUINONE, I, 383; IV, 698;
1,4-Naphthalenedione [130-15-4]

1,2-NAPHTHOQUINONE-4-SULFONATE, AMMONIUM SALT, III, 633;
1-Naphthalenesulfonic acid, 3,4-dihydro-3,4-dioxo-, ammonium salt [53684-60-9]

1,2-NAPHTHOQUINONE-4-SULFONATE POTASSIUM SALT, III, 633;
1-Naphthalenesulfonic acid, 3,4-dihydro-3,4-dioxo, potassium salt [5908-27-0]

NAPHTHORESORCINOL, III, 637;
1,3-Naphthalenediol [132-86-5]

α-NAPHTHYL ISOTHIOCYANATE, IV, 700;
Naphthalene, 1-isothiocyanato- [551-06-4]

β-NAPHTHYLMERCURIC CHLORIDE, II, 432;
Mercury, chloro-2-naphthalenyl- [39966-41-1]

N-β-NAPHTHYLPIPERIDINE, V, 816;
Piperidine, 1-(2-naphthalenyl)- [5465-85-0]

NEOPENTYL ALCOHOL, V, 818;
1-Propanol, 2,2-dimethyl- [75-84-3]

NEOPHYL CHLORIDE, IV, 702;
Benzene, (2-chloro-1,1-dimethylethyl)- [515-40-2]

NICOTINAMIDE-1-OXIDE, IV, 704;
3-Pyridinecarboxamide, 1-oxide [1986-81-8]

NICOTINIC ACID, I, 385;
3-Pyridinecarboxylic acid [59-67-6]

NICOTINIC ANHYDRIDE, V, 822;
3-Pyridinecarboxylic acid, anhydride [16837-38-0]

NICOTINONITRILE, IV, 706;
3-Pyridinecarbonitrile [100-54-9]

m-NITROACETOPHENONE, II, 434;
Ethanone, 1-(3-nitrophenyl)- [121-89-1]

o-NITROACETOPHENONE, IV, 708;
Ethanone, 1-(2-nitrophenyl)- [577-59-3]

1-NITRO-2-ACETYLAMINONAPHTHALENE, II, 438;
Acetamide, *N*-(1-nitro-2-naphthalenyl)- [5419-82-9]

o-NITROANILINE, I, 388;
Benzenamine, 2-nitro- [88-74-4]

9-NITROANTHRACENE, IV, 711;
Anthracene, 9-nitro- [602-60-8]

NITROANTHRONE, I, 390;
9(10*H*)-Anthracenone, 10-nitro- [6313-44-6]

NITROBARBITURIC ACID, II, 440;
2,4,6(1*H*,3*H*,5*H*)-Pyrimidinetrione, 5-nitro- [480-68-2]

o-NITROBENZALDEHYDE, III, 641; V, 825;
Benzaldehyde, 2-nitro- [552-89-6]

m-NITROBENZALDEHYDE DIMETHYLACETAL, III, 644;
Benzene, 1-(dimethoxymethyl)-3-nitro- [3395-79-7]

p-NITROBENZALDEHYDE, II, 441;
Benzaldehyde, 4-nitro- [555-16-8]

o-NITROBENZALDIACETATE, IV, 713;
Methanediol, (2-nitrophenyl)-, diacetate (ester) [6345-63-7]

p-NITROBENZALDIACETATE, IV, 713;
Methanediol, (4-nitrophenyl)-, diacetate (ester) [2929-91-1]

m-NITROBENZAZIDE, IV, 715;
Benzoyl azide, 3-nitro- [3532-31-8]

m-NITROBENZOIC ACID, I, 391;
Benzoic acid, 3-nitro- [121-92-6]

p-NITROBENZOIC ACID, I, 392;
Benzoic acid, 4-nitro- [62-23-7]

p-NITROBENZONITRILE, III, 646;
Benzonitrile, 4-nitro- [619-72-7]

p-NITROBENZOYL CHLORIDE, I, 394;
Benzoyl chloride, 4-nitro- [122-04-3]

p-NITROBENZOYL PEROXIDE, III, 649;
Peroxide, bis(4-nitrobenzoyl) [1712-84-1]

p-NITROBENZYL ACETATE, III, 650;
Benzenemethanol, 4-nitro-, acetate (ester) [619-90-9]

p-NITROBENZYL ALCOHOL, III, 652;
Benzenemethanol, 4-nitro- [619-73-8]

p-NITROBENZYL BROMIDE, II, 443;
Benzene, 1-(bromomethyl)-4-nitro- [100-11-8]

p-NITROBENZYL CYANIDE, I, 396;
  Benzeneacetonitrile, 4-nitro- [555-21-5]

m-NITROBIPHENYL, IV, 718;
  1,1'-Biphenyl, 3-nitro- [2113-58-8]

2-NITROCARBAZOLE, V, 829;
  9*H*-Carbazole, 2-nitro- [14191-22-1]

o-NITROCINNAMALDEHYDE, IV, 722;
  2-Propenal, 3-(2-nitrophenyl)- [1466-88-2]

m-NITROCINNAMIC ACID, I, 398;
  2-Propenoic acid, 3-(3-nitrophenyl)- [555-68-0]

2-NITRO-*p*-CYMENE, III, 653;
  Benzene, 1-methyl-4-(1-methylethyl)-2-nitro- [943-15-7]

5-NITRO-2,3-DIHYDRO-1,4-PHTHALAZINEDIONE, III, 656;
  1,4-Phthalazinedione, 2,3-dihydro-5-nitro- [3682-15-3]

m-NITRODIMETHYLANILINE, III, 658;
  Benzenamine, *N*,*N*-dimethyl-3-nitro- [619-31-8]

p-NITRODIPHENYL ETHER, II, 445;
  Benzene, 1-nitro-4-phenoxy- [620-88-2]

2-NITROETHANOL, V, 833;
  Ethanol, 2-nitro- [625-48-9]

2-NITROFLUORENE AND, 2-AMINOFLUORENE, II, 447;
  9*H*-Fluorene, 2-nitro- [607-57-8]
  9*H*-Fluoren-2-amine [153-78-6]

NITROGUANIDINE, I, 399;
  Guanidine, nitro- [556-88-7]

5-NITROINDAZOLE, III, 660;
  1*H*-Indazole, 5-nitro- [5401-94-5]

NITROMESITYLENE, II, 449;
  Benzene, 1,3,5-trimethyl-2-nitro- [603-71-4]

NITROMETHANE, I, 401;
  Methane, nitro- [75-52-5]

2-NITRO-4-METHOXYANILINE, III, 661;
  Benzenamine, 4-methoxy-2-nitro- [96-96-8]

N-NITROMORPHOLINE, V, 839;
  Morpholine, 4-nitro- [4164-32-3]

1-NITRO-2-NAPHTHOL, II, 451;
  2-Naphthalenol, 1-nitro- [550-60-7]

4-NITRO-1-NAPHTHYLAMINE, III, 664;
  1-Naphthalenamine, 4-nitro- [776-34-1]

1-NITROÖCTANE, IV, 724;
Octane, 1-nitro- [629-37-8]

m-NITROPHENOL, I, 404;
Phenol, 3-nitro- [554-84-7]

p-NITROPHENYLACETIC ACID, I, 406;
Benzeneacetic acid, 4-nitro- [104-03-0]

p-NITROPHENYLARSONIC ACID, III, 665;
Arsonic acid, (4-nitrophenyl)- [98-72-6]

1-(p-NITROPHENYL)-1,3-BUTADIENE, IV, 727;
Benzene, 1-(1,3-butadienyl)-4-nitro- [20264-89-5]

trans-o-NITRO-α-PHENYLCINNAMIC ACID, IV, 730;
Benzeneacetic acid, α-[(2-nitrophenyl)methylene]-, (E)- [19319-35-8]

m-NITROPHENYL DISULFIDE, V, 843;
Disulfide, bis(3-nitrophenyl) [537-91-7]

p-NITROPHENYL ISOCYANATE, II, 453;
Benzene, 1-isocyanato-4-nitro- [100-28-7]

p-NITROPHENYL SULFIDE, III, 667;
Benzene, 1,1'-thiobis[4-nitro- [1223-31-0]

o-NITROPHENYLSULFUR CHLORIDE, II, 455;
Benzenesulfenyl chloride, 2-nitro- [7669-54-7]

3-NITROPHTHALIC ACID, I, 408;
1,2-Benzenedicarboxylic acid, 3-nitro- [603-11-2]

4-NITROPHTHALIC ACID, II, 457;
1,2-Benzenedicarboxylic acid, 4-nitro- [610-27-5]

3-NITROPHTHALIC ANHYDRIDE, I, 410;
1,3-Isobenzofurandione, 4-nitro- [641-70-3]

4-NITROPHTHALIMIDE, II, 459;
1H-Isoindole-1,3(2H)-dione, 5-nitro- [89-40-7]

NITROSOBENZENE, III, 668;
Benzene, nitroso- [586-96-9]

N-NITROSOMETHYLANILINE, II, 460;
Benzenamine, N-methyl-N-nitroso- [614-00-6]

NITROSOMETHYLUREA, II, 461;
Urea, N-methyl-N-nitroso- [684-93-5]

NITROSOMETHYLURETHANE, II, 464; V, 842 (warning)
Carbamic acid, methylnitroso-, ethyl ester [615-53-2]

NITROSO-β-NAPHTHOL, I, 411;
2-Naphthalenol, 1-nitroso- [131-91-9]

β-NITROSTYRENE, I, 413;
Benzene, (2-nitroethenyl)- [102-96-5]

m-NITROSTYRENE, IV, 731;
Benzene, 1-ethenyl-3-nitro- [586-39-0]

2-NITROTHIOPHENE, II, 466;
Thiophene, 2-nitro- [609-40-5]

m-NITROTOLUENE, I, 415;
Benzene, 1-methyl-3-nitro- [99-08-1]

4-NITRO-2,2,4-TRIMETHYLPENTANE, V, 845;
Pentane, 2,2,4-trimethyl-4-nitro- [5342-78-9]

NITROUREA, I, 417;
Urea, nitro- [556-89-8]

6-NITROVERATRALDEHYDE, IV, 735;
Benzaldehyde, 4,5-dimethoxy-2-nitro- [20357-25-9]

2,4-NONANEDIONE, V, 848;
2,4-Nonanedione [6175-23-1]

2-NORBORNANONE, V, 852;
Bicyclo[2.2.1]heptan-2-one [497-38-1]

NORBORNYLENE, IV, 738;
Bicyclo[2.2.1]hept-2-ene [498-66-8]

NORCARANE, V, 855;
Bicyclo[4.1.0]heptane [286-08-8]

exo/endo-7-NORCARANOL, V, 859;
Bicyclo[4.1.0]heptan-7ol, (1α,6α,7α)- [13830-44-9]
Bicyclo[4.1.0]heptan-7ol, (1α,6α,7β)- [931-31-7]

NORTRICYCLANOL, V, 863;
Tricyclo[2.2.1.0$^{2.6}$]heptan-3-ol [695-04-5]

NORTRICYCLANONE, V, 866;
Tricyclo[2.2.1.0$^{2.6}$]heptan-3-one [695-05-6]

O

Δ$^{1(9)}$ OCTALONE-2, V, 872;
2(3H)-Naphthalenone, 4,4a,5,6,7,8-hexahydro- [1196-55-0]

OCTANAL, V, 872;
Octanal [124-13-0]

d- and l-OCTANOL-2, I, 418;
2-Octanol, (S)- [6169-06-8]
2-Octanol, (R)- [5978-70-1]

n-OCTYL BROMIDE, I, 30;
Octane, 1-bromo- [111-83-1]

OLEOYL CHLORIDE, IV, 739;
9-Octadecenoyl chloride, (Z)- [112-77-6]

OLEYL ALCOHOL, II, 468; III, 671;
9-Octadecen-1-ol, (Z)- [143-28-2]

ORTHANILIC ACID, II, 471;
Benzenesulfonic acid, 2-amino- [88-21-1]

2-OXA-7,7-DICHLORONORCARANE, V, 874;
2-Oxabicyclo[4.1.0]heptane, 7,7-dichloro- [7556-13-0]

OXALIC ACID (ANHYDROUS), I, 421;
Ethanedioic acid [144-62-7]

D-2-OXO-7,7-DIMETHYL-1-VINYLBICYCLO[2.2.1]HEPTANE, V, 877;
Bicyclo[2.2.1]heptan-2-one, 1-ethenyl-7,7-dimethyl-, (1S)- [53585-70-9]

OZONE, III, 673;
Ozone [10028-15-6]

P

PALLADIUM CATALYSTS, III, 685;
Palladium [7440-05-3]

PALLADIUM CATALYST FOR PARTIAL REDUCTION OF ACETYLENES, V, 880;

PALMITIC ACID, III, 605;
Hexadecanoic acid [57-10-3]

PARABANIC ACID, IV, 744;
Imidazolidinetrione [120-89-8]

[2.2]PARACYCLOPHANE, V, 883;
Tricyclo[8.2.2.2$^{4:7}$] hexadeca-4,6,10,12,13,15-hexaene [1633-22-3]

PELARGONIC ACID, II, 474;
Nonanoic acid [112-05-0]

2,3,4,5,6-PENTA-O-ACETYL-D-GLUCONIC ACID AND 2,3,4,5,6-PENTA-O-ACETYL-D-GLUCONYL CHLORIDE, V, 887;
D-Gluconic acid, pentaacetate [17430-71-6]
D-Gluconyl chloride, pentaacetate [53555-69-4]

PENTAACETYL d-GLUCONONITRILE, III, 690;
D-Gluconitrile, 2,3,4,5,6-pentaacetate [6272-51-1]

PENTACHLORBENZOIC ACID, V, 890;
Benzoic acid, pentachloro- [1012-84-6]

1,2,3,4,5-PENTACHLORO-5-ETHYLCYCLOPENTADIENE, V, 893;
1,3-Cyclopentadiene, 1,2,3,4,5-pentachloro-5-ethyl- [16177-48-3]

1,4-PENTADIENE, IV, 746;
1,4-Pentadiene [591-93-5]

PENTAERYTHRITOL, I, 425;
   1,3-Propanediol, 2,2-bis(hydroxymethyl)- [115-77-5]

PENTAERYTHRITYL BROMIDE AND IODIDE, II, 476;
   Pentaerythrityl bromide [3229-00-3]
   Pentaerythrityl iodide [1522-88-9]

PENTAERYTHRITYL TETRABROMIDE, IV, 753;
   Propane, 1,3-dibromo-2,2-bis(bromomethyl)- [3229-00-3]

PENTAMETHYLENE BROMIDE, I, 428; III, 692;
   Pentane, 1,5-dibromo- [111-24-0]

3,3-PENTAMETHYLENEDIAZIRINE, V, 897;
   1,2-Diazaspiro[2.5]oct-1-ene [930-82-5]

n-PENTANE, II, 478;
   Pentane [109-66-0]

1,5-PENTANEDIOL, III, 693;
   1,5-Pentanediol [111-29-5]

2-PENTENE, I, 430;
   2-Pentene [109-68-2]

3-PENTEN-2-OL, III, 696;
   3-Penten-2-ol [1569-50-2]

4-PENTEN-1-OL, III, 698;
   4-Penten-1-ol [821-09-0]

4-PENTYN-1-OL, IV, 755;
   4-Pentyn-1-ol [5390-04-5]

PERBENZOIC ACID, I, 431; V, 900, Warning; Same as PEROXYBENZOIC ACID, V,
   904;
   Benzenecarboperoxoic acid [93-59-4]

PERCHLOROFULVALENE, V, 901;
   1,3-Cyclopentadiene, 1,2,3,4-tetrachloro-5-(2,3,4,5-tetrachloro-2,4-cyclopentadien-1-yli-
   dene)- [6298-65-3]

PEROXYBENZOIC ACID, V, 904;
   Benzenecarboperoxoic acid [93-59-4]

PHENACYLAMINE HYDROCHLORIDE, V, 909;
   Ethanone, 2-amino-1-phenyl-, hydrochloride [5468-37-1]

PHENACYL BROMIDE, II, 480;
   Ethanone, 2-bromo-1-phenyl- [70-11-1]

PHENANTHRENE-9-ALDEHYDE, III, 701;
   9-Phenanthrenecarboxaldehyde [4707-71-5]

PHENANTHRENEQUINONE, IV, 757;
   9,10-Phenanthrenedione [84-11-7]

2- and 3-PHENANTHRENESULPHONIC ACIDS, II, 482;
2-Phenanthrenesulfonic acid [41105-40-2]
3-Phenanthrenesulfonic acid [2039-95-4]

PHENOLS; 6-METHOXY-2-NAPHTHOL, V, 918;
2-Naphthalenol, 6-methoxy- [5111-66-0]

PHENOXTHIN, II, 485;
Phenoxathiin [262-20-4]

γ-PHENOXYPROPYL BROMIDE, I, 435;
Benzene, (3-bromopropoxy) (588-63-6]

PHENYLACETAMIDE, IV, 760;
Benzeneacetamide [103-81-1]

PHENYLACETIC ACID, I, 436;
Benzeneacetic acid [103-82-2]

α-PHENYLACETOACETONITRILE, II, 487;
Benzeneacetonitrile, α-acetyl- [4468-48-8]

PHENYLACETYLENE, I, 438; IV, 763;
Benzene, ethynyl- [536-74-3]

dl-PHENYLALANINE, III, 705;
dl-β;PHENYLALANINE, II, 489;
DL-Phenylalanine [150-30-1]

γ-PHENYLALLYLSUCCINIC ACID, IV, 766;
Butanedioic acid, (3-phenyl-2-propenyl)- [5671-91-0]

1-PHENYL-3-AMINO-5-PYRAZOLONE, III, 708;
3H-Pyrazol-3-one, 5-amino-2,4-dihydro-2-phenyl- [4149-06-8]

PHENYLARSONIC ACID, II, 494;
Arsonic acid, phenyl- [98-05-5]

PHENYL AZIDE, III, 710;
Benzene, azido- [622-37-7]

p-PHENYLAZOBENZOIC ACID, III, 711;
Benzoic acid, 4-(phenylazo)- [1562-93-2]

p-PHENYLAZOBENZOYL CHLORIDE, III, 712;
Benzoyl chloride, 4-(phenylazo)- [104-24-5]

N-PHENYLBENZAMIDINE, IV, 769;
Benzenecarboximidamide, N-phenyl- [1527-91-9]

PHENYLBENZOYLDIAZOMETHANE, II, 496;
Ethanone, 2-diazo-1,2-diphenyl- [3469-17-8]

α-PHENYL-β-BENZOYLPROPIONITRILE, II, 498;
Benzenebutanenitrile, γ-oxo-α-phenyl- [6268-00-4]

PHENYLBROMOETHYNE, V, 921;
Benzene, (bromoethynyl)- [932-87-6]

trans-1-PHENYL-1,3-BUTADIENE, IV, 771;
  Benzene, 1,3-butadienyl-, (*E*)- [16939-57-4]

PHENYL *t*-BUTYL ETHER (METHOD I), V, 924;
  Benzene, (1,1-dimethylethoxy)- [6669-13-2]

PHENYL *t*-BUTYL ETHER (METHOD II), V, 926;
  Benzene, (1,1-dimethylethoxy)- [6669-13-2]

γ-PHENYLBUTYRIC ACID, II, 499;
  Benzenebutanoic acid [1821-12-1]

α-PHENYL-α-CARBETHOXYGLUTARONITRILE, IV, 776;
  Benzeneacetic acid, α-cyano-α-(2-cyanoethyl)-, ethyl ester [53555-70-7]

PHENYL CINNAMATE, III, 714;
  2-Propenoic acid, 3-phenyl-, phenyl ester [2757-04-2]

α-PHENYLCINNAMIC ACID, IV, 777;
  Benzeneacetic acid, α-(phenylmethylene)- [3368-16-9]

α-PHENYLCINNAMONITRILE, III, 715;
  Benzeneacetonitrile, α-(phenylmethylene)- [2510-95-4]

2-PHENYLCYCLOHEPTANONE, IV, 780;
  Cycloheptanone, 2-phenyl- [14996-78-2]

PHENYLCYCLOPROPANE, V, 929; Same as CYCLOPROPYLBENZENE, V, 328;
  Benzene, cyclopropyl- [873-49-4]

PHENYLDICHLOROPHOSPHINE, IV, 784;
  Phosphonous dichloride, phenyl- [644-97-3]

4-PHENYL-*m*-DIOXANE, IV, 786;
  1,3-Dioxane, 4-phenyl- [772-00-9]

o-PHENYLENE CARBONATE, IV, 788;
  1,3-Benzodioxol-2-one [2171-74-6]

o-PHENYLENEDIAMINE, II, 501;
  1,2-Benzenediamine [95-54-5]

α-PHENYLETHYLAMINE, II, 503; III, 717;
  Benzenemethanamine, α-methyl- [98-84-0]

d- AND *l*-α-PHENYLETHYLAMINE, II, 506;
  Benzenemethanamine, α-methyl-, (*R*)- [3886-69-9]
  Benzenemethanamine, α-methyl-, (*S*)- [2627-86-3]

R(+)- AND *S*(−)-α-PHENYLETHYLAMINE, V, 932;
  Benzenemethanamine, α-methyl-, (*R*)- [3886-69-9]
  Benzenemethanamine, α-methyl-, (*S*)- [2627-86-3]

β-PHENYLETHYLAMINE, III, 720;
  Benzeneethanamine [64-04-0]

2-PHENYLETHYL BENZOATE, V, 336;
  Benzoic acid, 2-phenylethyl ester [94-47-3]

β-PHENYLETHYLDIMETHYLAMINE, III, 723;
Benzeneethanamine, N,N-dimethyl- [1126-71-2]

PHENYLETHYLENE, I, 440;
Benzene, ethenyl- [100-42-5]

α-PHENYLGLUTARIC ANHYDRIDE, IV, 790;
2H-Pyran-2,6(3H)-dione, dihydro-3-phenyl- [2959-96-8]

PHENYLGLYOXAL, II, 509; V, 937;
Benzeneacetaldehyde, α-oxo- [1074-12-0]

PHENYLHYDRAZINE, I, 422;
Hydrazine, phenyl- [100-63-0]

β-PHENYLHYDROXYLAMINE, I, 445;
Benzenamine, N-hydroxy- [100-65-2]

2-PHENYLINDAZOLE, V, 941;
2H-Indazole, 2-phenyl- [3682-71-1]

2-PHENYLINDOLE, III, 725;
1H-Indole, 2-phenyl- [948-65-2]

PHENYL ISOTHIOCYANATE, I, 447;
Benzene, isothiocyanato- [103-72-0]

N-PHENYLMALEIMIDE, V, 944;
1H-Pyrrole-2,5-dione, 1-phenyl- [941-69-5]

PHENYLMETHYLGLYCIDIC ESTER, III, 727;
Oxiranecarboxylic acid, 3-methyl-3-phenyl-, ethyl ester [77-83-8]

1-PHENYLNAPHTHALENE, III, 729;
Naphthalene, 1-phenyl- [605-02-7]

PHENYLNITROMETHANE, II, 512;
Benzene, (nitromethyl)- [622-42-4]

2-PHENYL-5-OXAZOLONE, V, 946;
5(4H)-Oxazolone, 2-phenyl- [1199-01-5]

1-PHENYL-1-PENTEN-4-YN-3-OL, IV, 792;
1-Penten-4-yn-3-ol, 1-phenyl- [14604-31-0]

2-PHENYLPERFLUOROPROPENE, V, 949;
Benzene, [2,2-difluoro-1-(trifluoromethyl)ethenyl]- [1979-51-7]

9-PHENYLPHENANTHRENE, V, 952;
Phenanthrene, 9-phenyl- [844-20-2]

1-PHENYLPIPERIDINE, IV, 795;
Piperidine, 1-phenyl- [4096-20-2]

3-PHENYL-1-PROPANOL, IV, 798;
Benzenepropanol [122-97-4]

PHENYLPROPARGYL ALDEHYDE, III, 731;
2-Propynal, 3-phenyl- [2579-22-8]

PHENYLPROPARGYLALDEHYDE DIETHYL ACETAL, IV, 801;
Benzene, (3,3-diethoxy-1-propynyl)- [6142-95-6]

PHENYLPROPIOLIC ACID, II, 515;
2-Propynoic acid, 3-phenyl- [637-44-5]

α-PHENYLPROPIONALDEHYDE, III, 733;
Benzeneacetaldehyde, α-methyl- [93-53-8]

2-PHENYL-3-*n*-PROPYLISOXAZOLIDINE-4,5-*cis*-DICARBOXYLIC ACID *N*-PHENY-
LIMIDE, V, 957;
2*H*-Pyrrolo[3,4-*d*]isoxazole-4,6(3*H*,5*H*)-dione, dihydro-2,5-diphenyl-3-propyl- [53555-71-8]

2-PHENYLPYRIDINE, II, 517;
Pyridine, 2-phenyl- [1008-89-5]

PHENYLPYRUVIC ACID, II, 519;
Benzenepropanoic acid, α-oxo- [156-06-9]

4-PHENYLSEMICARBAZIDE, I, 450;
Hydrazinecarboxamide, *N*-phenyl- [537-47-3]

PHENYLSUCCINIC ACID, I, 451; IV, 804;
Butanedioic acid, phenyl- [635-51-8]

PHENYLSULFUR TRIFLUORIDE, V, 959;
Sulfur, trifluorophenyl- [672-36-6]

3-PHENYLSYDNONE, V, 962;
Sydnone, 3-phenyl- [120-06-9]

PHENYL THIENYL KETONE, II, 520;
Methanone, phenyl-2-thienyl- [135-00-2]

1-PHENYL-2-THIOBIURET, V, 966;
Urea, [(phenylamino)thioxomethyl]- [53555-72-9]

α-PHENYLTHIOUREA, III, 735;
Thiourea, phenyl- [103-85-5]

PHENYL(TRICHLOROMETHYL)MERCURY, V, 969;
Mercury, phenyl(trichloromethyl)- [3294-57-3]

PHENYLUREA, I, 453;
Urea, phenyl- [64-10-8]

PHLOROACETOPHENONE, II, 522;
Ethanone, 1-(2,4,6-trihydroxyphenyl)- [480-66-0]

PHLOROGLUCINOL, I, 455;
1,3,5-Benzenetriol [108-73-6]

o-PHTHALALDEHYDE, IV, 807;
1,2-Benzenecarboxaldehyde [643-79-8]

PHTHALALDEHYDIC ACID, II, 523; III, 737;
Benzoic acid, 2-formyl- [119-67-5]

PHTHALIDE, II, 526;
1(3*H*)-Isobenzofuranone [87-41-2]

PHTHALIMIDE, I, 457;
1*H*-Isoindole-1,3(2*H*)-dione [85-41-6]

α-PHTHALIMIDO-*o*-TOLUIC ACID, IV, 810;
Benzoic acid, 2-[(1,3-dihydro-1,3-dioxo-2*H*-isoindol-2-yl)methyl]- [53663-18-6]

sym. AND *unsym.o*-PHTHALYL CHLORIDES, II, 528;
1,2-Benzenedicarboyl dichloride [88-95-9]
1(3*H*)-Isobenzofuranone, 3,3-dichloro- [601-70-7]

N-PHTHALYL-L-β-PHENYLALANINE, V, 973;
2*H*-Isoindole-2-acetic acid, 1,3-dihydro-1,3-dioxo-α-(phenylmethyl)-, (*S*)- [5123-55-7]

PICOLINIC ACID HYDROCHLORIDE, III, 740;
2-Pyridinecarboxylic acid, hydrochloride [636-80-6]

PIMELIC ACID, II, 531;
Heptanedioic acid [111-16-0]

PINACOL HYDRATE, I, 459;
2,3-Butanediol, 2,3-dimethyl-, hexahydrate [6091-58-3]

PINACOLONE, I, 462;
2-Butanone, 3,3-dimethyl- [75-97-8]

PIPERONYLIC ACID, II, 538;
1,3-Benzodioxole-5-carboxylic acid [94-53-1]

PLATINUM CATALYST FOR REDUCTIONS, I, 463;
Platinum oxide (PtO$_2$) [1314-15-4]

POTASSIUM ANTHRAQUINONE-α-SULFONATE, II, 539;
1-Anthracenesulfonic acid, 9,10-dihydro-9,10-dioxo-, potassium salt [30845-78-4]

PROPIOLALDEHYDE, IV, 813;
2-Propynal [624-67-9]

PROPIONALDEHYDE, II, 541;
Propanal [123-38-6]

o-PROPIOPHENOL AND *p*-PROPIOPHENOL, II, 543;
1-Propanone, 1-(2-hydroxyphenyl)- [610-99-1]
1-Propanone, 1-(4-hydroxyphenyl)- [70-70-2]

n-PROPYLBENZENE, I, 471;
Benzene, propyl- [103-65-1]

iso-PROPYL BROMIDE, I, 37;
Propane, 2-bromo- [75-26-3]

n-PROPYL BROMIDE, I, 37;
Propane, 1-bromo- [106-94-5]

γ-*n*-PROPYLBUTYROLACTONE, III, 742;
2(3*H*)-Furanone, dihydro-5-propyl- [105-21-5]

l-PROPYLENE GLYCOL, II, 545;
1,2-Propanediol, (*R*)- [4254-14-2]

n-PROPYL SULFIDE, II, 547;
Propane, 1,1-thiobis- [111-47-7]

PROTOCATECHUALDEHYDE, II, 549;
Benzaldehyde, 3,4-dihydroxy- [139-85-5]

PROTOCATECHUIC ACID, III, 745;
Benzoic acid, 3,4-dihydroxy- [99-50-3]

PSEUDOIONONE, III, 747;
3,5,9-Undecatrien-2-one, 6,10-dimethyl- [141-10-6]

PSEUDOPELLETIERINE, IV, 816;
9-Azabicyclo[3.3.1]nonan-3-one, 9-methyl- [552-70-5]

PSEUDOTHIOHYDANTOIN, III, 751;
4(5*H*)-Thiazolone, 2-amino- [556-90-1]

PURIFICATION OF TETRAHYDROFURAN (*Warning*), V, 976;
Furan, tetrahydro- [109-99-9]

PUTRESCINE DIHYDROCHLORIDE, IV, 819;
1,4-Butanediamine, dihydrochloride [333-93-7]

PYOCYANINE, III, 753;
Phenazinium, 1-hydroxy-5-methyl-, hydroxide, inner salt [85-66-5]

2,3-PYRAZINEDICARBOXYLIC ACID, IV, 824;
2,3-Pyrazinedicarboxylic acid [89-01-0]

PYRIDINE-*N*-oxide, IV, 828;
Pyridine, 1-oxide [694-59-7]

4-PYRIDINESULFONIC ACID, V, 977;
4-Pyridinesulfonic acid [5402-20-0]

1-(α-PYRIDYL)-2-PROPANOL, III, 757;
2-Pyridineethanol, α-methyl- [5307-19-7]

PYROGALLOL 1-MONOMETHYL ETHER, III, 759;
1,2-Benzenediol, 3-methoxy- [934-00-9]

PYROMELLITIC ACID, II, 551;
1,2,4,5-Benzenetetracarboxylic acid [89-05-4]

α-PYRONE, V, 982;
2*H*-Pyran-2-one [504-31-4]

PYRROLE, I, 473;
1*H*-Pyrrole [109-97-7]

2-PYRROLEALDEHYDE, IV, 831;
1*H*-Pyrrole-2-carboxaldehyde [1003-29-8]

2-(1-PYRROLIDYL)PROPANOL, IV, 834;
1-Pyrrolidineethanol, β-methyl- [53663-19-7]

PYRUVIC ACID, I, 475;
Propanoic acid, 2-oxo- [127-17-3]

Q

QUINACETOPHENONE MONOMETHYL ETHER, IV, 836;
Ethanone, 1-(2-hydroxy-5-methoxyphenyl)- [705-15-7]

QUINIZARIN, I, 476;
9,10-Anthracenedione, 1,4-dihydroxy- [81-64-1]

QUINOLINE, I, 478;
Quinoline [91-22-5]

QUINONE, I, 482; II, 553;
2,5-Cyclohexadiene-1,4-dione [106-51-4]

p-QUINQUEPHENYL, V, 985;
1,1′:4′,1″:4″,1‴:4‴,1⁗-Quinquephenyl [3073-05-0]

3-QUINUCLIDONE HYDROCHLORIDE, V, 989;
1-Azabicyclo[2.2.2]octan-3-one, hydrochloride [1193-65-3]

R

REDUCTION OF CONJUGATED ALKENES WITH CHROMIUM(II) SULFATE: DIE-
THYL SUCCINATE, V, 993;
Butanedioic acid, diethyl ester [123-25-1]

REDUCTION OF ORGANIC HALIDES, CHLOROBENZENE TO BENZENE, V, 998;
Benzene [71-43-2]

REINECKE SALT, II, 555;
Chromate(1-), diamminetetrakis(thiocyanato-N)-, ammonium, (OC-6-11)- [13573-16-5]

RESACETOPHENONE, III, 761;
Ethanone, 1-(2,4-dihydroxyphenyl)- [89-84-9]

β-RESORCYLIC ACID, II, 557;
Benzoic acid, 2,4-dihydroxy- [89-86-1]

RHODANINE, III, 763;
4-Thiazolidinone, 2-thioxo- [141-84-4]

RUTHENOCENE, V, 1001;
Ruthenocene [1287-13-4]

S

SALICYL-o-TOLUIDE, III, 765;
Benzamide, 2-hydroxy-N-(2-methylphenyl)- [7133-56-4]

SEBACIL, IV, 838;
1,2-Cyclodecanedione [96-01-5]

SEBACOIN, IV, 840;
Cyclodecanone, 2-hydroxy- [96-00-4]

SEBACONITRILE, III, 768;
Decanedinitrile [1871-96-1]

SELENOPHENOL, III, 771;
Benzeneselenol [645-96-5]

SEMICARBAZIDE SULFATE, I, 485;
Hydrazinecarboxamide, sulfate [17464-82-3]

dl-SERINE, III, 774;
DL-Serine [302-84-1]

SODIUM AMIDE, III, 778;
Sodium amide [7782-92-5]

SODIUM p-ARSONO-N-PHENYLGLYCINAMIDE, I, 488;
Arsonic acid, [4-[(2-amino-2-oxoethyl)amino]phenyl]-, disodium salt [834-03-7]

SODIUM 2-BROMOETHANESULFONATE, II, 558;
Ethanesulfonic acid, 2-bromo-, sodium salt [4263-52-9]

SODIUM p-HYDROXYPHENYLARSONATE, I, 490;
Arsonic acid, (4-hydroxyphenyl)-, sodium salt [53663-20-0]

SODIUM NITROMALONALDEHYDE MONOHYDRATE, IV, 844; (Warning), V, 1004;
Propanedial, nitro-, ion(1-), sodium, monohydrate [53821-72-0]

SODIUM β-STYRENESULFONATE AND β-STYRENESULFONYL CHLORIDE, IV, 846;
Ethenesulfonic acid, 2-phenyl-, sodium salt [2039-44-3]
Ethenesulfonyl chloride, 2-phenyl- [4091-26-3]

SODIUM p-TOLUENESULFINATE, I, 492;
Benzenesulfinic acid, 4-methyl-, sodium salt [824-79-3]

SORBIC ACID, III, 783;
2,4-Hexadienoic acid, (E,E)- [110-44-1]

STEAROLIC ACID, III, 785; IV, 851;
9-Octadecynoic acid [506-24-1]

STEARONE, IV, 854;
18-Pentatriacontanone [504-53-0]

cis-STILBENE, IV, 857;
Benzene, 1,1′-(1,2-ethenediyl)bis-, (Z)- [645-49-8]

trans-STILBENE, III, 786;
Benzene, 1,1′-(1,2-ethenediyl)bis-, (E)- [103-30-0]

trans-STILBENE OXIDE, IV, 860;
Oxirane, 2,3-diphenyl-, trans- [1439-07-2]

STYRENE OXIDE, I, 494;
Oxirane, phenyl- [96-09-3]

STYRYLPHOSPHONIC DICHLORIDE, V, 1005;
Phosphonic dichloride, (2-phenylethenyl)- [4708-07-0]

SUCCINIC ANHYDRIDE, II, 560;
2,5-Furandione, dihydro- [108-30-5]

SUCCINIMIDE, II, 562;
2,5-Pyrrolidinedione [123-56-8]

o-SULFOBENZOIC ANHYDRIDE, I, 495;
3H-2,1-Benzoxathiol-3-one, 1,1-dioxide [81-08-3]

α-SULFOPALMITIC ACID, IV, 862;
Hexadecanoic acid, 2-sulfo- [1782-10-1]

SYRINGIC ALDEHYDE, IV, 866;
Benzaldehyde, 4-hydroxy-3,5-dimethoxy- [134-96-3]

T

dl-TARTARIC ACID, I, 497;
Butanedioic acid, 2,3-dihydroxy, (R^*,R^*)-($\pm$)- [133-37-9]

TAURINE, II, 563;
Ethanesulfonic acid, 2-amino- [107-35-7]

TEREPHTHALALDEHYDE, III, 788;
1,4-Benzenedicarboxaldehyde [623-27-8]

TEREPHTHALIC ACID, III, 791;
1,4-Benzenedicarboxylic acid [100-21-0]

TETRAACETYLETHANE, IV, 869;
2,5-Hexanedione, 3,4-diacetyl- [5027-32-7]

dl-4,4',6,6'-TETRACHLORODIPHENIC ACID, IV, 872;
[1,1'-Biphenyl]-2,2'-dicarboxylic acid, 4,4',6,6'-tetrachloro-, ($\pm$)- [53663-22-2]

TETRACYANOETHYLENE, IV, 877;
Ethenetetracarbonitrile [670-54-2]

TETRACYANOETHYLENE OXIDE, V, 1007;
Oxiranetetracarbonitrile [3189-43-3]

TETRAETHYLTIN, IV, 881;
Stannane, tetraethyl- [597-64-8]

1,2,3,4-TETRAHYDROCARBAZOLE, IV, 884;
1H-Carbazole, 2,3,4,9-tetrahydro- [942-01-8]

TETRAHYDROFURAN, II, 566;
Furan, tetrahydro- [109-99-9]

TETRAHYDROFURFURYL BROMIDE, III, 793;
Furan, 2-(bromomethyl)tetrahydro- [1192-30-9]

β-(TETRAHYDROFURYL)PROPIONIC ACID, III, 742;
2-Furanpropanoic acid, tetrahydro- [935-12-6]

ar-TETRAHYDRO-α-NAPHTHOL, IV, 887;
1-Naphthalenol, 5,6,7,8-tetrahydro- [529-35-1]

ac-TETRAHYDRO-β-NAPHTHYLAMINE, I, 499;
2-Naphthalenamine, 1,2,3,4-tetrahydro- [2954-50-9]

cis-Δ⁴-TETRAHYDROPHTHALIC ANHYDRIDE, IV, 890;
1,3-Isobenzofurandione, 3a,4,7,7a-tetrahydro-, cis- [935-79-5]

TETRAHYDROPYRAN, III, 794;
2H-Pyran, tetrahydro- [142-68-7]

TETRAHYDROTHIOPHENE, IV, 892;
Thiophene, tetrahydro- [110-01-0]

TETRAHYDROXYQUINONE, V, 1011;
2,5-Cyclohexadiene-1,4-dione, 2,3,5,6-tetrahydroxy- [319-89-1]

TETRAIODOPHTHALIC ANHYDRIDE, III, 796;
1,3-Isobenzofurandione, 4,5,6,7-tetraiodo- [632-80-4]

TETRALIN HYDROPEROXIDE, IV, 895;
Hydroperoxide, 1,2,3,4-tetrahydro-1-naphthalenyl [771-29-9]

α-TETRALONE, II, 569; III, 798; IV, 898;
1(2H)-Naphthalenone, 3,4-dihydro- [529-34-0]

β-TETRALONE, IV, 903;
2(1H)-Naphthalenone, 3,4-dihydro- [530-93-8]

TETRAMETHYLAMMONIUM 1,1,2,3,3-PENTACYANOPROPENIDE, V, 1013;
Methanaminium, N,N,N-trimethyl-, salt with 1-propene-1,1,2,3,3-pentacarbonitrile
(1:1) [53663-17-5]

TETRAMETHYLBIPHOSPHINE DISULFIDE, V, 1016;
Diphosphine, tetramethyl-, 1,2-disulfide [3676-97-9]

TETRAMETHYLENE CHLOROHYDRIN, II, 571;
1-Butanol, 4-chloro- [928-51-8]

2,3,4,6-TETRAMETHYL-d-GLUCOSE, III, 800;
D-Glucopyranose, 2,3,4,6-tetra-O-methyl- [7506-68-5]

2,2,6,6-TETRAMETHYLOLCYCLOHEXANOL, IV, 907;
1,1,3,3-Cyclohexanetetramethanol, 2-hydroxy- [5416-55-7]

TETRAMETHYL-p-PHENYLENEDIAMINE, V, 1018;
1,4-Benzenediamine, N,N,N',N'-tetramethyl- [100-22-1]

2,3,4.5-TETRAMETHYLPYRROLE, V, 1022;
1H-Pyrrole, 2,3,4,5-tetramethyl- [1003-90-3]

2,2,5,5-TETRAMETHYLTETRAHYDRO-3-KETOFURAN, V, 1024;
3(2H)-Furanone, dihydro-2,2,5,5-tetramethyl- [5455-94-7]

α,α,α',α'-TETRAMETHYLTETRAMETHYLENE GLYCOL, V, 1026;
2,6-Hexanediol, 2,6-dimethyl- [110-03-2]

2,4,5,7-TETRANITROFLUORENONE, V, 1029;
9H-Fluoren-9-one, 2,4,5,7-tetranitro- [746-53-2]

(+)- AND (−)-α-(2,4,5,7-TETRANITRO-9-FLUORENYLIDENEAMINOOXY)PROPIONIC
ACID, V, 1031;

Propanoic acid, 2-[[(2,4,5,7-tetranitro-9*H*-fluoren-9-ylidene)amino]-
oxy]-, (+)- [50996-73-1]
Propanoic acid, 2-[[(2,4,5,7-tetranitro-9*H*-fluoren-9-ylidene)amino]-
oxy]-, (−)- [50874-31-2]

TETRANITROMETHANE, III, 803;
Methane, tetranitro- [509-14-8]

TETRAPHENYLARSONIUM CHLORIDE HYDROCHLORIDE, IV, 910;
Arsonium, tetraphenyl-, (hydrogen dichloride), [21006-73-5]

TETRAPHENYLCYCLOPENTADIENONE, III, 806;
2,4-Cyclopentadien-1-one, 2,3,4,5-tetraphenyl- [479-33-4]

TETRAPHENYLETHYLENE, IV, 914;
Benzene, 1,1′,1″,1‴-(1,2-ethenediylidene)tetrakis- [632-51-9]

1,2,3,4-TETRAPHENYLNAPHTHALENE, V, 1037;
Naphthalene, 1,2,3,4-tetraphenyl- [751-38-2]

TETRAPHENYLPHTHALIC ANHYDRIDE, III, 807;
1,3-isobenzofurandione, 4,5,6,7-tetraphenyl- [4741-53-1]

TETROLIC ACID, V, 1043;
2-Butynoic acid [590-93-2]

2-THENALDEHYDE, III, 811; IV, 915;
2-Thiophenecarboxaldehyde [98-03-3]

3-THENALDEHYDE, IV, 918;
3-Thiophenecarboxaldehyde [498-62-4]

3-THENOIC ACID, IV, 919;
3-Thiophenecarboxylic acid [88-13-1]

3-THENYL BROMIDE, IV, 921;
Thiophene, 3-(bromomethyl)- [34846-44-1]

THIOBENZOIC ACID, IV, 924;
Benzenecarbothioic acid [98-91-9]

THIOBENZOPHENONE, II, 573; IV, 927;
Methanethione, diphenyl- [1450-31-3]

THIOBENZOYLTHIOGLYCOLIC ACID, V, 1046;
Acetic acid, [(phenylthioxomethyl)thio]- [942-91-6]

m-THIOCRESOL, III, 809; WARNING, V, 1050;
Benzenethiol, 3-methyl- [108-40-7]

p-THIOCYANODIMETHYLANILINE, II, 574;
Thiocyanic acid, 4-(dimethylamino)phenyl ester [7152-80-9]

β-THIODIGLYCOL, II, 576;
Ethanol, 2,2′-thiobis- [111-48-8]

2α-THIOHOMOPHTHALIMIDE, V, 1051;
3(2*H*)-Isoquinolinone, 1,4-dihydro-1-thioxo- [938-38-5]

THIOLACETIC ACID, IV, 928;
Ethanethioic acid [507-09-5]

THIOPHENE, II, 578;
Thiophene [110-02-1]

2-THIOPHENEALDEHYDE, III, 811; IV, 915;
2-Thiophenecarboxaldehyde [98-03-3]

THIOPHENOL, I, 504;
Benzenethiol [108-98-5]

THIOPHOSGENE, I, 506;
Carbonothioic dichloride [463-71-8]

THIOSALICYLIC ACID, II, 580;
Benzoic acid, 2-mercapto- [147-93-3]

dl-THREONINE, III, 813;
DL-Threonine [80-68-2]

THYMOQUINONE, I, 511;
2,5-Cyclohexadiene-1,4-dione, 2-methyl-5-(1-methylethyl)- [490-91-5]

o-TOLUALDEHYDE, III, 818; IV, 932;
Benzaldehyde, 2-methyl- [529-20-4]

p-TOLUALDEHYDE, II, 583;
Benzaldehyde, 4-methyl- [104-87-0]

o-TOLUAMIDE, II, 586; Warning, V, 1054;
Benzamide, 2-methyl- [527-85-5]

p-TOLUENESULFENYL CHLORIDE, IV, 934;
Benzenesulfenyl chloride, 4-methyl- [933-00-6]

p-TOLUENESULFINYL CHLORIDE, IV, 937;
Benzenesulfinyl chloride, 4-methyl- [10439-23-3]

p-TOLUENESULFONIC ANHYDRIDE, IV, 940;
Benzenesulfonic acid, 4-methyl-, anhydride [4124-41-8]

p-TOLUENESULFONYLHYDRAZIDE, V, 1055;
Benzenesulfonic acid, 4-methyl-, hydrazide [1576-35-8]

o-TOLUIC ACID, II, 588; III, 820;
Benzoic acid, 2-methyl- [118-90-1]

p-TOLUIC ACID, III, 822;
Benzoic acid, 4-methyl- [99-94-5]

o-TOLUIDINESULFONIC ACID, III, 824;
Benzenesulfonic acid, 4-amino-3-methyl- [98-33-9]

o-TOLUNITRILE, I, 514;
Benzonitrile, 2-methyl- [529-19-1]

p-TOLUNITRILE, I, 514;
Benzonitrile, 4-methyl- [104-85-8]

p-TOLUYL-*o*-BENZOIC ACID, I, 517;
Benzoic acid, 2-(4-methylbenzoyl)- [85-55-2]

m-TOLYLBENZYLAMINE, III, 827;
Benzenemethanamine, *N*-(3-methylphenyl)- [5405-17-4]

p-TOLYL CARBINOL, II, 590;
Benzenemethanol, 4-methyl- [589-18-4]

1-*p*-TOLYLCYCLOPROPANOL, V, 1058;
Cyclopropanol, 1-(4-methylphenyl)- [40122-37-0]

o-TOLYL ISOCYANIDE, V, 1060;
Benzene, 1-isocyano-2-methyl- [10468-64-1]

p-TOLYLMERCURIC CHLORIDE, I, 519;
Mercury, chloro(4-methylphenyl)- [539-43-5]

2-(*p*-TOLYLSULFONYL)-DIHYDROISOINDOLE, V, 1064;
1*H*-Isoindole, 2,3-dihydro-2-[(4-methylphenyl)sulfonyl]- [32372-83-1]

p-TOLYLSULFONYLMETHYLNITROSAMIDE, IV, 943;
Benzenesulfonamide, *N*,4-dimethyl-*N*-nitroso- [80-11-5]

1,3,5-TRIACETYLBENZENE, III, 829;
Ethanone, 1,1′,1″-(1,3,5-benzenetriyl)tris- [779-90-8]

2,4,5-TRIAMINONITROBENZENE, V, 1067;
1,2,4-Benzenetriamine, 5-nitro- [6635-35-4]

1,2,4-TRIAZOLE, V, 1070;
1*H*-1,2,4-Triazole [288-88-0]

TRIBIPHENYLCARBINOL, III, 831;
[1,1′-Biphenyl]-4-methanol, α,α-bis([1,1′-biphenyl]-4-yl)- [5341-14-0]

sym-TRIBROMOBENZENE, II, 592;
Benzene, 1,3,5-tribromo- [626-39-1]

2,4,6-TRIBROMOBENZOIC ACID, IV, 947;
Benzoic acid, 2,4,6-tribromo- [633-12-5]

1,2,3-TRIBROMOPROPANE, I, 521;
Propane, 1,2,3-tribromo- [96-11-7]

TRICARBALLYLIC ACID, I, 523;
1,2,3-Propanetricarboxylic acid [99-14-9]

TRICARBETHOXYMETHANE, II, 594;
Methanetricarboxylic acid, triethyl ester [6279-86-3]

TRICARBOMETHOXYMETHANE, II, 596;
Methanetricarboxylic acid, trimethyl ester [1186-73-8]

α,α,α-TRICHLOROACETANILIDE, V, 1074;
Acetamide, 2,2,2-trichloro-*N*-phenyl- [2563-97-5]

TRICHLOROETHYL ALCOHOL, II, 598;
Ethanol, 2,2,2-trichloro- [115-20-8]

TRICHLOROMETHYLPHOSPHONYL DICHLORIDE, IV, 950;
Phosphonic dichloride, (trichloromethyl)- [21510-59-8]

1,1,3-TRICHLORO-*n*-NONANE, V, 1076;
Nonane, 1,1,3-trichloro- [10575-86-7]

p-TRICYANOVINYL-*N,N*-DIMETHYLANILINE, IV, 953;
Ethenetricarbonitrile, [4-(dimethylamino)phenyl]- [6673-15-0]

TRIETHYLCARBINOL, II, 602;
3-Pentanol, 3-ethyl- [597-49-9]

TRIETHYLOXONIUM FLUOBORATE, V, 1080;
Oxonium, triethyl-, tetrafluoroborate(1-) [368-39-8]

TRIETHYL PHOSPHITE, IV, 955;
Phosphorous acid, triethyl ester [122-52-1]

1,1,1-TRIFLUOROHEPTANE, V, 1082;
Heptane, 1,1,1-trifluoro- [693-09-4]

m-TRIFLUOROMETHYL-*N,N*-DIMETHYLANILINE, V, 1085;
Benzenamine, *N,N*-dimethyl-3-(trifluoromethyl)- [329-00-0]

1,2,5-TRIHYDROXYPENTANE, III, 833;
1,2,5-Pentanetriol [14697-46-2]

1,2,3-TRIIODO-5-NITROBENZENE, II, 604;
Benzene, 1,2,3-triiodo-5-nitro- [53663-23-3]

TRIMETHYLACETIC ACID, I, 524;
Propanoic acid, 2,2-dimethyl- [75-98-9]

TRIMETHYLAMINE, I, 528;
Methanamine, *N,N*-dimethyl- [75-50-3]

TRIMETHYLAMINE HYDROCHLORIDE, I, 531;
Methanamine, *N,N*-dimethyl-, hydrochloride [593-81-7]

4,6,8-TRIMETHYLAZULENE, V, 1088;
Azulene, 4,6,8-trimethyl- [941-81-1]

2,6,6-TRIMETHYL-2,4-CYCLOHEXADIENONE, V, 1092;
2,4-Cyclohexadien-1-one, 2,6,6-trimethyl- [13487-30-4]

2,4,4-TRIMETHYLCYCLOPENTANONE, IV, 957;
Cyclopentanone, 2,4,4-trimethyl- [4694-12-6]

TRIMETHYLENE BROMIDE, I, 30;
Propane, 1,3-dibromo- [109-64-8]

TRIMETHYLENE CHLOROHYDRIN, I, 533;
1-Propanol, 3-chloro- [627-30-5]

TRIMETHYLENE CYANIDE, I, 536;
Pentanedinitrile [544-13-8]

TRIMETHYLENE OXIDE, III, 835;
Oxetane [503-30-0]

TRIMETHYLGALLIC ACID, I, 537;
Benzoic acid, 3,4,5-trimethoxy- [118-41-2]

TRIMETHYLOXONIUM FLUOBORATE, V, 1096;
Oxonium, trimethyl-, tetrafluoroborate (1-) [420-37-1]

TRIMETHYLOXONIUM 2,4,6-TRINITROBENZENESULFONATE, V, 1099;
Oxonium, trimethyl-, salt with 2,4,6-trinitrobenzenesulfonic acid (1:1) [13700-00-0]

2,2,4-TRIMETHYL-3-OXOVALERYL CHLORIDE, V, 1103;
Pentanoyl chloride, 2,2,4-trimethyl-3-oxo- [10472-34-1]

2,4,6-TRIMETHYLPYRYLIUM PERCHLORATE, V, 1106;
Pyrylium, 2,4,6-trimethyl-, perchlorate [940-93-2]

2,4,6-TRIMETHYLPYRYLIUM TETRAFLUOROBORATE, V, 1112;
Pyrylium, 2,4,6-trimethyl-, tetrafluoroborate(1-) [773-01-3]

2,4,6-TRIMETHYLPYRYLIUM TRIFLUOROMETHANESULFONATE, V, 1114;
Pyrylium, 2,4,6-trimethyl-, salt with trifluoromethanesulfonic acid (1:1) [40927-60-4]

TRIMYRISTIN, I, 538;
Tetradecanoic acid, 1,2,3-propanetriyl ester [555-45-3]

1,3,5-TRINITROBENZENE, I, 541;
Benzene, 1,3,5-trinitro- [99-35-4]

2,4,6-TRINITROBENZOIC ACID, I, 543;
Benzoic acid, 2,4,6-trinitro- [129-66-8]

2,4,7-TRINITROFLUORENONE, III, 837;
9H-Fluoren-9-one, 2,4,7-trinitro- [129-79-3]

TRIPHENYLALUMINUM, V, 1116;
Aluminum, triphenyl- [841-76-9]

TRIPHENYLAMINE, I, 544;
Benzenamine, N,N-diphenyl- [603-34-9]

TRIPHENYLCARBINOL, III, 839;
Benzenemethanol, α,α-diphenyl- [76-84-6]

TRIPHENYLCHLOROMETHANE, III, 841;
Benzene, 1,1′,1″-(chloromethylidyne)tris- [76-83-5]

TRIPHENYLENE, V, 1120;
Triphenylene [217-59-4]

TRIPHENYLETHYLENE, II, 606;
Benzene, 1,1′,1″-(1-ethenyl-2-ylidene)tris- [58-72-0]

2,3,5-TRIPHENYLISOXAZOLIDINE, V, 1124;
Isoxazolidine, 2,3,5-triphenyl- [13787-96-7]

TRIPHENYLMETHANE, I, 548;
Benzene, 1,1′,1″-methylidynetris- [519-73-3]

TRIPHENYLMETHYLSODIUM, II, 607;
Sodium, (triphenylmethyl)- [4303-71-3]

2,4,6-TRIPHENYLNITROBENZENE, V, 1128;
  1,1':3',1"-Terphenyl, 2'-nitro-5'-phenyl- [10368-47-5]

2,4,6-TRIPHENYLPHENOXYL, V, 1130;
  2,5-Cyclohexadien-1-one, 2,4,6-triphenyl-4-(5'-phenyl[1,1':3,1"-terphenyl]-2'-yloxy)-
  [10384-15-3]

α,β,β-TRIPHENYLPROPIONIC ACID, IV, 960;
  Benzenepropanoic acid, α,β-diphenyl- [53663-24-4]

α,α,β-TRIPHENYLPROPIONITRILE, IV, 962;
  Benzenepropanenitrile, α,α-diphenyl- [5350-82-3]

2,4,6-TRIPHENYLPYRYLIUM TETRAFLUOROBORATE, V, 1135;
  Pyrylium, 2,4,6-triphenyl-, tetrafluoroborate(1-) [448-61-3]

TRIPHENYLSELONIUM CHLORIDE, II, 240;
  Selenium, chlorotriphenyl-, (T-4)- [17166-13-1]

TRIPHENYLSTIBINE, I, 550;
  Stibine, triphenyl- [603-36-1]

TRIPTYCENE, IV, 964;
  9,10[1',2']-Benzenoanthracene, 9,10-dihydro- [477-75-8]

sym-TRITHIANE, II, 610;
  1,3,5-Trithiane [291-21-4]

TRITHIOCARBODIGLYCOLIC ACID, IV, 967;
  Acetic acid, 2,2'-[carbonothioylbis(thio)]bis- [6326-83-6]

TROPYLIUM FLUOBORATE, V, 1138;
  Cycloheptatrienylium, tetrafluoroborate(1-) [27081-10-3]

l-TRYPTOPHANE, II, 612;
  L-Tryptophan [73-22-3]

U

UNDECYL ISOCYANATE, III, 846;
  Undecane, 1-isocyanato- [2411-58-7]

10-UNDECYNOIC ACID, IV, 969;
  10-Undecynoic acid [2777-65-3]

UNSOLVATED n-BUTYLMAGNESIUM CHLORIDE, V, 1141;
  Magnesium, butylchloro- [693-04-9]

URAMIL, II, 617;
  2,4,6(1H,3H,5H)-Pyrimidinetrione, 5-amino- [118-78-5]

V

dl-VALINE, III, 848;
  DL-Valine [516-06-3]

VANILLIC ACID, IV, 972;
  Benzoic acid, 4-hydroxy-3-methoxy- [121-34-6]

VERATRALDEHYDE, II, 619;
   Benzaldehyde, 3,4-dimethoxy- [120-14-9]

VERATRONITRILE, II, 622;
   Benzonitrile, 3,4-dimethoxy- [2024-83-1]

VINYLACETIC ACID, III, 851;
   3-Butenoic acid [625-38-7]

VINYL CHLOROACETATE, III, 853;
   Acetic acid, chloro, ethenyl ester [2549-51-1]

VINYL LAURATE AND OTHER VINYL ESTERS, IV, 977;
   Dodecanoic acid, ethenyl ester [2146-71-6]

2-VINYLTHIOPHENE, IV, 980;
   Thiophene, 2-ethenyl- [1918-82-7]

VINYL TRIPHENYLPHOSPHONIUM BROMIDE, V, 1145;
   Phosphonium, ethenyltriphenyl-, bromide [5044-52-0]

X

XANTHONE, I, 552;
   9H-Xanthen-9-one [90-47-1]

XANTHYDROL, I, 554;
   9H-Xanthen-9-ol [90-46-0]

o-XYLYLENE DIBROMIDE, IV, 984;
   Benzene, 1,2-bis(bromomethyl)- [91-13-4]

CUMULATIVE CONTENTS INDEX

According to Chemical Abstracts Index Names

This second index has an alphabetized arrangement of the Chemical Abstracts Index Nmes first, followed by the CA Registry Numbers. Immediately below and set at the right are the Title Names in capital letters with citations to the *Collective Volume* numbers and pages.

A

1,2-Acenaphthalenedione [82-86-0];
    ACENAPHTHENEQUINONE, III, 1

5-Acenaphthylenebutanoic acid, 1,2-dihydro-γ-oxo- [16294-60-3];
    β-(3-ACENAPHTHOYL)PROPIONIC ACID, III, 6

1-Acenaphthylenol,1,2-dihydro- [6306-07-6]
    ACENAPHTHENOL-7, III, 3

Acetaldehyde, tribromo- [115-17-3];
    BROMAL, II, 87

Acetamide [60-35-5];
    ACETAMIDE, I, 3

Acetamide, *N*-bromo- [79-15-2];
    N-BROMOACETAMIDE, IV, 104

Acetamide, 2-chloro- [79-07-2];
    CHLOROACETAMIDE, I, 153

Acetamide, *N*-[4-(chloroacetyl)phenyl]- [140-49-8];
    p-CHLOROACETYLACETANILIDE, III, 183

Acetamide, 2-cyano- [107-91-5];
    CYANOACETAMIDE, I, 179

Acetamide,2,2-dichloro- [683-72-7];
    α,α-DICHLOROACETAMIDE, III, 260

Acetamide,*N*-(1-methyl-2-oxopropyl)- [6628-81-5];
    3-ACETAMIDO-2-BUTANONE, IV, 5

Acetamide, N-(1-nitro-2-naphthalenyl)- [5419-82-9];
1-NITRO-2-ACETYLAMINONAPHTHALENE, II, 438

Acetamide,N-[4-(4-oxo-2-thioxo-3-thiazolidinyl)phenyl]- [53663-36-8];
N-(p-ACETYLAMINOPHENYL)RHODANINE, IV, 6

Acetamide, 2,2,2-trichloro-N-phenyl- [2563-97-5];
α,α,α-TRICHLOROACETANILIDE, V, 1074

Acetic acid, (aminooxy)-, hydrochloride (2:1) [6921-14-4];
CARBOXYMETHOXYLAMINE HEMIHYDROCHLORIDE, III, 172

Acetic acid, arsono- [107-38-0];
ARSONOACETIC ACID, I, 73

Acetic acid, (4-arsonophenoxy)- [53663-15-3];
p-ARSONOPHENOXYACETIC ACID, I, 75

Acetic acid, bromo, ethyl ester [105-36-2];
ETHYL BROMOACETATE, III, 381

Acetic acid, 2,2'-[carbonothioylbis(thio)]bis- [6326-83-6];
TRITHIOCARBODIGLYCOLIC ACID, IV, 967

Acetic acid, chloro, ethenyl ester [2549-51-1];
VINYL CHLOROACETATE, III, 853

Acetic acid, chlorofluoro-, ethyl ester [401-56-9];
ETHYL CHLOROFLUOROACETATE, IV, 423

Acetic acid, cyanocyclohexylidene- [37107-50-9];
CYCLOHEXYLIDENECYANOACETIC ACID, IV, 234

Acetic acid, cyano, 1,1-dimethylethyl ester [1116-98-9];
t-BUTYL CYANOACETATE, V, 171

Acetic acid, cyano-, ethyl ester [105-56-6];
ETHYL CYANOACETATE, I, 254

Acetic acid, cyclohexylidene, ethyl ester [1552-92-7];
ETHYL CYCLOHEXYLIDENEACETATE, V, 547

Acetic acid, 2,2'-diarsenediyl)bis- [544-27-4];
ARSENOACETIC ACID, I, 73

Acetic acid, diazo-, 2-butenyl ester, (E)- [14746-03-3];
CROTYL DIAZOACETATE, V, 258

Acetic acid, diazo-, 1,1-dimethylethyl ester [35059-50-8];
t-BUTYL DIAZOACETATE, V, 179

Acetic acid, diazo, ethyl ester [623-73-4];
ETHYL DIAZOACETATE, III, 392; IV, 424

Acetic acid, dichloro- [79-43-6];
DICHLOROACETIC ACID, II, 181

Acetic acid, diethoxy, ethyl ester [6065-82-3];
ETHYL DIETHOXYACETATE, IV, 427

Acetic acid,1,1-dimethylethyl ester [540-88-5];
  t-BUTYL ACETATE, III, 141

Acetic acid, ethoxy- [627-03-2];
  ETHOXYACETIC ACID, II, 260

Acetic acid, ethoxy, ethyl ester [817-95-8];
  ETHYL ETHOXYACETATE, II, 260

Acetic acid, [[5-methyl-2-(1-methylethyl)cyclohexyl]oxy]-,[1R-(1a,2p,5a)]- [40248-63-3];
  l-MENTHOXYACETIC ACID, III, 544

Acetic acid, oxo, butyl ester [6295-06-3];
  n-BUTYL GLYOXYLATE, IV, 124

Acetic acid, [(phenylthioxomethyl)thio]- [942-91-6];
  THIOBENZOYLTHIOGLYCOLIC ACID, V, 1046

Acetonitrile, chloro-[107-14-2];
  CHLOROACETONITRILE, IV, 144

Acetonitrile, dibromo- [3252-43-5];
  DIBROMOACETONITRILE, IV, 254

Acetonitrile, (diethylamino)- [3010-02-4];
  DIETHYLAMINOACETONITRILE, III, 275

Acetonitrile, hydroxy- [107-16-4];
  GLYCOLONITRILE, III, 436

Acetonitrile, methoxy- [1738-36-9];
  METHOXYACETONITRILE, II, 387

Acetonitrile, (methyleneamino)- [109-82-0];
  METHYLENEAMINOACETONITRILE, I, 355

Acetyl chloride, [[5-methyl-2-(1-methylethyl)cyclohexyl]oxy]-, [1R-(1a,2p,5a)]- [15356-62-4];
  l-MENTHOXYACETYL CHLORIDE, III, 547

Acetyl isocyanate, chloro- [4461-30-7];
  α-CHLOROACETYL ISOCYANATE, V, 204

9-Acridinamine [90-45-9];
  9-AMINOACRIDINE, III, 53

9(10*H*)-Acridinone [578-95-0];
  ACRIDONE, II, 15

β-Alanine [107-95-9];
  β-ALANINE, II, 19; III, 34

DL-Alanine [302-72-7];
  dl-ALANINE, I, 21

β-Alanine, *N*-(3-ethoxy-3-oxopropyl)-*N*-methyl-,ethyl ester [6315-60-2];
  DI-β-CARBETHOXYETHYLMETHYLAMINE, III, 258

Alanine, 2-methyl- [62-57-7];
  α-AMINOISOBUTYRIC ACID, II, 29

L-Alanine, 3-sulfo-, hemihydrate [53643-49-5];
   CYSTEIC ACID MONOHYDRATE, III, 226

Aluminum, triphenyl- [841-76-9];
   TRIPHENYLALUMINUM, V, 1116

Androst-5-ene-17-carboxylic acid, 3-(acetyloxy)-,(3β,17β)- [7150-18-7];
   3β-ACETOXYETIENIC ACID, V, 8

9-Anthracenecarboxaldehyde [642-31-9];
   9-ANTHRALDEHYDE, III, 98

Anthracene, 9-chloro- [716-53-0];
   9-CHLOROANTHRACENE, V, 206

Anthracene, 9,10-dibromo-[523-27-3]
   9,10-DIBROMOANTHRACENE, I, 207

Anthracene, 9,10-dihydro- [613-31-0];
   9,10-DIHYDROANTHRACENE, V, 398

9,10-Anthracenedione, 1-bromo-4-(methylamino)- [128-93-8];
   1-METHYLAMINO-4-BROMOANTHRAQUINONE, III, 575

9,10-Anthracenedione, 1-chloro- [82-44-0];
   α-CHLOROANTHRAQUINONE, II, 128

9,10-Anthracenedione, 1,4-dihydroxy- [81-64-1];
   QUINIZARIN, I, 476

9,10-Anthracenedione, 2,3-dimethyl- [6531-35-7];
   2,3-DIMETHYLANTHRAQUINONE, III, 310

9,10-Anthracenedione, 2-methyl- [84-54-8];
   β-METHYLANTHRAQUINONE, I, 353

9,10-Anthracenedione, 1-(methylamino)- [82-38-2];
   1-METHYLAMINOANTHRAQUINONE, III, 573

Anthracene, 9-nitro- [602-60-8];
   9-NITROANTHRACENE, IV, 711

1-Anthracenesulfonic acid, 9,10-dihydro-9,10-dioxo-,potassium salt [30845-78-4];
   POTASSIUM ANTHRAQUINONE-α-SULFONATE, II, 539

9(10H)-Anthracenone [90-44-8];
   ANTHRONE, I, 60

9(10H)-Anthracenone, 10-nitro- [6313-44-6];
   NITROANTHRONE, I, 390

D-Arabinose [10323-20-3];
   D-ARABINOSE, III, 101

L-Arabinose [5328-37-0];
   l-ARABINOSE, I, 67

L-Arginine, monohydrochloride [1119-34-2];
   d-ARGININE HYDROCHLORIDE, II, 49

Arsonic acid, [4-[(2-amino-2-oxoethyl)amino]phenyl]-,disodium salt [834-03-7];
   SODIUM p-ARSONO-N-PHENYLGLYCINAMIDE, I, 488

Arsonic acid,(4-aminophenyl)- [98-50-0];
   ARSANILIC ACID, I, 70

Arsonic acid, (4-hydroxyphenyl)-, sodium salt [53663-20-0];
   SODIUM p-HYDROXYPHENYLARSONATE, I, 490

Arsonic acid, (4-nitrophenyl)- [98-72-6];
   p-NITROPHENYLARSONIC ACID, III, 665

Arsonic acid, phenyl- [98-05-5];
   PHENYLARSONIC ACID, II, 494

Arsonium, tetraphenyl-, (hydrogen dichloride) [21006-73-5];
   TETRAPHENYLARSONIUM CHLORIDE HYDROCHLORIDE, IV, 910

DL-Aspartic acid [617-45-8];
   DL-ASPARTIC ACID, IV, 55

9-Azabicyclo[3.3.1]nonan-3-one, 9-methyl- [552-70-5];
   PSEUDOPELLETIERINE, IV, 816

1-Azabicyclo[2.2.2]octan-3-one, hydrochloride [1193-65-3];
   3-QUINUCLIDONE HYDROCHLORIDE, V, 989

2H-Azepine, 3,4,5,6-tetrahydro-7-methoxy- [2525-16-8];
   O-METHYLCAPROLACTIM, IV, 588

2H-Azepin-2-one, 1,3-dihydro-3,5,7-trimethyl- [936-85-6];
   1,3-DIHYDRO-3,5,7-TRIMETHYL-2H-AZEPIN-2-ONE; V, 408

2H-Azepin-2-one, hexahydro- [105-60-2];
   2-KETOHEXAMETHYLENIMINE, II, 371

Azetidinesulfonyl chloride, 2,2-dimethyl-4-oxo [17174-96-8]
2-Azetidinone, 4,4-dimethyl- [4879-95-2] ;
   β-ISOVALEROLACTAM-N-SULFONYL CHLORIDE, AND β-ISOVALEROLAC-
   TAM, V, 673

Aziridine [151-56-4];
   ETHYLENIMINE, IV, 433

Aziridine, 1-ethyl-2-methylene- [872-39-9];
   N-ETHYLALLENIMINE, V, 541

Aziridine, 2-phenyl-3-(phenylmethyl)-, cis- [1605-08-9];
   cis-2-BENZYL-3-PHENYLAZIRIDINE, V, 83

2-Azobicyclo[2.2.2]octan-3-one [3306-69-2];
   3-ISOQUINUCLIDONE, V, 670

Azulene, 4,6,8-trimethyl- [941-81-1];
   4,6,8-TRIMETHYLAZULENE, V, 1088

B

Benzaldehyde, 2-amino [529-23-7];
o-AMINOBENZALDEHYDE, III, 56

Benzaldehyde, 4-amino- [556-18-3];
p-AMINOBENZALDEHYDE, IV, 31

Benzaldehyde, 4-bromo-,[1122-91-4];
p-BROMOBENZALDEHYDE, II, 89

Benzaldehyde, 2-bromo-4-methyl- [824-54-4];
2-BROMO-4-METHYLBENZALDEHYDE, V, 139

Benzaldehyde, 3-chloro- [587-04-2];
m-CHLOROBENZALDEHYDE, II, 130

Benzaldehyde, 4-chloro- [104-88-1];
p-CHLOROBENZALDEHYDE, II, 133

Benzaldehyde, 3,4-dihydroxy- [139-85-5];
PROTOCATECHUALDEHYDE, II, 549

Benzaldehyde, 3,4-dimethoxy- [120-14-9];
VERATRALDEHYDE, II, 619

Benzaldehyde, 4,5-dimethoxy-2-nitro- [20357-25-9];
6-NITROVERATRALDEHYDE, IV, 735

Benzaldehyde, 4-(dimethylamino)- [100-10-7];
p-DIMETHYLAMINOBENZALDEHYDE, I, 214; IV, 331

Benzaldehyde, 2,4-dinitro- [528-75-6];
2,4-DINITROBENZALDEHYDE, II, 223

Benzaldehyde, 3-hydroxy- [100-83-4];
m-HYDROXYBENZALDEHYDE, III, 453

Benzaldehyde, 4-hydroxy-3,5-dimethoxy- [134-96-3];
SYRINGIC ALDEHYDE, IV, 866

Benzaldehyde, 2-methoxy- [135-02-4];
o-ANISALDEHYDE, V, 46

Benzaldehyde, 3-methoxy- [591-31-1];
m-METHOXYBENZALDEHYDE, III, 564

Benzaldehyde, 2-methyl- [529-20-4];
o-TOLUALDEHYDE, III, 818; IV, 932

Benzaldehyde, 4-methyl- [104-87-0];
p-TOLUALDEHYDE, II, 583

Benzaldehyde, 2-nitro- [552-89-6];
o-NITROBENZALDEHYDE, III, 641; V, 825

Benzaldehyde, 4-nitro- [555-16-8];
p-NITROBENZALDEHYDE, II, 441

Benzaldehyde, 2,4,6-trimethyl- [487-68-3];
MESITALDEHYDE, III, 549; V, 49

Benzamide, N-hydroxy- [495-18-1];
BENZOHYDROXAMIC ACID II, 67

Benzamide, 2-hydroxy-N-(2-methylphenyl)- [7133-56-4];
SALICYL-o-TOLUIDE, III, 765

Benzamide, 2-methyl- [527-85-5];
o-TOLUAMIDE, II, 586; Warning, V, 1054

Benzamide, N-phenyl- [93-98-1];
BENZANILIDE, I, 82

7H-Benz[de]anthracen-7-one [82-05-3];
BENZANTHRONE, II, 62

Benzenamine, 4,4'-azobis- [538-41-0];
4,4'-DIAMINOAZOBENZENE, V, 341

Benzenamine, 2-bromo-4-methyl- [583-68-6];
3-BROMO-4-AMINOTOLUENE, I, 111

Benzenamine, 4-chloro-N-ethyl- [13519-75-0];
N-ETHYL-p-CHLOROANILINE, IV, 420

Benzenamine, 2,6-dibromo- [608-30-0];
2,6-DIBROMOANILINE, III, 262

Benzenamine, 2,6-dichloro- [608-31-1];
2,6-DICHLOROANILINE, III, 262

Benzenamine, 4-(2,2-dichloroethenyl)-N,N-dimethyl- [6798-58-9];
β,β-DICHLORO-p-DIMETHYLAMINOSTYRENE, V, 361

Benzenamine, 2,6-diiodo-4-nitro- [5398-27-6];
2,6-DIIODO-p-NITROANILINE, II, 196

Benenamine, 3,4-dimethoxy- [6315-89-5];
4-AMINOVERATROLE, II, 44

Benzenamine,3-(dimethoxymethyl)- [53663-37-9];
m-AMINOBENZALDEHYDE DIMETHYLACETAL, III, 59

Benzenamine, 3,4-dimethyl- [95-64-7];
3,4-DIMETHYLANILINE, III, 307

Benzenamine, N,N-dimethyl-3-nitro- [619-31-8];
m-NITRODIMETHYLANILINE, III, 658

Benzenamine, N,N-dimethyl-3-(trifluoromethyl)- [329-00-0];
m-TRIFLUOROMETHYL-N,N-DIMETHYLANILINE, V, 1085

Benzenamine, 2,4-dinitro- [97-02-9];
2,4-DINITROANILINE, II, 221

Benzenamine, 2,6-dinitro- [606-22-4];
2,6-DINITROANILINE, IV, 364

Benzenamine, N,N-diphenyl- [603-34-9];
TRIPHENYLAMINE, I, 544

Benzenamine, 4,4'-dithiobis- [722-27-0];
p-AMINOPHENYL DISULFIDE, III, 86

Benzenamine, N-ethyl-3-methyl- [102-27-2];
N-ETHYL-m-TOLUIDINE, II, 290

Benzenamine, N-hydroxy- [100-65-2];
β-PHENYLHYDROXYLAMINE, I, 445

Benzenamine, N-hydroxy-N-nitroso-, ammonium salt [135-20-6];
CUPFERRON, I, 177

Benzenamine, 4-iodo- [540-37-4];
p-IODOANILINE, II, 347

Benzenamine, N,N'-methanetetraylbis- [622-16-2];
DIPHENYLCARBODIIMIDE(METHODS I AND II), V, 501; 504

Benzenamine, 4-methoxy-2-nitro- [96-96-8];
2-NITRO-4-METHOXYANILINE, III, 661

Benzenamine, 2-methyl-5-(methylethyl)- [2051-53-8];
2-AMINO-p-CYMENE, III, 63

Benzenamine, 2-methyl-6-nitro- [570-24-1];
2-AMINO-3-NITROTOLUENE, IV, 42

Benzenamine, N-methyl-N-nitroso- [614-00-6];
N-NITROSOMETHYLANILINE, II, 460

Benzenamine, 2-nitro- [88-74-4];
o-NITROANILINE, I, 388

Benzenamine, N-(phenylmethylene)- [538-51-2];
BENZALANILINE, I, 80

Benzenamine, 4,4'-sulfonylbis- [80-08-0];
4,4'-DIAMINODIPHENYLSULFONE, III, 239

Benzenamine, 4-(triphenylmethyl)- [22948-06-7];
p-AMINOTETRAPHENYLMETHANE, IV, 47

Benzene [71-43-2];
REDUCTION OF ORGANIC HALIDES, CHLOROBENZENE TO BENZENE, V, 998

Benzeneacetaldehyde, α-methyl- [93-53-8];
α-PHENYLPROPIONALDEHYDE, III, 733

Benzeneacetaldehyde, α-oxo- [1074-12-0];
PHENYLGLYOXAL, II, 509; V, 937

Benezeneacetaldehyde, α-phenyl- [947-91-1];
DIPHENYLACETALDEHYDE, IV, 375

Benzeneacetamide [103-81-1];
PHENYLACETAMIDE, IV, 760

Benzeneacetamide, α-hydroxy- [4410-31-5];
MANDELAMIDE, III, 536

Benzeneacetic acid [103-82-2]
PHENYLACETIC ACID, I, 436

Benzeneacetic acid, α- acetyl-, ethyl ester [5413-05-8];
ETHYL α-PHENYLACETOACETATE, II, 284

Benzeneacetic acid, α-(acetyloxy)- [5438-68-6];
ACETYLMANDELIC ACID,I, 12

Benzeneacetic acid, α-amino-(±)- [2835-06-5];
dl-α-AMINOPHENYLACETIC ACID, III, 84

Benzeneacetic acid, 4-amino- [1197-55-3];
p-AMINOPHENYLACETIC ACID, I, 52

Benzeneacetic acid, α-amino-α-methyl-(±) [6945-32-0];
dl-α-AMINO-α-PHENYLPROPIONIC ACID, III, 88

Benzeneacetic acid, 4-bromo-α-hydroxy- [6940-50-7];
p-BROMOMANDELIC ACID, IV, 110

Benzeneacetic acid, 2-carboxy- [89-51-0];
HOMOPHTHALIC ACID, V, 612

Benzeneacetic acid, α-chloro- [4755-72-0];
α-CHLOROPHENYLACETIC ACID, IV, 169

Benzeneacetic acid, 4-chloro-α-(4-chlorophenyl)- [83-05-6];
DI-(p-CHLOROPHENYL)ACETIC ACID, III, 270; Correction,V, 370

Benzeneacetic acid, α-cyano-α-(2-cyanoethyl)-, ethyl ester [53555-70-7];
α-PHENYL-α-CARBETHOXYGLUTARONITRILE, IV, 776

Benzeneacetic acid, α-cyano-, ethyl ester [4553-07-5];
ETHYL PHENYLCYANOACETATE, IV, 461

Benzeneacetic acid, 3,4-dimethoxy- [93-40-3];
HOMOVERATRIC ACID, II, 333

Benzeneacetic acid, 4-(dimethylamino)-, ethyl ester [17078-29-4];
ETHYL p-DIMETHYLAMINOPHENYLACETATE, V, 552

Benzeneacetic acid, ethyl ester [101-97-3];
ETHYL PHENYLACETATE, I, 270

Benzeneacetic acid, α-hydroxy- [90-64-2]
MANDELIC ACID, I, 336; III, 538

Benzeneacetic acid, α-hydroxy-2,5-dimethyl- [5766-40-5];
2,5-DIMETHYLMANDELIC ACID, III, 326

Benzeneacetic acid, α-hydroxy-α-methyl- [515-30-0];
ATROLACTIC ACID, IV, 58

Benzeneacetic acid, α-hydroxy-α-phenyl- [76-93-7]
BENZILIC ACID, I, 89

Benzeneacetic acid, 4-nitro- [104-03-0]
  p-NITROPHENYLACETIC ACID, I, 406

Benzeneacetic acid, α-[(2-nitrophenylmethylene]-, (E)- [19319-35-8];
  trans-o-NITRO-α-PHENYLCINNAMIC ACID, IV, 730

Benzeneacetic acid, α-oxo- [611-73-4]
  BENZOYLFORMIC ACID, III, 114

Benzeneacetic acid, α-oxo-, ethyl ester [1603-79-8];
  ETHYL BENZOYLFORMATE, I, 241

Benzeneacetic acid, α-phenyl- [117-34-0];
  DIPHENYLACETIC ACID, I, 224

Benzeneacetic acid, α-(phenylmethylene)- [3368-16-9];
  α-PHENYLCINNAMIC ACID, IV, 777

Benzeneacetic acid, 2,4,6-trimethyl- [4408-60-0];
  MESITYLACETIC ACID, III, 557

Benzeneacetonitrile [140-29-4];
  BENZYL CYANIDE, I, 107

Benzeneactonitrile, α-acetyl- [4468-48-8];
  α-PHENYLACETOACETONITRILE, II, 487

Benzeneacetonitrile, α-cyclohexyl- [3893-23-0];
  α-CYCLOHEXYLPHENYLACETONITRILE, III, 219

Benzeneacetonitrile, α-(dimethylamino)- [827-36-1];
  α-N,N-DIMETHYLAMINOPHENYLACETONITRILE, V, 437

Benzeneacetonitrile, α-(diphenylmethylene)- [6304-33-2];
  α,β-DIPHENYLCINNAMONITRILE, IV, 387

Benzeneacetonitrile, 4-methoxy- [104-47-2];
  p-METHOXYPHENYLACETONITRILE, IV, 576

Benzeneacetonitrile, 4-nitro- [555-21-5];
  p-NITROBENZYL CYANIDE, I, 396

Benzeneacetonitrile, α-oxo- [613-90-1];
  BENZOYL CYANIDE, III, 112

Benzeneacetonitrile, α-phenyl- [86-29-3];
  DIPHENYLACETONITRILE, III, 347

Benzeneacetonitrile, α-(phenylmethylene)- [2510-95-4];
  α-PHENYLCINNAMONITRILE, III, 715

Benzeneacetyl chloride,α-(acetyloxy)- [1638-63-7];
  ACETYLMANDELYL CHLORIDE, I, 12

Benzene, azido- [622-37-7];
  PHENYL AZIDE, III, 710

Benzene, 1,2-bis(bromomethyl)- [91-13-4];
  o-XYLYLENE DIBROMIDE, IV, 984

Benzene, 1,2-bis(butylthio)- [53663-38-0];
   1,2-BIS(n-BUTYLTHIO)BENZENE, V, 107

Benzene, 1-bromo-2-chloro- [694-80-4];
   o-CHLOROBROMOBENZENE, III, 185

Benzene, 4-bromo-1,2-dimethyl- [583-71-1];
   4-BROMO-o-XYLENE, III, 138

Benzene, (bromoethynyl)- [932-87-6];
   PHENYLBROMOETHYNE, V, 921

Benzene, 1-bromo-2-fluoro- [1072-85-1];
   1-BROMO-2-FLUOROBENZENE; V, 133

Benzene, 1-bromo-2-methyl- [95-46-5];
   o-BROMOTOLUENE, I, 135

Benzene, 1-bromo-3-methyl- [591-17-3];
   m-BROMOTOLUENE, I, 133

Benzene, 1-bromo-4-methyl- [106-38-7];
   p-BROMOTOLUENE, I, 136

Benzene, 1-(bromomethyl)-4-nitro- [100-11-8];
   p-NITROBENZYL BROMIDE, II, 443

Benzene, 1-bromo-3-nitro- [585-79-5];
   m-BROMONITROBENZENE, I, 123

Benzene, 1-bromo-4-(phenylmethyl)- [2116-36-1];
   p-BROMODIPHENYLMETHANE, V, 130

Benzene (3-bromo-1-propenyl)- [4392-24-9];
   CINNAMYL BROMIDE, V, 249

Benzene, (3-bromopropoxy) [588-63-6];
   γ-PHENOXYPROPYL BROMIDE, I, 435

Benzene, 2-bromo-1,3,5-trimethyl- [576-83-0];
   BROMOMESITYLENE, II, 95

Benzene, 1,1'-(4,3-butadiene-1,4-diyl)bis-[886-65-7];
   1,4-DIPHENYL-1,3-BUTADIENE, II, 229; V, 499

Benzene, 1,3-butadienyl-, (E)- [16939-57-4];
   trans-1-PHENYL-1,3-BUTADIENE, IV, 771

Benzene, 1-(1,3-butadienyl)-4-nitro- [20264-89-5];
   1-(p-NITROPHENYL)-1,3-BUTADIENE, IV, 727

Benzene, 1,1'-(1,3-butadiyne-1,4-diyl)bis- [886-66-8];
   DIPHENYLDIACETYLENE, V, 517

Benzenebutanenitrile, α-(4-chlorophenyl)-β-oxo [35741-47-0];
   α-(4-CHLOROPHENYL)-γ-PHENYLACETOACETONITRILE, IV, 174

Benzenebutanenitrile, γ-oxo- α-phenyl- [6268-00-4];
   α-PHENYL-β-BENZOYLPROPIONITRILE, II, 498

Benzenebutanoic acid [1821-12-1];
γ-PHENYLBUTYRIC ACID, II, 499

Benzenebutanoic acid, β-ethyl-β-methyl- [53663-16-4];
3-BENZYL-3-METHYLPENTANOIC ACID, IV, 93

Benzenebutanoic acid, γ-oxo- [2051-95-8];
β-BENZOYLPROPIONIC ACID, II, 81

Benzenebutanoic acid, α-phenyl-, ethyl ester [53608-81-4];
ETHYL 2,4-DIPHENYLBUTANOATE, V, 559

Benzene, 1-butoxy-2-nitro- [7252-51-9];
o-n-BUTOXYNITROBENZENE, III, 140

Benzene, butyl- [104-51-8];
n-BUTYLBENZENE, III, 157

Benzenecarboperoxoic acid [93-59-4];
PEROXYBENZOIC ACID, V, 904; I, 431; (Warning) V, 900

Benezenecarboperoxoic acid, 2-carboxy- [2311-91-3];
MONOPERPHTHALIC ACID, III, 619; V, 805

Benzenecarbothioic acid [98-91-9];
THIOBENZOIC ACID, IV, 924

1,2-Benzenecarboxaldehyde [643-79-8];
o-PHTHALALDEHYDE, IV, 807

Benzenecarboximidamide, N,N'-diphenyl- [2556-46-9];
N,N-DIPHENYLBENZAMIDINE, IV, 383

Benzenecarboximidamide, N-phenyl- [1527-91-9];
N-PHENYLBENZAMIDINE, IV, 769

Benzene, 1-chloro-4-(chloromethoxy)- [21151-56-4];
p-CHLOROPHENOXYMETHYL CHLORIDE, V, 221

Benzene, (2-chloro-1,1-dimethylethyl)- [515-40-2];
NEOPHYL CHLORIDE, IV, 702

Benzene, 2-chloro-1,3-dinitro- [606-21-3];
1-CHLORO-2,6-DINITROBENZENE, IV, 160

Benzene, 1-chloro-3-ethenyl- [2039-85-2];
m-CHLOROSTYRENE, III, 204

Benzene, 1,1'-[(2-chloroethoxy)methylene]bis- [32669-06-0];
BENZHYDRYL β-CHLOROETHYL ETHER, IV, 72

Benzene, 1-chloro-4-isothiocyanato- [2131-55-7];
p-CHLOROPHENYL ISOTHIOCYANATE, I, 165; V,223

Benzene, 1-chloro-2-methyl- [95-49-8];
o-CHLOROTOLUENE, I, 170

Benzene, 1-chloro-4-methyl- [106-43-4];
p-CHLOROTOLUENE, I, 170

Benzene, 1,1',1"-(chloromethylidyne)tris- [76-83-5];
TRIPHENYLCHLOROMETHANE, III, 841

Benzene, 1-chloro-3-nitro- [121-73-3];
m-CHLORONITROBENZENE, I, 162

Benzene, cyclohexyl- [827-52-1];
CYCLOHEXYLBENZENE, II, 151

Benzene, cyclopropyl- [873-49-4];
CYCLOPROPYLBENZENE, V, 328; same as PHENYLCYCLOPROPANE, V, 929

Benzene, 1,1'-cyclopropylidenebis- [3282-18-6];
1,1-DIPHENYLCYCLOPROPANE, V, 509

1,4-Benzenediacetic acid, 2,5-dihydroxy- [5488-16-4];
2,5-DIHYDROXY-p-BENZENEDIACETIC ACID, III, 286

1,2-Benzenediamine [95-54-5];
o-PHENYLENEDIAMINE, II, 501

1,3-Benzenediamine, 4-methyl- [95-80-7];
2,4-DIAMINOTOLUENE, II, 160

1,2-Benzenediamine, 4-nitro- [99-56-9];
1,2-DIAMINO-4-NITROBENZENE, III, 242

1,4-Benzenediamine, N,N,N',N'-tetramethyl- [100-22-1];
TETRAMETHYL-p-PHENYLENEDIAMINE, V, 1018

Benzene, 1,1'-(diazomethylene)bis- [883-40-9];
DIPHENYLDIAZOMETHANE, III, 351

Benzenediazonium, 2-carboxy-, hydroxide, inner salt [1608-42-0];
BENZENEDIAZONIUM-2-CARBOXYLATE, V, 54

1,3-Benzenedicarboxaldehyde [626-19-7];
ISOPHTHALALDEHYDE, V, 668

1,4-Benzenedicarboxaldehyde [623-27-8];
TEREPHTHALALDEHYDE, III, 788

1,4-Benzenedicarboxylic acid [100-21-0];
TEREPHTHALIC ACID, III, 791

1,3-Benzenedicarboxylic acid, 2-hydroxy- [606-19-9];
2-HYDROXYISOPHTHALIC ACID, V, 617

1,2-Benzenedicarboxylic acid, 3-nitro- [603-11-2];
3-NITROPHTHALIC ACID, I, 408

1,2-Benzenedicarboxylic acid, 4-nitro- [610-27-5];
4-NITROPHTHALIC ACID, II, 457

1,2-Benzenedicarbonyl dichloride [88-95-9];
Sym.O-PHTHALYL CHLORIDE, II, 528

Benzene, (dichloroiodo)- [932-72-9];
IODOBENZENE DICHLORIDE, III, 482

Benzene, 1,3-dichloro-2-nitro- [601-88-7];
2,6-DICHLORONITROBENZENE, V, 367

Benzene, (3,3-diethoxy-1-propynyl)- [6142-95-6];
PHENYLPROPARGYLALDEHYDE DIETHYL ACETAL, IV, 801

Benzene, (2,2-difluoroethenyl)- [405-42-5];
β,β-DIFLUOROSTYRENE, V, 390

Benzene, (difluoromethyl)- [455-31-2];
α α-DIFLUOROTOLUENE, V, 396

Benzene, [2,2-difluoro-1-(trifluoromethyl)ethenyl]- [1979-51-7];
2-PHENYLPERFLUOROPROPENE, V, 949

Benzene, 1-(dimethoxymethyl)-3-nitro- [3395-79-7];
m-NITROBENZALDEHYDE DIMETHYLACETAL, III, 644

Benzene, (1,1-dimethylethoxy)- [6669-13-2];
PHENYL t-BUTYL ETHER (METHOD I), V, 924

Benzene, (1,1-dimethylethoxy)- [6669-13-2];
PHENYL t-BUTYL ETHER(METHOD II), V, 926

Benzene, 1,4-dinitro- [100-25-4];
p-DINITROBENZENE, II, 225

1,2-Benzenediol [120-80-9];
CATECHOL, I, 149

1,3-Benzenediol, 4-bromo- [6626-15-9];
4-BROMORESORCINOL, II, 100

1,4-Benzenediol, diacetate [1205-91-0];
HYDROQUINONE DIACETATE, III, 452

1,2-Benzenediol, 3-methoxy- [934-00-9];
PYROGALLOL 1-MONOMETHYL ETHER, III, 759

1,2-Benzenedithiol [17534-15-5];
1,2-DIMERCAPTOBENZENE, V, 419

Benzeneethanamine [64-04-0];
β-PHENYLETHYLAMINE, III, 720

Benzeneethanamine, α,α-dimethyl- [122-09-8];
α,α-DIMETHYL-β-PHENETHYLAMINE, V, 471

Benzeneethanamine, N,N-dimethyl- [1126-71-2];
β-PHENYLETHYLDIMETHYLAMINE, III, 723

Benzeneethanamine, N-methyl-α-phenyl- [53663-25-5]

Benzeneethanamine, N-methyl-α-phenyl-, hydrochloride [7400-77-3];
N-METHYL-1,2-DPHENYLETHYLAMINE AND HYDROCHLORIDE, IV, 605;

Benzene, 1,1'-(1,2-ethanediyl)bis[4-nitro- [736-30-1];
p,p'-DINITROBIBENZYL, IV, 367

Benzeneethanimidoyl chloride, N-hydroxy-α-oxo- [4937-87-5];
Ω-CHLOROISONITROSOACETOPHENONE, III, 191

Benzene, 1,1'-(1,2-ethenediyl)bis-, (E)- [103-30-0];
trans-STILBENE, III, 786

Benzene, 1,1'-(1,2-ethenediyl)bis-, (Z)- [645-49-8];
cis-STILBENE, IV, 857

Benzene, 1,1'-(1,2-ethenediyl)bis[4-methoxy-,(E)- [15638-14-9];
trans-4,4'-DIMETHOXYSTILBENE, V, 428

Benzene, 1,1',1'',1'''-(1,2-ethenediylidene)tetrakis- [632-51-9];
TETRAPHENYLETHYLENE, IV, 914

Benzene, ethenyl- [100-42-5];
PHENYLETHYLENE, I, 440 (STYRENE)

Benzene, 1,1'-ethenylidenebis- [530-48-3];
1,1-DIPHENYLETHYLENE, I, 226

Benzene, 1-ethenyl-3-nitro- [586-39-0];
m-NITROSTYRENE, IV, 731

Benzene, 1,1',1''-(1-ethenyl-2-ylidene)tris- [58-72-0];
TRIPHENYLETHYLENE, II, 606

Benzene, 1,1'-ethylidenebis [4-methyl-] [530-45-0];
1,1-DI-p-TOLYLETHANE, I, 229

Benzene, 1,1'-(1,2-ethynediyl)bis- [501-65-5];
DIPHENYLACETYLENE, III, 350; IV, 377

Benzene, ethynyl- [536-74-3];
PHENYLACETYLENE, I, 438; IV, 763

Benzene, fluoro- [462-06-6];
FLUOROBENZENE, II, 295

Benzene, hexamethyl- [87-85-4];
HEXAMETHYLBENZENE, IV, 520

Benzenehexol [608-80-0];
HEXAHYDROXYBENZENE, V, 595

Benzene, iodo- [591-50-4];
IODOBENZENE, I, 323; II, 351

Benzene, 4-iodo-1,2-dimethoxy- [5460-32-2];
4-IODOVERATROLE, IV, 547

Benzene, 1-iodo-2,4-dinitro- [709-49-9];
2,4-DINITROIODOBENZENE, V, 478

Benzene, iodosyl- [536-80-1];
IODOSOBENZENE, III, 483; V, 658

Benzene, iodoxy- [696-33-3];
IODOXYBENZENE, III, 485; V, 665

Benzene, 1-isocyanato-4-nitro- [100-28-7];
p-NITROPHENYL ISOCYANATE, II, 453

Benzene, 1-isocyano-2-methyl- [10468-64-1];
o-TOLYL ISOCYANIDE, V, 1060

Benzene, isothiocyanato- [103-72-0];
PHENYL ISOTHIOCYANATE, I, 447

Benzenemethanamine, 2,3-dimethoxy-N-methyl- [53663-28-8];
N-METHYL-2,3-DIMETHOXYBENZYLAMINE, IV, 603

Benzenemethanamine, α-methyl- [98-84-0];
α-PHENYLETHYLAMINE, II, 503; III, 717

Benzenemethanamine, α-methyl-, (R)- [3886-69-9];
Benzenemethanamine, α-methyl-, (S)- [2627-86-3];
d-AND l-α-PHENYLETHYLAMINE, II, 506 R(+)-AND S(−)-α-PHENYLETHY-
LAMINE, V, 932

Benzenemethanamine, N-(3-methylphenyl)- [5405-17-4];
m-TOLYLBENZYLAMINE, III, 827

Benzenemethanamine, N-phenyl- [103-32-2];
BENZYLANILINE, I, 102

Benzenemethanamine,N,N,2-trimethyl- [4525-48-8];
2-METHYLBENZYLDIMETHYLAMINE, IV, 585

Benzenemethanaminium, N,N,N-trimethyl-, ethoxide [27292-06-4];
BENZYLTRIMETHYLAMMONIUM ETHOXIDE, IV, 98

Benzenemethanimine, α-phenyl- [1013-88-3];
DIPHENYL KETIMINE, V, 520

Benzenemethanimine, α-phenyl-, hydrochloride [5319-67-5];
DIPHENYLMETHANE IMINE HYDROCHLORIDE, II, 234

Benzenemethanol, 2-amino- [5344-90-1];
o-AMINOBENZYL ALCOHOL, III, 60

Benzenemethanol, 3-chloro-α-methyl [6939-95-3];
m-CHLOROPHENYLMETHYLCARBINOL, III, 200

Benzenemethanol, α,α-diphenyl- [76-84-6];
TRIPHENYLCARBINOL, III, 839

Benzenemethanol, 2-methyl- [89-95-2];
o-METHYLBENZYL ALCOHOL, IV, 582

Benzenemethanol, 4-methyl- [589-18-4];
p-TOLYL CARBINOL, II, 590

Benzenemethanol, 4-nitro- [619-73-8];
p-NITROBENZYL ALCOHOL, III, 652

Benzenemethanol, 4-nitro-, acetate (ester) [619-90-9];
p-NITROBENZYL ACETATE, III, 650

Benzenemethanol, α-phenyl-[91-01-0];
BENZOHYDROL, I, 90

Benzene, methoxy- [100-66-3];
ANISOLE, I, 58

Benzene, 1-methoxy-3,5-dinitro- [5327-44-6];
3,5-DINITROANISOLE, I, 219

Benzene, 1-methoxy-2-phenoxy- [1695-04-1];
2-METHOXYDIPHENYL ETHER, III, 566

Benzene, 1,1'-methylenebis- [101-81-5];
DIPHENYLMETHANE, II, 232

Benzene, 1,1'-methylenebis[2,4,6-trimethyl- [733-07-3];
DIMESITYLMETHANE, V, 422

Benzene, 1,1',1''-methylidynetris- [519-73-3];
TRIPHENYLMETHANE, I, 548

Benzene, 1-methyl-3-(1-methylethyl)- [535-77-3];
m-CYMENE, V, 332

Benzene, 1-methyl-4-(1-methylethyl)-2-nitro- [943-15-7];
2-NITRO-p-CYMENE, III, 653

Benzene, 1-methyl-4-(methylsulfonyl)- [3185-99-7];
METHYL-p-TOLYL SULFONE, IV, 674

Benzene, 1-methyl-3-nitro- [99-08-1];
m-NITROTOLUENE, I, 415

Benzene, (methylsulfinyl)- [1193-82-4];
METHYL PHENYL SULFOXIDE, V, 791

Benzene, (2-nitroethenyl)- [102-96-5];
β-NITROSTYRENE, I, 413

Benzene, (nitromethyl)- [622-42-4];
PHENYLNITROMETHANE, II, 512

Benzene, 1-nitro-4-phenoxy- [620-88-2];
p-NITRODIPHENYL ETHER, II, 445

Benzene, nitroso- [586-96-9];
NITROSOBENZNE, III, 668

Benzene, pentyl [538-68-1];
n-AMYLBENZENE, II, 47

Benzene, 1, 1'-pentylidenebis- [1726-12-1];
1,1-DIPHENYLPENTANE, V, 523

Benzenepropanamide, β-oxo-N-phenyl- [959-66-0];
BENZOYLACETANILIDE, III, 108; IV, 80

Benzenepropanenitrile, α,α-diphenyl- [5350-82-3];
α,α,β-TRIPHENYLPROPIONITRILE, IV, 962

Benzenepropanoic acid [501-52-0];
HYDROCINNAMIC ACID, I, 311

Benzenepropanoic acid, α-acetyl-2,3-dimethoxy-, ethyl ester [53608-80-3];
ETHYL α-ACETYL-β-(2,3-DIMETHOXYPHENYL)PROPIONATE, IV, 408

Benzenepropanoic acid, β-amino-(±)-[3646-50-2];
dl-β-AMINO-β-PHENYLPROPIONIC ACID, III, 91

Benzenepropanoic acid, 2-carboxy- [776-79-4];
β-(o-CARBOXYPHENYL) PROPIONIC ACID, IV, 136

Benzenepropanoic acid, β-cyano-α-oxo-, ethyl ester [6362-63-6];
ETHYL PHENYLCYANOPYRUVATE, II, 287

Benzenepropanoic acid, α,α-dimethyl-β-oxo-, ethyl ester [25491-42-3];
ETHYL BENZOYLDIMETHYLACETATE, II, 268

Benzenepropanoic acid, α,β-diphenyl- [53663-24-4];
α,β,β-TRIPHENYLPROPIONIC ACID, IV, 960

Benzenepropanoic acid, β-hydroxy-, ethyl ester [5764-85-2];
ETHYL β-PHENYL-β-HYDROXYPROPIONATE, III, 408

Benzenepropanoic acid, 4-hydroxy-α-oxo- [156-39-8];
p-HYDROXYPHENYLPYRUVIC ACID, V, 627

Benzenepropanoic acid, β-hydroxy-β-phenyl-, ethyl ester [894-18-8];
ETHYL β-HYDROXY-β,β-DIPHENYLPROPIONATE, V, 564

Benzenepropanoic acid, α-oxo- [156-06-9];
PHENYLPYRUVIC ACID, II, 519

Benzenepropanoic acid, β-oxo, ethyl ester [94-02-0];
ETHYL BENZOYLACETATE, II, 266; III, 379; IV, 415

Benzenepropanoic acid, α-phenyl-[3333-15-1];
α,β-DIPHENYLPROPIONIC ACID, V, 526

Benzenepropanol [122-97-4];
3-PHENYL-1-PROPANOL, IV, 798

Benzenepropanoic acid, α,β-dibromo-, ethyl ester [5464-70-0];
ETHYL α,β-DIBROMO-β-PHENYLPROPIONATE, II, 270

Benzene, propyl- [103-65-1];
n-PROPYLBENZENE, I, 471

Benzene, 1,1'-selenobis-[1132-39-4];
DIPHENYL SELENIDE, II, 238

Benzeneselenol [645-96-5];
SELENOPHENOL, III, 771

Benzenesulfenyl chloride, 2,4-dinitro- [528-76-7];
2,4-DINITROBENZENESULFENYL CHLORIDE, V, 474

Benzenesulfenyl chloride, 4-methyl- [933-00-6];
p-TOLUENESULFENYL CHLORIDE, IV, 934

Benzenesulfenyl chloride, 2-nitro- [7669-54-7];
  o-NITROPHENYLSULFUR CHLORIDE, II, 455

Benzenesulfinic acid, 4-(acetylamino)- [710-24-7];
  p-ACETAMINOBENZENESUFINIC ACID, I, 7

Benzenesulfinic acid, methyl ester [670-98-4];
  METHYL BENZENESULFINATE, V, 723

Benzenesulfinic acid, 4-methyl-, sodium salt [824-79-3];
  SODIUM *p*-TOLUENESULFINATE, I, 492

Benzenesulfinyl chloride, 4-methyl- [10439-23-3];
  p-TOLUENESULFINYL CHLORIDE, IV, 937

Benzenesulfonic acid, 2-amino- [88-21-1];
  ORTHANILIC ACID, II, 471

Benzenesulfonic acid, 4-amino-3-methyl- [99-33-9];
  o-TOLUIDINESULFONIC ACID, III, 824

Benzenesulfonic acid, 4-methyl-, anhydride [4124-41-8];
  p-TOLUENESULFONIC ANHYDRIDE, IV, 940

Benzenesulfonic acid, 4-methyl-, butyl ester [778-28-9];
  n-BUTYL *p*-TOLUENESULFONATE, I, 145

Benzenesulfonic acid, 4-methyl-, dodecyl ester [10157-76-3];
  n-DODECYL *p*-TOLUENESULFONATE, III, 366

Benzenesulfonic acid, 4-methyl-, hydrazide [1576-35-8];
  p-TOLUENESULFONLHYDRAZIDE, V, 1055

Benzenesulfonyl chloride [98-09-9];
  BENZENESULFONYL CHLORIDE I, 84

Benzenesulfonyl chloride, 4-(acetylamino)- [121-60-8];
  p-ACETAMINOBENZENESULFONYL CHLORIDE, I, 8

1,2,4,5-Benzenetetracarboxylic acid [89-05-4];
  PYROMELLITIC ACID, II, 551

Benzene, 1,2,3,5-tetramethyl- [527-53-7];
  ISODURENE, II, 360

Benzene, 1,2,4,5-tetramethyl- [95-93-2];
  DURENE, II, 248

Benzene, 1,1'-thiobis- [139-66-2];
  DIPHENYL SULFIDE, II, 242

Benzene, 1,1'-thiobis[4-nitro- [1223-31-0];
  p-NITROPHENYL SULFIDE, III, 667

Benzenethiol [108-98-5];
  THIOPHENOL, I, 504

Benzenethiol, 3-methyl- [108-40-7];
  m-THIOCRESOL, III, 809; *WARNING*, V, 1050

1,2,4-Benzenetriamine, 5-nitro- [6635-35-4];
2,4,5-TRIAMINONITROBENZENE, V, 1067

Benzene, 1,3,5-tribromo- [626-39-1];
sym-TRIBROMOBENZENE, II, 592

Benzene, 1,2,3-triiodo-5-nitro- [53663-23-3];
1,2,3-TRIIODO-5-NITROBENZENE, II, 604

Benzene, 1,2,3-trimethyl- [526-73-8];
HEMIMELLITENE, IV, 508

Benzene, 1,3,5-trimethyl- [108-67-8];
MESITYLENE, I, 341

Benzene, 1,3,5-trimethyl-2-nitro- [603-71-4];
NITROMESITYLENE, II, 449

Benzene, 1,3,5-trinitro- [99-35-4];
1,3,5-TRINITROBENZENE, I, 541

1,3,5-Benzenetriol [108-73-6];
PHLOROGLUCINOL, I, 455

1,2,4-Benzenetriol, triacetate [613-03-6];
HYDROXYHYDROQUINONE TRIACETATE, I, 317

9,10[1′,2′]-Benzenoanthracene, 9,10-dihydro- [477-75-8];
TRIPTYCENE, IV, 964

1H/Benzimidazole [51-17-2];
BENZIMIDAZOLE, II, 65

2H/Benzimidazole-2-thione, 1,3-dihydro- [583-39-1];
2-MERCAPTOBENZIMIDAZOLE, IV, 569

1,3-Benzodioxole-5-carboxylic acid [94-53-1];
PIPERONYLIC ACID, II, 538

1,3-Benzodioxol-2-one [2171-74-6];
o-PHENYLENE CARBONATE, IV, 788

Benzofuran [271-89-6];
COUMARONE, V, 251

2-Benzofurancarboxylic acid [496-41-3];
COUMARILIC ACID, III, 209

Benzofuran, 3-methyl- [21535-97-7];
3-METHYLCOUMARONE, IV, 590

Benzofurazan, 1-oxide [480-96-6];
BENZOFURAZAN OXIDE, IV, 74

Benzoic acid, 4-acetyl-, methyl ester [3609-53-8];
METHYL p-ACETYLBENZOATE, IV, 579

Benzoic acid, 4-amino-, ethyl ester [94-09-7];
ETHYL p-AMINOBENZOATE, I, 240

Benzoic acid, 2-amino-5-iodo- [5326-47-6];
5-IODOANTHRANILIC ACID, II, 349

Benzoic acid, anhydride [93-97-0];
BENZOIC ANHYDRIDE, I, 91

Benzoic acid, 2-bromo-3-methyl- [53663-39-1];
2-BROMO-3-METHYLBENZOIC ACID, IV, 114

Benzoic acid, 2-bromo-3-nitro- [573-54-6];
2-BROMO-3-NITROBENZOIC ACID, I, 125

Benzoic acid, 5-[(3-carboxy-4-hydroxyphenyl)(3-carboxy-4-oxo-2,5-cyclohexadien-
1-ylidene)methyl]-2-hydroxy-, triammonium salt [569-58-4];
AMMONIUM SALT OF AURIN TRICARBOXYLIC ACID, I, 54

Benzoic acid, 2-chloro- [118-91-2];
o-CHLOROBENZOIC ACID, II, 135

Benzoic acid, 2-(cyanomethyl)- [6627-91-4];
o-CARBOXYPHENYLACETONITRILE, III, 174

Benzoic acid, 2-[(1,3-dihydro-1,-1,3-dioxo-2H-isoindol-2-yl)methyl]- [53663-18-6];
α-PHTHALIMIDO-o-TOLUIC ACID, IV, 810

Benzoic acid, 2,4-dihydroxy- [89-86-1];
β-RESORCYLIC ACID, II, 557

Benzoic acid, 3,4-dihydroxy- [99-50-3];
PROTOCATECHUIC ACID, III, 745

Benzoic acid, 3,5-dihydroxy- [99-10-5];
3,5-DIHYDROXYBENZOIC ACID, III, 288

Benzoic acid, 2-[[4-(dimethylamino)phenyl]azo]- [493-52-7];
METHYL RED, I, 374

Benzoic acid, 2,5-dinitro- [610-28-6];
2,5-DINITROBENZOIC ACID, III, 334

Benzoic acid, 3,5-dinitro- [99-34-3];
3,5-DINITROBENZOIC ACID III, 337

Benzoic acid, 4-fluoro- [456-22-4];
p-FLUOROBENZOIC ACID, II, 299

Benzoic acid, 2-formyl- [119-67-5];
PHTHALALDEHYDIC ACID, II, 523; III, 737

Benzoic acid, 4-hydroxy- [99-96-7];
p-HYDROXYBENZOIC ACID, II, 341

Benzoic acid, 2-hydroxy-, 4-chlorophenyl ester [2944-58-3];
p-CHLOROPHENYL SALICYLATE, IV, 178

Benzoic acid, 2-hydroxy-3,5-diiodo- [133-91-5];
2-HYDROXY-3,5-DIIODOBENZOIC ACID, II, 343

Benzoic acid, 4-hydroxy-3-methoxy- [121-34-6];
VANILLIC ACID, IV, 972

Benzoic acid, 3-iodo- [618-51-9];
   m-IODOBENZOIC ACID, II, 353

Benzoic acid, 4-iodo- [619-58-9];
   p-IODOBENZOIC ACID, I, 325

Benzoic acid, 2-mercapto- [147-93-3];
   THIOSALICYLIC ACID, II, 580

Benzoic acid, 2-methyl- [118-90-1];
   o-TOLUIC ACID, II, 588; III, 820

Benzoic acid, 4-methyl- [99-94-5];
   p-TOLUIC ACID, III, 822

Benzoic acid, 2-(4-methylbenzoyl)- [85-55-2];
   p-TOLUYL-o-BENZOIC ACID, I, 517

Benzoic acid, 3-nitro- [121-92-6];
   m-NITROBENZOIC ACID, I, 391

Benzoic acid, 4-nitro- [62-23-7];
   p-NITROBENZOIC ACID, I, 392

Benzoic acid, 3-nitro-, methyl ester [618-95-1];
   METHYL m-NITROBENZOATE, I, 372

Benzoic acid, pentachloro- [1012-84-6];
   PENTACHLORBENZOIC ACID, V, 890

Benzoic acid, 4-(phenylazo)- [1562-93-2];
   p-PHENYLAZOBENZOIC ACID, III, 711

Benzoic acid, 2-phenylethyl ester [94-47-3];
   2-PHENYLETHYL BENZOATE, V, 336

Benzoic acid, phenylmethyl ester [120-51-4];
   BENZYL BENZOATE, I, 104

Benzoic acid, 2-(1H-pyrrol-1-yl)-, methyl ester [10333-67-2];
   1-(2-METHOXYCARBONYLPHENYL)PYRROLE, V, 716

Benzoic acid, 2-sulfo-, monoammonium salt [6939-89-5];
   ACID AMMONIUM o-SULFOBENZOATE, I, 14

Benzoic acid, 2,4,6-tribromo- [633-12-5];
   2,4,6-TRIBROMOBENZOIC ACID, IV, 947

Benzoic acid, 3,4,5-trimethoxy- [118-41-2];
   TRIMETHYLGALLIC ACID, I, 537

Benzoic acid, 2,4,6-trimethyl- [480-63-7];
   MESITOIC ACID, III, 553; V, 706

Benzoic acid, 2,4,6-trinitro- [129-66-8];
   2,4,6-TRINITROBENZOIC ACID, I, 543

Benzonitrile, 2,6-dimethoxy- [16932-49-3];
   2,6-DIMETHOXYBENZONITRILE, III, 293

Benzonitrile, 3,4-dimethoxy- [2024-83-1];
VERATRONITRILE, II, 622

Benzonitrile, 2-methyl- [529-19-1];
o-TOLUNITRILE, I, 514

Benzonitrile, 4-methyl- [104-85-8];
p-TOLUNITRILE, I, 514

Benzonitrile, 2-methyl-3,5-dinitro- [948-31-2];
3,5-DINITRO-o-TOLUNITRILE, V, 480

Benzonitrile, 4-nitro- [619-72-7];
p-NITROBENZONITRILE, III, 646

2H-1-Benzopyran-3-carboxylic acid, 2-oxo-, ethyl ester [1846-76-0];
3-CARBETHOXYCOUMARIN, III, 165

1H-2-Benzopyran-1,3(4H)-dione [703-59-3];
HOMOPHTHALIC ACID AND ANHYDRIDE, III, 449

2H-1-Benzopyran-2-one, 6,7-dihydroxy-4-methyl- [529-84-0];
4-METHYLESCULETIN, I, 360

4H-1-Benzopyran-4-one, 2-ethyl- [14736-30-2];
2-ETHYLCHROMONE, III, 387

2H-1-Benzopyran-2-one, 4-methyl- [607-71-6];
4-METHYLCOUMARIN, III, 581

4H-1-Benzopyran-4-one, 2-phenyl- [525-82-6];
FLAVONE, IV, 478

1H.5H-Benzo[ij]quinolizine, 2,3,6,7-tetrahydro- [479-59-4];
JULOLIDINE, III, 504

1,2,3-Benzothiadiazole, 1,1-dioxide [37150-27-9];
1,2,3-BENZOTHIADIAZOLE 1,1-DIOXIDE, V, 60

2-Benzothiazolamine, 6-methyl- [2536-91-6];
2-AMINO-6-METHYLBENZOTHIAZOLE, III, 76

1H-Benzotriazole [95-14-7];
1,2,3-BENZOTRIAZOLE, III, 106

3H-2,1-Benzoxamercurol-3-one, 7-nitro- [53663-14-2];
ANHYDRO-2-HYDROXYMERCURI-3-NITROBENZOIC ACID, I, 56

3H-2,1-Benzoxathiol-3-one, 1,1-dioxide [81-08-3];
o-SULFOBENZOIC ANHYDRIDE, I, 495

2H-3,1-Benzoxazine-2,4(1H)-dione [118-48-9];
ISATOIC ANHYDRIDE, III, 488

Benzoyl azide, 3-nitro- [3532-31-8];
m-NITROBENZAZIDE, IV, 715

Benzoyl chloride, 2-chloro- [609-65-4];
o-CHLOROBENZOYL CHLORIDE, I, 155

Benzoyl chloride, 4-nitro- [122-04-3];
    p-NITROBENZOYL CHLORIDE, I, 394

Benzoyl chloride, 4-(phenylazo)- [104-24-5];
    p-PHENYLAZOBENZOYL CHLORIDE, III, 712

Benzoyl chloride, 2,4,6-trimethyl- [938-18-1];
    MESITOIC ACID AND MESITOYL CHLORIDE, III, 555

Benzoyl fluoride [455-32-3];
    BENZOYL FLUORIDE, V, 66

Bicyclo[2.2.1]hepta-2,5-diene,7-(1,1-dimethylethoxy)-[877-06-5];
    7-t-BUTOXYNORBORNADIENE, V, 151

Bicyclo[4.1.0]heptane [286-08-8];
    NORCARANE, V, 855

Bicyclo[2.2.1]heptane-1-carboxylic acid, 7,7-dimethyl-, (±)-[53684-59-6];
    D,L,-KETOPINIC ACID, V, 689

Bicyclo[2.2.1]heptane-1-methanesulfonic acid, 7,7-dimethyl-2-oxo-, (±)- [5872-08-2];
    D,L-10-CAMPHORSULFONIC ACID(REYCLER'S ACID), V, 194

Bicyclo[2.2.1]heptane-1-methanesulfonyl chloride, 7,7-dimethyl-2-oxo-, (±)- [6994-93-0];
    D,L-10-CAMPHORSULFONYL CHLORIDE, V, 196

Bicyclo[4.1.0]heptan-7ol, (1α,6α,7α)- [13830-44-9];
Bicyclo [4.1.0]heptan-7ol, (1α,6α,7β)- [931-31-7];
    exo/endo-7-NORCARANOL, V, 859

Bicyclo[2.2.1]heptan-2-one [497-38-1];
    2-NORBORNANONE, V, 852

Bicyclo[2.2.1]heptan-2-one, 1-ethenyl-7,7-dimethyl-, (1S)- [53585-70-9];
    D-2-OXO-7,7-DIMETHYL-1-VINYLBICYCLO[2.2.1]HEPTANE, V, 877

Bicyclo[2.2.1]hept-2-ene [498-66-8];
    NORBORNYLENE, IV, 738

Bicyclo[2.2.1]hept-2-ene, 7,7-dimethoxy- [875-04-7];
    7,7-DIMETHOXYBICYCLO[2.2.1]HEPTENE, V, 424

Bicyclo[2.2.1]hept-2-en-7-one [694-71-3];
    BICYCLO[2.2.1]HEPTEN-7-ONE, V, 91

[1,1'-Bicyclohexyl]-1,1'dicarbonitrile [18341-40-7];
    1,1'-DICYANO-1,1'-BICYCLOHEXYL, IV, 273

Bicyclo[2.1.0]pentane [185-94-4];
    BICYCLO[2.1.0]PENTANE, V, 96

Bicyclo[4.2.0]octa-1,3,5-triene-7-carbonitrile [6809-91-2];
    1-CYANOBENZOCYCLOBUTENE, V, 263

1,1'-Biphenyl,4-bromo- [92-66-0]
    p-BROMOBIPHENYL, I, 113

[1,1'-Biphenyl]-2-carboxylic acid, 2'-formyl- [6720-26-9];
DIPHENALDEHYDIC ACID, V, 493

[1,1'-Biphenyl]-4,4'-diamine,*N*,*N*'-diethyl- [6290-86-4];
N,*N*'-DIETHYLBENZIDINE, IV, 283

1,1'-Biphenyl,4,4'-dibromo- [92-86-4];
4,4'-DIBROMOBIPHENYL, IV, 256

[1,1'-Biphenyl]-2,2'-dicarboxaldehyde [1210-05-5];
DIPHENALDEHYDE, V, 489

[1,1'-Biphenyl]-2,2'-dicarboxylic acid [482-05-3];
DIPHENIC ACID, I, 222

[1,1'-Biphenyl]-2,2'-dicarboxylic acid, 4,4',6,6'-tetrachloro-, (±)- [53663-22-2];
dl-4,4',6,6'-TETRACHLORODIPHENIC ACID, IV, 872

1,1'-Biphenyl, 3,4-dichloro [2974-92-7]
ARYLBENZENES: 3,4-DICHLOROBIPHENYL, V, 51

1,1'-Biphenyl, 4,4'-difluoro- [398-23-2];
4,4'-DIFLUOROBIPHENYL, II, 188

1,1'-Biphenyl, 3,3'-dimethoxy- [6161-50-8];
3,3'-DIMETHOXYBIPHENYL, III, 295

1,1'-Biphenyl, 3,3'-dimethyl- [612-75-9];
3,3'-DIMETHYLBIPHENYL, III, 295

1,1'-Biphenyl, 2,2'-dinitro- [2436-96-6];
2,2'-DINITROBIPHENYL, III, 339

[1,1'-Biphenyl]-3,3'-diol [612-76-0];
3,3'-DIHYDROXYBIPHENYL, V, 412

Biphenylene, [259-79-0];
BIPHENYLENE, V, 54

[1,1'-Biphenyl]-4-methanol, α,α-bis([1,1'-biphenyl]-4-yl)- [5341-14-0];
TRIBIPHENYLCARBINOL, III, 831

1,1'-Biphenyl, 3-nitro- [2113-58-8];
m-NITROBIPHENYL, IV, 718

2,2'-Bipyridine, [366-18-7];
2,2'-BIPYRIDINE, V, 102

[5,5'-Bipyrimidine]-2,2',4,4',6,6'(1*H*,1'*H*,3*H*,3'*H*,5*H*,5'*H*)-hexonedihydrate [6011-27-4];
ALLOXANTIN DIHYDRATE, III, 42; IV, 25

Boric acid (H₃BO₃), tributyl ester [688-74-4];
n-BUTYL BORATE, II, 106

Boroxin, triphenyl- [3262-89-3];
BENZENEBORONIC ANHYDRIDE, IV, 68

1,3-Butadiene, [106-99-0];
1,3-BUTADIENE, II, 102

1,4-Butadiene, 1-chloro-1,4,4-trifluoro- [764-14-7];
   1-CHLORO-1,4,4-TRIFLUOROBUTADIENE, V, 235

1,3-Butadiene, 2,3-dimethyl- [513-81-5];
   2,3-DIMETHYL-1,3-BUTADIENE, III, 312

2,3-Butadienoic acid, methyl ester [18913-35-4];
   METHYL BUTADIENOATE, V, 734

Butanal, 2,2,3-trichloro- [76-36-8];
   BUTYRCHLORAL, IV, 130

Butanamide, 3-oxo-N-phenyl- [102-01-2];
   ACETOACETANILIDE, III, 10

1-Butanamine N-methyl- [110-68-9];
   N-METHYLBUTYLAMINE, V, 736

Butane, 1-bromo- [109-65-9];
   n-BUTYL BROMIDE, I, 28,37

Butane, 2-bromo- [78-76-2];
   sec-BUTYL BROMIDE, I, 38

Butane, 1-bromo-3-methyl- [107-82-4];
   iso-AMYL BROMIDE, I, 27

Butane, 1-chloro- [109-69-3];
   n-BUTYL CHLORIDE, I, 142

1,4-Butanediamine, dihydrochloride [333-93-7];
   PUTRESCINE DIHYDROCHLORIDE, IV, 819

Butane, 1,4-diiodo- [628-21-7];
   1,4-DIIODOBUTANE, IV, 321

Butanedinitrile, 2,3-diphenyl- [5424-86-2];
   2,3-DIPHENYLSUCCINONITRILE, IV, 392

Butane, 1,4-dinitro- [4286-49-1];
   1,4-DINITROBUTANE, IV, 368

Butanedioic acid, 2,3-dibromo- [526-78-3];
   α,β-DIBROMOSUCCINIC ACID, II, 177

Butanedioic acid, diethyl ester [123-25-1];
   DIETHYL SUCCINATE, V, 993

Butanedioic acid, 2,2-difluoro- [665-31-6];
   2,2-DIFLUOROSUCCINIC ACID, V, 393

Butanedioic acid, 2,3-dihydroxy, (R*,R*)-($\pm$)- [133-37-9];
   dl-TARTARIC ACID, I, 497

Butanedioic acid, diphenyl ester [621-14-7];
   DIPHENYL SUCCINATE, IV, 390

Butanedioic acid, (diphenylmethylene)-, 1-ethyl ester [5438-22-2];
   β-CARBETHOXY-γ,γ-DIPHENYLVINYLACETIC ACID, IV, 132

Butanedioic acid, 2-ethyl-2-methyl- [631-31-2];
α-ETHYL-α-METHYLSUCCINIC ACID; V, 572

Butanedioic acid, methyl- [498-21-5]
METHYLSUCCINIC ACID, III, 615

Butanedioic acid, methylene- [97-65-4];
ITACONIC ACID, II, 368

Butanedioic acid, (1-oxoheptyl)-, diethyl ester [41117-78-6];
ETHYL ENANTHYLSUCCINATE, IV, 430

Butanedioic acid, phenyl- [635-51-8];
PHENYLSUCCINIC ACID, I, 451; IV, 804

Butanedioic acid, (3-phenyl-2-propenyl)- [5671-91-0];
γ-PHENYLALLYLSUCCINIC ACID, IV, 766

2,3-Butanediol, 2,3-dimethyl-, hexahydrate [6091-58-3];
PINACOL HYDRATE, I, 459

2,3-Butanedione, dioxime [95-45-4];
DIMETHYLGLYOXIME, II, 204

Butanedioyl dichloride, methylene- [1931-60-8];
ITACONYL CHLORIDE, IV, 554

Butane, 1,1'-[(1-methylethylidene)bis(oxy)]bis- [141-72-0];
ACETONE DIBUTYL ACETAL, V, 5

Butanenitrile, 4-chloro- [628-20-6];
γ-CHLOROBUTYRONITRILE, I, 156

Butane, 1,1'-oxybis[4-chloro- [6334-96-9];
4,4'-DICHLORODIBUTYL ETHER, IV, 266

Butanoic acid, 2-acetyl-3-methyl-, ethyl ester [1522-46-9];
ETHYL α-ISOPROPYLACETOACETATE, III, 405

Butanoic acid, 2-acetyl-3-oxo-, ethyl ester [603-69-0];
ETHYL DIACETYLACETATE, III, 390

Butanoic acid, 4-amino- [56-12-2];
γ-AMINOBUTYRIC ACID, II, 25

Butanoic acid, 2-amino-2-ethyl- [2566-29-2];
α-AMINODIETHYLACETIC ACID, III, 66

Butanoic acid, 4-bromo, ethyl ester [2969-81-5];
ETHYL γ-BROMOBUTYRATE, V, 545

Butanoic acid, 2-bromo-3-methyl-, [565-74-2];
α-BROMOISOVALERIC ACID, II, 93

Butanoic acid, butyl ester, [109-21-7];
n-BUTYL, n-BUTYRATE, I, 138

Butanoic acid, 4-chloro-4-oxo, methyl ester [1490-25-1];
β-CARBOMETHOXYPROPIONYL CHLORIDE, III, 169

Butanoic acid, 3-oxo-, 1,1-dimethylethyl ester [1694-31-1];
    t-BUTYL ACETOACETATE, V, 155

Butanoic acid, 2-methyl-, (±)- [600-07-7];
    dl-METHYLETHYLACETIC ACID, I, 361

Butanoic acid, 2-nitro-, ethyl ester [2531-81-9];
    ETHYL α-NITROBUTYRATE, IV, 454

Butanoic acid, 3-oxo-, ethyl ester [141-97-9];
    ETHYL ACETOACETATE, I, 235

1-Butanol, 4-chloro- [928-51-8];
    TETRAMETHYLENE CHLOROHYDRIN, II, 571

1-Butanol, 4-chloro-, benzoate [946-02-1];
    4-CHLOROBUTYL BENZOATE, III, 187

2-Butanol, 3-methyl- [598-75-4];
    METHYL ISOPROPYL CARBINOL, II, 406

2-Butanone, 3,4-dibromo-4-phenyl- [6310-44-7];
    BENZALACETONE DIBROMIDE, III, 105

2-Butanone, 3,3-dimethyl- [75-97-8];
    PINACOLONE, I, 462

2-Butanone, 4-(diethylamino)- [3299-38-5];
    1-DIETHYLAMINO-3-BUTANONE, IV, 281

2-Butanone, 3-methyl- [563-80-4];
    METHYL ISOPROPYL KETONE, II, 408

Butanoyl chloride [141-75-3];
    n-BUTYRYL CHLORIDE, I, 147

2-Butenedinitrile, (*E*)- [764-42-1];
    FUMARONITRILE, IV, 486

2-Butenedinitrile, 2,3-diamino, (*Z*)- [1187-42-4];
    DIAMINOMALEONITRILE(HYDROGEN CYANIDE TETRAMER), V, 344

2-Butenedioic acid (*E*)- [110-17-8];
    FUMARIC ACID, II, 302

2-Butenedioic acid, 2-methyl-, (*E*)- [498-24-8];
    MESACONIC ACID, II, 382

2-Butenedioic acid, 2-methyl, (*Z*)- [498-23-7];
    CITRACONIC ACID, II, 140

2-Butene-1,4-dione, 1,4-diphenyl-, (*E*)- [959-28-4];
    trans-DIBENZOYLETHYLENE, III, 248

2-Butenedioyl dichloride, (*E*)- [627-63-4];
    FUMARYL CHLORIDE, III, 422

3-Butenenitrile [109-75-1];
    ALLYL CYANIDE, I, 46

3-Butenoic acid [625-38-7];
VINYLACETIC ACID, III, 851

2-Butenoic acid, 2-cyano-3-phenyl-, ethyl ester [18300-89-5];
ETHYL(1-PHENYLETHYLIDENE)CYANOACETATE, IV, 463

2-Butenoic acid, 2,3-dibromo-4-oxo-, (Z)- [488-11-9];
MUCOBROMIC ACID, III, 621; IV, 688

2-Butenoic acid, 3-methyl- [541-47-9];
β,β-DIMETHYLACRYLIC ACID, III, 302

2-Butenoic acid, 4-oxo-4-phenyl- [583-06-2];
β-BENZOYLACRYLIC ACID, III, 109

2-Butenoic acid, 3-(phenylamino)-, ethyl ester [31407-07-5];
ETHYL β-ANILINOCROTONATE, III, 374

3-Buten-2-one, 3-bromo-4-phenyl- [31207-17-7]
α-BROMOBENZALACETONE, III, 125

2-Buten-1-one, 1,3-diphenyl- [495-45-4];
DYPNONE, III, 367

3-buten-2-one, 4-(2-furanyl)- [623-15-4];
2-FURFURALACETONE, I, 283

3-Buten-2-one, 4-phenyl- [122-57-6];
BENZALACETONE, I, 77

1-Buten-3-yne [689-97-4];
MONOVINYLACETYLENE, IV, 683

2-Butynedioic acid [142-45-0];
ACETYLENEDICARBOXYLIC ACID, II, 10

2-Butynedioic acid, dimethyl ester [762-42-5];
DIMETHYL ACETYLENECARBOXYLATE, IV, 329

2-Butynoic acid [590-93-2];
TETROLIC ACID, V, 1043

2-Butyn-1-ol [764-01-2];
2-BUTYN-1-OL, IV, 128

3-Butyn-2-ol, 2-methyl- [115-19-5];
DIMETHYLETHYNYLCARBINOL, III, 320

C

Carbamic acid, butyl ester [592-35-8];
n-BUTYL CARBAMATE, I, 140

Carbamic acid, 1,1-dimethylethyl ester [4248-19-5];
t-BUTYL CARBAMATE, V, 162

Carbamic acid, methyl-, ethyl ester [105-40-8];
ETHYL N-METHYLCARBAMATE, II, 278

Carbamic acid, methylnitroso-, ethyl ester [615-53-2];
  NITROSOMETHYLURETHANE, II, 464; V, 842

Carbamimidic acid, methyl ester, monohydrochloride [5329-33-9];
  METHYLISOUREA HYDROCHLORIDE, IV, 645

Carbamimidothioic acid, methyl ester, sulfate (2:1) [867-44-7];
  S-METHYL ISOTHIOUREA SULFATE, II, 411

Carbamothioic chloride, diethyl- [88-11-9];
  DIETHYLTHIOCARBAMYL CHLORIDE, IV, 307

9H-Carbazole, 2-nitro- [14191-22-1];
  2-NITROCARBAZOLE, V, 829

1H-Carbazole, 2,3,4,9-tetrahydro- [942-01-8];
  1,2,3,4-TETRAHYDROCARBAZOLE, IV, 884

Carbonazidic acid, 1,1-dimethylethyl ester [1070-19-5];
  t-BUTYL AZIDOFORMATE, V, 157

Carbonic acid, compound, with hydrazinecarboximidamide(1:1) [2582-30-1];
  AMINOGUANIDINE BICARBONATE, III, 73

Carbonochloridic acid, phenylmethyl ester [501-53-1];
  CARBOBENZOXY CHLORIDE, AND DERIVATIVES, III, 167

Carbonothioic dichloride [463-71-8];
  THIOPHOSGENE, I, 506

Casein [9000-71-9];
  CASEIN, II, 120

Cholan-24-al [26606-02-0];
  CHOLANE-24-AL, V, 242

Chol-23-ene-3,12-diol, 24,24-diphenyl-, diacetate, (3α, 5β, 12α)- [53608-88-1];
  3,12-DIACETOXY-bisnor-CHOLANYLDIPHENYLETHYLENE, III, 237

Cholestane, 3-methoxy-, (3β, 5α)- [1981-90-4];
  CHOLESTANYL METHYL ETHER, V, 245

Cholestan-3-ol (3β, 5α) [80-97-7];
  DIHYDROCHOLESTEROL, II, 191

Cholestan-3-one, (5a)- [566-88-1];
  CHOLESTANONE, II, 139

Cholest-4-ene-3,6-dione [984-84-9];
  Δ⁴-CHOLESTEN-3,6-DIONE, IV, 189

Cholest-5-en-3-ol(3β)- [57-88-5];
  CHOLESTEROL, IV, 195

Cholest-4-en-3-one [601-57-0];
  CHOLESTENONE, III, 207; Δ⁴-CHOLEST-3-ONE, IV, 192; 195

Cholest-5-en-3-one [601-54-7];
  Δ⁵-CHOLESTEN-3-ONE, IV, 195

Chromate(1-), diamminetetrakis(thiocyanato-*N*)-, ammonium, (OC-6-11)- [13573-16-5];
REINECKE SALT, II, 555

Chromium copper oxide, (CuCr₂O₄) [12018-10-9];
COPPER CHROMITE CATALYST, II, 142

Cyanamide, (2-chlorophenyl)- [45765-25-1];
o-CHLOROPHENYLCYANAMIDE, IV, 172

Cyanamide, di-2-propenyl- [538-08-9];
DIALLYLCYANAMIDE, I, 203

Cyanamide, methyl-1-naphthalenyl- [53663-33-5];
N-METHYL-1-NAPHTHYLCYANAMIDE, III, 608

Cyanogen bromide, [506-68-3];
CYANOGEN BROMIDE, II, 150

Cyclobutanamine [2516-34-9];
CYCLOBUTYLAMINE, V, 273

Cyclobutanecarboxylic acid [3721-95-7];
CYCLOBUTANECARBOXYLIC ACID, III, 213

1,1-Cyclobutanedicarboxylic acid [5445-51-2];
1,1-CYCLOBUTANEDICARBOXYLIC ACID, III, 213

1,1-Cyclobutanedicarboxylic acid, diethyl ester [3779-29-1];
DIETHYL 1,1-CYCLOBUTANEDICARBOXYLATE, IV, 288

1,2-Cyclobutanedicarboxylic acid, 3-methylene-, dimethyl ester [53684-52-9];
DIMETHYL 3-METHYLENECYCLOBUTANE-1,2-DICARBOXYLATE, V, 459

Cyclobutane, 1,2-diethenyl, *cis*- [16177-46-1];
cis-1,2-DIVINYLCYCLOBUTANE, V, 528

Cyclobutane, 1,2-diethenyl-, *trans*- [6553-48-6];
trans-1,2-DIVINYLCYCLOBUTANE, V, 528

Cyclobutene, 3,4-dichloro-1,2,3,4-tetramethyl- [1194-30-5];
3,4-DICHLORO-1,2,3,4-TETRAMETHYLCYCLOBUTENE, V, 370

1,2-Cyclodecanediol [21014-77-7];
1,2-CYCLODECANEDIOL, IV, 216

1,2-Cyclodecanedione [96-01-5];
SEBACIL, IV, 838

Cyclodecanone [1502-06-3];
CYCLODECANONE, IV, 218; V, 277

Cyclodecanone, 2-hydroxy- [96-00-4];
SEBACOIN, IV, 840

Cyclododecene, (Z)- [1129-89-1];
cis-CYCLODODECENE, V, 281

Cycloheptanone [502-42-1];
CYCLOHEPTANONE, IV, 221

Cycloheptanone, 2-phenyl- [14996-78-2];
  2-PHENYLCYCLOHEPTANONE, IV, 780

Cycloheptatrienylium, tetrafluoroborate(1-) [27081-10-3];
  TROPYLIUM FLUOBORATE, V, 1133

1,3-Cyclohexadiene [592-57-4];
  1,3-CYCLOHEXADIENE, V, 285

2,5-Cyclohexadiene-1-carboxylic acid [4794-04-1];
  1,4-DIHYDROBENZOIC ACID, V, 400

2,5-Cyclohexadiene-1,4-dione [106-51-4];
  QUINONE, I, 482; II, 553

2,5-Cyclohexadiene-1,4-dione, 2-chloro- [695-99-8];
  CHLORO-p-BENZOQUINONE, IV, 148

2,5-Cyclohexadiene-1,4-dione, 2-methyl-5-(1-methylethyl)- [490-91-5];
  THYMOQUINONE, I, 511

2,5-Cyclohexadiene-1,4-dione, 2,3,5,6-tetrahydroxy- [319-89-1];
  TETRAHYDROXYQUINONE, V, 1011

2,5-Cyclohexadiene-1,4-dione, 2,3,5,6-tetramethyl- [527-17-3];
  DUROQUINONE, II, 254

2,5-Cyclohexadien-1-one, 2,6-dibromo-4-(chloroimino)- [537-45-1];
  2,6-DIBROMOQUINONE-4-CHLOROIMIDE, II, 175

2,4-Cyclohexadien-1-one, 2,3,4,5,6,6-hexamethyl- [3854-96-4];
  2,3,4,5,6,6-HEXAMETHYL-2,4-CYCLOHEXADIEN-1-ONE, V, 598

2,4-Cyclohexadien-1-one, 2,6,6-trimethyl- [13487-30-4];
  2,6,6-TRIMETHYL-2,4-CYCLOHEXADIENONE, V, 1092

2,5-Cyclohexadien-1-one, 2,4,6-triphenyl-4-(5'-phenyl[1,1':3,1''-terphenyl]-2'-yloxy)- [10384-15-3];
  2,4,6-TRIPHENYLPHENOXYL, V, 1130

Cyclohexanamine, 1-methyl- [6526-78-9];
  1-AMINO-1-METHYLCYCLOHEXANE, V, 35

Cyclohexane, 1,1-bis(2-propenyloxy)- [53608-84-7];
  CYCLOHEXANONE DIALLYL ACETAL, V, 292

Cyclohexane, (2-bromo-2-propenyl)- [53608-85-8];
  3-CYCLOHEXYL-2-BROMOPROPENE, I, 186

Cyclohexanecarbonitrile, 1,1'-azobis- [2094-98-6];
  1,1'-AZO-bis-1-CYCLOHEXANENITRILE, IV, 66

Cyclohexanecarbonitrile, 1,1'-hydrazobis- [17643-01-5];
  1,2-DI-1(1-CYANO)CYCLOHEXYLHYDRAZINE, IV, 274

Cyclohexanecarboxylic acid, 1-methyl- [1123-25-7];
  1-METHYLCYCLOHEXANECARBOXYLIC ACID, V, 739

Cyclohexanecarboxylic acid, 3,4,5-trihydroxy-, (1α, 3α, 4α, 5α)- [53796-39-7];
Cyclohexanecarboxylic acid, 3,4,5-tris(acetyloxy)-, 1α, 3α, 4α, 5α)- [53796-40-0];
HEXAHYDROGALLIC ACID AND HEXAHYDROGALLIC ACID TRIACETATE, V,
591

Cyclohexane, cyclohexylidene- [4233-18-5];
CYCLOHEXYLIDENECYCLOHEXANE, V, 297

Cyclohexane, 1,2-dibromo- [5401-62-7];
1,2-DIBROMOCYCLOHEXANE, II, 171

1,2-Cyclohexanedicarboxylic acid, diethyl ester, cis- [17351-07-4];
DIETHYL cis-HEXAHYDROPHTHALATE, IV, 304

1,3-Cyclohexanedicarboxylic acid, 2-oxo-, dimethyl ester [25928-05-6];
DIMETHYL CYCLOHEXANONE-2,6-DICARBOXYLATE, V, 439

1,2-Cyclohexanediol, trans- [1460-57-7];
trans-1,2-CYCLOHEXANEDIOL, III, 217

1,3-Cyclohexanedione [504-02-9];
DIHYDRORESORCINOL, III, 278

1,4-Cyclohexanedione [637-88-7];
1,4-CYCLOHEXANEDIONE, V, 288

1,3-Cyclohexanedione, 5,5-dimethyl- [126-81-8];
5,5-DIMETHYL-1.3-CYCLOHEXANEDIONE, II, 200

1,2-Cyclohexanedione, dioxime [492-99-9];
1,2-CYCLOHEXANEDIONE DIOXIME, IV, 229

1,3-Cyclohexanedione, 2-methyl- [1193-55-1];
2-METHYL-1,3-CYCLOHEXANEDIONE, V, 743

Cyclohexane, iodo- [626-62-0];
IODOCYCLOHEXANE, IV, 543

Cyclohexane, isocyano- [931-53-3];
CYCLOHEXYL ISOCYANIDE, V, 300

Cyclohexanemethanamine, N,N-dimethyl- [16607-80-0];
N,N-DIMETHYLCYCLOHEXYLMETHYLAMINE, IV, 339

Cyclohexanemethanol [100-49-2];
CYCLOHEXYLCARBINOL, I, 188

Cyclohexane, methylene- [1192-37-6];
METHYLENECYCLOHEXANE, IV, 612; V, 751

Cyclohexane, 2-propynyl- [17715-00-3];
3-CYCLOHEXYLPROPINE, I, 191

1,1,3,3-Cyclohexanetetramethanol, 2-hydroxy- [5416-55-7];
2,2,6,6-TETRAMETHYLOLCYCLOHEXANOL, IV, 907

Cyclohexanimine, N-chloro- [6681-70-5];
N-CHLOROCYCLOHEXYLIDENEIMINE, V, 208

Cyclohexanol, 2-chloro- [1561-86-0];
   2-CHLOROCYCLOHEXANOL, I, 158

Cyclohexanol, 4-(1,1-dimethylethyl, trans- [21862-63-5];
   trans-4-t-BUTYLCYCLOHEXANOL, V, 175

Cyclohexanol, 1,1'-(1,2-thynediyl)bis- [78-54-6];
   1,1'-ETHYNYLENE-bis-CYCLOHEXANOL, IV, 471

Cyclohexanol, 1-ethynyl- [78-27-3];
   1-ETHYNYLCYCLOHEXANOL, III, 416

Cyclohexanone, 2-butyl-2-methyl- [1197-78-0];
   2-n-BUTYL-2-METHYLCYCLOHEXANONE, V, 187

Cyclohexanone, 2-chloro- [822-87-7];
   2-CHLOROCYCLOHEXANONE, III, 188

Cyclohexanone, 2-chloro-2-methyl- [10409-46-8];
2-Cyclohexen-1-one, 2-methyl- [1121-18-2];
   2-CHLORO-2-METHYLCYCLOHEXANONE AND 2-METHYL-2-CYCLOHEXA-
   NONE, IV, 162

Cyclohexanone, 5-methyl-2-(1-methylethyl)-, (2S-trans)- [14073-97-3];
   l-MENTHONE, I, 340

Cyclohexanone, 2-(2-propenyl)- [94-66-6];
   2-ALLYLCYCLOHEXANONE, III, 44; V, 25

Cyclohexene [110-83-8];
   CYCLOHEXENE, I, 183

1-Cyclohexene-1-acetonitrile [6975-71-9];
   l-CYCLOHEXENYLACETONITRILE, IV, 234

1-Cyclohexene-1-carboxaldehyde, 2-chloro- [1680-73-5];
   2-CHLORO-1-FORMYL-1-CYCLOHEXENE, V, 215

2-Cyclohexene-1-carboxylic acid, 2,6-dimethyl-4-oxo-, ethyl ester [6102-15-4];
   3,5-DIMETHYL-4-CARBETHOXY-2-CYCLOHEXEN-1-ONE, III, 317

4-Cyclohexene-1,2-dicarboxylic acid, diethyl ester, cis- [4841-85-4];
   DIETHYL cis-Δ⁴-TETRAHYDROPHTHALATE, IV, 304

2-Cyclohexen-1-ol, benzoate [3352-93-0];
   3-BENZOYLOXYCYCLOHEXENE, V, 70

2-Cyclohexen-1-one [930-68-7];
   2-CYCLOHEXENONE, V, 294

2-Cyclohexen-1-one, 3,5-dimethyl- [1123-09-7];
   3,5-DIMETHYL-2-CYCLOHEXEN-1-ONE, III, 317

2-Cyclohexen-1-one, 3-ethoxy- [5323-87-5];
   3-ETHOXY-2-CYCLOHEXENONE, V, 539

1,2-Cyclononadiene [1123-11-1];
   1,2-CYCLONONADIENE, V, 306

Cyclooctanecarboxylic acid, 2-oxo-, ethyl ester [4017-56-5];
2-CARBETHOXYCYCLOOCTANONE, V, 198

Cyclooctanone [502-49-8];
CYCLOÖCTANONE, V, 310

Cyclooctene, (E)- [931-89-5];
transCYCLOOCTENE, V, 315

1,3-Cyclopentadiene [542-92-7];
CYCLOPENTADIENE, IV, 238

1,3-Cyclopentadiene, 1,2,3,4,5-pentachloro-5-ethyl- [16177-48-3];
1,2,3,4,5-PENTACHLORO-5-ETHYLCYCLOPENTADIENE, V, 893

1,3-Cyclopentadiene, 1,2,3,4-tetrachloro-5-(2,3,4,5-tetrachloro-2,4-cyclopentadien-1-ylidene)-
[6298-65-3];
PERCHLOROFULVALENE, V, 901

2,4-Cyclopentadien-1-one, 2,3,4,5-tetraphenyl- [479-33-4]
TETRAPHENYLCYCLOPENTADIENONE, III, 806

Cyclopentanecarboxaldehyde [872-53-7];
CYCLOPENTANECARBOXALDEHYDE, V, 320

Cyclopentanecarboxylic acid, methyl ester [4630-80-2];
METHYL CYCLOPENTANECARBOXYLATE, IV, 594

Cyclopentanecarboxylic acid, 2-oxo-, ethyl ester [611-10-9];
2-CARBETHOXYCYCLOPENTANONE, II, 116

1,3-Cyclopentanedione, 2-methyl- [765-69-5];
2-METHYLCYCLOPENTANE-1,3-DIONE, V, 747

Cyclopentanol, 2-chloro-, trans- [20377-80-4];
trans-2-CHLOROCYCLOPENTANOL, IV, 157

Cyclopentanone [120-92-3];
CYCLOPENTANONE, I, 192

Cyclopentanone, 2-(phenylmethyl)- [2867-63-2];
2-BENZYLCYCLOPENTANONE, V, 76

Cyclopentanone, 2,4,4-trimethyl- [4694-12-6];
2,4,4-TRIMETHYLCYCLOPENTANONE, IV, 957

Cyclopentene, 3-chloro- [96-40-2];
3-CHLOROCYCLOPENTENE, IV, 238

2-Cyclopentene-1,4-diol [4157-01-1];
3,5-DIHYDROXYCYCLOPENTENE, V, 414

3-Cyclopentene-1,2-diol, cis- [694-29-1];
DIHYDROXYCYCLOPENTENE, V, 414

4-Cyclopentene-1,3-dione [930-60-9];
2-CYCLOPENTENE-1,4-DIONE, V, 324

2-Cyclopenten-1-one [930-30-3];
2-CYCLOPENTENONE, V, 326

Cyclopropane, bromo- [4333-56-6];
BROMOCYCLOPROPANE, V, 126

Cyclopropanecarbonitrile [5500-21-0];
CYCLOPROPYL CYANIDE: III, 223

Cyclopropanecarboxylic acid [1759-53-1];
CYCLOPROPANECARBOXYLIC ACID, III, 221

Cyclopropanol, 1-(4-methylphenyl)- [40122-37-0];
1-p-TOLYLCYCLOPROPANOL, V, 1058

2-Cyclopropen-1-one, 2,3-diphenyl- [886-38-4];
DIPHENYLCYCLOPROPENONE, V, 514

L-Cystine [56-89-3];
l-CYSTINE, I, 194

D

1,10-Decanediamine [646-25-3];
DECAMETHYLENEDIAMINE, III, 229

Decane, 1,10-dibromo- [4101-68-2]
DECAMETHYLENE BROMIDE, III, 227

Decanedinitrile [1871-96-1];
SEBACONITRILE, III, 768

Decanedioic, acid, monoethyl ester [693-55-0]
ETHYL HYDROGEN SEBACATE, II, 276

1,10-Decanediol [112-47-0];
DECAMETHYLENE GLYCOL, II, 154

2,5-Decanediol, 2-methyl- [53731-34-3];
2-METHYL-2,5-DECANEDIOL, IV, 601

Decanoic acid, 10-amino-10-oxo-, methyl ester [53663-35-7];
METHYL SEBACAMATE, III, 613

1,2-Diazaspiro[2.5]oct-1-ene [930-82-5];
3,3-PENTAMETHYLENEDIAZIRINE, V, 897

Diazenecarbothioic acid, phenyl-,2-phenylhydrazide [60-10-6];
DITHIZONE, III, 360

Diazenedicarboxylic acid, bis(1,1-dimethylethyl) ester [870-50-8];
t-BUTYL AZODIFORMATE, V, 160

Diazenedicarboxylic acid, diethyl ester [1972-28-7];
ETHYL AZODICARBOXYLATE, III, 375; IV, 411; Warning, V, 544

Diazene, diphenyl- [103-33-3];
AZOBENZENE, III, 103

Diazene, diphenyl-, 1-oxide [495-48-7];
AZOXYBENZENE, II, 57

Dibenz[c,e][1,2]azaborine, 5,6-dihydro-6-methyl- [15813-13-5];
10-METHYL-10,9-BORAZAROPHENANTHRENE, V, 727

1,3-Dioxane-5,5-dimethanol, 2-phenyl- [2425-41-4];
MONOBENZALPENTAERYTHRITOL, IV, 679

1,3-Dioxane, 4-phenyl- [772-00-9];
4-PHENYL-*m*-DIOXANE, IV, 786

1,3-Dioxolane-4-methanol, 2,2-dimethyl-, (±)- [22323-83-7];
dl-ISOPROPYLIDENEGLYCEROL, III, 502

Diphosphine, tetramethyl-, 1,2-disulfide [3676-97-9];
TETRAMETHYLBIPHOSPHINE DISULFIDE, V, 1016

Disulfide, bis(2-nitrophenyl) [1155-00-6];
DI-*o*-NITROPHENYL DISULFIDE, I, 220

Disulfide, bis(3-nitrophenyl) [537-91-7];
m-NITROPHENYL DISULFIDE, V, 843

Disulfide, dibenzoyl [644-32-6];
BENZOYL DISULFIDE, III, 116

1,4-Dithiane [505-29-3];
p-DITHIANE, IV, 396

Docosanedioic acid [505-56-6];
DOCOSANEDIOIC ACID, V, 533

13-Docosenoic acid, (Z)- [112-86-7];
ERUCIC ACID, II, 258

1-Dodecanamine, *N*-methyl- [7311-30-0];
LAURYLMETHYLAMINE, IV, 564

Dodecane, 1-bromo- [143-15-7];
n-DODECYL BROMIDE, I, 29; II, 246

1-Dodecanethiol [112-55-0];
n-DODECYL MERCAPTAN, III, 363

Dodecanoic acid, ethenyl ester [2146-71-6];
VINYL LAURATE AND OTHER VINYL ESTERS, IV, 977

Dodecanoic acid, 2-methylene- [52756-21-5];
2-METHYLENEDODECANOIC ACID, IV, 616

1-Dodecanol [112-53-8];
LAURYL ALCOHOL, II, 372

2-Dodecenoic acid, (E)- [32466-54-9];
trans-2-DODECENOIC ACID, IV, 398

2-Dodecenoic acid, 2-methyl-, (E)- [53663-29-9];
trans-2-METHYL-2-DODECENOIC ACID, IV, 608

E

Ethanamine, 2-bromo-hydrobromide [2576-47-8];
    β-BROMOETHYLAMINE HYDROBROMIDE, II, 91

Ethanamine, 2-chloro-N,N-dimethyl-, hydrochloride [4584-46-7];
    β-DIMETHYLAMINOETHYL CHLORIDE HYDROCHLORIDE, IV, 333

Ethanamine, 2,2-diethoxy- [645-36-3];
    AMINOACETAL, III, 50

Ethanamine, N-Methyl- [624-78-2];
    N-METHYLETHYLAMINE, V, 758

Ethanaminium, 2-(benzoyloxy)-N,N,N-trimethyl-, chloride [2964-09-2];
    BENZOYLCHOLINE CHLORIDE, IV, 84

Ethanaminium, 2-(benzoyloxy)-N,N,N-trimethyl-, iodide [17518-43-3];
    BENZOYLCHOLINE IODIDE, IV, 84

Ethanaminium, 2-chloro-N,N,N-trimethyl-2-oxo-, chloride [53684-57-4];
    N-CHLOROBETAINYL CHLORIDE, IV, 154

Ethanaminium, 2-hydrazino-N,N,N-trimethyl-2-oxo-chloride [123-46-6];
    BETAINE HYDRAZIDE HYDROCHLORIDE, II, 85

1,2-Ethandiol, 1,1,2,2-tetraphenyl- [464-72-2];
    BENZOPINACOL, II, 71

Ethane, bromo- [74-96-4];
    ETHYL BROMIDE, I, 29, 36

Ethane, 2-bromo-1,1-diethoxy- [2032-35-1];
    BROMOACETAL, III, 123

Ethane, 1-bromo-2-ethoxy- [592-55-2];
    β-ETHOXYETHYL BROMIDE, III, 370

Ethane, 2-chloro-1-ethoxy-1,1,2-trifluoro- [310-71-4];
    2-CHLORO-1,1,2-TRIFLUOROETHYL ETHYL ETHER, IV, 184

Ethane, 1-chloro-2-(methylthio)- [542-81-4];
    β-CHLOROETHYL METHYL SULFIDE, II, 136

1,2-Ethanediamine,N,N-dibutyl- [3529-09-7];
    β-DI-n-BUTYLAMINOETHYLAMINE, III, 254

Ethane, 1,1-diethoxy [105-57-7];
    ACETAL, I, 1

Ethanedioic acid [144-62-7];
    OXALIC ACID (ANHYDROUS), I, 421

Ethanedioic acid, diethyl ester [95-92-1];
    ETHYL OXALATE, I, 261

Ethanedioic acid, dimethyl ester [553-90-2];
    METHYL OXALATE, II, 414

Ethanedione, bis[4-(dimethylamino)phenyl]- [17078-27-2];
4,4'BIS(DIMETHYLAMINO)BENZIL, V, 111

Ethanedione, diphenyl- [134-81-6];
BENZIL, I, 87

1,2-Ethanedisulfonic acid, 1,2-dihydroxy- [18381-20-9];
GLYOXAL BISULFITE, III, 438

1,2-Ethanedithiol [540-63-6];
ETHANEDITHIOL, IV, 401

Ethane, isocyano- [624-79-3];
ETHYL ISOCYANIDE, IV, 438

Ethane,1,1',1",1'''-[methanetetrayltetrakis(oxy)]-tetrakis- [78-09-1];
ETHYL ORTHOCARBONATE, IV, 457

Ethane,1,1',1"-[methylidynetris(oxy)]tris- [122-51-0];
ETHYL ORTHOFORMATE, I, 258

Ethanesulfonic acid, 2-amino- [107-35-7];
TAURINE, II, 563

Ethanesulfonic acid, 2-bromo-, sodium salt [4263-52-9];
SODIUM 2-BROMOETHANESULFONATE, II, 558

Ethanethioic acid [507-09-5];
THIOLACETIC ACID, IV, 928

Ethanethiol, 2,2-diethoxy- [53608-94-9];
DIETHYL MERCAPTOACETAL, IV, 295

Ethanimidamide, monohydrochloride [124-42-5];
ACETAMIDINE HYDROCHLORIDE, I, 5

Ethanol, 2-bromo- [540-51-2];
2-BROMOETHANOL, I, 117

Ethanol, 2-(cyclohexyloxy)- [1817-88-5];
2-CYCLOHEXYLOXYETHANOL, V, 303

Ethanol, 2,2-dichloro- [598-38-9];
2,2-DICHLOROETHANOL, V, 271

Ethanol, 2-(diethylamino)- [100-37-8];
β-DIETHYLAMINOETHYL ALCOHOL, II, 183

Ethanol, 2-[(1-methylethyl)amino]- [109-56-8];
2-ISOPROPYLAMINOETHANOL, III, 501

Ethanol, 2-(methylthio)- [5271-38-5];
β-HYDROXYETHYL METHYL SULFIDE, II, 345

Ethanol, 2-nitro- [625-48-9];
2-NITROETHANOL, V, 833

Ethanol, 2,2'-thiobis- [111-48-8];
β-THIODIGLYCOL, II, 576

Ethanol, 2,2,2-trichloro- [115-20-8];
TRICHLOROETHYL ALCOHOL, II, 598

Ethanone, 2-(acetyloxy)-1,2-diphenyl- [574-06-1];
BENZOIN ACETATE, II, 69

Ethanone, 2-amino-1-phenyl-, hydrochloride [5468-37-1];
PHENACYLAMINE HYDROCHLORIDE, V, 909

Ethanone, 1-(9-anthracenyl)- [784-04-3];
9-ACETYLANTHRACENE, IV, 8

Ethanone, 1,1',1''-(1,3,5-benzenetriyl)tris- [779-90-8];
1,3,5-TRIACETYLBENZENE, III, 829

Ethanone, 1,2-bis(4-methoxyphenyl)- [120-44-5];
DEOXYANISOIN, V, 339

Ethanone, 2-bromo-1-(4-bromophenyl)- [99-73-0];
p-BROMOPHENACYL BROMIDE, I, 127

Ethanone, 1-(3-bromophenyl)- [2142-63-4];
3-BROMOACETOPHENONE, V, 117

Ethanone, 1-(4-bromophenyl)- [99-90-1];
p-BROMOACETOPHENONE, I, 109

Ethanone, 2-bromo-1-phenyl- [70-11-1];
PHENACYL BROMIDE, II, 480

Ethanone, 2-chloro-1,2-diphenyl- [447-31-4];
DESYL CHLORIDE, II, 159

Ethanone, 1-(1-cyclohexen-1-yl)- [932-66-1];
1-ACETYLCYCLOHEXENE, III, 22

Ethanone, 1-cyclohexyl- [823-76-7];
CYCLOHEXYL METHYL KETONE, V, 775

Ethanone, 1-cyclopropyl- [765-43-5];
METHYL CYCLOPROPYL KETONE, IV, 597

Ethanone, 2-diazo-1,2-diphenyl- [3469-17-8];
PHENYLBENZOYLDIAZOMETHANE, II, 496

Ethanone, 1-(2',5'-dihydroxy[1,1'-biphenyl]-4-yl)- [3948-13-8];
2-p-ACETYLPHENYLHYDROQUINONE IV, 15

Ethanone, 1-(2,4-dihydroxyphenyl)- [89-84-9];
RESACETOPHENONE, III, 761

Ethanone, 1.(2,5-dihydroxyphenyl)- [490-78-8];
2,5-DIHYDROXYACETOPHENONE, III, 280

Ethanone, 1-(2,6-dihydroxyphenyl)- [699-83-2];
2,6-DIHYDROXYACETOPHENONE, III, 281

Ethanone, 1,2-diphenyl- [451-40-1];
DESOXYBENZOIN, II, 156

Ethanone, 1-(9H-fluoren-2-yl)- [781-73-7];
2-ACETYLFLUORENE, III, 23

Ethanone, 1-(1-hydroxycyclohexyl)- [1123-27-9];
1-ACETYLCYCLOHEXANOL, IV, 13

Ethanone, 2-hydroxy-1,2-diphenyl- [119-53-9];
BENZOIN, I, 94

Ethanone, 1-(2-hydroxy-5-methoxyphenyl)- [705-15-7];
QUINACETOPHENONE MONOMETHYL ETHER, IV, 836

Ethanone, 2-methoxy-1-phenyl- [4079-52-1];
ω-METHOXYACETOPHENONE, III, 562

Ethanone, 1-[2-methyl-5-(1-methylethyl)phenyl]- [1202-08-0];
ACETO p-CYMENE, II, 3

Ethanone, 1-(2-nitrophenyl)- [577-59-3];
o-NITROACETOPHENONE, IV, 708

Ethanone, 1-(3-nitrophenyl)- [121-89-1];
m-NITROACETOPHENONE, II, 434

Ethanone, 1-(9-phenanthrenyl)- [2039-77-2];
9-ACETYLPHENANTHRENE, III, 26

Ethanone, tetraphenyl- [466-37-5];
β-BENZOPINACOLONE, II, 73

Ethanone, 1-(2-thienyl)- [88-15-3];
2-ACETOTHIENONE, II, 8; III, 14

Ethanone, 1-(2,3,4-trihydroxyphenyl)- [528-21-2];
GALLACETOPHENONE, II, 304

Ethanone, 1-(2,4,6-trihydroxyphenyl)- [480-66-0];
PHLOROACETOPHENONE, II, 522

Ethenamine, 1,2,2-trichloro-N,N-diethyl- [686-10-2];
N,N-DIETHYL-1,2,2-TRICHLOROVINYLAMINE, V, 387

Ethene, 1,1-dichloro-2,2-difluoro- [79-35-6];
1,1-DICHLORO-2,2-DIFLUOROETHYLENE, IV, 268

Ethene, 1,1-diethoxy- [2678-54-8];
KETENE DIETHYLACETAL, III, 506

Ethenesulfonic acid, 2-phenyl-, sodium salt [2039-44-3];
Ethenesulfonyl chloride, 2-phenyl- [4091-26-3];
    SODIUM β-STYRENESULFONATE AND β-STYRENESULFONYL CHLORIDE, IV, 846

Ethenetetracarbonitrile [670-54-2];
TETRACYANOETHYLENE, IV, 877

Ethenetetracarboxylic acid, tetraethyl ester [6174-95-4];
   ETHYL ETHYLENETETRACARBOXYLATE, II, 273

Ethenetricarbonitrile, [4-(dimethylamino)phenyl]- [6673-15-0];
   p-TRICYANOVINYL-N,N-DIMETHYLANILINE, IV, 953

Ethenone [463-51-4];
   KETENE, I, 330; V, 679

Ethenone, diphenyl [525-06-4];
   DIPHENYLKETENE, III, 356

Ethyne, ethoxy- [927-80-0];
   ETHOXYACETYLENE, IV, 404

F

Ferrocene [102-54-5];
   FERROCENE, IV, 473

Ferrocene, (cyanomethyl)- [1316-91-2];
   FERROCENYLACETONITRILE, V, 578

Ferrocene, (hydroxymethyl)- [1273-86-5];
   HYDROXYMETHYLFERROCENE, V, 621

9H-Fluoren-2-amine [153-78-6];
   2-AMINOFLUORENE, II, 447; V, 30

9H-Fluorene-9-carboxylic acid [1989-33-9]
   9-FLUORENECARBOXYLIC ACID, IV, 482

9H-Fluorene-2-carboxylic acid, 9-oxo- [784-50-9];
   FLUORENONE-2-CARBOXYLIC ACID, III, 420

9H-Fluorene, 9-methyl- [2523-37-7];
   9-METHYLFLUORENE, IV, 623

9H-Fluorene, 2-nitro [607-57-8];
   2-NITROFLUORENE, II, 447

9H-Fluoren-9-one, 2,4,5,7-tetranitro- [746-53-2];
   2,4,5,7-TETRANITROFLUORENONE, V, 1029

9H-Fluoren-9-one, 2,4,7-trinitro- [129-79-3];
   2,4,7-TRINITROFLUORENONE, III, 837

Formamide, N-methyl-N-phenyl- [93-61-8];
   N-METHYLFORMANILIDE, III, 590

Furan [110-00-9];
   FURAN, I, 274

Furan, 2-(bromomethyl)tetrahydro- [1192-30-9];
   TETRAHYDROFURFURYL BROMIDE, III, 793

2-Furancarboxaldehyde [98-01-1];
   FURFURAL, I, 280

2-Furancarboxaldehyde, 5-methyl- [620-02-0];
5-METHYLFURFURAL, II, 393

2-Furancarboxylic acid [88-14-2];
2-FUROIC ACID, IV, 493

2-Furancarboxylic acid, -3-methyl [4412-96-8];
3-METHYL-2-FUROIC ACID, IV, 628

2-Furancarboxylic acid, 3-methyl-, methyl ester [6141-57-7];
METHYL 3-METHYL-2-FUROATE, IV, 649

Furan, 2,5-dihydro-2,5-dimethoxy- [332-77-4];
2,5-DIHYDRO-2,5-DIMETHOXYFURAN, V, 403

2,5-Furandione, 3,4-bis(acetyloxy)dihydro-, (3S-trans)- [19523-83-2];
DIACETYL d-TARTARIC ANHYDRIDE, IV, 242

2,5-Furandione, dihydro- [108-30-5];
SUCCINIC ANHYDRIDE, II, 560

2,5-Furandione, dihydro-3-methylene- [2170-03-8]
ITACONIC ANHYDRIDE, II, 368

2,5-Furandione, 3-methyl- [616-02-4];
CITRACONIC ANHYDRIDE, II, 140

2-Furanmethanamine, N,N,5-trimethyl- [14496-35-6];
5-METHYLFURFURYLDIMETHYLAMINE, IV, 626

2-Furanmethanethiol [98-02-2];
2-FURFURYL MERCAPTAN, IV, 491

2-Furanmethanol [98-00-0];
2-FURYLCARBINOL, I, 276

2-Furanmethanol, acetate [623-17-6];
2-FURYLMETHYL ACETATE, I, 285

Furan, 3-methyl- [930-27-8];
3-METHYLFURAN, IV, 628

3-Furanol, tetrahydro- [453-20-3];
3-HYDROXYTETRAHYDROFURAN, IV, 534

2(5H)-Furanone [497-23-4];
γ-CROTONOLACTONE, V, 255

2(3H)-Furanone, 3-acetyl-5-(chloromethyl)dihydro- [3154-75-4];
α-ACETYL-δ-CHLORO-γ-VALEROLACTONE, IV, 10

2(3H)-Furanone, dihydro-5-propyl- [105-21-5];
γ-n-PROPYLBUTYROLACTONE, III, 742

3(2H)-Furanone, dihydro-2,2,5,5-tetramethyl- [5455-94-7];
2,2,5,5-TETRAMETHYLTETRAHYDRO-3-KETOFURAN, V, 1024

2(3H)-Furanone, 5-phenyl-3(phenylmethylene)- [4361-96-0];
α-BENZYLIDENE-γ-PHENYL-Δβ,γ-BUTENOLIDE, V, 80

2-Furanpropanoic acid, tetrahydro- [935-12-6];
   β-(TETRAHYDROFURYL)PROPIONIC ACID, III, 742

Furan, tetrahydro- [109-99-9];
   TETRAHYDROFURAN, II, 566; PURIFICATION OF TETRAHYDROFURAN (Warning), V, 976

Furan, tetrahydro-2,2-dimethyl-5-pentyl- [53684-53-0];
   5,5-DIMETHYL-2-n-PENTYLTETRAHYDROFURAN, IV, 350

Furazan, dimethyl- [4975-21-7];
   DIMETHYLFURAZAN, IV, 342

2-Furoic acid [88-14-2];
   2-FURANCARBOXYLIC ACID, I, 276

                                    G

D-Gluconic acid, pentaacetate [17430-71-6];
   2,3,4,5,6-PENTA-O-ACETYL-D-GLUCONIC ACID, V, 887

D-Glucononitrile, 2,3,4,5,6-pentaacetate [6272-51-1];
   PENTAACETYL d-GLUCONONITRILE, III, 690

D-Gluconyl chloride, pentaacetate [53555-69-4];
   2,3,4,5,6-PENTA-O-ACETYL-D-GLUCONYL CHLORIDE, V, 887

β-D-Glucopyranose, 1,2,3,4-tetraacetate [13100-46-4];
   β-d-GLUCOSE-1,2,3,4-TETRAACETATE, III, 432

β-D-Glucopyranose, 2,3,4,6-tetraacetate [3947-62-4];
   β-d-GLUCOSE-2,3,4,6-TETRAACETATE, III, 434

β-D-Glucopyranose, 6-O-(2,3,4,6-tetra-O-acetyl-β-D-glucopyranosyl)-tetraacetate [4613-78-9];
   β-GENTIOBIOSE OCTAACETATE, III, 428

D-Glucopyranose, 2,3,4,6-tetra-O-methyl- [7506-68-5];
   2,3,4,6-TETRAMETHYL-d-GLUCOSE, III, 800

α-D-Glucopyranoside, methyl [97-30-3];
   α-METHYL d-GLUCOSIDE, I, 364

α-D-Glucopyranosyl bromide, tetraacetate [572-09-8];
   ACETOBROMOGLUCOSE, III, 11

α-D-Glucopyranosyl chloride, 2-(acetylamino)-2-deoxy-,3,4,6-triacetate [3068-34-6];
   2-ACETAMIDO-3,4,6-TRI-O-ACETYL-2-DEOXY-α-D-GLUCOPYRANOSYL CHLORIDE, V, 1

D-Glucose, 2-amino-2-deoxy-, hydrochloride [66-84-2];
   d-GLUCOSAMINE HYDROCHLORIDE, III, 430

D-Glucose, 4-O-β-D-glucopyranosyl- [528-50-7];
   CELLOBIOSE, II, 122

D-Glucose, 4-O-α-D-glucopyranosyl-, octaacetate [5346-90-7];
   α-CELLOBIOSE OCTAACETATE, II, 124

D-Glutamic acid [6893-26-1];
   d-GLUTAMIC ACID, I, 286

Glycine [56-40-6];
   GLYCINE, I, 298

Glycine, N-acetyl- [543-24-8];
   ACETYLGLYCINE, II, 11

Glycine, N-(aminoiminomethyl)- [352-97-6];
   GUANIDOACETIC ACID, III, 440

Glycine, N-benzoyl- [495-69-2];
   HIPPURIC ACID, II, 328

Glycine, N-(carboxymethyl)-N-methyl- [4408-64-4];
   METHYLIMINODIACETIC ACID, II, 397

Glycine, 1,1-dimethylethyl ester [6456-74-2];
   GLYCINE t-BUTYL ESTER, V, 586

Glycine, ethyl ester, hydrochloride [623-33-6];
   GLYCINE ETHYL ESTER HYDROCHLORIDE, II, 310

Guanidine, mononitrate [506-93-4];
   GUANIDINE NITRATE, I, 302; Warning, V, 589

Guanidine, nitro- [556-88-7];
   NITROGUANIDINE, I, 399

D-Gulonic acid, γ-lactone [6322-07-2];
   D-GULONIC-γ-LACTONE, IV, 506

H

Heptanal, 2-bromo- [16486-84-3];
   α-BROMOHEPTALDEHYDE, III, 127

Heptanal, oxime [629-31-2];
   HEPTALDOXIME, II, 313

Heptanamide [628-62-6];
   n-HEPTAMIDE, IV, 513

1-Heptanamine [111-68-2];
   n-HEPTYLAMINE, II, 318

Heptane, 1-bromo-2-fluoro- [1786-32-9];
   1-BROMO-2-FLUOROHEPTANE, V, 136

Heptanedioic acid [111-16-0];
   PIMELIC ACID, II, 531

Heptanedioic acid, 3-oxo-, diethyl ester [40420-22-2];
   DIETHYL β-KETOPIMELATE, V, 384

Heptanedioic acid, 4-oxo-, diethyl ester [6317-49-3];
   DIETHYL γ-OXOPIMELATE, IV, 302

Heptane, 1,1,1-trifluoro- [693-09-4];
  1,1,1-TRIFLUOROHEPTANE, V, 1082

Heptanoic acid [111-14-8];
  n-HEPTANOIC ACID, II, 315

Heptanoic acid, 2-fluoro- [1578-58-1];
  2-FLUOROHEPTANOIC ACID, V, 580

Heptanoic acid, 3-methyl- [53663-30-2];
  3-METHYLHEPTANOIC ACID, V, 762

Heptanoic acid, 6-oxo- [3128-07-2];
  δ-ACETYL-n-VALERIC ACID, IV, 19

1-Heptanol [111-70-6];
  n-HEPTYL ALCOHOL, I, 304

2-Heptanol [543-49-7];
  2-HEPTANOL, II, 317

2-Heptanone [110-43-0]
  METHYL n-AMYL KETONE, I, 351

4-Heptanone [123-19-3];
  4-HEPTANONE, V, 589

2-Heptene, 4-bromo- [22118-57-6];
  4-BROMO-2-HEPTENE, IV, 108

1-Hepten-4-ol, 4,6-dimethyl- [32189-75-6];
  4,6-DIMETHYL-1-HEPTEN-4-OL, V, 452

1-Hepten-3-one, 1-chloro-6-methyl- [18378-90-0];
  β-CHLOROVINYL ISOAMYL KETONE, IV, 186

Hexadecane [544-76-3];
  n-HEXADECANE, II, 320

Hexadecane, 1-iodo- [544-77-4];
  n-HEXADECYL IODIDE, II, 322

Hexadecanoic acid [57-10-3];
  PALMITIC ACID, III, 605

Hexadecanoic acid, methyl ester [112-39-0];
  METHYL PALMITATE, III, 605

Hexadecanoic acid, 2-sulfo- [1782-10-1];
  α-SULFOPALMITIC ACID, IV, 862

1,5-Hexadiene, [592-42-7];
  BIALLYL, III, 121

2,4-Hexadienedioic acid [505-70-4];
  MUCONIC ACID, III, 623

2,4-Hexadienoic acid, (E,E)- [110-44-1];
  SORBIC ACID, III, 783

Hexanal [66-25-1];
n-HEXALDEHYDE, II, 323

Hexane, 1,6-diiodo- [629-09-4];
1,6-DIIODOHEXANE, IV, 323

Hexane, 1,6-diisocyanato- [822-06-0];
HEXAMETHYLENE DIISOCYANATE, IV, 521

Hexanedioic acid [124-04-9];
ADIPIC ACID, I, 18

Hexanedioic acid, diethyl ester [141-28-6];
ETHYL ADIPATE, II, 264

1,6-Hexanediol [629-11-8];
HEXAMETHYLENE GLYCOL, II, 325

2,6-Hexanediol, 2,6-dimethyl- [110-03-2];
$\alpha,\alpha,\alpha',\alpha'$-TETRAMETHYLTETRAMETHYLENE GLYCOL, V, 1026

2,5-Hexanedione, 3,4-diacetyl- [5027-32-7];
TETRAACETYLETHANE, IV, 869

1,6-Hexanedione, 1,6-diphenyl- [3375-38-0];
1,4-DIBENZOYLBUTANE, II, 169

Hexane, 1-fluoro- [373-14-8];
n-HEXYL FLUORIDE, IV, 525

Hexanenitrile, 2-ethyl- [4528-39-6];
2-ETHYLHEXANONITRILE, IV, 436

Hexanoic acid, 2-acetyl-, ethyl ester [1540-29-0];
ETHYL n-BUTYLACETOACETATE, I, 248

Hexanoic acid, 6-amino- [60-32-2];
ϵ-AMINOCAPROIC ACID, II, 28; IV, 39

Hexanoic acid, anhydride [2051-49-2];
n-CAPROIC ANHYDRIDE, III, 164

Hexanoic acid, 6-(benzoylamino)- [956-09-2];
ϵ-BENZOYLAMINOCAPROIC ACID, II, 76

Hexanoic acid, 6-(benzoylamino)-2-bromo- [1700-05-6];
ϵ-BENZOYLAMINO-α-BROMOCAPROIC ACID, II, 74

Hexanoic acid, 2-bromo- [616-05-7];
α-BROMO-n-CAPROIC ACID, I, 115

Hexanoic acid, 2-cyano-, ethyl ester [7391-39-1];
ETHYL n-BUTYLCYANOACETATE, III, 385

1-Hexanol [111, 27-3];
n-HEXYL ALCOHOL, I, 306

1-Hexanol, 6-chloro- [2009-83-8];
HEXAMETHYLENE CHLOROHYDRIN, III, 446

2-Hexanone, 1,1-dimethoxy-5-methyl- [53684-58-5];
β-KETOISOOCTALDEHYDE DIMETHYL ACETAL, IV, 558

1,3,5-Hexatriene [2235-12-3];
1,3,5-HEXATRIENE, V, 608

3-Hexene, 3,4-dinitro- [53684-54-1];
3,4-DINITRO-3-HEXENE, IV, 372

5-Hexen-2-one, 5-methyl- [3240-09-3];
5-METHYL-5-HEXEN-2-ONE, V, 767

1-Hexyne [693-02-7];
n-BUTYLACETYLENE, IV, 117

L-Histidine, monohydrochloride [645-35-2];
l-HISTIDINE MONOHYDROCHLORIDE, II, 330

Hydrazinecarboxamide, 2-(2-amino-1-methylethylidene)-, monohydrochloride [10469-70-2];
AMINOACETONE SEMICARBAZONE HYDROCHLORIDE, V, 27

Hydrazinecarboxamide, N-phenyl- [537-47-3];
4-PHENYLSEMICARBAZIDE, I, 450

Hydrazinecarboxamide, sulfate [7464-82-3];
SEMICARBAZIDE SULFATE, I, 485

Hydrazinecarboxylic acid, 1,1-dimethylethyl ester [870-46-2];
t-BUTYL CARBAZATE, V, 166

Hydrazinecarboxylic acid, ethyl ester [4114-31-2];
ETHYL HYDRAZINECARBOXYLATE, III, 404

Hydrazine, 1,2-dimethyl-, dihydrochloride [306-37-6];
sym-DIMETHYLHYDRAZINE DIHYDROCHLORIDE, II, 208

Hydrazine, 1,1-dimethyl-, monohydrochloride [593-82-8];
unsym-DIMETHYLHYDRAZINE HYDROCHLORIDE, II, 211

Hydrazine, (2,4-dinitrophenyl)- [119-26-6];
2,4-DINITROPHENYLHYDRAZINE, II, 228

Hydrazine, 1-methyl-1-phenyl-[618-40-6];
α-METHYL-α-PHENYLHYDRAZINE, II, 418

Hydrazine, methyl-, sulfate (1:1) [302-15-8];
METHYLHYDRAZINE SULFATE, II, 395

Hydrazin, phenyl- [100-63-0];
PHENYLHYDRAZINE, I, 442

Hydrazine, sulfate (1:1) [10034-93-2];
HYDRAZINE SULFATE, I, 309

Hydrobromic acid [10035-10-6];
HYDROGEN BROMIDE (ANHYDROUS), II, 338

Hydrocyanic acid [74-90-8];
HYDROGEN CYANIDE (ANHYDROUS), I, 314

Hydroperoxide, 1,2,3,4-tetrahydro-1-naphthalenyl [771-29-9];
TETRALIN HYDROPEROXIDE, IV, 895

Hydroxylamine, hydrochloride [5470-11-1];
HYDROXYLAMINE HYDROCHLORIDE, I, 318

Hypochlorous acid, 1,1-dimethylethyl ester [507-40-4];
t-BUTYL HYPOCHLORITE, IV, 125; V, 184; *Warning*, V, 183

I

1*H*-Imidazole [288-32-4];
IMIDAZOLE, III, 471

1*H*-Imidazole, 1,1'-carbonylbis- [530-62-1];
1,1'-CARBONYLDIIMIDAZOLE, V, 201

1*H*-Imidazole-4-methanol, monohydrochloride [32673-41-9];
4(5)-HYDROXYMETHYLIMIDAZOLE HYDROCHLORIDE, III, 460

Imidazolidene, 2-(1,3-diphenyl-2-imidazolidinylidene)-1,3-diphenyl- [2179-89-7];
BIS(1,3-DIPHENYLIMIDAZOLIDINYLIDENE-2). V, 115

2,4-Imidazolidinedione, 5,5-dimethyl- [77-71-4];
5,5-DIMETHYLHYDANTOIN, III, 323

2-Imidazolidinethione [96-45-7];
ETHYLENE THIOUREA, III, 394

Imidazolidinetrione [120-89-8];
PARABANIC ACID, IV, 744

4*H*-Imidazol-4-one, 2-amino-1,5-dihydro-1-methyl- [60-27-5];
CREATININE, I, 172

2*H*-Imidazol-2-one, 1,5-dihydro-4,5-diphenyl- [53684-56-3];
4,5-DIPHENYLGLYOXALONE, II, 231

Imidodicarbonic dihydrazide [4375-11-5];
DIAMINOBIURET, III, 404

1*H*-Indazole [271-44-3];
INDAZOLE, III, 475; IV, 536; V, 650

1*H*-Indazole, 5-nitro- [5401-94-5];
5-NITROINDAZOLE, III, 660

2*H*-Indazole, 2-phenyl- [3682-71-1];
2-PHENYLINDAZOLE, V, 941

1*H*-Indene-2-carboxylic acid, 67-dimethoxy-3-methyl-, ethyl ester [5415-54-3];
ETHYL 6,7-DIMETHOXY-3-METHYLINDENE-2-CARBOXYLATE, V, 550

1*H*-Indene, 2,3-dihydro-1-methyl-3-phenyl- [6416-39-3];
1-METHYL-3-PHENYLINDANE, IV, 665

1*H*-Inden-1-one, 2,3-dihydro- [83-33-0];
α-HYDRINDONE, II, 336

2H-Inden-2-one, 1,3-dihydro- [615-13-4];
  2-INDANONE, V, 647

1H-Inden-1-one, 2,3-diphenyl- [1801-42-9];
  2,3-DIPHENYLINDONE (2,3-DIPHENYL-1-INDENONE), III, 353

1H-Indole [120-72-9];
  INDOLE, III, 479

1H-Indole-3-acetic acid [87-51-4];
  INDOLE-3-ACETIC ACID, V, 654

1H-Indole-3-carbonitrile [5457-28-3];
  INDOLE-3-CARBONITRILE, V, 656

1H-Indole-3-carboxaldehyde [487-89-8];
  INDOLE-3-ALDEHYDE, IV, 539

1H-Indole-2-carboxylic acid, ethyl ester [3770-50-1];
  ETHYL INDOLE-2-CARBOXYLATE, V, 567

1H-Indole-2,3-dione [91-56-5];
  ISATIN, I, 327

1H-Indole, 1-methyl- [603-76-9];
  1-METHYLINDOLE, V, 769

1H-Indole, 2-methyl- [95-20-5];
  2-METHYLINDOLE, III, 597

1H-Indole, 2-phenyl- [948-65-2];
  2-PHENYLINDOLE, III, 725

2H-Indol-2-one, 3-acetyl-1,3-dihydro- [17266-70-5];
  3-ACETYLOXINDOLE, V, 12

2H-Indol-2-one, 1,3-dihydro-3-methyl- [1504-06-9];
  3-METHYLOXINDOLE, IV, 657

2H-Indol-2-one, 3-ethyl-1,3-dihydro-1-methyl- [2525-35-1];
  1-METHYL-3-ETHYLOXINDOLE, IV, 620

Iodine, bis(acetate-O)phenyl- [3240-34-4];
  IODOSOBENZENE DIACETATE, V, 660

Iodine cyanide, I(CN) [506-78-5];
  CYANOGEN IODIDE, IV, 207

Iodonium, diphenyl, iodide [2217-79-0];
  DIPHENYLIODONIUM IODIDE, III, 355

Iron, chloro[7,12-diethenyl-3,8,13,17-tetramethyl-21H,23H-porphine-2,18-dipropanoato(2-)-
    $N^{21},N^{22},N^{23},N^{24}$]-, (SP-5-13)- [15489-47-1];
  HEMIN, III, 442

1,3-Isobenzofurandione, 4-nitro- [641-70-3];
  3-NITROPHTHALIC ANHYDRIDE, I, 410

1,3-Isobenzofurandione, 3a,4,7,7a-tetrahydro-, cis- [935-79-5];
   cis-Δ⁴-TETRAHYDROPHTHALIC ANHYDRIDE, IV, 890

1,3-Isobenzofurandione, 4,5,6,7-tetraiodo- [632-80-4];
   TETRAIODOPHTHALIC ANHYDRIDE, III, 796

1,3-Isobenzofurandione, 4,5,6,7-tetraphenyl- [4741-53-1];
   TETRAPHENYLPHTHALIC ANHYDRIDE, III, 807

1(3H)-Isobenzofuranone [87-41-2];
   PHTHALIDE, II, 526

1(3H)-Isobenzofuranone, 3-bromo- [6940-49-4];
   3-BROMOPHTHALIDE, V, 145

1(3H)-Isobenzofuranone, 3,3-dichloro- [601-70-7];
   unsym. o-PHTHALYL CHLORIDE, II, 528

1(3H)-Isobenzofuranone, 3-(phenylmethylene)- [575-61-1];
   BENZALPHTHALIDE, II, 61

2H-Isoindole-2-acetic acid, 1,3-dihydro-1,3-dioxo-α-(phenylmethyl)-, (S)- [5123-55-7];
   N-PHTHALYL-L-β-PHENYLALANINE, V, 973

1H-Isoindole, 2,3-dihydro- [496-12-8];
   1,3-DIHYDROISOINDOLE, V, 406

1H-Isoindole, 2,3-dihydro-2-[(4-methylphenyl)sulfonyl]- [32372-83-1];
   2-(-TOLYLSULFONYL)-DIHYDROISOINDOLE, V, 1064

1H-Isoindole-1,3(2H)-dione [85-41-6];
   PHTHALIMIDE, I, 457

1H-Isoindole-1,3(2H)-dione, 2-(2-bromoethyl)- [574-98-1];
   β-BROMOETHYLPHTHALIMIDE, I, 119; IV, 106

1H-Isoindole-1,3(2H)-dione, 5-nitro- [89-40-7];
   4-NITROPHTHALIMIDE, II, 459

1H-Isoindole-1,3(2H)-dione, 2-(phenylmethyl)- [2142-01-0];
   BENZYL PHTHALIMIDE, II, 83

DL-Isoleucine [443-79-8];
   dl-ISOLEUCENE, III, 495

Isoquinoline, 1-methyl- [1721-93-3];
   1-METHYLISOQUINOLINE, IV, 641

3(2H)-Isoquinolinone, 1,4-dihydro-1-thioxo- [938-38-5];
   2α-THIOHOMOPHTHALIMIDE, V, 1051

1(2H)-Isoquinolinone, 2-hydroxy-3-methyl- [7114-79-6];
   2-HYDROXY-3-METHYLISOCARBOSTYRYL, V, 623

Isoxazolidine, 2,3,5-triphenyl- [13787-96-7];
   2,3,5-TRIPHENYLISOXAZOLIDINE, V, 1124

L

DL-Leucine [328-39-2];
  dl-LEUCINE, III, 523

DL-Lysine, dihydrochloride [617-68-5];
  dl-LYSINE HYDROCHLORIDES, DI-, II, 374

DL-Lysine, monohydrochloride [70-53-1];
  dl-LYSINE HYDROCHLORIDES, MONO-, II, 374

M

Magnesium, butylchloro- [693-04-9];
  UNSOLVATED *n*-BUTYLMAGNESIUM CHLORIDE, V, 1141

D-Mannose [3458-28-4];
  D-MANNOSE, III, 541

α-D-Mannoside, methyl [27939-30-6];
  α-METHYL MANNOSIDE, I, 371

Mercury, bis(4-methylphenyl)- [537-64-4];
  DI-*p*-TOLYLMERCURY, I, 231

Mercury, (4-carboxyphenyl)chloro- [59-85-8];
  p-CHLOROMERCURIBENZOIC ACID, I, 159

Mercury, chloro(2-hydroxyphenyl)- [90-03-9];
  O-CHLOROMERCURIPHENOL, I, 161

Mercury, chloro(4-methylphenyl)- [539-43-5];
  p-TOLYLMERCURIC CHLORIDE, I, 519

Mercury, chloro-2-naphthalenyl- [39966-41-1];
  β-NAPHTHYLMERCURIC CHLORIDE, II, 432

Mercury, di-2-naphthalenyl- [19510-26-0];
  MERCURY DI-β-NAPHTHYL, II, 381

Mercury, diphenyl- [587-85-9];
  DIPHENYLMERCURY, I, 228

Mercury, phenyl(trichloromethyl)- [3294-57-3];
  PHENYL(TRICHLOROMETHYL)MERCURY, V, 969

Methanamine, 1-(2,4-cyclopentadien-1-ylidene)-*N*,*N*-dimethyl- [696-68-4];
  6-(DIMETHYLAMINO)FULVENE, V, 431

Methanamine, *N*,*N*-dimethyl- [75-50-3];
  TRIMETHYLAMINE, I, 528

Methanamine, *N*,*N*-dimethyl-, hydrochloride [593-81-7];
  TRIMETHYLAMINE HYDROCHLORIDE, I, 531

Methanamine, hydrochloride [593-51-1];
  METHYLAMINE HYDROCHLORIDE, I, 347

Methanamine, *N*-hydroxy-*N*-methyl-, hydrochloride [16645-06-0];
  N,*N*-DIMETHYLHYDROXYLAMINE HYDROCHLORIDE, IV, 612

Methanaminium, 1-ferrocenyl-*N*,*N*,*N*-trimethyl-, iodide [12086-40-7];
  N,*N*-DIMETHYLAMINOMETHYLFERROCENE METHIODIDE, V, 434

Methanaminium, *N*,*N*,*N*-trimethyl-, salt with 1-propene-1,1,2,3,3-pentacarbonitrile(1:1)
  [53663-17-5];
  TETRAMETHYLAMMONIUM 1,1,2,3,3-PENTACYANOPROPENIDE, V, 1013

Methane, chloromethoxy- [107-30-2];
  MONOCHLOROMETHYL ETHER, I, 377

Methane, diazo- [334-88-3];
  DIAZOMETHANE, II, 165; III, 244; IV, 250; V, 351

Methane, dibromo- [74-95-3];
  METHYLENE BROMIDE, I, 357

Methane, dichloromethoxy- [4885-02-3];
  DICHLOROMETHYL METHYL ETHER, V, 365

Methane, diiodo- [75-11-6];
  METHYLENE IODIDE, I, 358

Methanediol, 2-furanyl, diacetate [613-75-2];
  FURFURAL DIACETATE, IV, 489

Methanediol, (2-nitrophenyl)-, diacetate (ester) [6345-63-7];
  o-NITROBENZALDIACETATE, IV, 713

Methanediol, (4-nitrophenyl)-, diacetate (ester) [2929-91-1];
  p-NITROBENZALDIACETATE; V, 713

Methane, iodo- [74-88-4];
  METHYL IODIDE, II, 399

Methane, isocyano- [593-75-9];
  METHYL ISOCYANIDE, V, 772

Methane, isothiocyanato- [556-61-6];
  METHYL ISOTHIOCYANATE, III, 599

Methane, nitro- [75-52-5];
  NITROMETHANE, I, 401

Methane, oxybis[chloro- [542-88-1];
  BISCHLOROMETHYL ETHER, IV, 101; (*Hazard Note*), V, 218

Methanesulfinyl chloride [676-85-7];
  METHANESULFINYL CHLORIDE, V, 709

Methanesulfonyl chloride [124-63-0];
  METHANESULFONYL CHLORIDE, IV, 571

Methane, tetranitro- [509-14-8];
  TETRANITROMETHANE, III, 803

Methanethione, diphenyl- [1450-31-3];
    THIOBENZOPHENONE, II, 573; IV, 927

Methanetricarboxylic acid, triethyl ester [6279-86-3];
    TRICARBETHOXYMETHANE, II, 594

Methanetricarboxylic acid, trimethyl ester [1186-73-8];
    TRICARBOMETHOXYMETHANE, II, 596

Methanimidamide, monoacetate [3473-63-0];
    FORMAMIDINE ACETATE, V, 582

Methanimidic acid, N-phenyl-, ethyl ester [6780-49-0];
    ETHYL N-PHENYLFORMIMIDATE, IV, 464

Methanone, (2-aminophenyl)phenyl- [2835-77-0];
    2-AMINOBENZOPHENONE, IV, 34

Methanone, dicyclopropyl- [1121-37-5];
    DICYCLOPROPYL KETONE, IV, 278

Methanone, [4-(dimethylamino)phenyl]phenyl- [530-44-9];
    p-DIMETHYLAMINOBENZOPHENONE, I, 217

Methanone, diphenyl- [119-61-9];
    BENZOPHENONE, I, 95

Methanone, diphenyl-, oxime [574-66-3];
    BENZOPHENONE OXIME, II, 70

Methanone, phenyl-3-pyridinyl- [5424-19-1];
    3-BENZOYLPYRIDINE IV, 88

Methanone, phenyl-2-thienyl- [135-00-2];
    PHENYL THIENYL KETONE, II, 520

DL-Methionine [59-51-8];
    dl-METHIONINE, II, 384

Morpholine, 4-(1-cyclohexen-1-yl(- [670-80-4];
    1-MORPHOLINO-1-CYCLOHEXENE, V, 808

Morpholine, 4-nitro- [4164-32-3];
    N-NITROMORPHOLINE, V, 839

N

1-Naphthalenamine, 4-nitro- [776-34-1];
    4-NITRO-1-NAPHTHYLAMINE, III, 664

2-Naphthalenamine, 1,2,3,4-tetrahydro- [2954-50-9];
    ac-TETRAHYDRO-β-NAPHTHYLAMINE, I, 499

Naphthalene, 1-bromo-, [90-11-9];
    α-BROMONAPHTHALENE I, 121

Naphthalene, 2-bromo- [580-13-2];
    2-BROMONAPHTHALENE, V, 142

1-Naphthalenecarbonitrile [86-53-3];
α-NAPHTHONITRILE, III, 631

1-Naphthalenecarboxaldehyde [66-77-3]
1-NAPHTHALDEHYDE, IV, 690

2-Naphthalenecarboxaldehyde [66-99-9];
β-NAPHTHALDEHYDE, III, 626

1-Naphthalenecarboxaldehyde, 2-ethoxy- [19523-57-0];
2-ETHOXY-1-NAPHTHALDEHYDE, III, 98

1-Naphthalenecarboxaldehyde, 2-hydroxy- [708-06-5];
2-HYDROXY-1-NAPHTHALDEHYDE, III, 463

1-Naphthalenecarboxylic acid [86-55-5];
α-NAPHTHOIC ACID, II, 425

2-Naphthalenecarboxylic acid [93-09-4];
β-NAPHTHOIC ACID, II, 428

2-Naphthalenecarboxylic acid, 3-amino- [5959-52-4];
3-AMINO-2-NAPHTHOIC ACID, III, 78

2-Naphthalenecarboxylic acid, 3,4-dihydro-6,7-dimethoxy- [53684-50-7];
6,7-DIMETHOXY-3,4-DIHYDRO-2-NAPHTHOIC ACID, III, 300

1-Naphthalenecarboxylic acid, ethyl ester [3007-97-4];
ETHYL α-NAPHTHOATE, II, 282

Naphthalene, 1-(chloromethyl)- [86-52-2];
1-CHLOROMETHYLNAPHTHALENE, III, 195

2,3-Naphthalenedicarboxylic acid [2169-87-1];
2,3-NAPHTHALENEDICARBOXYLIC ACID, V, 810

2,6-Naphthalenedicarboxylic acid [1141-38-4];
2,6-NAPHTHALENEDICARBOXYLIC ACID, V, 813

Naphthalene, 1,4-dinitro- [6921-26-2];
1,4-DINITRONAPHTHALENE, III, 341

1,3-Naphthalenediol [132-86-5];
NAPHTHORESORCINOL, III, 637

1,2-Naphthalenedione [524-42-5];
1,2-NAPHTHOQUINONE, II, 430

1,4-Naphthalenedione [130-15-4];
1,4-NAPHTHOQUINONE, I, 383; IV, 698

1,4-Naphthalenedione, 2-hydroxy- [83-72-7];
2-HYDROXY-1,4-NAPHTHOQUINONE, III, 465

1,6(2H,7H)-Naphthalenedione, 3,4,8,8a-tetrahydro-8a-methyl- [20007-72-1];
1,6-DIOXO-8a-METHYL-1,2,3,4,6,7,8,8a-OCTAHYDRONAPHTHALENE, V, 486

1,3-Naphthalenedisulfonic acid, 6,6'-[(3,3'-dimethyl[1,1'-biphenyl]-4,4'-diyl)bis(azo)]bis[4-amino-5-hydroxy-, tetrasodium salt [314-13-6];
COUPLING OF *o*-TOLUIDINE AND CHICAGO ACID, II, 145

1,5-Naphthalenedisulfonyl dichloride [1928-01-4];
NAPHTHALENE-1,5-DISULFONYL CHLORIDE, IV, 693

1,5-Naphthalenedithiol [5325-88-2];
1,5-NAPHTHALENEDITHIOL, IV, 695

Naphthalene, 1-isothiocyanato- [551-06-4];
α-NAPHTHYL ISOTHIOCYANATE, IV, 700

Naphthalene, 1-phenyl- [605-02-7];
1-PHENYLNAPHTHALENE, III, 729

1-Naphthalenesulfonic acid, 4-amino-3-hydroxy- [116-63-2];
1-AMINO-2-NAPHTHOL-4-SULFONIC ACID, II, 42

1-Naphthalenesulfonic acid, 3,4-dihydro-3,4-dioxo-, ammonium salt [53684-60-9];
1,2-NAPHTHOQUINONE-4-SULFONATE, AMMONIUM SALT, III, 633

1-Naphthalenesulfonic acid, 3,4-dihydro-3,4-dioxo-, potassium salt [5908-27-0];
1,2-NAPHTHOQUINONE-4-SULFONATE POTASSIUM SALT, III, 633

Naphthalene, 1,2,3,4-tetraphenyl- [751-38-2];
1,2,3,4-TETRAPHENYLNAPHTHALENE, V, 1037

1-Naphthalenol, 4-amino-, hydrochloride [5959-56-8];
1,4-AMINONAPHTHOL HYDROCHLORIDE, I, 49; II, 39

2-Naphthalenol, 1-amino-, hydrochloride [1198-27-2];
1,2-AMINONAPHTHOL HYDROCHLORIDE, II, 33

2-Naphthalenol, 1-(aminophenylmethyl)- [481-82-3];
β-NAPHTHOL PHENYLAMINOMETHANE, I, 381

2-Naphthalenol, 6-bromo- [15231-91-1]
6-BROMO-2-NAPHTHOL, III, 132

2-Naphthalenol, 6-methoxy- [5111-66-0];
PHENOLS: 6-METHOXY-2-NAPHTHOL, V, 918

2-Naphthalenol, 1-nitro- [550-60-7];
1-NITRO-2-NAPHTHOL, II, 451

2-Naphthalenol, 1-nitroso- [131-91-9];
NITROSO-β-NAPHTHOL, I, 411

1-Naphthalenol, 5,6,7,8-tetrahydro- [529-35-1];
ar-TETRAHYDRO-α-NAPHTHOL, IV, 887

1(2*H*)-Naphthalenone, 3,4-dihydro- [529-34-0];
α-TETRALONE, II, 569; III, 798; IV, 898

2(1*H*)-Naphthalenone, 3,4-dihydro- [530-93-8];
β-TETRALONE, IV, 903

2(3*H*)-Naphthalenone, 4,4a,5,6,7,8-hexahydro- [1196-55-0];
Δ$^{1(9)}$ OCTALONE-2, V, 869

Naphtho[1,2-*c*]furan-1,3-dione [5343-99-7];
1,2-NAPHTHALIC ANHYDRIDE, II, 423

Naphtho[1,2-*c*]furan-1,3-dione, 4,5-dihydro- [37845-14-0];
3,4-DIHYDRO-1,2-NAPHTHALIC ANHYDRIDE, II, 194

Naphtho[1,2-*d*]thiazol-2(1*H*)-imine, 1-methyl- [53663-31-3];
1-METHYL-2-IMINO-β-NAPHTHOTHIAZOLINE, III, 595

Nickel [7440-02-0];
CATALYST, RANEY NICKEL, W-2, III, 181

CATALYST, RANEY NICKEL, W-6, III, 176

Nitric acid, methyl ester [598-58-3];
METHYL NITRATE, II, 412

Nitridotricarbonic acid, triethyl ester [3206-31-3];
ETHYL *N*-TRICARBOXYLATE, III, 415

Nitrous acid, butyl ester [544-16-1];
n-BUTYL NITRITE, II, 108

Nitroxide, bis(1,1-dimethylethyl)- [2406-25-9];
DI-*t*-BUTYL NITROXIDE, V, 355

Nonanedinitrile [1675-69-0];
AZELANITRILE, IV, 62

Nonanedioic acid [123-99-9];
AZELAIC ACID, II, 53

2,4-Nonanedione [6175-23-1];
2,4-NONANEDIONE, V, 848

4,6-Nonanedione, 2,8-dimethyl- [7307-08-6];
DIISOVALERYLMETHANE, III, 291

Nonane, 1,1,3-trichloro- [10575-86-7];
1,1,3-TRICHLORO-*n*-NONANE, V, 1076

Nonanoic acid [112-05-0];
PELARGONIC ACID, II, 174

Nonanoic acid, 9-cyano- [5810-19-5];
ω-CYANOPELARGONIC ACID, III, 768

Nonanoic acid, 9-cyano-, methyl ester [53663-26-6];
METHYL ω-CYANOPELARGONATE, III, 584

5-Nonanol [623-93-8];
DI-*n*-BUTYLCARBINOL, II, 179

24-Norcholan-23-oic acid, 3,12-dihydroxy-, (3α, 5β, 12α)- [53608-86-9];
nor-DESOXYCHOLIC ACID, III, 234

Norleucine [5157-09-5];
    α-AMINO-n-CAPROIC ACID, I, 48

O

9,12-Octadecadienoic acid (Z,Z)- [60-33-3];
    LINOLEIC ACID, III, 526

Octadecanedioic acid, diethyl ester [1472-90-8];
    ETHYL 1,16-HEXADECANEDICARBOXYLATE, III, 401

Octadecanedioic acid, dimethyl ester [1472-93-1];
    DIMETHYL OCTADECANEDIOATE, V, 463

Octadecanoic acid, 9,10-dihydroxy- [120-87-6];
    9,10-DIHYDROXYSTEARIC ACID, IV, 317

9,12,15-Octadecatrienoic acid, (Z,Z,Z)- [463-40-1];
    LINOLENIC ACID, III, 531

9-Octadecen-1-ol, (Z)- [143-28-2];
    OLEYL ALCOHOL, II, 468; III, 671

9-Octadecenoyl chloride, (Z)- [112-77-6];
    OLEOYL CHLORIDE, IV, 739

9-Octadecynoic acid [506-24-1];
    STEAROLIC ACID, III, 785; IV, 851

Octanal [124-13-0];
    OCTANAL, V, 872

Octane, 1-bromo- [111-33-1];
    n-OCTYL BROMIDE, I, 30

Octane, 2,7-dimethyl-2,7-dinitro- [53684-51-8];
    2,7-DIMETHYL-2,7-DINITROÖCTANE, V, 445

Octanenitrile, 8-oxo- [13050-09-4];
    7-CYANOHEPTANAL, V, 266

Octane, 1-nitro- [629-37-8];
    1-NITROÖCTANE, IV, 724

Octanoic acid, 7-methyl-4-oxo-, methyl ester [53663-32-4];
    METHYL 4-KETO-7-METHYLOCTANOATE, III, 601

Octanoic acid, 2-methyl-3-oxo-, 1-methylpropyl ester [53663-40-4];
    sec-BUTYL-α-n-CAPROYLPROPIONATE, IV, 120

2-Octanol [123-96-6];
    METHYL-n-HEXYLCARBINOL, I, 366

2-Octanol, (S)- [6169-06-8];
2-Octanol, (R)- [5978-70-1];
    d- and l-OCTANOL-2, I, 418

4-Octanone, 5-hydroxy- [496-77-5];
BUTYROIN, II, 114

2-Octenoic acid, 4-ethyl-2-methyl- [6975-97-9];
4-ETHYL-2-METHYL-2-OCTENOIC ACID, IV, 444

7-Oxabicyclo[4.1.0]heptane [286-20-4];
CYCLOHEXENE OXIDE, I, 185

2-Oxabicyclo[4.1.0]heptane, 7,7-dichloro-[7556-13-0];
2-OXA-7,7-DICHLORONORCARANE, V, 874

7-Oxabicyclo[4.1.0]heptan-2-one, 4,4,6-trimethyl- [10276-21-8];
ISOPHORONE OXIDE, IV, 552

1-Oxaspiro[2.5]octane [185-70-6];
METHYLENECYCLOHEXANE OXIDE, V, 755

1-Oxaspiro[2.5]octane-2-carboxylic acid, ethyl ester [6975-17-3];
ETHYL β,β-PENTAMETHYLENEGLYCIDATE, IV, 459

1,2-Oxathiane, 2,2-dioxide [1633-83-6];
4-HYDROXY-1-BUTANESULFONIC ACID SULTONE, IV, 529

Oxaziridine, 2-(1,1-dimethylethyl)-3-phenyl- [7731-34-2];
2-t-BUTYL-3-PHENYLOXAZIRANE, V, 191

5(4H)-Oxazolone, 4-[(3,4-dimethoxyphenyl)methylene]-2-phenyl- [541-56-5];
AZLACTONE OF α-BENZOYLAMINO-β-(3,4-DIMETHOXYPHENYL)-ACRYLIC ACID, II, 55

5(4H)-Oxazolone, 2-phenyl- [1199-01-5];
2-PHENYL-5-OXAZOLONE, V, 946

Oxepin, 2,7-dimethyl- [1487-99-6];
2,7-DIMETHYLOXEPIN, V, 467

Oxetane [503-30-0];
TRIMETHYLENE OXIDE, III, 835

2-Oxetanone, 3,3-dimethyl-4-(1-methylethylidene)- [3173-79-3];
DIMETHYLKETENE β-LACTONE DIMER, V, 456

2-Oxetanone, 4-methylene- [674-82-8]
KETENE DIMER, III, 508

Oxirane, 2.3-bis(2-chlorophenyl)- [53608-92-7];
2,2'-DICHLORO-α,α-EPOXYBIBENZYL, V, 358

Oxirane, (bromomethyl)- [3132-64-7];
EPIBROMOHYDRIN, II, 256

Oxiranecarboxylic acid, 3-methyl-3-phenyl-, ethyl ester [77-83-8];
PHENYLMETHYLGLYCIDIC ESTER, III, 727

Oxirane, (chloromethyl)- [106-89-8];
EPICHLOROHYDRIN, I, 233; II, 256

Oxirane, 2,3-diphenyl-, *trans*- [1439-07-2];
    trans-STILBENE OXIDE, IV, 860

Oxirane, phenyl- [96-09-3];
    STYRENE OXIDE, I, 494

Oxiranetetracarbonitrile [3189-43-3];
    TETRACYANOETHYLENE OXIDE, V, 1007

Oxonium, triethyl-, tetrafluoroborate (1-) [368-39-8];
    TRIETHYLOXONIUM FLUOBORATE, V, 1080

Oxonium, trimethyl-, salt with 2,4,6-trinitrobenzenesulfonic acid (1:1) [13700-00-0];
    TRIMETHYLOXONIUM 2,4,6-TRINITROBENZENESULFONATE, V, 1099

Oxonium, trimethyl-, tetrafluoroborate (1-) [420-37-1];
    TRIMETHYLOXONIUM FLUOBORATE, V, 1096

Ozone [10028-15-6];
    OZONE, III, 673

P

Palladium [7440-05-3];
    PALLADIUM CATALYSTS, III, 685

1,4-Pentadiene [591-93-5];
    1,4-PENTADIENE, IV, 746

1,4-Pentadien-3-one, 1,5-diphenyl- [538-58-9];
    DIBENZALACETONE, II, 167

Pentaerythrityl bromide [3229-00-3];
Pentaerythrityl iodide [1522-88-9];
    PENTAERYTHRITYL BROMIDE AND IODIDE, II, 476

1-Pentalenecarboxylic acid, octahydro-,)1α, 3$a\alpha$,6$a\alpha$)- [18209-43-3];
    exo-cis-BICYCLO[3.3.0]OCTANE-2-CARBOXYLIC ACID, V, 93

Pentanal, 5-hydroxy- [4221-03-8];
    5-HYDROXYPENTANAL, III, 470

Pentane [109-66-0];
    n-PENTANE, II, 478

Pentane, 1,5-dibromo- [111-24-0];
    PENTAMETHYLENE BROMIDE, I, 428; III, 692

Pentanedinitrile [544-13-8];
    TRIMETHYLENE CYANIDE, I, 536

Pentanedinitrile, 3-hydroxy- [13880-89-2];
    3-HYDROXYGLUTARONITRILE, V, 614

Pentanedioic acid [110-94-1];
    GLUTARIC ACID, I, 289; IV, 496

Pentanedioic acid, 3,3-dimethyl- [4839-46-7];
    β,β-DIMETHYLGLUTARIC ACID, IV, 345

Pentanedioic acid, 3-ethyl-3-methyl [5345-01-7];
β-ETHYL-β-METHYLGLUTARIC ACID, IV, 441

Pentanedioic acid, 3-methyl- [626-51-7];
β-METHYLGLUTARIC ACID, III, 591

Pentanedioic acid, 3-methyl-3-phenyl- [4160-92-3]
β-METHYL-β-PHENYLGLUTARIC ACID, IV, 664

Pentanedioic acid, 2-oxo- [328-50-7];
α-KETOGLUTARIC ACID, III, 510; V, 687

Pentanedioic acid, 2-oxo-, diethyl ester [5965-53-7];
ETHYL ETHOXALYLPROPINATE, II, 272

Pentanedioic acid, 3-oxo- [542-05-2];
ACETONEDICARBOXYLIC ACID, I, 10

Pentanedioic acid, 3-oxo-, diethyl ester [105-50-0];
ETHYL ACETONEDICARBOXYLATE, I, 237

1,5-Pentanediol [111-29-5];
1,5-PENTANEDIOL, III, 693

1,5-Pentanediol, 3-methyl- [4457-71-0];
3-METHYL-1,5-PENTANEDIOL, IV, 660

2,4-Pentanedione [123-54-6];
ACETYLACETONE, III, 16

2,4-Pentanedione, 3-methyl- [815-57-6];
3-METHYLPENTANE-2,4-DIONE, V, 785

Pentane, 2,2,4-trimethyl-4-nitro- [5342-78-9];
4-NITRO-2,2,4-TRIMETHYLPENTANE, V, 845

1,2,5-Pentanetriol [14697-46-2];
1,2,5-TRIHYDROXYPENTANE, III, 833

1,3,5-Pentanetrione, 1-(4-methoxyphenyl)-5-phenyl- [1678-17-7];
1-(p-METHOXYPHENYL)-5-PHENYL-1,3,5-PENTANETRIONE, V, 718

Pentanoic acid, 5-bromo-, methyl ester [5454-83-1];
METHYL 5-BROMOVALERATE, III, 578

Pentanoic acid, 2,4-dioxo-, ethyl ester [615-79-2];
ETHYL ACETOPYRUVATE, I, 238

Pentanoic acid, 3-methyl [105-43-1];
3-METHYLPENTANOIC ACID, II, 416

Pentanoic acid, 4-methyl-4-nitro-, methyl ester [16507-02-1];
METHYL γ-METHYL-γ-NITROVALERATE, IV, 652

Pentanoic acid, 4-oxo- [123-76-2];
LEVULINIC ACID, I, 335

3-Pentanol, 3-ethyl- [597-49-9];
TRIETHYLCARBINOL, II, 602

2-Pentanone, 4-amino-4-methyl-, ethanedioate (2:1) [53608-87-0];
DIACETONAMINE HYDROGEN OXALATE, I, 196

2-Pentanone, 4-hydroxy-4-methyl- [123-42-2];
DIACETONE ALCOHOL, I, 199

Pentanoyl chloride, 2,2,4-trimethyl-3-oxo- [10472-34-1];
2,2,4-TRIMETHYL-3-OXOVALERYL CHLORIDE, V, 1103

18-Pentatriacontanone [504-53-0];
STEARONE, V, 854

2-Pentene [109-68-2];
2-PENTENE, I, 430

2-Pentenoic acid, 2-cyano-3-ethyl-, ethyl ester [868-04-2];
ETHYL(1-ETHYLPROPYLIDENE)CYANOACETATE, III, 399

3-Pentenoic acid, 2-cyano-3-ethyl-2-methyl-, ethyl ester [53608-83-6];
ETHYL (1-ETHYLPROPENYL)METHYLCYANOACETATE, III, 397

3-Penten-2-ol [1569-50-2];
3-PENTEN-2-OL, III, 696

4-Penten-1-ol [821-9-0];
4-PENTEN-1-OL, III, 698

1-Penten-3-one, 4,4-dimethyl-1-phenyl- [538-44-3];
BENZALPINACOLONE, I, 81

3-Penten-2-one, 4-methyl- [141-79-7];
MESITYL OXIDE, I, 345

1-Penten-4-yn-3-ol, 1-phenyl- [14604-31-0];
1-PHENYL-1-PENTEN-4-YN-3-OL, IV, 792

4-Pentyn-1-ol [5390-04-5];
4-PENTYN-1-OL, IV, 755

Peroxide, bis(4-nitrobenzoyl) [1712-84-1];
p-NITROBENZOYL PEROXIDE, III, 649

Phenanthrene, 9-bromo- [573-17-1];
9-BROMOPHENANTHRENE, III, 134

9-Phenanthrenecarbonitrile [2510-55-6];
9-CYANOPHENANTHRENE, III, 212

9-Phenanthrenecarboxaldehyde [4707-71-5];
PHENANTHRENE-9-ALDEHYDE, III, 701

1-Phenanthrenecarboxylic acid, 1,2,3,4,4a,4b,5,6,10,10a-decahydro-1,4a-dimethyl-7-(1-methylethyl)-, [1R-(1α, 4aβ, 4bα, 10aα)]- [514-10-3];
ABIETIC ACID, IV, 1

1-Phenanthrenecarboxylic acid, 1,2,3,4,4a,4b,5,9,10,10a-decahydro-1,4a-dimethyl-7-(1-methylethyl)-, [1R-(1α, 4aβ, 4bα, 10aα)]- [79-54-9];
LEVOPIMARIC ACID, V, 699

4-Phenanthrenecarboxylic acid, 5-formyl- [5684-15-1];
  5-FORMYL-4-PHENANTHROIC ACID, IV, 484

Phenanthrene, 9,10-dihydro- [776-35-2];
  9,10-DIHYDROPHENANTHRENE,IV, 313

9,10-Phenanthrenedione [84-11-7];
  PHENANTHRENEQUINONE, IV, 757

Phenanthrene, 9-phenyl- [844-20-2];
  9-PHENYLPHENANTHRENE, V, 952

2-Phenanthrenesulfonic acid [41105-40-2];
3-Phenanthrenesulfonic acid [2039-95-4];
  2- and 3-PHENANTHRENESULPHONIC ACIDS, II, 482

Phenazinium, 1-hydroxy-5-methyl-, hydroxide, inner salt [85-66-5];
  PYOCYANINE, III, 753

Phenol, 2-amino-4-nitro- [99-57-0];
  2-AMINO-4-NITROPHENOL, III, 82

Phenol, 2-bromo- [95-56-7];
  o-BROMOPHENOL, II, 97

Phenol, 4-bromo- [106-41-2];
  p-BROMOPHENOL, I, 128

Phenol, 2-bromo-4-methyl- [6627-55-0];
  3-BROMO-4-HYDROXYTOLUENE, III, 130

Phenol, 2-(chloromethyl)-4-nitro- [2973-19-5];
  2-HYDROXY-5-NITROBENZYL CHLORIDE, III, 468

Phenol, 2,6-dibromo-4-nitro- [99-28-5];
  2,6-DIBROMO-4-NITROPHENOL, II, 173

Phenol, 2,6-dichloro- [87-65-0];
  2,6-DICHLOROPHENOL, III, 267

Phenol, 2-heptyl- [5284-22-0];
  o-n-HEPTYLPHENOL, III, 444

Phenol, 2-iodo- [533-58-4];
  o-IODOPHENOL, I, 326

Phenol, 4-iodo- [540-38-5];
  p-IODOPHENOL, II, 355

Phenol, 2-methoxy-4-methyl- [93-51-6];
  CREOSOL, IV, 203

Phenol, 2-methoxy-6-(2-propenyl)- [579-60-2];
  o-EUGENOL, III, 418

Phenol, 4-methyl- [106-44-5];
  p-CRESOL, I, 175

Phenol, 3-nitro- [554-84-7];
    m-NITROPHENOL, I, 404

Phenoxathiin [262-20-4]
    PHENOXTHIN, II, 485

DL-Phenylalanine [150-30-1];
    dl-PHENYLALANINE, III, 705
    dl-β-PHENYLALANINE, II, 489

Phosphine, (4-bromophenyl)diphenyl- [734-59-8];
    DIPHENYL-*p*-BROMOPHENYLPHOSPHINE, V, 496

Phosphinous chloride, bis(1-methylethyl)- [40244-90-4];
    CHLORODIISOPROPYLPHOSPHINE, V, 211

1*H*-Phosphole, 2,3-dihydro-4-methyl-1-phenyl-, 1-oxide [707-61-9];
    3-METHYL-1-PHENYLPHOSPHOLENE OXIDE, V, 787

Phosphonic acid, methyl, bis(1-methylethyl) ester [1445-75-6];
    DIISOPROPYL METHYLPHOSPHONATE, IV, 325

Phosphonic dichloride, (2-phenylethyl)- [4708-07-0];
    STYRYLPHOSPHONIC DICHLORIDE, V, 1005

Phosphonic dichloride, (trichloromethyl)- [21510-59-8];
    TRICHLOROMETHYLPHOSPHONYL DICHLORIDE, IV, 950

Phosphonium, ethenyltriphenyl-, bromide [5044-52-0];
    VINYL TRIPHENYLPHOSPHONIUM BROMIDE, V, 1145

Phosphonothioic dichloride, (chloromethyl)- [1983-27-3];
    CHLOROMETHYLPHOSPHONOTHIOIC DICHLORIDE, V,218

Phosphonous dichloride, phenyl- [644-97-3];
    PHENYLDICHLOROPHOSPHINE, IV, 784

Phosphorane, (dichloromethylene)triphenyl- [6779-08-4];
    DICHLOROMETHYLENETRIPHENYLPHOSPHORANE, V, 361

Phosphoric acid, tributyl ester [126-73-8];
    n-BUTYL PHOSPHATE, II, 109

Phosphorous acid, triethyl ester [122-52-1];
    TRIETHYL PHOSPHITE, IV, 955

Phosphorous triamide, hexamethyl [1608-26-0];
    HEXAMETHYLPHOSPHOROUS TRIAMIDE, V, 602

1,4-Phthalazinedione, 5-amino-2,3-dihydro- [521-31-3];
    5-AMINO-2,3-DIHYDRO-1,4-PHTHALAZINEDIONE, III, 69

1,4-Phthalazinedione, 2,3-dihydro-5-nitro- [3682-15-3];
    5-NITRO-2,3-DIHYDRO-1,4-PHTHALAZINEDIONE, III, 656

Piperazine, 1-(phenylmethyl)- [2759-28-6];
    1-BENZYLPIPERAZINE, V, 88

Piperidine, 1-benzoyl- [776-75-0];
  BENZOYL PIPERIDINE, I, 99

3,5-Piperidinedicarbonitrile, 4-methyl-2,6-dioxo-4-phenyl- [6936-95-4];
  β-METHYL-β-PHENYL-α,α'-DICYANOGLUTARIMIDE, IV, 662

2,6-Piperidinedione [1121-89-7];
  GLUTARIMIDE, IV, 496

Piperidine, 1-ethyl- [766-09-6];
  N-ETHYLPIPERIDINE, V, 575

Piperidine, 1-(2-naphthalenyl)- [5465-85-0];
  N-β-NAPHTHYLPIPERIDINE, V, 816

Piperidine, 1-phenyl- [4096-20-2];
  1-PHENYLPIPERIDINE, IV, 795

Platinum oxide (PtO₂) [1314-15-4];
  PLATINUM CATALYST FOR REDUCTIONS, I, 463

Pregn-5-en- 18-oic acid, 3-(acetyloxy)-20-hydroxy-, γ-lactone, (3β, 20R)- [3020-10-8];
  18,20-LACTONE OF 3β-ACETOXY-20β-HYDROXY-5-PREGNENE-18-OIC ACID, V,
  692

1,2-Propadiene [463-49-0];
  ALLENE, V, 22

Propanal [123-38-6];
  PROPIONALDEHYDE, II, 541

Propanal, 2,3-dihydroxy, (+)- [56-82-6];
  dl-GLYCERALDEHYDE, II, 305

Propanal, 2-oxo-, 1-(phenylhydrazone) [5391-74-2]
  METHYLGLYOXAL-ω-PHENYLHYDRAZONE, IV, 633

Propanamide, 2-hydroxy- [2043-43-8];
  LACTAMIDE, III, 516

Propanamide, 2-methyl- [563-83-7];
  ISOBUTYRAMIDE, III, 490

2-Propanamine, 2-methyl-, [75-64-9];
  t-BUTYLAMINE, III, 148

2-Propanamine, 2-methyl, hydrochloride [10017-37-5];
  t-BUTYLAMINE HYDROCHLORIDE, III, 153

1-Propanaminium, 3-[(ethylcarbonimidoyl)amino]-N,N,N-trimethyl-, iodide [22572-40-3];
  1-ETHYL-3-(3-DIMETHYLAMINO)PROPYLCARBODIIMIDE METHIODIDE, V,
  555

Propane, 1-bromo- [106-94-5];
  n-PROPYL BROMIDE, I, 37

Propane, 2-bromo- [75-26-3];
  ISOPROPYL BROMIDE, I, 37

Propane, 1-bromo-2-methyl- [78-77-3];
ISOBUTYL BROMIDE, II, 358

Propane, 2-bromo-2-methyl- [507-19-7];
t-BUTYL BROMIDE, I, 38

Propane, 3-chloro-1,1-diethoxy- [35573-93-4];
β-CHLOROPROPIONALDEHYDE ACETAL, II, 137

Propane, 2-chloro-2-methyl- [507-20-0];
t-BUTYL CHLORIDE, I, 144

Propanedial, nitro-, ion(1-), sodium, monohydrate [53821-72-0];
SODIUM NITROMALONALDEHYDE MONOHYDRATE, IV, 844; (Warning), V,
1004

1,3-Propanediamine, N,N-dibutyl- [102-83-0];
γ-DI-nBUTYLAMINOPROPYLAMINE, III, 256

1,3-Propanediamine, N'-(ethylcarbonimidoyl)-N,N-dimethyl-, monohydrochloride
[25952-53-8];
1-ETHYL-3-(3-DIMETHYLAMINO)PROPYLCARBODIIMIDE HYDROCHLORIDE,
V, 555

Propane, 1,3-dibromo- [109-64-8];
TRIMETHYLENE BROMIDE, I, 30

Propane, 1,3-dibromo-2,2-bis(bromomethyl)- [3229-00-3];
PENTAERYTHRITYL TETRABROMIDE, IV, 753

Propanedinitrile [109-77-3];
MALONONITRILE, II, 379; III, 535

Propanedinitrile, amino-, mono(4-methylbenzenesulfonate) [5098-14-6];
AMINOMALONONITRILE p-TOLUENESULFONATE, V, 32

Propanedinitrile, 1,3-dioxolan-2-ylidene- [5694-65-5];
DICYANOKETENE ETHYLENE ACETAL, IV, 276

Propanedioic acid [141-82-2];
MALONIC ACID, II, 376

Propanedioic acid, (acetylamino)-, diethyl ester [1068-90-2];
DIETHYL ACETAMIDOMALONATE, V, 373

Propanedioic acid, acetyl-, diethyl ester [570-08-1];
ETHYL ACETOSUCCINATE, II, 262

Propanedioic acid, amino-, diethyl ester, hydrochloride [13433-00-6];
DIETHYL AMINOMALONATE HYDROCHLORIDE, V, 376

Propanedioic acid, benzoyl-, diethyl ester [1087-97-4];
DIETHYL BENZOYLMALONATE, IV, 285

Propanedioic acid, (benzoyloxy)ethyl-, diethyl ester [6259-78-5];
DIETHYL[O-BENZOYL]ETHYLTARTRONATE, V, 379

Propanedioic acid, bis(1,1-DIMETHYLETHYL) ESTER [541-16-2];
DI-t-BUTYL MALONATE, IV, 261

Propanedioic acid, bis(hydroxymethyl)-, diethyl ester [20605-01-0];
DIETHYL BIS(HYDROXYMETHYL)MALONATE, V, 381

Propanedioic acid, bromo-, diethyl ester [685-870];
ETHYL BROMOMALONATE, I, 245

Propanedioic acid, butyl, diethyl ester [133-08-4];
ETHYL n-BUTYLMALONATE, I, 250

Propanedioic acid, 2-cyclopenten-1-yl-, diethyl ester [53608-93-8];
DIETHYL Δ^2-CYCLOPENTENYLMALONATE, IV, 291

Propanedioic acid, (1,3-dihydro-1,3-dioxo-2H-isoindol-2-yl)-diethyl ester, [5680-61-5];
ETHYL PHTHALIMIDOMALONATE, I, 271

Propanedioic acid, 1,1-dimethylethyl ethyl ester [32864-38-3];
ETHYL $tert$-BUTYL MALONATE, IV, 417

Propanedioic acid, (ethoxymethylene)-, diethyl ether [87-13-8];
ETHYL ETHOXYMETHYLENEMALONATE, III, 395

Propanedioic acid, ethylidene-, diethyl ester [1462-12-0];
DIETHYL ETHYLIDENEMALONATE, IV, 293

Propanedioic acid, hexadecyl-, diethyl ester [41433-81-2];
CETYLMALONIC ESTER, IV, 141

Propanedioic acid, methyl-, diethyl ester [609-08-5];
ETHYL METHYLMALONATE, II, 279

Propanedioic acid, methylene-, diethyl ester [3377-20-6];
DIETHYL METHYLENEMALONATE, IV, 298

Propanedioic acid, oxo-, monoethyl ester [27728-17-2];
ETHYL OXOMALONATE, I, 266

Propanedioic acid, phenyl-, diethyl ester [83-13-6];
ETHYL PHENYLMALONATE, II, 288

Propanedioic acid, (phenylmethylene)-, diethyl ester [5292-53-5];
ETHYL BENZALMALONATE, III, 377

1,2-Propanediol, (R)- [4254-14-2];
l-PROPYLENE GLYCOL, II, 545

1,3-Propanediol, 2,2-bis(hydroxymethyl)- [115-77-5];
PENTAERYTHRITOL, I, 425

1,3-Propanediol, 2-(bromomethyl)-2-(hydroxymethyl)- [19184-65-7];
MONOBROMOPENTAERYTHRITOL, IV, 681

1,2-Propanediol, 3-chloro- [96-24-2];
GLYCEROL α-MONOCHLOROHYDRIN, I, 294

1,2-Propanediol, 3,3-diethoxy-, ($\pm$)- [10487-05-5];
dl-GLYCERALDEHYDE DIETHYL ACETAL, II, 307

1,2-Propanediol, 3-phenoxy- [538-43-2];
  α-GLYCERYL PHENYL ETHER, I, 296

1,3-Propanedione, 1,3-diphenyl- [120-46-7];
  DIBENZOYLMETHANE, I, 205; III, 251

1,2-Propanedione, 1-phenyl- [579-07-7];
  ACETYLBENZOYL, III, 20

1,2-Propanedione, 1-phenyl-, 2-oxime [119-51-7];
  ISONITROSOPROPIOPHENONE, II, 363

Propane, 1,1,1,2,2,3,3-heptachloro- [594-89-8];
  unsym-HEPTACHLOROPROPANE, II, 312

Propanenitrile, 3-amino- [151-18-8];
  β-AMINOPROPIONITRILE, III, 93

Propanenitrile, 3-[(2-chlorophenyl)amino]- [94-89-3];
  3-(o-CHLOROANILINO)PROPIONITRILE, IV, 146

Propanenitrile, 3-ethoxy- [2141-62-0];
  β-ETHOXYPROPIONITRILE, III, 372

Propanenitrile, 3-hydroxy- [109-78-4];
  ETHYLENE CYANOHYDRIN, I, 256

Propanenitrile, 3-[(2-hydroxyethyl)thio]- [15771-37-6];
  β-(2-HYDROXYETHYLMERCAPTO)PROPIONITRILE, III, 458

Propanenitrile, 2-hydroxy-2-methyl- [75-86-5];
  ACETONE CYANOHYDRIN, II, 7

Propanenitrile, 3,3'-iminobis- [111-94-4];
  bis-(β-CYANOETHYL)AMINE, III, 93

Propanenitrile, 2-methyl- [78-82-0];
  ISOBUTYRONITRILE, III, 493

Propanenitrile, 3-(phenylamino)- [1075-76-9];
  N-2-CYANOETHYLANILINE, IV, 205

1,1,2,3-Propanetetracarboxylic acid, tetraethyl ester [635-03-0];
  ETHYL PROPANE-1,1,2,3-TETRACARBOXYLATE, I, 272

Propane, 1,1-thiobis- [111-47-7];
  n-PROPYL SULFIDE, II, 547

Propane, 1,2,3-tribromo- [96-11-7];
  1,2,3-TRIBROMOPROPANE, I, 521

1,2,3-Propanetricarboxylic acid [99-14-9];
  TRICARBALLYLIC ACID, I, 523

Propane, 1,1,3-triethoxy- [7789-92-6];
  β-ETHOXYPROPIONALDEHYDE ACETAL, III, 371

Propanetrione, diphenyl- [643-75-4];
  DIPHENYL TRIKETONE, II, 244

Propanoic acid, 3-bromo-, [590-92-1];
  β-BROMOPROPIONIC ACID, I, 131

Propanoic acid, 3-bromo-, ethyl ester [539-74-2];
  ETHYL β-BROMOPROPIONATE, I, 246

Propanoic acid, 3-bromo-, methyl ester [3395-91-3];
  METHYL β-BROMOPROPIONATE, III, 576

Propanoic acid, 3-chloro- [107-94-8];
  β-CHLOROPROPIONIC ACID, I, 166

Propanoic acid, 3-chloro-2,2,3-trifluoro- [425-97-8];
  3-CHLORO-2,2,3-TRIFLUOROPROPIONIC ACID, V, 239

Propanoic acid, 2,2-dimethyl- [75-98-9];
  TRIMETHYLACETIC ACID, I, 524

Propanoic acid, 3-hydroxy- [503-66-2];
  β-HYDROXYPROPIONIC ACID, I, 321

Propanoic acid, 2-hydroxy-, 1-methylethyl ester [617-51-6];
  ISOPROPYL LACTATE, II, 365

Propanoic acid, 2-hydroxy-, 2-propenyl ester [5349-55-3];
  ALLYL LACTATE, III, 46

Propanoic acid, 2-oxo- [127-17-3];
  PYRUVIC ACID, I, 475

Propanoic acid, 2-oxo-, ethyl ester [617-35-6];
  ETHYL PYRUVATE, IV, 467

Propanoic acid, 2-oxo-, methyl ester [600-22-6];
  METHYL PYRUVATE, III, 610

Propanoic acid, 2-[[(2,4,5,7-tetranitro-9H-fluoren-9-ylidene)amino]oxy]-, (+)- [50996-73-1];
Propanoic acid, 2-[[(2,4,5,7-tetranitro-9H-fluoren-9-ylidene)amino]oxy]-, (−)- [50874-31-2];
  (+)-AND (−)-α-(2,4,5,7-TETRANITRO-9-FLUORENYLIDENEAMINOOXY)-
  PROPIONIC ACID, V, 1031

Propanoic acid, 3,3'-thiobis-, dimethyl ester [4131-74-2];
  METHYL β-THIODIPROPIONATE, IV, 669

1-Propanol, 3-chloro- [627-30-5];
  TRIMETHYLENE CHLOROHYDRIN, I, 533

1-Propanol, 3-chloro-, acetate [628-09-1];
  γ-CHLOROPROPYL ACETATE, III, 203

2-Propanol, 1,3-dibromo- [96-21-9];
  GLYCEROL α,γ-DIBROMOHYDRIN, II, 308

2-Propanol, 1,3-dichloro- [96-23-1];
  GLYCEROL α,γ-DICHLOROHYDRIN, I, 292

1-Propanol, 2,2-dimethyl- [75-84-3];
  NEOPENTYL ALCOHOL, V, 818

2-Propanol, 2-methyl, aluminum salt [556-91-2];
ALUMINUM *t*-BUTOXIDE, III, 48

2-Propanone, 1-bromo- [598-31-2];
BROMOACETONE, II, 88

2-Propanone, 1-chloro-3-phenyl- [937-38-2];
BENZYL CHLOROMETHYL KETONE, III, 119

2-Propanone, 1-(4-chlorophenyl)-3-phenyl- [35730-03-1];
1-(*p*-CHLOROPHENYL)-3-PHENYL-2-PROPANONE, IV, 176

2-Propanone, 1,3-dichloro- [534-07-6];
α,γ-DICHLOROACETONE, I, 211

1-Propanone, 3-(dimethylamino)-1-phenyl- hydrochloride [879-72-1];
β-DIMETHYLAMINOPROPIOPHENONE HYDROCHLORIDE, III, 305

1-Propanone, 1,3-diphenyl- [1083-30-3];
BENZYLACETOPHENONE, I, 101

2-Propanone, 1,1-diphenyl- [781-35-1];
α,α-DIPHENYLACETONE, III, 343

2-Propanone, 1-hydroxy- [116-09-6];
ACETOL, II, 5

1-Propanone, 1-(2-hydroxyphenyl)- [610-99-1];
1-Propanone, 1-(4-hydroxyphenyl)- [70-70-2];
o-PROPIOPHENOL AND *p*-PROPIOPHENOL, II, 543

2-Propanone, 1-(2-methoxyphenyl)- [5211-62-1];
o-METHOXYPHENYLACETONE, IV, 573

2-Propanone, 1-phenyl- [103-79-7];
METHYL BENZYL KETONE, II, 389

1-Propanone, 1,3,3-triphenyl- [606-86-0];
β,β-DIPHENYLPROPIOPHENONE, II, 236

2-Propenal [107-02-8];
ACROLEIN, I, 15

2-Propenal, 3-(2-nitrophenyl)- [1466-88-2];
o-NITROCINNAMALDEHYDE, IV, 722

2-Propenamide, 2-methyl- [79-39-0];
METHACRYLAMIDE, III, 560

2-Propenamide, *N*-(phenylmethyl)- [13304-62-6];
N-BENZYLACRYLAMIDE, V, 73

2-Propen-1-amine [107-11-9];
ALLYLAMINE, II, 24

2-Propen-1-amine, 2-bromo- [6943-51-7];
2-BROMOALLYLAMINE, V, 121

2-Propen-1-amine, 2-bromo-*N*-ethyl- [871-23-8];
    N-(2-BROMOALLYL)ETHYLAMINE, V, 124

2-Propen-1-amino, *N*-2-propenyl- [124-02-7];
    DIALLYLAMINE, I, 201

1-Propene, 3-bromo- [106-95-6];
    ALLYL BROMIDE, I, 27

1-Propene, 2,3-dibromo- [513-31-5];
    2,3-DIBROMOPROPENE, I, 209

1-Propene, 3,3-diethoxy- [3054-95-3];
    ACROLEIN ACETAL, II, 17; IV, 21

2-Propenenitrile, 3-(2-furanyl)- [7187-01-1];
    3-(2-FURYL)ACRYLONITRILE, V, 585

1-Propene-1,2,3-tricarboxylic acid [499-12-7];
    ACONITIC ACID, II, 12

2-Propenoic acid [79-10-7];
    ACRYLIC ACID III, 30

2-Propenoic acid, 2-(acetylamino)-3-phenyl- [5469-45-4];
    α-ACETAMINOCINNAMIC ACID, II, 1

2-Propenoic acid, butyl ester [141-32-2];
    n-BUTYL ACRYLATE, III, 146

2-Propenoic cid, 2-cyano-3-phenyl- [1011-92-3];
    α-CYANO-β-PHENYLACRYLIC ACID, I, 181

2-Propenoic acid, 3-(2,3-dimethoxyphenyl)- [7461-60-1];
    2,3-DIMETHOXYCINNAMIC ACID, IV, 327

2-Propenoic acid, 3-(2-furanyl)- [539-47-9];
    FURYLACRYLIC ACID, III, 425

2-Propenoic acid, 3-(3-nitrophenyl)- [555-68-0]
    m-NITROCINNAMIC ACID, I, 398

2-Propenoic acid, 3-phenyl-, ethyl ester [103-36-6];
    ETHYL CINNAMATE, I, 252

2-Propenoic acid, 3-phenyl-, phenyl ester [2757-04-2];
    PHENYL CINNAMATE, III, 714

2-Propen-1-ol [107-18-6];
    ALLYL ALCOHOL, I, 42

2-Propen-1-one, 1,3-diphenyl- [94-41-7];
    BENZALACETOPHENONE, I, 78

1-Propen-1-one, 2-methyl- [598-26-5];
    DIMETHYLKETENE, IV, 348

2-Propenone, oxime [127-06-0];
    ACETOXIME, I, 318

2-Propynal [624-67-9];
PROPIOLALDEHYDE, IV, 813

2-Propynal, 3-phenyl- [2579-22-8];
PHENYLPROPARGYL ALDEHYDE, III, 731

2-Propynoic acid, 3-phenyl- [637-44-5];
PHENYLPROPIOLIC ACID, II, 515

2H-Pyran-3-carboxylic acid, 4,6-dimethyl-2-oxo- [33953-26-3];
2H-Pyran-5-carboxylic acid, 4,6-dimethyl-2-oxo-, ethyl ester [3385-34-0];
ISODEHYDROACETIC ACID AND ETHYL ISODEHYDROACETATE, IV, 549

2H-Pyran-5-carboxylic acid, 2-oxo- [500-05-0];
COUMALIC ACID, IV, 201

4H-Pyran-2,6-dicarboxylic acid, 4-oxo- [99-32-1];
CHELIDONIC ACID, II, 126

2H-Pyran, 3,4-dihydro- [110-87-2];
2,3-DIHYDROPYRAN, III, 276

2H-Pyran, 3,4-dihydro-2-methoxy-4-methyl- [53608-95-0];
3,4-DIHYDRO-2-METHOXY-4-METHYL-2H-PYRAN, IV, 311

2H-Pyran-2,4(3H)-dione, 3-acetyl-6-methyl- [520-45-6];
DEHYDROACETIC ACID, III, 231

2H-Pyran-2,6(3H)-dione, dihydro-4-methyl- [4166-53-4];
β-METHYLGLUTARIC ANHYDRIDE, IV, 630

2H-Pyran-2,6(3H)-dione, dihydro-3-phenyl- [2959-96-8];
α-PHENYLGLUTARIC ANHYDRIDE, IV, 790

2H-Pyran-2-one [504-31-4];
α-PYRONE, V, 982

2H-Pyran-2-one, 4,6-dimethyl- [675-09-2];
4,6-DIMETHYLCOUMALIN, IV, 337

4H-Pyran-4-one, 2,6-dimethyl-3,5-diphenyl- [33731-54-3];
2,6-DIMETHYL-3,5-DIPHENYL-4H-PYRAN-4-ONE, V, 450

4H-Pyren-4-one, 2-(4-methoxyphenyl)-6-phenyl- [14116-43-9];
2-(p-METHOXYPHENYL)-6-PHENYL-4-PYRONE, V, 721

2H-Pyran-2-one, tetrahydro-4-methyl- [1121-84-2];
β-METHYL-δ-VALEROLACTONE, IV, 677

2H-Pyran, tetrahydro- [142-68-7];
TETRAHYDROPYRAN, III, 794

2,3-Pyrazinedicarboxylic acid [89-01-0];
2,3-PYRAZINEDICARBOXYLIC ACID, IV, 824

1H-Pyrazol-3-amine [1820-80-0];
3(5)-AMINOPYRAZOLE, V, 39

1*H*-Pyrazole, 3,5-dimethyl- [67-51-6];
3,5-DIMETHYLPYRAZOLE, IV, 351

3*H*-Pyrazol-3-one, 5-amino-2,4-dihydro-2-phenyl- [4149-06-8];
1-PHENYL-3-AMINO-5-PYRAZOLONE, III, 708

Pyrene 1-bromo- [1714-29-0];
3-BROMOPYRENE V, 147

1-Pyrenol [5315-79-7];
3-HYDROXYPYRENE, V, 632

3-Pyridinamine [462-08-8];
3-AMINOPYRIDINE, IV, 45

2-Pyridinamine, *N*-(phenylmethyl)- [6935-27-9];
2-BENZYLAMINOPYRIDINE, IV, 91

2-Pyridineacetic acid, ethyl ester [2739-98-2];
ETHYL 2-PYRIDYLACETATE, III, 413

Pyridine, 2-bromo- [109-04-6];
2-BROMOPYRIDINE, III, 136

3-Pyridinecarbonitrile [100-54-9];
NICOTINONITRILE, IV, 706

3-Pyridinecarbonitrile, 2-chloro- [6602-54-6];
2-CHLORONICOTINONITRILE, V, 166

3-Pyridinecarbonitrile, 1,2-dihydro-6-methyl-2-oxo- [4241-27-4];
3-CYANO-6-METHYL-2(1)-PYRIDONE, IV, 210

2-Pyridinecarbonitrile, 6-methyl- [1620-75-3];
2-CYANO-6-METHYLPYRIDINE, IV, 269

3-Pyridinecarboxamide, 1-oxide [1986-81-8];
NICOTINAMIDE-1-OXIDE, IV, 704

3-Pyridinecarboxylic acid [59-67-6];
NICOTINIC ACID, I, 385

3-Pyridinecarboxylic acid, anhydride [16837-38-0];
NICOTINIC ANHYDRIDE, V, 822

3-Pyridinecarboxylic acid, 1,6-dihydro-6-oxo- [5006-66-6];
6-HYDROXYNICOTINIC ACID, IV, 532

2-Pyridinecarboxylic acid, hydrochloride [636-80-6];
PICOLINIC ACID HYDROCHLORIDE, III, 740

2,3-Pyridinediamine [452-58-4];
2,3-DIAMINOPYRIDINE, V, 346

Pyridine, 2,6-dimethyl- [108-48-5];
2,6-DIMETHYLPYRIDINE, II, 214

2-Pyridineethanol, α-methyl- [5307-19-7];
1-(α-PYRIDYL)-2-PROPANOL, III, 757

Pyridine, 4-ethyl- [536-75-4];
  4-ETHYLPYRIDINE, III, 410

Pyridine, 5-ethyl-2-methyl- [104-90-5];
  5-ETHYL-2-METHYLPYRIDINE, IV, 451

Pyridine, 3-methyl-4-nitro, 1-oxide [1074-98-2];
  3-METHYL-4-NITROPYRIDINE-1-OXIDE, IV, 654

Pyridine, 2-phenyl- [1008-89-5];
  2-PHENYLPYRIDINE, II, 517

Pyridine, 1-oxide [694-59-7];
  PYRIDINE-N-oxide, IV, 828

4-Pyridinesulfonic acid [5402-20-0];
  4-PYRIDINESULFONIC ACID, V, 977

Pyridinium, 1-amino-, iodide [6295-87-0];
  1-AMINOPYRIDINIUM IODIDE, V, 43

2(1H)-Pyridinone, 1-methyl- [694-85-9];
  1-METHYL-2-PYRIDONE, II, 419

2-Pyrimidinamine, N,N-dimethyl- [5621-02-3];
  2-(DIMETHYLAMINO)PYRIMIDINE, IV, 336

4-Pyrimidinamine, 2,6-dimethyl- [461-98-3];
  4-AMINO-2,6-DIMETHYLPYRIMIDINE, III, 71

5-Pyrimidinecarbonitrile, 1heptyl-1,2,3,6-tetrahydro-6-imino-2-oxo- [53608-90-5];
  3-n-HEPTYL-5-CYANOCYTOSINE, IV, 515

5-Pyrimidinecarbonitrile, 1,2,3,4-tetrahydro-4-oxo-2-thioxo- [23945-49-5];
  2-MERCAPTO-4-HYDROXY-5-CYANOPYRIMIDINE, IV, 566

5-Pyrimidinecarboxylic acid, 4-amino-1,2-dihydro-2-thioxo-, ethyl ester [774-07-2];
  2-MERCAPTO-4-AMINO-5-CARBETHOXYPYRIMIDINE, IV, 566

Pyrimidine, 2-chloro- [1722-12-9];
  2-CHLOROPYRIMIDINE, IV, 182

2,4(1H, 3H)-Pyrimidinedione, 4,5-diamino-, monohydrochloride [53608-89-2];
  DIAMINOURACIL HYROCHLORIDE, IV, 247

2,4(1H, 3H)-Pyrimidinedione, 6-methyl- [626-48-2];
  6-METHYLURACIL, II, 422

Pyrimidine, 4-methyl- [3438-46-8];
  4-METHYLPYRIMIDINE, V, 794

2(1H)-Pyrimidinethione [1450-85-7];
  2-MERCAPTOPYRIMIDINE, V, 703

2,4,6(1H, 3H, 5H)-Pyrimidinetrione [67-52-7];
  BARBITURIC ACID, II, 60

2,4,6(1H, 3H, 5H)-Pyrimidinetrione, 5-amino- [118-78-5];
  URAMIL, II, 617

2,4,6(1*H*, 3*H*, 5*H*)-Pyrimidinetrione, 5,5-dihydroxy- [3237-50-1];
   ALLOXAN MONOHYDRATE, III, 37; IV, 23

2,4,6(1*H*, 3*H*, 5*H*)-Pyrimidinetrione, 5-nitro- [480-68-2];
   NITROBARBITURIC ACID, II, 440

4(1*H*)-Pyrimidinone, 2,6-diamino- [56-06-4];
   2,4-DIAMINO-6-HYDROXYPYRIMIDINE, IV, 245

4(1*H*)-Pyrimidinone, 6-methyl- [3524-87-6];
   4-METHYL-6-HYDROXYPYRIMIDINE, IV, 638

1*H*-Pyrrole [109-97-7];
   PYRROLE, I, 473

1*H*-Pyrrole-2-carboxaldehyde [1003-29-8];
   2-PYRROLEALDEHYDE, IV, 831

1*H*-Pyrrole-2-carboxylic acid, 3,5-dimethyl-, ethyl ester [2199-44-2];
   2,4-DIMETHYL-5-CARBETHOXYPYRROLE, II, 198

1*H*-Pyrrole-2,4-dicarboxylic acid, 3,5-dimethyl-, diethyl ester [2436-79-5];
   2,4-DIMETHYL-3,5-DICARBETHOXYPYRROLE, II, 202

1*H*-Pyrrole, 2,3-dihydro-1-methyl-5-(methylthio)- [25355-52-6];
   2-METHYLMERCAPTO-*N*-METHYL-Δ^2-PYRROLINE, V, 780

1*H*-Pyrrole, 2,4-dimethyl- [625-82-1];
   2,4-DIMETHYLPYRROLE, II, 217

1*H*-Pyrrole, 2,5-dimethyl- [625-84-3];
   2,5-DIMETHYLPYRROLE, II, 219

1*H*-Pyrrole-2,5-dione, 1-phenyl- [941-69-5];
   N-PHENYLMALEIMIDE, V, 944

1*H*-Pyrrole, 2,4-diphenyl- [3274-56-4];
   2,4-DIPHENYLPYRROLE, III, 358

1*H*-Pyrrole, 3-ethyl-2,4-dimethyl- [517-22-6];
   KRYPTOPYRROLE, III, 513

1*H*-Pyrrole, 2,3,4,5-tetramethyl- [1003-90-3];
   2,3,4.5-TETRAMETHYLPYRROLE, V, 1022

1-Pyrrolidineacetic acid, α-methyl-, ethyl ester [26846-86-6];
   ETHYL α-(1-PYRROLIDYL)PROPIONATE, IV, 466

Pyrrolidine, 1-butyl- [767-10-2];
   1-*n*-BUTYLPYRROLIDINE, III, 159

Pyrrolidine, 2,2-dimethyl- [35018-15-6];
   2,2-DIMETHYLPYRROLIDINE, IV, 354

2,5-Pyrrolidinedione [123-56-8];
   SUCCINIMIDE, II, 562

2,5-Pyrrolidinedione, 1-iodo- [516-12-1];
   N-IODOSUCCINIMIDE, V, 663

1-Pyrrolidineethanol, β-methyl- [53663-19-7];
  2-(1-PYRROLIDYL)PROPANOL, IV, 834

2-Pyrrolidinone, 1,5-dimethyl- [5075-92-3];
  1,5-DIMETHYL-2-PYRROLIDONE, III, 328

2-Pyrrolidinone, 5,5-dimethyl- [5165-28-6];
  5,5-DIMETHYL-2-PYRROLIDONE, IV, 357

2H-Pyrrolo[3,4-d]isoxazole-4,6(3H, 5H)-dione, dihydro-2,5-diphenyl-3-propyl- [53555-71-8];
  2-PHENYL-3-n-PROPYLISOXAZOLIDINE-4,5-cis-DICARBOXYLIC ACID N-PHEN-
  YLIMIDE, V, 957

Pyrylium, 2,4,6-trimethyl-, perchlorate [940-93-2];
  2,4,6-TRIMETHYLPYRYLIUM PERCHLORATE, V, 1106

Pyrylium, 2,4,6-trimethyl-, salt with trifluoromethanesulfonic acid (1:1) [40927-60-4];
  2,4,6-TRIMETHYLPYRYLIUM TRIFLUOROMETHANESULFONATE, V, 1114

Pyrylium, 2,4,6-trimethyl-, tetrafluoroborate(1-) [773-01-3];
  2,4,6-TRIMETHYLPYRYLIUM TETRAFLUOROBORATE, V, 1112

Pyrylium, 2,4,6-triphenyl-, tetrafluoroborate(1-) [448-61-3];
  2,4,6-TRIPHENYLPYRYLIUM TETRAFLUOROBORATE, V, 1135

Q

2,4(1H, 3H)-Quinazolinedione [86-96-4];
  BENZOYLENE UREA, II, 79

Quinoline [91-22-5];
  QUINOLINE, I, 478

4-Quinolinecarboxylic acid, 1,2-dihydro-2-oxo- [15733-89-8];
  2-HYDROXYCINCHONINIC ACID, III, 456

Quinoline, 2-chloro-4-methyl- [634-47-9];
  2-CHLOROLEPIDINE, III, 194

Quinoline, 4,7-dichloro- [86-98-6];
  4,7-DICHLOROQUINOLINE, III, 272

Quinoline, 2,4-dimethyl- [1198-37-4];
  2,4-DIMETHYLQUINOLINE, III, 329

Quinoline, 6-methoxy-8-nitro- [85-81-4];
  6-METHOXY-8-NITROQUINOLINE, III, 568

Quinoline, 4-methyl- [491-35-0];
  LEPIDINE, III, 519

3-Quinolinol [580-18-7];
  3-HYDROXYQUINOLINE, V, 635

4-Quinolinol, 2-methyl- [607-67-0];
  2-METHYL-4-HYDROXYQUINOLINE, III, 593

2(1*H*)-Quinolinone, 4-methyl- [607-66-9];
    4-METHYLCARBOSTYRIL, III, 580

1,1':4',1":4",1''':4''',1''''-Quinquephenyl [3073-05-0];
    p-QUIQUEPHENYL, V, 985

R

Ruthenocene [1287-13-4];
    RUTHENOCENE, V, 1001

S

Selenium, chlorotriphenyl-, (*T*-4)- [17166-13-1];
    TRIPHENYLSELONIUM CHLORIDE, II, 240

Selenium, dichlorodiphenyl- [2217-81-4];
    DIPHENYLSELENIUM DICHLORIDE, II, 240

Selenourea, *N,N*-dimethyl- [5117-16-8];
    N,N-DIMETHYLSELENOUREA, IV, 359

DL-Serine [302-84-1];
    dl-SERINE, III, 774

Sodium amide [7782-92-5];
    SODIUM AMIDE, III, 778

Sodium, (triphenylmethyl)- [4303-71-3];
    TRIPHENYLMETHYLSODIUM, II, 607

Stannane, dibutyldiethenyl- [7330-43-0];
    DI-*n*-BUTYLDIVINYLTIN, IV, 258

Stannane, tetraethyl- [597-64-8];
    TETRAETHYLTIN, IV, 881

Stibine, triphenyl- [603-36-1];
    TRIPHENYLSTIBINE, I, 550

Sulfuric acid, dibutyl ester [625-22-9];
    n-BUTYL SULFATE, II, 111

Sulfur, trifluorophenyl- [672-36-6];
    PHENYLSULFUR TRIFLUORIDE, V, 959

Sulfurous acid, dibutyl ester [626-85-7];
    n-BUTYL SULFITE, II, 112

Sulfuryl chloride isocyanate [1189-71-5];
    CHLOROSULFONYL ISOCYANATE, V, 226

Sydnone, 3-phenyl- [120-06-9]
    3-PHENYLSYDNONE, V, 962

T

1,1':3',1"-Terphenyl, 2'-nitro-5'-phenyl- [10368-47-5];
    2,4,6-TRIPHENYLNITROBENZENE, V, 1128

1,1':2',1''-Terphenyl, 3',4',5',6'-tetraphenyl- [992-04-1];
  HEXAPHENYLBENZENE, V, 604

Tetradecanoic acid [544-63-8];
  MYRISTIC ACID, I, 379; III, 605

Tetradecanoic acid, methyl ester [124-10-7];
  METHYL MYRISTATE, III, 605

Tetradecanoic acid, 1,2,3-propanetriyl ester [555-45-3];
  TRIMYRISTIN, I, 538

2,5,7,10-Tetraoxaundecane, 6-mthylene- [5130-02-9];
  KETENE DI(2-METHOXYETHYL) ACETAL, V, 684

7-Thiabicyclo[4.1.0]heptane [286-28-2];
  CYCLOHEXENE SULFIDE, IV, 232

2-Thiazolamine, 4-methyl- [1603-91-4];
  2-AMINO-4-METHYLTHIAZOLE, II, 31

Thiazole, 2,4-dimethyl- [541-58-2];
  2,4-DIMETHYLTHIAZOLE, III, 332

4-Thiazolidinone, 2-thioxo- [141-84-4];
  RHODANINE, III, 763

4(5H)-Thiazolone, 2-amino- [556-90-1];
  PSEUDOTHIOHYDANTOIN, III, 751

Thiirane [420-12-2];
  ETHYLENE SULFIDE, V, 562

Thiirane, chloro-, 1,1-dioxide [10038-13-8];
  2-CHLOROTHIIRANE, 1,1-DIOXIDE, V, 231

Thiocyanic acid, 4-(dimethylamino)phenyl ester [7152-80-9];
  p-THIOCYANODIMETHYLANILINE, II, 574

Thiocyanic acid 1-methylethyl ester [625-59-2];
  ISOPROPYL THIOCYANATE, II, 366

Thiophene [110-02-1];
  THIOPHENE, II, 578

Thiophene, 3-bromo- [872-31-1];
  3-BROMOTHIOPHENE, V, 149

Thiophene, 3-(bromomethyl)- [34846-44-1];
  3-THENYL BROMIDE, IV, 921

2-Thiophenecarboxaldehyde [98-03-3];
  2-THENALDEHYDE; 2-THIOPHENEALDEHYDE, III, 811; IV, 915

3-Thiophenecarboxaldehyde [498-62-4];
  3-THENALDEHYDE, IV, 918

3-Thiophenecarboxylic acid [88-13-1];
  3-THENOIC ACID, IV, 919

Thiophene, 2-(chloromethyl)- [765-50-4];
2-CHLOROMETHYLTHIOPHENE, III, 197

3,4-Thiophenedicarbonitrile, 2,5-diamino [17989-89-8];
2,5-DIAMINO-3,4-DICYANOTHIOPHENE, IV, 243

Thiophene, 2,5-dihydro-3-methyl-, 1,1-dioxide [1193-10-8];
ISOPRENE CYCLIC SULFONE, III, 499

Thiophene, 2-ethenyl- [1918-82-7];
2-VINYLTHIOPHENE, IV, 980

Thiophene, 2-iodo- [3437-95-4];
2-IODOTHIOPHENE, II, 357; IV, 545

Thiophene, 3-methyl- [616-44-4];
3-METHYLTHIOPHENE, IV, 671

Thiophene, 2-(methylthio)- [5780-36-9];
METHYL 2-THIENYL SULFIDE, IV, 667

Thiophene, 2-nitro- [609-40-5];
2-NITROTHIOPHENE, II, 466

Thiophene-2-ol [17236-58-7];
2-HYDROXYTHIOPHENE, V, 642

Thiophene, tetrahydro- [110-01-0];
TETRAHYDROTHIOPHENE, IV, 892

Thiourea, (aminoiminomethyl)- [2114-02-5];
GUANYLTHIOUREA, IV, 502

Thiourea, (2-chlorophenyl)- [5344-82-1];
o-CHLOROPHENYLTHIOUREA, IV, 180

Thiourea, methyl- [598-52-7];
METHYLTHIOUREA, III, 617

Thiourea, N-methyl-N-1-naphthalenyl- [53663-34-6];
1-METHYL-1-(1-NAPHTHYL)-2-THIOUREA, III, 609

Thiourea, phenyl- [103-85-5];
α-PHENYLTHIOUREA, III, 735

DL-Threonine [80-68-2];
dl-THREONINE, III, 813

1-Triazene, 1,3-diphenyl [136-35-6];
DIAZOAMINOBENZENE, II, 163

1-Triazene, 1-methyl-3-(4-methylphenyl)- [21124-13-0];
1-METHYL-3-p-TOLYLTRIAZENE, V, 797

1,3,5-Triazine-2,4-diamine, 6-(chloromethyl-N-phenyl [30355-60-3];
2-AMINO-4-ANILINO-6-(CHLOROMETHYL)-s-TRIAZINE, IV, 29

1,3,5-Triazine-2,4-diamine, 6-phenyl- [91-76-9];
BENZOGUANAMINE, IV, 78

1,3,5-Triazine, hexahydro-1,3,5-tris(1-oxopropyl)- [30805-19-7];
   HEXAHYDRO-1,3,5-TRIPROPIONYL-s-TRIAZINE, IV, 518

1H-1,2,4-Triazol-3-amine [61-82-5];
   3-AMINO-IH-1,2,4-TRIAZOLE, III, 95

4H-1,2,4-Triazol-4-amine [584-13-4];
   4-AMINO-4H-1,2,4-TRIAZOLE, III, 96

1H-1,2,3-Triazol-5-amine, 1,4-diphenyl- [29704-63-0];
1H-1,2,3-Triazol-4-amine, N,5-diphenyl- [53684-55-2];
   1,4-DIPHENYL-5-AMINO-1,2,3-TRIAZOLE, AND 4-PHENYL-5-ANILINO-1,2,3-
   TRIAZOLE, IV, 380

1H-1,2,4-Triazole [288-88-0];
   1,2,4-TRIAZOLE, V, 1070

12-Tricosanone [540-09-0];
   LAURONE, IV, 560

Tricyclo[3.3.1.1$^{3.7}$]decane [281-23-2];
   ADAMANTANE, V, 16

Tricyclo[3.3.1.1$^{3.7}$]decane-1-carboxylic acid [828-51-3];
   1-ADAMANTANECARBOXYLIC ACID, V, 20

Tricyclo[2.2.1.0$^{2.6}$]heptan-3-ol [695-04-5];
   NORTRICYCLANOL, V, 863

Tricyclo[2.2.1.0$^{2.6}$]heptan-3-one [695-05-6];
   NORTRICYCLANONE, V, 866

Tricyclo[8.2.2.2$^{4.7}$]hexadeca-4,6,10,12,13,15-hexaene [1633-22-3];
   [2,2]PARACYCLOPHANE, V, 883

Tridecanoic acid, ethyl ester [28267-29-0];
   ETHYL n-TRIDECYLATE, II, 292

Triphenylene [217-59-4];
   TRIPHENYLENE, V, 1120

1,3,5-Trithiane [291-21-4];
   sym-TRITHIANE, II, 610

L-Tryptophan [73-22-3];
   l-TRYPTOPHANE, II, 612

DL-Tyrosine, 3-hydroxy-N-methyl- [53663-27-7];
   N-METHYL-3,4-DIHYDROXYPHENYLALANINE, III, 586
                              U
Undecanedioic acid [1852-04-6];
   HENDECANEDIOIC ACID, IV, 510

Undecanedioic acid, monomethyl ester [3927-60-4];
   METHYL HYDROGEN HENDECANEDIOATE, IV, 635

Undecanedioic acid, 6-oxo- [3242-53-3];
   6-KETOHENDECANEDIOIC ACID, IV, 555

Undecane, 1-isocyanato- [2411-58-7];
UNDECYL ISOCYANATE, III, 846

3,5,9-Undecatrien-2-one, 6,10-dimethyl- [141-10-6];
PSEUDOIONONE, III, 747

10-Undecynoic acid [2777-65-3];
10-UNDECYNOIC ACID, IV, 969

Urea, (4-bromophenyl)- [1967-25-5];
p-BROMOPHENYLUREA, IV, 49

Urea, N-cyano-N'-phenyl- [41834-91-7];
1-CYANO-3-PHENYLUREA, IV, 213

Urea, N,N-dimethyl- [598-94-7];
asym-DIMETHYLUREA, IV, 361

Urea, (2,5-dioxo-4-imidazolidinyl)- [97-59-6];
ALLANTOIN, II, 21

Urea, N,N'-diphenyl- [102-07-8];
sym-DIPHENYLUREA, I, 453

Urea, (4-ethoxyphenyl)- [150-69-6];
p-ETHOXYPHENYLUREA, IV, 52

Urea, hydroxy- [127-07-1];
HYDROXYUREA, V, 645

Urea, N-methyl-N-nitroso- [684-93-5];
NITROSOMETHYLUREA, II, 461

Urea, nitro- [556-89-8];
NITROUREA, I, 417

Urea, phenyl- [64-10-8];
PHENYLUREA, I, 453

Urea, [(phenylamino)thioxomethyl]- [53555-72-9];
1-PHENYL-2-THIOBIURET, V, 966

V

DL-Valine [516-06-3];
dl-VALINE, III, 848

X

9H-Xanthen-9-ol [90-46-0];
XANTHYDROL, I, 554

9H-Xanthen-9-one [90-47-1];
XANTHONE, I, 552

Z

Zinc, diethyl- [557-20-0];
DIETHYL ZINC, II, 184

TYPE OF REACTION INDEX

This index lists the preparations contained in the five *Collective Volumes* in accordance with general types of reactions. Only those preparations which can be classified under the selected heading with some definiteness are included. The arrangement of types and of preparations is alphabetical.

A valuable source of information concerning types of reactions described in *Organic Syntheses* is *Reaction Index of Organic Syntheses* by Shigehiko Sugasawa and Seijiro Nakai. The English edition of this book was published jointly in 1966 by Hirokawa Publishing Co., Inc., Tokyo, and John Wiley & Sons, Inc., New York. The procedures are classified under 31 main headings with numerous subdivisions according to the types of functional groups undergoing change. Each preparation is demonstrated by equations, with solvents, catalysts, experimental condition over the arrow and percentage yield beside the product. Names of compounds are only used occasionally, structural formulas are preferred. Citations to *Annual Volumes* 1 through 45 and *Collective Volumes* I through IV of *Organic Syntheses* are given.

ACETAL AND KETAL FORMATION
ACETONE DIBUTYL ACETAL, **V**,5
ACROLEIN ACETAL, **IV**, 21
BROMOACETAL, **III**,123
CYCLOHEXANONE DIALLYL ACETAL, **V**, 292
2-CYCLOHEXYLOXYETHANOL, **V**, 303
DICYANOKETENE ETHYLENE ACETAL, **IV**, 276
7,7-DIMETHOXYBICYCLO(2.2.1)-
HEPTENE, **V**, 424
β-ETHOXYPROPIONALDEHYDE ACETAL, **III**, 371
ETHYL DIETHOXYACETATE, **IV**, 427
dl-ISOPROPYLIDENE GLYCEROL, **III**, 502
β-KETOISOÖCTALDEHYDE DIMETHYL ACETAL, **IV**, 558
MONOBENZALPENTAERYTHRITOL, **IV**, 681
m-NITROBENZALDEHYDE DI-

METHYL ACETAL, III, 644
PHENYLPROPARGYL ALDE-
HYDE, III, 731
PHENYLPROPARGYLALDE-
HYDE DIETHYLACETAL,
IV, 801
ACYLATION (see also Replacement Reac-
tions; (J) Metals by Various
Groups)
3-ACETAMIDO-2-BUTANONE, IV,
5
ACETANILIDE, I, 332
ACETOBROMOGLUCOSE, III, 11
p-ACETOTOLUIDIDE, I, 111
3β-ACETOXYETIENIC ACID, V,
8
ACETYLGLYCINE, II, 11
ACETYLMANDELIC ACID, I, 12
2-AMINO-3-NITROTOLUENE,IV,
42
BENZANILIDE, I, 82
BENZOHYDROXAMIC ACID, II,
67
BENZOIN ACETATE, II, 69
BENZOYLACETANILIDE, IV, 80
ε-BENZOYLAMINOCAPROIC
ACID, II, 76
BENZOYLCHOLINE CHLORIDE,
IV, 84
BENZOYL PIPERIDINE, I, 99
4,4'-BIS(DIMETHYLAMINO)
BENZIL, V, 111
β-BROMOETHYLPHTHALIM-
IDE, IV, 106
t-BUTL ACETATE, III, 141
t-BUTYL AZODIFORMATE, V,
160
2-CARBETHOXYCY-
CLOÖCTANONE, V, 198
1,1'-CARBONYLDIIMIDAZOLE,
V, 201
α-CELLOBIOSE OCTAACETATE,
II, 124
α-CHLOROACETYL ISOCYAN-
ATE, V, 204
p-CHLOROPHENYL ISOTHIO-

CYANATE, V, 223
p-CHLOROPHENYL SALICY-
LATE, IV, 178
CHOLESTERYL ACETATE, II,
193
CYCLOHEXYLUREA, V, 801
sym-DIBENZOYLHYDAZINE, II,
208
DIETHYL ACETAMIDOMALON-
ATE, V, 373
DIETHYL BENZOYLMALON-
ATE, IV, 285
N,N-DIMETHYLCYCLOHEXYL-
METHYLAMINE, IV, 339
2,6-DIMETHYLPHENYLTHI-
OUREA, V, 802
DIPHENYL SUCCINATE, IV, 390
DOCOSANEDIOIC ACID, V, 533
ETHYL AZODICARBOXYLATE,
IV, 411
ETHYL BENZOYLACETATE, IV,
415
ETHYL DIACETYLACETATE,
III, 390
ETHYL N-TRICARBOXYLATE,
III, 415
1-ETHYL-3-(3-DIMETHYL-
AMINO)PROPYLCARBO-
DIIMIDE HYDROCHLOR-
IDE AND METHIODIDE,
V, 555
FLAVONE, IV, 478
FURFURAL DIACETATE, IV, 489
2-FURYLMETHYL ACETATE, I,
285
GALLACETOPHENONE, II, 304
β-GENTIOBIOSE OCTAACE-
TATE, III, 428
β-d-GLUCOSE-1,2,3,4-TETRAACE-
TATE, III, 432
n-HEPTAMIDE, IV, 513
HEXAMETHYLENE DIISO-
CYANATE, IV, 521
HEXAMETHYLPHOSPHOROUS
TRIAMIDE, V, 602
HIPPURIC ACID, II, 328

HYDROQUINONE DIACETATE,
    III, 452
HYDROXYHYDROQUINONE
    TRIACETATE, **I**, 317
6-HYDROXYNICOTINIC ACID,
    IV, 532
HYDROXYUREA, **V**, 645
INDAZOLE, **V**, 650
ISATOIC ANHYDRIDE, **III**, 488
α-KETOGLUTARIC ACID, **V**, 687
1-(p-METHOXYPHENYL)-5-
    PHENYL-1,3,5-PENTANE-
    TRIONE, **V**, 718
2-(p-METHOXYPHENYL)-6-
    PHENYL-4-PYRONE, **V**, 721
trans-2-METHYL-2-DODECENOIC
    ACID, **IV**, 608
2-METHYLENEDODECANOIC
    ACID, **IV**, 616
1-METHYL-3-ETHYLOXINDOLE,
    IV, 620
3-METHYLOXINDOLE, **IV**, 657
o-NITROACETOPHENONE, **IV**,
    708
o- and p-NITROBENZALDIACE-
    TATE, **IV**, 713
2-NITRO-4-METHOXYANILINE,
    III, 661
PARABANIC ACID, **IV**, 744
PENTAERYTHRITYL TETRA-
    BROMIDE, **IV**, 753
2-PHENYLCYCLOHEPTANONE,
    IV, 780
o-PHENYLENE CARBONATE, **IV**,
    788
2-PHENYLETHYL BENZOATE,
    V, 336
PHENYLGLYOXAL, **V**, 937
N-PHENYLMALEIMIDE, **V**, 944
1-PHENYL-2-THIOBIURET, **V**, 966
N-PHTHALYL-1-β-PHENYLALA-
    NINE, **V**, 973
PUTRESCINE DIHYDROCHLOR-
    IDE, **IV**, 819
2α-THIOHOMOPHTHALIMIDE,
    V, 1051

dl-THREONINE, **III**, 813
p-TOLUENESULFONYLHYDRA-
    ZIDE, **V**, 1055
p-TOLUENESULFONYLME-
    THYLNITROSAMIDE, **IV**,
    943
1,2,4-TRIAZOLE, **V**, 1070
α,α,α-TRICHLOROACETANI-
    LIDE, **V**, 1074
TRIETHYL PHOSPHITE, **IV**, 955
ADDITION (See Grignard Reaction; Hy-
    drolysis; Reduction). The re-
    agents which add are included in
    brackets after the name of the
    preparation.
    A. to C=C
β-AMINOPROPIONITRILE, **III**, 93
BENZALACETOPHENONE DI-
    BROMIDE, [Br—Br], **I**, 205
exo-cis-BICYCLO(3.3.0)OCTANE-2-
    CARBOXYLIC ACID, **V**, 93
BROMOACETAL, **III**, 123
1-BROMO-2-FLUOROHEPTANE,
    V, 136
α-BROMOHEPTALDEHYDE, **III**,
    127
BUTYRCHLORAL, **IV**, 130
trans-2-CHLOROCYCLOPEN-
    TANOL, **IV**, 157
2-CHLOROCYCLOHEXANOL,
    [HO—Cl], **I**, 158
α-CHLOROHYDRINDENE, [Cl—
    Cl], **II**, 336
2-CHLORO-1,1,2-TRIFLUORO-
    ETHYL ETHYL ETHER, **IV**,
    184
3-CHLORO-2,2,3-TRIFLUOROPRO-
    PIONIC ACID, **V**, 239
CHOLESTEROL, **IV**, 195
7-CYANOHEPTANAL, **V**, 266
CYCLOHEXYLBENZENE,
    [H—C$_6$H$_5$], **II**, 151
1,2-CYCLONONADIENE, **V**, 306
1,2-DIBROMOCYCLOHEXANE,
    [Br—Br], **II**, 171
α,β-DIBROMOSUCCINIC ACID,

[Br—Br], **II**, 177
3,4-DICHLORO-1,2,3,4-TETRAME-
THYLCYCLOBUTENE, **V**,
370
9,10-DIHYDROXYSTEARIC
ACID, **IV**, 317
DIPHENYLACETYLENE, **III**, 350
ETHYL α,β-DIBROMO-β-PHEN-
YLPROPIONATE [Br—Br],
II, 270
ETHYL ENANTHYLSUCCIN-
ATE, **IV**, 430
unsym-HEPTACHLOROPROPANE
[Cl—CHCl₂], **II**, 312
β-(2-HYDROXYETHYLMER-
CAPTO)-PROPIONITRILE,
III, 458
2-INDANONE, **V**, 647
IODOCYCLOHEXANE, **IV**, 543
LINOLEIC ACID, **III**, 526
LINOLENIC ACID, **III**, 531
NEOPENTYL ALCOHOL, **V**, 818
NORTRICYCLANOL, **V**, 863
STEAROLIC ACID, **III**, 785; **IV**,
851
STYRYLPHOSPHONIC DI-
CHLORIDE, **V**, 1005
2,2,5,5-TETRAMETHYLTETRA-
HYDRO-3-KETOFURAN, **V**,
1024
1,2,3-TRIBROMOPROPANE[Br—
Br], **I**, 521
1,1,3-TRICHLORO-*n*-NONANE, **V**,
1076
10-UNDECYNOIC ACID, **IV**, 969
B. To C≡C
1-ACETYLCYCLOHEXANOL, **IV**,
13
β-CHLOROVINYL ISOAMYL KE-
TONE, **IV**, 186
2,2,5,5-TETRAMETHYLTETRA-
HYDRO-3-KETOFURAN, **V**,
1024
C. To C≡N
1-METHYL-2-PYRIDONE[CH₃—
SO₄CH₃], **II**, 419

D. To C≡N (See also Grignard and
Related Reactions)
ACETAMIDINE
HYDROCHLORIDE[H—
OC₂H₅], **I**, 5
BENZAMIDINE
HYDROCHLORIDE[H—
OC₂H₅], **I**, 6
DIPHENYL KETIMINE, **V**, 520
N,N-DIMETHYLSELENOUREA,
IV, 359
ETHYL α-PHENYLACETOACE-
TATE
[H—OC₂H₅], **II**, 284
GUANIDINE NITRATE[H—NH₂],
I, 302
GUANYLTHIOUREA, **IV**, 502
METHYLISOUREA HYDROCH-
LORIDE, **IV**, 645
N-PHENYLBENZAMIDINE, **IV**,
769
PHLOROACETOPHENONE
[H—C₆H₂(OH)₃], **II**, 522
E. to C=O
ACETONE CYANOHYDRIN
[H—CN], **II**, 7
dl-ALANINE, [H—CN], **I**, 21
ATROLACTIC ACID, **IV**, 58
p-BROMODIPHENYLMETHANE,
V, 130
p-BROMOPHENYLUREA, **IV**, 49
1-CYANO-3-PHENYLUREA, **IV**, 213
CYCLODECANONE, **V**, 277
1,2-CYCLOHEXANEDIONE DIOX-
IME, **IV**, 229
CYCLOHEXYL METHYL KE-
TONE, **V**, 775
CYCLOHEXYLUREA, **V**, 801
1,2-DI-1-(1-CYANO)CYCLO-
HEXYLHYDRAZINE, **IV**, 274
DIETHYLAMINOACETONITRILE,
III, 275
α-*N,N*-DIMETHYLAMINOPHENY-
LACETONITRILE, **V**, 437
DIMETHYLETHYNYLCARBINOL,
III, 320

DIPHENYLACETYLENE, **IV,** 377

p-ETHOXYPHENYLUREA, **IV,** 52

1-ETHYL-3-(3-DIMETHYL-
AMINO)-PROPYLCARBODI-
IMIDE, **V,** 555

ETHYL *β*-HYDROXY-*β,β*-DIPHEN-
YLPROPIONATE, **V,** 564

1-ETHYNYLCYCLOHEXANOL, **III,**
416

1,1'-ETHYNYLENE-*bis*-CYCLO-
HEXANOL, **IV,** 471

GLYCOLONITRILE, **III,** 436

MANDELIC ACID, [H—CN], **I,** 336

METHOXYACETONITRILE,
[H—CN], **II,** 387

N-METHYL-2,3-DIMETHOXYBEN-
ZYLAMINE, **IV,** 603

N-METHYL-,2-DIPHENYLETHYL-
AMINE, **IV,** 605

1-MORPHOLINO-1-CYCLOHEX-
ENE, **V,** 808

1,2,3,4-TETRAHYDROCARBA-
ZOLE, **IV,** 884

β-TETRALONE, **IV,** 903

2,2,6,6-TETRAMETHYLOLCY-
CLOHEXANOL, **IV,** 907

THIOBENZOPHENONE, **IV,** 927

F. To C=S

DITHIZONE, **III,** 360

ETHYLENE THIOUREA, **III,** 394

METHYL ISOTHIOCYANATE, **III,**
599

RHODANINE, **III,** 763

G. To N=C=O

BENZOYLENE UREA
[H—NHC₆H₅], **II,** 79

H. To C=C—C=C (See also Cyclization,
B)

3-CHLOROCYCLOPENTENE, **IV,**
238

2,5-DIHYDRODIMETHOXY-2,5-
FURAN, **V,** 403

2,3-DIMETHYLANTHRAQUI-
NONE, **III,** 310

ISOPRENE CYCLIC SULFONE, **III,**
499

NORBORNYLENE, **IV,** 738

cis-Δ⁴-TETRAHYDROPHTHALIC
ANHYDRIDE, **IV,** 890

TETRAPHENYLPHTHALIC AN-
HYDRIDE, **III,** 807

TRIPTYCENE, **IV,** 964

I. To C=C—C=O (See also Grignard and
Related Reactions, Michael Reaction)

2-*p*-ACETYLPHENYLHYDROQUI-
NONE, **IV,** 15

dl-*β*-AMINO-*β*-PHENYLPROPIONIC
ACID, **III,** 91

BENZALACETONE DIBROMIDE,
III, 105

β-CHLOROPROPIONIC ACID
[H-Cl], **I,** 166

COUMARILIC ACID, **III,** 209

DIACETONAMINE HYDROGEN
OXALATE [H—NH₂], **I,** 196

DI-*β*-CARBETHOXYETHYLME-
THYLAMINE, **III,** 258

3,4-DIHYDRO-2-METHOXY-4-
METHYL-2*H*-PYRAN, **IV,** 311

2,5-DIHYDROXY-*p*-BENZENEDI-
ACETIC ACID, **III,** 286

5,5-DIMETHYL-1,3-CYCLOHEXA-
NEDIONE
[H—CH(CO₂C₂H₅)₂], **II,** 200

β-*β*-DIPHENYLPROPIO-
PHENONE[H-C₆H₅], **II,** 236

β-ETHOXYPROPIONALDEHYDE
ACETAL, **III,** 371

ETHYL *α,β*-DICYANO-*β*-PHENYL-
PROPIONATE
[H—CN], **I,** 452

ETHYL PROPANE-1,1,2,3-
TETRACARBOXYLATE[H-
CH(CO₂C₂H₅)₂], **I,** 272

HYDROXYHYDROQUINONE TRI-
ACETATE
[H—OCOCH₃], **I,** 317

METHYL *β*-BROMOPROPIONATE,
III, 576

β-METHYLGLUTARIC ANHY-
DRIDE, **IV,** 630

METHYL γ-METHYL-γ-NITRO-
VALERATE, **IV,** 652
β-METHYL-β-PHENYL-α,α'-DI-
CYANOGLUTARIMIDE, **IV,**
662
METHYLSUCCINIC ACID, **III,** 615
METHYL β-THIODIPROPIONATE,
IV, 669
γ-PHENYLALLYLSUCCINIC
ACID, **IV,** 766
α-PHENYL-β-BENZOYL-
PROPIONITRILE[H—CN], **II,**
498
PHENYLPROPARGYL ALDE-
HYDE, **III,** 731
PHENYLSUCCINIC ACID, **IV,** 804
dl-SERINE, **III,** 774
dl-THREONINE, **III,** 813
J. MISCELLANEOUS
ACETOACETANILIDE, **III,** 10
2-AMINO-6-METHYLBENZOTHIA-
ZOLE, **III,** 76
t-BUTYL CARBAMATE, **V,** 162
o-CHLOROPHENYLTHIOUREA,
IV, 180
2,5-DIAMINO-3,4-DICYANOTHIO-
PHENE, **IV,** 243
2,3-DIPHENYLSUCCINONITRILE,
IV, 392
α-PHENYLTHIOUREA, **III,** 735
VINYL CHLOROACETATE, **III,** 853
ALCOHOLYSIS
BUTYL OLEATE **II,** 469
ALKYLATION (See also, Condensation;
Friedel-Crafts Reaction; Grig-
nard Reactions; Replacement
Reactions) (The reagent em-
ployed is given in brackets)
A. C-ALKYLATION
α-ACETYL-δ-CHLORO-γ-VALERO-
LACTONE
[NaOC₂H₅], **IV,** 10
2-ALLYLCYCLOHEXANONE
[NaNH₂] **III,** 44
p-AMINOTETRAPHENYLME-
THANE, **IV,** 47

DL-ASPARTIC ACID [NaOC₂H₅], **IV,**
55
2-BENZYLCYCLOPENTANONE
[Na], **V,** 76
p-BROMODIPHENYLMETHANE
[H₂SO₄], **V,** 130
n-BUTYLACETYLENE [NaNH₂],
IV, 117
2-*n*-BUTYL-2-METHYLCYCLO-
HEXANONE [KNH₂], **V,** 187
1,1-CYCLOBUTANEDICARBOX-
YLIC ACID [NaOC₂H₅], **III,**
213
α-CYCLOHEXYLPHENYLACE-
TONITRILE [NaNH₂], **III,** 219
CYCLOPROPANECARBOXYLIC
ACID [NaOH], **III,** 221
CYCLOPROPYL CYANIDE
[NaNH₂], **III,** 223
DIETHYL 1,1-CYCLOBUTANEDI-
CARBOXYLATE
[NaOC₂H₅], **IV,** 288
DIETHYL Δ²-CYCLOPENTENYL-
MALONATE [NaOC₂H₅], **IV,**
291
DIETHYL β-KETOPIMELATE [Na],
V, 384
DIMESITYLMETHANE [HCO₂H],
V, 422
3,4-DINITRO-3-HEXENE [KOH],
IV, 372
1,6-DIOXO-8*a*-METHYL-1,2,3,4,6,-
7,8*a*-OCTAHYDRONA-
PHTHALENE, **V,** 486
DIPHENYLMETHANE, **II,** 232
1,1-DIPHENYLPENTANE,
[NaNH₂], **V,** 523
α-β-DIPHENYLPROPIONIC ACID
[NaNH₂], **V,** 526
ETHYL α-ACETOGLUTARATE, **II,**
263
ETHYL ACETOSUCCINATE, **II,**
262
ETHYL *n*-BUTYLACETOACE-
TATE, **I,** 248
ETHYL *n*-BUTYLMALONATE, **I,**
250

ETHYL *sec.*-BUTYLMALONATE,
II, 417
ETHYL 2,4-DIPHENYLBUTAN-
OATE [NaNH₂], V, 559
ETHYL (1-ETHYLPROPENYL)
METHYL CYANOACETATE
[NaOC₂H₅], III, 397
ETHYL α-ISOPROPYLACETOACE-
TATE [BF₃], III, 405
ETHYL ISOPROPYLMALONATE,
II, 94
ETHYL METHYLMALONATE, II,
279
HEXAMETHYLBENZENE [Al₂O₃],
IV, 520
INDOLE-3-ACETIC ACID [KOH],
V, 654
dl-ISOLEUCINE [NaOC₂H₅], III, 495
9-METHYLFLUORENE
[NaOCH₃], IV, 623
5-METHYL-5-HEXEN-2-ONE
[K₂CO₃], V, 767
1-METHYLISOQUINOLINE
[LiC₆H₅], IV, 641
3-METHYLPENTANE-2,4-DIONE
[K₂CO₃], V, 785
2,4-NONANEDIONE [NaNH₂], V,
848
Δ¹(9)-OCTALONE-2, V, 869
PELARGONIC ACID, II, 474
1,2,3,4,5-PENTACHLORO-5-
ETHYLCYCLOPENTA-
DIENE, V, 893
dl-PHENYLALANINE [NaOC₂H₅],
III, 705
n-PROPYLBENZENE, I, 471
TETRAACETYLETHANE
[NaOH], IV, 869
p-TRICYANOVINYL-*N,N*-DIME-
THYLANILINE, IV, 953
2,6,6-TRIMETHYL-2,4-CYCLO-
HEXADIENONE [Li], V, 1092
α,α,β-TRIPHENYLPROPIONI-
TRILE, [KNH₂], IV, 962
B. *N*-ALKYLATION

BENZOYLCHOLINE CHLORIDE,
IV, 84
2-BENZYLAMINOPYRIDINE, IV, 91
BENZYLANILINE, I, 102
BENZYLPHTHALIMIDE, II, 83
1-BENZYLPIPERAZINE, V, 88
BENZYLTRIMETHYLAM-
MONIUM ETHOXIDE, IV, 98
trans-CYCLOOCETENE, V, 315
sym,-DIBENZOYLDIMETHYLHY-
DRAZINE, II, 209
γ-DI-*n*-BUTYLAMINOPROPYLAM-
INE, III, 256
N,N'-DIETHYLBENZIDINE, IV,
283
N,N-DIMETHYLAMINOMETHYL-
FERROCENE METHIODIDE,
V, 434
N,N'-DIPHENYLBENZAMIDINE,
IV, 383
N-ETHYL-*p*-CHLOROANILINE, IV,
420
ETHYL *p*-DIMETHYLAMINO-
PHENYLACETATE, V, 552
1-ETHYL-3-(3-DIMETHYLAMINO)-
PROPYLCARBODIIMIDE
METHIODIDE, V, 555
ETHYL *N*-PHENYLFORMIMI-
DATE, IV, 464
N-ETHYLPIPERIDINE, V, 575
ETHYL α-(1-PYRROLIDYL)PRO-
PIONATE, IV, 466
N-ETHYL-*m*-TOLUIDINE, II, 290
1,3,5-HEXATRIENE, V, 608
JULOLIDINE, III, 504
METHYLAMINE HYDRO-
CHLORIDE, I, 347
o-METHYLBENZYL ALCOHOL,
IV, 582
2-METHYLBENZYLDIMETHY-
LAMINE, IV, 585
N-METHYLBUTYLAMINE, V, 736
N-METHYLETHYLAMINE, V, 758
METHYLHYDRAZINE
SULFATE, II, 395

1-METHYLINDOLE, **V**, 769
o-NITROBENZALDEHYDE, **V**, 825
(2.2)PARACYCLOPHANE, **V**, 883
1-PHENYLPIPERIDINE, **IV**, 795
α-PHTHALIMIDO-*o*-TOLUIC ACID,
   IV, 810
PYOCYANINE, **III**, 753
3-QUINUCLIDONE HYDRO-
   CHLORIDE, **V**, 989
TETRAMETHYL-*p*-PHENYLENE-
   DIAMINE, **V**, 1018
2(*p*-TOLYLSULFONYL)-
   DIHYDROISOINDOLE, **V**,
   1064
m-TRIFLUOROMETHYL-*N*,*N*-DI-
   METHYLANILINE, **V**, 1085
TRIMETHYLAMINE, **I**, 528
TRIMETHYLAMINE HYDRO-
   CHLORIDE, **I**, 531
2-VINYLTHIOPHENE, **IV**, 980
C. *O*-ALKYLATION
ALLOXANTIN DIHYDRATE, **IV**, 25
ANISOLE, **I**, 58
o-*n*-BUTOXYNITROBENZENE
   [K$_2$CO$_3$], **III**, 140
CARBOXYMETHOXYLAMINE HE-
   MIHYDROCHLORIDE [N-
   aOH], **III**, 172
CHOLESTANYL METHYL ETHER,
   V, 245
COUMARONE [NaOH], **V**, 251
7-CYANOHEPTANAL [(C$_2$H$_5$)$_3$N], **V**,
   266
2-CYANO-6-METHYLPYRIDINE, **V**,
   269
o-EUGENOL [K$_2$CO$_3$], **III**, 418
β-*d*-GLUCOSE-1,2,3,4-TETRAACE-
   TATE, [C$_5$H$_5$N], **III**, 432
KETENE DI(2-METHOXY-
   ETHYL)ACETAL [Na], **V**, 684
l-MENTHOXYACETIC ACID [Na],
   III, 544
METHOXYACETONITRILE, **II**, 387
m-METHOXYBENZALDEHYDE
   [NaOH], **III**, 564
O-METHYLCAPROLACTIM

[K$_2$CO$_3$], **IV**, 588
3-METHYLCOUMARONE
   [NaOH], **IV**, 590
α-METHYL-*d*-GLUCOSIDE, **I**, 364
METHYL *β*-NAPHTHYL ETHER, **I**,
   59
QUINACETOPHENONE MONO-
   METHYL ETHER [K$_2$CO$_3$],
   IV, 836
2,3,4,6-TETRAMETHYL-*d*-GLU-
   COSE [NaOH], **III**, 800
o-TOLUALDEHYDE [NaOC$_2$H$_5$], **IV**,
   932
TRIMETHYLGALLIC ACID, **I**, 537
TRIETHYLOXONIUM FLUOBOR-
   ATE, **V**, 1080
TRIMETHYLOXONIUM FLUO-
   BORATE, **V**, 1096
TRIMETHYLOXONIUM 2,4,6-
   TRINITROBENZENESUL-
   FONATE, **V**, 1099
VERATRALDEHYDE, **II**, 619
D. S-ALKYLATION
1,2-BIS(*n*-BUTYLTHIO)-
   BENZENE(Cu$_2$O), **V**, 107
p-DITHIANE, **IV**, 396
n-DODECYL MERCAPTAN, **III**,
   363
ETHANEDITHIOL, **IV**, 401
2-FURFURYLMERCAPTAN, **IV**,
   491
GUANIDOACETIC ACID, **III**, 440
S-METHYL ISOTHIOUREA SUL-
   FATE, **II**, 411
2-METHYLMERCAPTO-*N*-
   METHYL-Δ^{2}PYRROLINE,
   V, 780
METHYL *p*-TOLYL SULFONE,
   IV, 674
n-PROPYL SULFIDE, **II**, 547
PSEUDOTHIOHYDANTOIN, **III**,
   751
RHODANINE, **III**, 763
AMINATION
1-AMINO-1-METHYLCYCLO-
   HEXANE, **V**, 35

1-AMINOPYRIDINIUM IODIDE,
V, 43
N-BENZYLACRYLAMIDE, V, 73
α,α-DIMETHYL-β-PHENETHY-
LAMINE, V, 471
ARBUSOV REACTION
DIISOPROPYL METHYLPHOS-
PHONATE, IV, 325
ARSONATION
ARSANILIC ACID, I, 70
SODIUM p-HYDROXYPHENYL-
ARSONATE, I, 490
ARYLATION (See also Cyclization A; B)
2-p-ACETYLPHENYLHYDRO-
QUINONE, IV, 15
3,4-DICHLOROBIPHENYL, V, 51
2,4-DINITROBENZENESUL-
FENYL CHLORIDE, V, 474
2-METHOXYDIPHENYL ETHER,
III, 566
m-NITROBIPHENYL, IV, 718
o-NITRODIPHENYL ETHER, II,
446
p-NITRODIPHENYL ETHER, II,
445
PHENYL t-BUTYL ETHER
(METHOD II), V, 926
N-PHENYLCARBAZOLE, I, 547
2-PHENYLPYRIDINE, II, 517
4-PYRIDINESULFONIC ACID, V,
977
TETRAPHENYLARSONIUM
CHLORIDE HYDRO-
CHLORIDE, IV, 910
1,2,3,4-TETRAPHENYLNA-
PHTHALENE, V, 1037
TRIPHENYLAMINE, I, 544
BART REACTION
p-NITROPHENYLARSONIC
ACID, III, 665
BECKMANN REACTION
7-CYANOHEPTANAL, V, 266
BLAISE REACTION
sec-BUTYL α-n-CAPROYLPRO-
PIONATE, IV, 120

BOORD SYNTHESIS
1,4-PENTADIENE, IV, 746
BOUVEAULT-BLANC REDUCTION (See
Reduction, F
CO₂R→CH₂OH[Na + ROH]
CANNIZZARO REACTION (See Reduc-
tion, C)
CHO → CH₂OH[CH₂O + KOH]).
CARBONATION
β-RESORCYLIC ACID, II, 557
CARBOXYLATION
1-ADAMANTANECARBOXYLIC
ACID, V, 20
2-CARBETHOXYCY-
CLOÖCTANONE, V, 198
DIMETHYL CYCLOHEXANONE-
2,6-DICARBOXYLATE, V,
439
MESITOIC ACID, V, 706
1-METHYLCYCLOHEXANECAR-
BOXYLIC ACID, V, 739
TETROLIC ACID, V, 1043
CATALYST PREPARATION
CATALYST, RANEY NICKEL,. W-
6, III, 176, W-2, III, 181
CATALYSTS, PALLADIUM, III,
685
CHLOROETHYLATION
2-VINYLTHIOPHENE, IV, 980
CHLOROMETHYLATION
1-CHLOROMETHYLNAPHTHAL-
ENE, III, 195
2-CHLOROMETHYLTHIO-
PHENE, III, 197
2-HYDROXY-5-NITROBENZYL
CHLORIDE, III, 468
MESITYLACETIC ACID, III, 557
CLEAVAGE (See also Hydrolysis)
A. DEALKYLATION
3,3′-DIHYDROXYBIPHENYL, V,
412
TETRAMETHYL-p-PHENYLENE-
DIAMINE, V, 1018
B. HYDROLYTIC (See also Hydrolysis)
DIHYDROXYCYCLOPENTENE, V,
414

ETHYL γ-BROMOBUTYRATE, **V**, 545

3-HYDROXYGLUTARONITRILE, **V**, 614

α-KETOGLUTARIC ACID, **V**, 687

NEOPENTYL ALCOHOL, **V**, 818

exo/endo-7-NORCARANOL, **V**, 859

C. KETONIC

t-BUTYL DIAZOACETATE, **V**, 179

2-*n*-BUTYL-2-METHYLCYCLO-HEXANONE, **V**, 187

DIETHYL β-KETOPIMELATE, **V**, 384

DOCOSANEDIOIC ACID, **V**, 533

2-METHYLCYCLOPENTANE-1,3-DIONE, **V**, 747

5-METHYL-5-HEXEN-2-ONE, **V**, 767

D. OXIDATIVE

o-ANISALDEHYDE, **V**, 46

1,2,4-TRIAZOLE, **V**, 1070

E. REDUCTIVE

2-CYCLOHEXYLOXYETHANOL, **V**, 303

1,2-DIMERCAPTOBENZENE, **V**, 419

F. MISCELLANEOUS

2,4-DINITROBENZENESUL-FENYL CHLORIDE, **V**, 474

2,2,4-TRIMETHYL-3-OXOVAL-ERYL CHLORIDE, **V**, 1103

CLEMMENSEN REDUCTION, (See Reduction)

$C=O \rightarrow CH_2[Zn + HCl]$, **II**, 499

CONDENSATION. The term "condensation" is used here in a restricted sense and applies to those reactions in which a carbon–carbon bond is formed by the elimination of a simple molecule. *Cyclization* and *Dehydration* reactions are listed separately. Many other reactions that produce a carbon–carbon bond are listed under other headings. (See *Addition*: Alkylation, Diazotization: *Friedel-Crafts Reaction*: *Grignard Reaction*: *Rearrangement*: *Reduction*.)

The subheadings illustrate the types of reactions leading to the compounds listed. Reagents used are included in brackets.

A. CARBONYL-ACTIVE METHYL CONDENSATIONS

ANISALACETONE [NaOH], **I**, 78

BENZALACETONE [NaOH], **I**, 77

BENZALACETOPHENONE [NaOH], **I**, 78

BENZALPINACOLONE [NaOH], **I**, 81

DIACETONE ALCOHOL [Ba(OH)$_2$], **I**, 199

DIBENZALACETONE [NaOH], **II**, 167

ETHYL CINNAMATE [NaOC$_2$H$_5$], **I**, 252

2-FURFURALACETONE [NaOH], **I**, 283

m-NITROCINNAMIC ACID [CH$_3$CO$_2$Na], **I**, 398

NITROSTYRENE [NaOH], **I**, 413

PENTAERYTHRITOL [Ca(OH)$_2$], **I**, 425

ETHYL α-ACETYL-β-(2,3-DIME-THOXYPHENYL)-PROPIONATE [C$_5$H$_{11}$N], **IV**, 408

ETHYL BENZALMALONATE [C$_5$H$_{11}$N], **III**, 377

ETHYL *n*-BUTYLCYANOACE-TATE, **III**, 385

ETHYL (1-ETHYL-PROPYL-IDENE)CYANOACETATE [CH$_3$CO$_2$NH$_4$], **III**, 399

β-ETHYL-β-METHYLGLUTARIC ACID [CH$_3$CO$_2$NH$_4$], **IV**, 441

α-ETHYL-α-METHYLSUCCINIC ACID [KCN], **V**, 572

ETHYL(1-PHENYLETHYL-IDENE)CYANOACETATE

[$CH_3CO_2NH_4$], **IV,** 463
ETHYL PROPANE-1,1,3,3-TETRA-
    CARBOXYLATE
    [$(C_2H_5)_2NH$], **I,** 290
FURYLACRYLIC ACID [C_5H_5N],
    III, 425
3-(2-FURYL)ACRYLONITRILE
    [$CH_3CO_2NH_4$; C_5H_5N], **V,** 585
p-HYDROXYPHENYLPYRUVIC
    ACID [$C_5H_{11}N$], **V,** 627
o-METHOXYPHENYLACETONE
    [$C_4H_9NH_2$], **IV,** 573
trans-o-NITRO-α-PHENYLCIN-
    NAMIC ACID [$(C_2H_5)_3N$,
    $(CH_3CO)_2O$], **IV,** 730
N-METHYL-3,4-DIHYDROXY-
    PHENYLALANINE [Heat],
    III, 586
β-METHYLGLUTARIC ACID
    [$C_5H_{11}N$], **III,** 591
2-NITROETHANOL [KOH], **V,** 833
m-NITROSTYRENE [C_5H_5N], **IV,**
    731
α-PHENYLCINNAMIC ACID
    [$(C_2H_5)_3N$], **IV,** 777
α-PHENYLCINNAMONITRILE
    [$NaOC_2H_5$], **III,** 715
SORBIC ACID [C_5H_5N], **III,** 783
B. CARBONYL-ACTIVE METHYL-
    ENE CONDENSATIONS
3-BENZYL-3-METHYLPENTAN-
    OIC ACID [β-ALANINE],
    IV, 93
AZLACTONE OF α-ACETAMI-
    NOCINNAMIC ACID
    [CH_3CO_2Na-$(CH_3CO)_2O$), **II,** 1
AZLACTONE OF α-BENZOY-
    LAMINOCINNAMIC ACID
    [CH_3CO_2Na-$(CH_3CO)_2O$], **II,**
    490
AZLACTONE OF α-BENZOY-
    LAMINO-β-(3,4-DIME-
    THOXYPHENYL)ACRYLIC
    ACID [CH_3CO_2Na-
    $(CH_3CO)_2O$], **II,** 55

α-BENZYLIDINE-γ-PHENYL-$\Delta^{\beta,\gamma}$-
    BUTENOLIDE [CH_3CO_2Na],
    V, 80
p-BROMODIPHENYLMETHANE
    [KOH], **V,** 130
3-CARBETHOXYCOUMARIN
    [$C_5H_{11}N$], **III,** 165
β-CARBETHOXY-γ,γ-DIPHENYL-
    VINYLACETIC ACID
    [$KOC(CH_3)_3$], **IV.** 132
α-CYANO-β-PHENYLACRYLIC
    ACID [NaOH], **I,** 181
CYCLOHEPTANONE [$NaOC_2H_5$],
    IV, 221
CYCLOHEXYLIDENECY-
    ANOACETIC ACID
    [$CH_3CO_2NH_4$], **IV,** 234
DICYCLOPROPYL KETONE
    [$NaOCH_3$], **IV,** 278
DIETHYL BIS(HYDROXY-
    METHYL)MALONATE,
    [$KHCO_3$], **V,** 381
DIETHYL ETHYLIDENEMA-
    LONATE [$(CH_3CO)_2O$], **IV,**
    293
1,4-DIHYDRO-3,5-DICARBETH-
    OXY-2,6-DIMETHYLPYRI-
    DINE [$(C_2H_5)_2NH$], **II,** 214
2,3-DIMETHOXYCINNAMIC
    ACID [C_5H_5N, $C_5H_{11}N$], **IV,**
    327
6-(DIMETHYLAMINO)FULVENE
    [Na], **V,** 431
3,5-DIMETHYL-2-CYCLOHEXEN-
    1-ONE [$C_5H_{11}N$], **III,** 317
1,4-DIPHENYLBUTADIENE
    [$(CH_3CO)_2O$-PbO], **II,** 229
α,β-DIPHENYLCINNAMONI-
    TRILE [$NaNH_2$], **IV,** 387
2,3-DIPHENYLSUCCINONI-
    TRILE [NaCN], **IV,** 392
C. CARBONYL-YLIDE
DICHLOROMETHYLENETRI-
    PHENYLPHOSPHORANE
    [KOC_4H_9], **V,** 361

β,β-DIFLUOROSTYRENE, **V**, 390
1,4-DIPHENYL-1,3-BUTADIENE
  [LiOC₂H₅], **V**, 499
1,1-DIPHENYLCYCLOPROPANE
  [NaH], **V**, 509
ETHYL CYCLOHEXYLIDENE-
  ACETATE [NaH], **V**, 547
METHYLENECYCLOHEXANE
  [LiC₄H₉], **V**, 751
2-PHENYLPERFLUOROPRO-
  PENE, **V**, 949
p-QUINQUEPHENYL [LiOC₂H₅],
  V, 985

D. ESTER ACTIVE METHYLENE
2-CARBETHOXYCYCLOPEN-
  TANONE [Na], **II**, 116
ETHYL ETHOXYALYLPRO-
  PIONATE [NaOC₂H₅], **II**, 272
ETHYL INDOLE-2-CARBOXY-
  LATE [C₂H₅OK], **V**, 567
ETHYL 2-KETOCYCLOHEXYL-
  GLYOXALATE [NaOC₂H₅],
  II, 531
ETHYL PHENYLCYANOPYRU-
  VATE [NaOC₂H₅], **II**, 287
ETHYL PHENYLMALONATE
  [NaOC₂H₅], **II**, 288
α-PHENYLACETOACETONITRILE
  [NaOC₂H₅], **II**, 487

E. ESTER-ESTER CONDENSATIONS
BUTYROIN [Na], **II**, 114
CETYLMALONIC ESTER
  [NaOC₂H₅], **IV**, 141
3,4-DIHYDRO-1,2-NAPHTHALIC
  ANHYDRIDE [H₂SO₄], **II**, 194
6,7-DIMETHOXY-3,4-DIHYDRO-2-
  NAPHTHOIC ACID
  [NaOC₂H₅], **III**, 300
ETHYL ACETOACETATE
  [NaOC₂H₅], **I**, 235
ETHYL BENZOYLACETATE
  [NaOC₂H₅], **III**, 379
ISOBUTYROIN [Na], **II**, 115
α-KETOGLUTARIC ACID
  [KOC₂H₅], **III**, 510
PIVALOIN [Na], **II**, 115

PROPIONOIN [Na], **II**, 115
F. ESTER-KETONE CONDENSA-
  TIONS
ACETYLACETONE [NaOC₂H₅],
  III, 17
3-CYANO-6-METHYL-2(1)-PYRI-
  DONE [NaOCH₃], **IV**, 210
DIBENZOYLMETHANE
  [NaOC₂H₅], **III**, 251
DIISOVALERYLMETHANE
  [NaNH₂], **III**, 291
2-ETHYLCHROMONE [Na], **III**, 387
ETHYL β-HYDROXY-β,β-DIPHEN-
  YLPROPIONATE
  [LiNH₂], **V**, 564
INDAZOLE [NaOC₂H₅], **IV**, 536
1,3,5-TRIACETYLBENZENE [Na],
  III, 829

G. MISCELLANEOUS
ACETYLACETONE [BF₃], **III**, 17
ACRIDONE [H₂SO₄], **II**, 15
AMINOACETONE SEMICARBA-
  ZONE HYDROCHLORIDE
  [C₅H₅N], **V**, 27
BENZALPHTHALIDE
  [CH₃CO₂Na], **II**, 61
BENZANTHRONE [H₂SO₄], **II**, 62
BENZOIN [NaCN], **I**, 94
2,2'-BIPYRIDINE [Ni], **V**, 102
BIS(1,3-DIPHENYLIMIDAZOLI-
  DINYLIDENE-2) [heat], **V**,
  115
α-(4-CHLOROPHENYL)-γ-PHEN-
  YLACETOACETONITRILE
  [NaOC₂H₅], **IV**, 174
CYCLOPENTANONE [Ba(OH)₂], **I**,
  192
DIAMINOMALEONITRILE
  [NaCN], **V**, 344
p-DIMETHYLAMINOBENZO-
  PHENONE [POCl₃], **I**, 217
5,5-DIMETHYL-1,3-CYCLOHEXA-
  NEDIONE [NaOC₂H₅], **II**, 200
2,4-DIMETHYL-3,5-DICARBETH-
  OXYPYRROLE, **II**, 202
DYPNONE [Al(OC(CH₃)₃)₃], **III**, 367

ETHYL ACETONEDIOXALATE [NaOC$_2$H$_5$], **II,** 126
ETHYL ACETOPYRUVATE [NaOC$_2$H$_5$], **I,** 238
ETHYL ETHYLENETETRACAR-BOXYLATE [Na$_2$CO$_3$], **II,** 273
ETHYL PHENYLCYANOACE-TATE [NaOC$_2$H$_5$], **IV,** 461
ETHYL PROPANE-1,1,2,3-TETRA-CARBOXYLATE [NaOC$_2$H$_5$], **I,** 272
FURYLACRYLIC ACID [CH$_3$CO$_2$K], **III,** 426
4-HEPTANONE [Fe], **V,** 589
3-n-HEPTYL-5-CYANOCYTO-SINE, **IV,** 515
MESITYLENE [H$_2$SO$_4$], **I,** 341
p-PHENYLAZOBENZOIC ACID, **III,** 711
PSEUDOIONONE [NaOC$_2$H$_5$], **III,** 747
TETRACYANOETHYLENE, **IV,** 877
TETRAPHENYLCYCLOPENTA-DIENONE [KOH], **III,** 806
TETRAPHENYLETHYLENE, **IV,** 914
p-TOLUALDEHYDE [Cu$_2$Cl$_2$—AlCl$_3$], **II,** 583
m-TOLYLBENZYLAMINE, **III,** 827

COUPLING
4,4'-DIAMINOAZOBENZENE, **V,** 341
2,2'-DICHLORO-α,α'-EPOXIBI-BENZYL, **V,** 1026
DIPHENYLDIACETYLENE, **V,** 517
$\alpha,\alpha.\alpha',\alpha'$-TETRAMETHYLTETRA-METHYLENE GLYCOL, **V,** 1026

CYANATION
2-CYANO-6-METHYLPYRIDINE, **V,** 269
α-ETHYL-α-METHYLSUCCINIC ACID, **V,** 572

CYANOETHYLATION
3(5)AMINOPYRAZOLE, **V,** 39
3-(o-CHLOROANILINO)-PROPIONITRILE, **IV,** 146
N-2-CYANOETHYLANILINE, **IV,** 205
α-PHENYL-α-CARBETHOXY-GLUTARONITRILE, **IV,** 776
TETRAMETHYLAMMONIUM 1,-1,2,3,3-PENTACYANOPRO-PENIDE, **V,** 1013

CYCLIZATIONS. These reactions include a variety of methods which are employed here for the formation of cyclic systems. The reagent used is included in brackets.

A. CONDENSATION
α-ACETYL-δ-CHLORO-γ-VALER-OLACTONE, (NaOC$_2$H$_5$), **IV,** 10
ACRIDONE (H$_2$SO$_4$), **II,** 15
BENZANTHRONE (H$_2$SO$_4$ + Cu), **II,** 62
3-CARBETHOXYCOUMARIN (C$_5$H$_{11}$N + CH$_3$CO$_2$H), **III,** 165
2-CARBETHOXYCYCLOPEN-TANONE(Na), **II,** 116
CHELIDONIC ACID (C$_2$H$_5$ONa), **II,** 126
COUMALIC ACID (H$_2$SO$_4$, SO$_3$), **IV,** 201
COUMARONE (CH$_3$CO$_2$Na), **V,** 251
3-CYANO-6-METHYL-2-(1)-PYRI-DONE (PIPERIDINE ACE-TATE), **IV,** 210
1,1-CYCLOBUTANEDICARBOX-YLIC ACID (C$_2$H$_5$ONa), **III,** 213
1,4-CYCLOHEXANEDIONE (NaOC$_2$H$_5$), **V,** 288
CYCLOPENTANONE (Ba(OH)$_2$), **I,** 192
CYCLOPROPANECARBOXYLIC ACID (NaOH), **III,** 221

CYCLOPROPYL CYANIDE
(NaNH₂), III, 223
DEHYDROACETIC ACID
(NaHCO₃), III, 231
2,2'-DICHLORO-α,α'-EPOXYBI-
BENZIL((CH₃)₂N)₃P), V, 358
αα-DICYANO-β-ETHYL-β-
METHYLGLUTARIMIDE-
(CH₃CO₂NH₄), IV, 441
DICYCLOPROPYLKETONE
(NaOH), IV, 278
DIETHYL 1,1-CYCLOBUTANEDI-
CARBOXYLATE, (NaOC₂H₅),
IV, 288
3,4-DIHYDRO-1,2-NAPHTHALIC
ANHYDRIDE (H₂SO₄), II,
194
6,7-DIMETHOXY-3,4-DIHYDRO-
2-NAPHTHOIC ACID
(H₂SO₄ + H₃PO₄), III, 300
3,5-DIMETHYL-4-CARBETHOXY-
2-CYCLOHEXEN-1-
ONE(H₂SO₄ + CH₃CO₂H) III,
318
5,5-DIMETHYL-1,3-CYCLOHEX-
ANEDIONE(C₂H₅ONa), II, 200
2,6-DIMETHYL-3,5-DIPHENYL-
4H-PYRAN-4-ONE(poly-
phosphoric acid), V, 450
1,6-DIOXO-8a-METHYL-1,2,3,4,6,-
7,8a-OCTAHYDRONA-
PHTHALENE(C₄H₉N), V, 486
DIPHENYLCYCLOPROPEN-
ONE((C₂H₅)₃N), V, 514
2,3-DIPHENYLINDONE (Heat) III,
353
2-ETHYLCHROMONE (HCl+
CH₃CO₂H) III, 387
ETHYL 6,7-DIMETHOXY-3-
METHYLINDENE-2-CAR-
BOXYLATE (Polyphosphoric
Acid), V, 550
ETHYL ISODEHYDROACE-
TATE, (H₂SO₄), IV, 549
FLAVONE, (Heat), IV, 478
FLAVONE, (Heat), IV, 479

FLUORENECARBOXYLIC ACID
(AlCl₃), IV, 482
ISODEHYDROACETIC ACID
(H₂SO₄), IV, 549
JULOLIDINE (Heat) III, 504
KETENE DIMER (Acetone + cool-
ing) IV, 508
6-METHOXY-8-NITROQUINO-
LINE (H₂SO₄ + As₂O₅) III,
568
β-METHYLANTHROQUINONE
(H₂SO₄·SO₃), I, 353
4-METHYLCARBOSTYRIL
(H₂SO₄) III, 580
4-METHYLCOUMARIN (AlCl₃) III,
581
3-METHYLCOUMARONE,
(H₂SO₄), IV, 590
METHYLENECYCLOHEXANE
OXIDE (NaH), V, 755
2-METHYLCYCLOPENTANE-1,3-
DIONE (NaOC₂H₅), V, 747
METHYL CYCLOPROPYL KE-
TONE (NaOH). IV, 597
4-METHYLESCULETIN (H₂SO₄),
I, 360
1-METHYL-3-ETHYLOXINDOLE
(AlCl₃), IV, 620
4-METHYL-7-HYDROXYCOU-
MARIN (H₂SO₄), III, 282
2-METHYL-4-HYDROXYQUINO-
LINE (Heat), III, 593
1-METHYL-3-PHENYLINDANE
(H₂SO₄), IV, 665
NAPHTHORESORCINOL (H₂SO₄),
III, 637
NORTRICYCLANOL
(BF₃·(C₂H₅)₂O), V, 863
Δ¹⁽⁹⁾-OCTALONE-2, V, 869
PHENYLPHENANTHRENE (Pho-
tochemical), V, 952
QUINIZARIN [H₂SO₄ + H₃BO₃], I,
476
QUINOLINE (H₂SO₄), I, 478
SEBACOIN (Na), IV, 840
α-TETRALONE (AlCl₃), II, 569;

(Poly H$_3$PO$_4$) **III**, 798; (AlCl$_3$)
IV, 898 (SnCl$_4$) **IV**, 900
TETRAPHENYLCYCLOPENTA-
DIENONE (KOH), **III**, 806
1-*p*-TOLYLCYCLOPROPANOL
(BrMgC$_2$H$_5$; FeCl$_3$), **V**, 1058
1,3,5-TRIACETYLBENZENE
(NaOC$_2$H$_5$; CH$_3$CO$_2$H), **III**,
829
XANTHONE (Heat), **I**, 552
B. CYCLOADDITIONS
BICYCLO(2.1.0)PENTANE, **V**, 96
BIPHENYLENE, **V**, 54
1-CHLORO-1,4,4-TRIFLUOROBU-
TADIENE, **V**, 235
3,4-DICHLORO-1,2,3,4-TETRAME-
THYLCYCLOBUTENE, **V**,
370
2,2-DIFLUOROSUCCINIC ACID,
V, 393
3,4-DIHYDRO-2-METHOXY-4-
METHYL-2*H*-PYRAN, (heat),
IV, 311
7,7-DIMETHOXYBICYCLO
(2.2.1)HEPTENE, **V**, 424
2,3-DIMETHYLANTHRAQUI-
NONE (Heat), **III**, 310
DIMETHYL-3-METHYLENECY-
CLOBUTANE-1,2-DICAR-
BOXYLATE, **V**, 459
DIPHENYLCARBODIIMIDE
(Method II), **V**, 504
1,1-DIPHENYLCYCLOPROPANE,
V, 509
cis- and *trans*-1,2-DIVINYLCYCLO-
BUTANE, **V**, 528
HEXAPHENYLBENZENE, **V**, 604
ISOPRENE CYCLIC SULFONE
(Heat), **III**, 499
β-ISOVALEROLACTAM-*N*-SUL-
FONYL CHLORIDE and *β*-
ISOVALEROLACTAM, **V**,
673
NORBORNYLENE (Heat), **IV**, 738
(2,2)PARACYCLOPHANE, **V**, 883
2-PHENYL-3-*n*-PROPYLISOAZO-
LIDINE-4,5-*cis*-DICARBOX-
YLIC ACID *N*-PHENYLIM-
IDE, **V**, 957
p-QUINQUEPHENYL, **V**, 985
STYRENE OXIDE (C$_6$H$_5$CO$_3$H), **I**,
494
Cis-Δ^4-TETRAHYDROPHTHALIC
ANHYDRIDE, **IV**, 890
1,2,3,4-TETRAPHENYLNA-
PHTHALENE, **V**, 1037
TETRAPHENYLPHTHALIC AN-
HYDRIDE (Heat), **III**, 807
TRIPHENYLENE, **V**, 1120
2,3,5-TRIPHENYLISOXAZOLI-
DINE, **V**, 1124
C. FREE RADICAL; CARBENOID;
NITRENOID;
exo-cis-BICYCLO(3.3.0)-OCTANE-
2-CARBOXYLIC ACID
((C$_6$H$_5$CO)$_2$O$_2$), **V**, 93
BICYCLO(2.1.0)PENTANE(Heat),
V, 96
CYCLOPROPYLBENZENE(Zn—
Cu), **V**, 328
METHYLENECYCLOHEXANE
OXIDE (NaH), **V**, 755
2-NITROCARBAZOLE (Heat), **V**,
829
NORCARANE (Zn—Cu), **V**, 855
exo/endo-7-NORCARANOL
(CH$_3$Li), **V**, 859
2-OXA-7,7-DICHLORONORCAR-
ANE (NaOCH$_3$), **V**, 874
D-2-OXO-7,7-DIMETHYL-1-
VINYLBICYCLO(2.2.1)-
HEPTANE ((C$_2$H$_5$)$_3$N), **V**, 877
2-PHENYLCYCLOHEPTANONE
(K$_2$CO$_3$)**IV**, 780
PHENYLINDAZOLE ((C$_2$H$_5$O)$_3$P),
V, 941
PSEUDOPELLETIERINE, **IV**, 816
TRIPTYCENE (Mg), **IV**, 964
D. OXIDATIVE
BENZOFURAZAN OXIDE
(NaOCl), **IV**, 74; (Heat); **IV**, 75
18,20-LACTONE of 3*β*-ACETOXY-

20β-HYDROXY-5-PREG-
NENE-18-OIC ACID
(Pb(OCOCH₃)₄; I₂), **V**, 692
PYROMELLITIC ACID (H₂SO₄ +
Hg), **II**,551
trans-STILBENEOXIDE
(CH₃CO₃H) **IV**, 860
TETRAHYDROXYQUINONE (O₂;
Na₂SO₃), **V**, 1011
E. REDUCTIVE
cis-2-BENZYL-3-PHENYLAZIRI-
DINE (LiAlH₄), **V**, 83
ETHYL INDOLE-2-CARBOXY-
LATE (H₂/Pt), **V**, 567
PHENACYLAMINE HYDROCH-
LORIDE (NaOCH₃), **V**, 909
2,3,4,5-TETRAMETHYLPYRROLE
(Zn; CH₃CO₂H), **V**, 1022
F. MISCELLANEOUS
"ACETONE-ANIL" (I₂ + heat), **III**,
329
3-ACETOXINDOLE (KNH₂), **V**, 12
3(5)-AMINOPYRAZOLE (H₂SO₄), **V**,
39
N-(*p*-ACETYLAMINO-
PHENYL)RHODAMINE,
(Heat), **IV**, 6
2-AMINO-4-ANILINO-6-(CHLOR-
OMETHYL)-*s*-TRIAZINE, **IV**,
29
4-AMINO-2,6-DIMETHYLPYRIM-
IDINE (CH₃OK), **III**, 71
2-AMINO-6-METHYLBENZO-
THIAZOLE (SO₂Cl₂), **III**, 76
2-AMINO-4-METHYLTHIAZOLE
(NaOH), **II**, 31
3-AMINO-1*H*-1,2,4-TRIAZOLE
(Heat), **III**, 95
4-AMINO-4*H*-1,2,4-TRIAZOLE
(Heat), **III**, 96
ANHYDRO-2-HYDROXYMER-
CURI-3-NITROBENZOIC
ACID (Hg(OCOCH₃)₂), **I**, 56
AZLACTONE OF α-ACETAMI-
NOCINNAMIC ACID (Ac₂O
+ CH₃CO₂Na), **II**, 1

AZLACTONE OF α-BENZOY-
LAMINOCINNAMIC ACID
(Ac₂O + CH₃CO₂Na), **II**, 490
AZLACTONE OF α-BENZOY-
LAMINO-b-(3,4-DIME-
THOXYPHENYL)-
ACRYLIC ACID (Ac₂O +
CH₃CO₂Na), **II**, 55
BARBITURIC ACID (C₂H₅ONa),
II, 60
BENZENEBORONIC ANHY-
DRIDE (Heat) **IV**, 68
BENZIMIDAZOLE (Heat) **II**, 65
BENZOGUANAMINE (KOH), **IV**,
78
1,2,3-BENZOTHIADIAZOLE 1,1-
DIOXIDE (HNO₂), **V**, 61
1,2,3-BENZOTRIAZOLE
(CH₃CO₂H + NaNO₂), **III**, 106
BENZOYLENE UREA (NaOH), **II**,
79
α-BENZYLIDENE-γ-PHENYL-
Δβ,γ-BUTENOLIDE (NaO-
COCH₃; (CH₃CO)₂O), **V**, 80
BIS(1,3-DIPHENYLIMIDAZOLI-
DINYLIDENE-2) (Heat), **V**,
115
t-BUTYLPHTHALIMIDE (Heat),
III, 152
1-*n*-BUTYLPYRROLIDINE
(H₂SO₄) **III**, 159
ε-CAPROLACTAM (2-Ketohexame-
thylene imine) **II**, 77, 371
7-CHLORO-4-HYDROXY-3-QUIN-
OLINECARBOXYLIC ACID
(Heat), **III**, 273
CHLOROTHIIRANE 1,1-DIOX-
IDE ((C₂H₅)₃N), **V**, 231
COUMARILIC ACID (KOH), **III**,
209
CREATININE (HCl), **I**, 172
1-CYANOBENZOCYCLOBUTENE
(NaNH₂), **V**, 263
CYCLOHEXENE OXIDE (NaOH),
I, 185
1,2-CYCLONONADIENE

(KOC$_4$H$_9$), **V**, 306

DIACETYL-*d*-TARTARIC ANHY-
DRIDE [(CH$_3$CO)$_2$O, H$_2$SO$_4$],
IV, 242

2,5-DIAMINO-3,4-DICYANO-
THIOPHENE, (C$_5$H$_5$N), **IV**,
243

2,4-DIAMINO-6-HYDROXYPYR-
IMIDINE, (NaOC$_2$H$_5$), **IV**,
245

DIAMINOURACIL HYDROCH-
LORIDE, (NaOC$_2$H$_5$) **IV**, 247

DICYANOKETENE ETHYLENE
ACETAL (Urea), **IV**, 276

2,3-DIHYDROPYRAN (Al$_2$O$_3$) **III**,
276

2,4-DIMETHYL-3-ACETYL-5-
CARBETHOXYPYRROLE
(Zn + AcOH) **III**, 513

2,4-DIMETHYL-3,5-DICARBETH-
OXYPYRROLE (CH$_3$CO$_2$H +
Zn) **II**, 202

2,2-DIMETHYLENEIMINE
(NaOH) **III**, 148

DIMETHYLFURAZAN, (Heat),
IV, 342

5,5-DIMETHYLHYDANTOIN
((NH$_4$)$_2$CO$_3$), **III**, 323

5,5-DIMETHYL-2-*n*-PENTYLTE-
TRAHYDROFURAN
(H$_3$PO$_4$), **IV**, 350

3,5-DIMETHYLPYRAZOLE, **IV**,
351

2,6-DIMETHYLPYRIDINE
((C$_2$H$_5$)$_2$NH), **II**, 214

2,5-DIMETHYLPYRROLE
((NH$_4$)$_2$CO$_3$), **II**, 219

1,5-DIMETHYL2-PYRROLIDONE
[Ni+H$_2$], **III**, 328

5,5-DIMETHYL-2-PYRROLIDONE,
[Ni+H$_2$], **IV**, 357

2,4-DIMETHYLTHIAZOLE, [P$_2$S$_5$],
III, 332

1,4-DIPHENYL-5-AMINO-1,2,3-
TRIAZOLE, (NaOCH$_3$), **IV**,
380

p-DITHIANE, [NaOC$_2$H$_5$], **IV**, 396

4,5-DIPHENYLGLYOXALONE,
[CH$_3$CO$_2$H + Heat], **II**, 231

2,4-DIPHENYLPYRROLE, [Ni +
H$_2$], **III**, 358

EPIBROMOHYDRIN, [Ca(OH)$_2$],
II, 257

EPICHLOROHYDRIN, [NaOH], **I**,
233; [Ca(OH)$_2$], **II**, 256

N-ETHYLALLENIMINE, [NaNH$_2$]
V, 541

ETHYLENE THIOUREA, [HCl],
III, 394

ETHYLENIMINE, [NaOH], **IV**, 433

5-ETHYL-2-METHYLPYRIDINE,
[CH$_3$CO$_2$NH$_4$], **IV**, 451

FURFURAL, [H$_2$SO$_4$·NaCl], **I**, 280

GLUTARIMIDE, [Heat], **IV**, 496

D-GULONIC-γ-LACTONE, **IV**, 506

3-*n*-HEPTYL-5-CYANOCYTOSINE
[NaOCH$_3$], **IV**, 515

HEXAHYDRO-1,3,5-TRIPRO-
PIONYL-*s*-TRIAZINE,
[H$_2$SO$_4$], **IV**, 518

HOMOPHTHALIC ANHYDRIDE,
[Ac$_2$O], **III**, 449

4-HYDROXY-1-BUTANESUL-
FONIC ACID SULTONE,
[Heat], **IV**, 529

2-HYDROXYCINCHONINIC
ACID, [NaOH], **III**, 456

4(5)-HYDROXYMETHYLIMIDA-
ZOLE HYDROCHLORIDE
[Cuco$_3$], **III**, 460

3-HYDROXYTETRAHYDRO-
FURAN, [p-C$_7$H$_7$SO$_3$H], **IV**,
534

IMIDAZOLE, [CH$_2$O + NH$_3$], **III**,
471

INDAZOLE, [HCl], **III**, 475; **IV**,
536; [Heat], **V**, 650

INDOLE [KOC(CH$_3$)$_3$], **III**, 479

ISATIN [H$_2$SO$_4$], **I**, 327

ISATOIC ANHYDRIDE, [COCl$_2$],
III, 488

dl-ISOPROPYLIDINEGLY-

CEROL, [CH₃C₆H₄SO₃H(*p*)],
III, 502
3-ISOQUINUCLIDONE, [Heat], V,
670
ITACONIC ANHYDRIDE, [Heat],
II, 368
2-MERCAPTO-4-AMINO-5-CAR-
BETHOXYPYRIMIDINE
AND 2-MERCAPTO-4-HY-
DROXY-5-CYANOPYRIMI-
DINE, [NaOC₂H₅], IV, 566
MERCAPTOBENZIMIDAZOLE,
IV, 569
2-MERCAPTOPYRIMIDINE,
[HCl], V, 703
MESITYLENE, [H₂SO₄], I, 341
1-(2-METHOXYCARBONYL-
PHENYL)-PYRROLE,
[CH₃CO₂H], V, 716
10-METHYL-10,9-BORAZARO-
PHENANTHRENE, [AlCl₃],
V, 727
5-METHYLFURFURAL, [HCl], II,
393
β-METHYLGLUTARIC ANHY-
DRIDE, [(CH₂CO)₂O], IV, 630
4-METHYL-6-HYDROXYPYRIMI-
DINE, [NaOCH₃], IV, 638
1-METHYL-2-IMINO-β-NA-
PHTHOTHIAZOLINE, [Br₂
+ AcOH], III, 595
2-METHYLINDOLE [NaNH₂], III,
597
METHYL-3-METHYL-2-FU-
ROATE [Heat], IV, 649
3-METHYLOXINDOLE, [CaH₂],
IV, 657
β-METHYL-β-PHENYL-α,α′-DI-
CYANOGLUTARIMIDE,
[NaOC₂H₅], IV, 662
3-METHYL-1-PHENYLPHOSPHO-
LENE 1-Oxide, V, 787
4-METHYLPYRIMIDINE,
[NH₄Cl], V, 794
3-METHYLTHIOPHENE, [P₄S₇],
IV, 671

6-METHYLURACIL, [HCl], II, 422
β-METHYL-γ-VALEROLAC-
TONE, [CuCr₂O₄], IV, 677
MONOBENZALPENTAERY-
THRITOL, [HCl], IV, 679
5-NITRO-2,3-DIHYDRO-1,4-
PHTHALAZINEDIONE,
[N₂H₂·H₂SO₄], III, 656
5-NITROINDAZOLE, [AcOH +
NaNO₂], III, 660
3-NITROPHTHALIC ANHY-
DRIDE, [Ac₂O], I, 410
PARABANIC ACID, [NaOCH₃],
IV, 744
3,3-PENTAMETHYLENEDIAZI-
RINE, V, 897
PHENOXTHIN, [AlCl₃], II, 485
1-PHENYL-3-AMINO-5-PYRAZO-
LONE, [C₂H₅ONa], III, 708
PHENYLCYCLOPROPANE, V,
929
4-PHENYL-*m*-DIOXANE, [H₂SO₄],
IV, 786
o-PHENYLENE CARBONATE,
[NaOH], IV, 788
α-PHENYLGLUTARIC ANHY-
DRIDE, [Ac₂O], IV, 790
2-PHENYLINDOLE, [ZnCl₂], III,
725
N-PHENYLMALEIMIDE,
[(CH₃CO)₂O], V, 944
PHENYLMETHYLGLYCIDIC
ESTER, [NaNH₂], III, 727
2-PHENYL-5-OXAZOLONE,
[(CH₃CO)₂O], V, 946
3-PHENYLSYDNONE,
[(CH₃CO)₂O], V, 962
PHTHALHYDRAZIDE,
[H₂NNH₂·H₂O], III, 153
PHTHALIDE, [HCl], II, 526
PHTHALIMIDE, [NH₄OH; Heat],
I, 457
unsym-o-PHTHALYL CHLORIDE,
[AlCl₃], II, 528
PSEUDOTHIOHYDANTOIN,
[Heat], III, 751

PYOCYANINE, **III**, 753
2,3-PYRAZINEDICARBOXYLIC
ACID, **IV**, 824
PYRROLE, [NH₄OH; Heat], **I**, 473
QUINOXALINE, **IV**, 824
3-QUINUCLIDONE HYDRO-
CHLORIDE, [KOC₂H₅], **V**, 989
RHODANINE, [HCl], **III**, 763
SUCCINIC ANHYDRIDE,
[CH₃COCl or POCl₃], **II**, 560
SUCCINIMIDE, [Heat], **II**, 562
o-SULFOBENZOIC ANHY-
DRIDE, [SOCl₂], **I**, 495
1,2,3,4-TETRAHYDROCARBA-
ZOLE, [CH₃CO₂H], **IV**, 884
TETRAHYDROTHIOPHENE, [H-
CON(CH₃) + Heat], **IV**, 892
2,2,5,5-TETRAMETHYLTETRA-
HYDRO-3-KETOFURAN,
[H₂SO₄], **V**, 1024
2α-THIOHOMOPHTHALIMIDE,
[AlCl₃], **V**, 1051
THIOPHENE, [P₂S₃], **II**, 578
2-(p-TOLYLSULFONYL)-
DIHYDROISOINDOLE,
[NaH], **V**, 1064
1,2,4-TRIAZOLE, [NaOH], **V**, 1070
TRIMETHYLENE OXIDE, [KOH],
III, 835
2,4,6-TRIMETHYLPYRYLIUM
PERCHLORATE, [HClO₄],
V, 1106
2,4,6-TRIMETHYLPYRYLIUM
TETRAFLUOROBORATE,
[HBF₄], **V**, 1112
2,4,6-TRIMETHYLPYRYLIUM
TRIFLUOROMETHANE-
SULFONATE, [CF₃SO₃H], **V**,
1114
2,4,6-TRIPHENYLNITROBEN-
ZENE, [(C₂H₅)₃N], **V**, 1128
2,4,6-TRIPHENYLPYRYLIUM TE-
TRAFLUOROBORATE,
[HBF₄], **V**, 1135
sym.-TRITHIANE, [HCl], **II**, 610

DARZENS REACTION
ETHYL β,β-PENTAMETHYLE-
NEGLYCIDATE, **IV**, 459
METHYL 3-METHYL-2-FURO-
ATE, **IV**, 649
PHENYLMETHYLGLYCIDIC
ESTER, **III**, 727
DEAMINATION
2-PHENYLETHYL BENZOATE, **V**,
336
DECARBONYLATION
CETYLMALONIC ESTER, **IV**, 141
CYCLOHEXYLIDENECYCLO-
HEXANE, **V**, 297
2,4,4-TRIMETHYLCYCLOPENTAN-
ONE, **IV**, 957
DECARBOXYLATION
3-ACETAMIDO-2-BUTANONE, **IV**,
5
ACETONEDICARBOXYLIC ACID,
I, 10
ANHYDRO-o-HYDROXYMER-
CURIBENZOIC ACID, **I**, 57
ANHYDRO-2-HYDROXYMER-
CURI-3-NITROBENZOIC
ACID, **I**, 56
α-BROMO-n-CAPROIC ACID, **II**,
95
α-BROMOISOCAPROIC ACID, **II**,
95
α-BROMOISOVALERIC ACID, **II**,
93
α-BROMO-β-METHYLVALERIC
ACID, **II**, 95
4-BROMORESORCINOL, **II**, 100
n-CAPROIC ACID, **II**, 417, 474
COUMALIC ACID, **IV**, 201
COUMARONE, **V**, 251
CYCLOBUTANECARBOXYLIC
ACID, **III**, 213
CYCLODECANONE, **V**, 277
1,4-CYCLOHEXANEDIONE, **V**,
288
1-CYCLOHEXENYLACETO-
NITRILE, **IV**, 234
CYCLOPENTANONE, **I**, 192

DIALLYLAMINE, **I,** 201
DIBROMOACETONITRILE, **IV,**
254
DI-*n*-BUTYLAMINE, **I,** 202
2,6-DICHLOROPHENOL, **III,** 267
4,7-DICHLOROQUINOLINE, **III,**
272
DIETHYL BENZOYLMALON-
ATE, **IV,** 285
4,6-DIMETHYLCOUMALIN, **IV,**
337
3,5-DIMETHYL-2-CYCLOHEXEN-
1-ONE, **III,** 317
2,5-DIMETHYLMANDELIC
ACID, **III,** 326
2,6-DIMETHYLPYRIDINE, **II,** 214
2,4-DIMETHYLPYRROLE, **II,** 217
α-ETHYL-*α*-METHYLSUCCINIC
ACID, **V,** 572
FURAN, **I,** 274
n-HEPTAMIDE, **IV,** 513
3-HYDROXYQUINOLINE, **V,** 635
IMIDAZOLE, **III,** 471
dl-ISOLEUCINE, **III,** 495
ITACONIC ANHYDRIDE, **II,** 368
α-KETOGLUTARIC ACID, **III,** 510
LAURONE, **IV,** 560
dl-METHIONINE, **II,** 384
METHYL *n*-AMYL KETONE, **I,**
351
METHYL BENZYL KETONE, **II,**
391
METHYL ISOBUTYL KETONE, **I,**
352
METHYL *n*-BUTYL KETONE, **I,**
352
3-METHYLCOUMARONE, **IV,** 590
METHYL CYCLOPROPYL KE-
TONE, **IV,** 597
3-METHYLFURAN, **IV,** 628
4-METHYLHEXANONE-2, **I,** 352
3-METHYLPENTANOIC ACID,
II, 416
β-METHYL-*β*-PHENYLGLU-
TARIC ACID, **IV,** 664
NAPHTHORESORCINOL, **III,** 637

o-NITROACETOPHENONE,
IV,708
NITROMETHANE, **I,** 401
m-NITROSTYRENE, **IV,** 731
PELARGONIC ACID, **II,** 474
dl-PHENYLALANINE, **III,** 705
PHENYLETHYLENE, **I,** 440
PHENYLNITROMETHANE, **II,**
512
PHLOROGLUCINOL, **I,** 455
α-PYRONE, **V,** 982
PYRUVIC ACID, **I,** 475
cis-STILBENE, **IV,** 857
1,2,3,4-TETRAPHENYLNA-
PHTHALENE, **V,** 1037
TRICARBALLYLIC ACID, **I,** 523
1,3,5-TRINITROBENZENE, **I,** 541
DEHALOGENATION
ALLENE, **V,** 22
3-BROMOTHIOPHENE, **V,** 149
CHOLESTEROL, **IV,** 195
1,1-DICHLORO-2,2-DIFLUOROE-
THYLENE, **IV,** 268
2,2-DIFLUOROSUCCINIC ACID,
V, 393
7,7-DIMETHOXYBICYCLO(2.2.1)-
HEPTENE, **V,** 424
DIMETHYLKETENE, **IV,** 348
PERCHLOROFULVALENE, **V,**
901
TRIPTYCENE, **IV,** 964
DEHYDRATION (See also, Condensation;
Esterification; Replacement).
The agents used are included in
brackets after the names of the
preparations.
ACETAMIDE [Heat], **I,** 3
ACETONE DICARBOXYLIC
ACID [Fuming H₂SO₄], **I,** 10
ACONITIC ACID [H₂SO₄], **II,** 12
ACROLEIN [KHSO₄ + H₂SO₄], **I,**
15
ALLYL ALCOHOL [Heat], **I,** 42
2-AMINO-4-METHYLTHIOZOLE,
II, 31
AZELANITRILE [SiO₂], **IV,** 62

BENZALAZINE, **II**, 395

BENZALANILINE, **I**, 80

BENZHYDRYL β-CHLORO-
ETHYL ETHER [H₂SO₄], **IV**,
72

BENZIL HYDRAZONE, **II**, 497

BENZIMIDAZOLE, **II**, 65

BENZOIC ANHYDRIDE
[(CH₃CO)₂O], **I**, 91

BENZOPHENONE OXIME, **II**, 70

BENZOYLENE UREA, **II**, 79

BENZYLIDENEARGININE, **II**, 49

BISCHLORMETHYL ETHER
[ClSO₃H], **IV**, 101

BUTYRCHLORAL [HCl], **IV**, 130

CHELIDONIC ACID, **II**, 126

CHLOROACETONITRILE [P₂O₅],
IV, 144

2-CHLORONICOTINONITRILE
[PCl₅], **IV**, 166

m-CHLOROSTYRENE [KHSO₄], **III**,
204

COUMALIC ACID [H₂SO₄, SO₃], **IV**,
201

CREATININE [HCl], **I**, 172

CYCLOHEXANONE OXIME, **II**, 76,
314

CYCLOHEXENE [H₃PO₄], **II**, 152;
[H₂SO₄], **I**, 183

2-CYCLOHEXENONE [H₂SO₄], **V**,
294

CYCLOHEXYL ISOCYANIDE
[POCl₃; C₅H₅N], **V**, 300

CYCLOPENTENE [H₃PO₄], **II**, 153

2-CYCLOPENTENONE
[*p*-CH₃C₆H₄SO₃H], **V**, 326

DIACETYL-*d*-TARTARIC ANHY-
DRIDE [(CH₃CO)₂O, H₂SO₄],
IV, 242

4,4'-DICHLORODIBUTYL ETHER
[H₂SO₄], **IV**, 266

2,3-DIMETHYL-1,3-BUTADIENE
[HBr], **III**, 312

2,3-DIMETHYL-1-3-BUTADIENE
[Al₂O₃], **III**, 313

DIMETHYLFURAZAN [MALEIC

ANHYDRIDE], **IV**, 342

5,5-DIMETHYL-2-*n*-PENTYL-
TETRAHYDROFURAN
[H₃PO₄], **IV**, 350

2,5-DIMETHYLPYRROLE, **II**, 219

2,4-DINITROBENZALDEHYDE, **II**,
223

4,5-DIPHENYLGLYOXALONE, **II**,
231

1,1-DIPHENYLETHYLENE [Heat],
I, 226

2,3-DIPHENYLINDONE [Heat], **III**,
353

1-ETHYL-3-(3-DIMETHYL-
AMINO)PROPYLCARBODI-
IMIDE HYDROCHLORIDE
and METHIODIDE
[*p*-CH₃C₆H₄SO₂Cl; (C₂H₅)₃N],
V, 555

2-ETHYLHEXANONITRILE
[SOCl₂], **IV**, 436

FUMARONITRILE [P₂O₅], **IV**, 486

FURFURAL [H₂SO₄], **I**, 280

GLUTARIMIDE [Heat], **IV**, 496

HEPTALDOXIME, **II**, 313

INDOLE-3-CARBONITRILE
[(NH₄)₂-HPO₄], **V**, 656

ISOBUTYRONITRILE [P₂O₅], **III**,
493

ITACONIC ANHYDRIDE [Heat], **II**,
368

MALONONITRILE [PCl₅—POCl₃],
II, 379; [POCl₃], **III**, 535

MESITYL OXIDE [I₂], **I**, 345

METHACRYLAMIDE [H₂SO₄], **III**,
560

2-METHYLBENZIMIDAZOLE, **II**,
66

METHYL-ω-CYANOPELARGON-
ATE [P₂O₅], **III**, 584

β-METHYLGLUTARIC ANHY-
DRIDE [(CH₃CO)₂O], **IV**, 630

METHYL ISOCYANIDE
[*p*-CH₃C₆H₄SO₂Cl; C₉H₇N], **V**,
772

6-METHYLURACIL [HCl], **II**, 422

NICOTINIC ANHYDRIDE
[COCl₂; (C₂H₅)₃N], **V**, 822
NICOTINONITRILE [P₂O₅], **IV**, 706
3-NITROPHTHALIC ANHYDRIDE
[(CH₃CO)₂O], **I**, 410
OXALIC ACID [Heat], **I**, 421
PENTAACETYL-d-GLUCONONI-
TRILE [(CH₃CO)₂O], **III**, 690
2-PENTENE [H₂SO₄], **I**, 430
1-PHENYLNAPHTHALENE
[(CH₃CO)₂O], **III**, 729
trans-1-PHENYL-1,3-BUTADIENE
[H₂SO₄], **IV**, 771
α-PHENYLGLUTARIC ANHY-
DRIDE [(CH₃CO)₂O], **IV**, 790
PINACOLONE [H₂SO₄], **I**, 462
PYRROLE [Heat], **I**, 473
PYRUVIC ACID [KHSO₄], **I**, 475
SEBACONITRILE [Heat], **III**, 768
SUCCINIC ANHYDRIDE
[CH₃COCl] [POCl₃], **II**, 560
p-TOLUENESULFONIC ANHY-
DRIDE [P₂O₅], **IV**, 940
o-TOLYL ISOCYANIDE [POCl₃;
KOC₄H₉], **V**, 1060
TRIPHENYLETHYLENE [H₂SO₄],
II, 606
sym.-TRITHIANE [HCl], **II**, 610
VERATRONITRILE [(CH₃CO)₂O],
II, 622
DEHYDROGENATION (See also Oxida-
tion.) The reagent used is in-
cluded in brackets after the
name of the preparations.
2,2'-BIPYRIDINE [Ni], **V**, 102
trans-4,4'-DIMETHOXYSTILBENE
[DDQ], **V**, 428
2,4-DIPHENYLPYRROLE [Se], **III**,
358
INDAZOLE [Pd], **IV**, 536
β-METHYL-δ-VALEROLACTONE
[CuCr₂O₄], **IV**, 677
1,2-NAPHTHALIC ANHYDRIDE
[S], **II**, 423
PHENOXTHIN [S], **II**, 485
PHENYLBENZOYLDIAZOME-

THANE [HgO], **II**, 496
1-PHENYLNAPHTHALENE [S], **III**,
729
TETRAPHENYLPHTHALIC AN-
HYDRIDE [Br₂], **III**, 807
DEHYDROHALOGENATION. The reagent
used is included in brackets
after the names of the prepara-
tions.
α-BROMOBENZALACETONE
[CH₃CO₂Na], **III**, 125
2-BUTYN-1-OL [NaNH₂], **IV**, 128
N-CHLOROCYCLOHEXYLIDE-
NEIMINE [KOAc], **V**, 208
γ-CROTONOLACTONE [Et₃N], **V**,
255
1,3-CYCLOHEXADIENE
[NaOCH(CH₃)₂], **V**, 285
DI-(p-CHLOROPHENYL)ACETIC
ACID [KOH], **III**, 270
2,7-DIMETHYLOXEPIN
[KOC(CH₃)₃], **V**, 467
DIPHENYLACETYLENE [KOH],
III, 350
DIPHENYLCARBODIIMIDE
[Method II], [(C₂H₅)₃N], **V**, 504
DIPHENYLCYCLOPROPENONE
[(C₂H₅)₃N], **V**, 514
3,4-DINITRO-3-HEXENE [KOH],
IV, 372
trans-2-DODECENOIC ACID
[KOC(CH₃)₃], **IV**, 398
ETHOXYACETYLENE [NaNH₂],
IV, 404
KETENE DIETHYLACETAL
[KOC(CH₃)₃], **III**, 506
6-KETOHENDECANEDIOIC ACID
[(C₂H₅)₃N], **IV**, 555
LAURONE [(C₂H₅)₃N], **IV**, 560
2-METHYL-2-CYCLOHEXENONE
[collidine or LiCl in
HCON(CH₃)₂], **IV**, 162
trans-2-METHYL-2-DODECENOIC
ACID [C₉H₇N], **IV**, 608
2-METHYLENEDODECANOIC
ACID [NaOH], **IV**, 616

MONOVINYLACETYLENE
[KOH], **IV,** 683
MUCONIC ACID [KOH], **III,** 623
9-NITROANTHRACENE [NaOH],
IV, 711
4-PENTYN-1-OL [NaNH₂], **IV,** 755
PHENACYLAMINE HYDROCH-
LORIDE [NaOCH₃], **V,** 909
PHENYLACETYLENE [NaNH₂],
IV, 763
PHENYLPROPARGYL ALDEHYDE
[K₂CO₃], **III,** 731; [KOH], **III,**
731
STEAROLIC ACID [KOH], **III,** 785;
[NaNH₂], **IV,** 851
10-UNDECYNOIC ACID [NaNH₂],
IV, 969

DELEPINE REACTION
2-BROMOALLYLAMINE, **V,** 121

DESULFURIZATION. The reagent used is
included in brackets.
o-CHLOROPHENYLCYANA-
MIDE [(CH₃CO₂)₂Pb], **IV,** 172
4-METHYL-6-HYDROXYPYRIMI-
DINE [Ni—H₂], **IV,** 638

DIAZOTIZATION
A. PREPARATION OF DIAZONIUM
SALTS (warning: Hazardous.
Observe safety precautions).
BENZENEDIAZONIUM-2-CAR-
BOXYLATE, **V,** 54
4,4′-BIPHENYLENE-*bis*-DIAZON-
IUM FLUOBORATE, **II,** 188
o-BROMOBENZENE DIAZONIUM
HEXAFLUOROPHOSPHATE,
V, 133
p-CARBETHOXYBENZENEDI-
AZONIUM FLUOBORATE,
II, 299
β-NAPHTHALENEDIAZONIUM
CHLORIDE-MERCURIC
CHLORIDE COMPOUND, **II,**
432
o-NITROBENZENEDIAZONIUM
FLUOBORATE, **II,** 226

p-NITROBENZENEDIAZONIUM
FLUOBORATE, **II,** 225
B. COUPLING REACTIONS (Reten-
tion of N₂)
AZO DYE FROM *o*-TOLIDINE AND
CHICAGO ACID, **II,** 145
4-BENZENEAZO-1-NAPHTHOL, **I,**
49
1,2,3-BENZOTHIADIAZOLE 1,1-
DIOXIDE, **V,** 61
DIAZOAMINOBENZENE, **II,** 163
METHYL RED, **I,** 374
1-METHYL-3-*p*-TOLYLTRIAZENE,
V, 797
ORANGE I, **II,** 39
ORANGE II, **II,** 36
C. REDUCTION OF AZO AND DIA-
ZONIUM GROUPS
1,2-AMINONAPHTHOL HY-
DROCHLORIDE, **II,** 35
1,4-AMINONAPHTHOL HY-
DROCHLORIDE, **I,** 49; **II,** 39
PHENYLHYDRAZINE, **I,** 442
D. REPLACEMENT OF DIAZON-
IUM GROUP (Loss of N₂)
1. By Aldehyde
2-BROMO-4-METHYLBEN-
ZALDEHYDE, **V,** 139
2. By Aryl
p-BROMOBIPHENYL, **I,** 113
m-NITROBIPHENYL, **IV,** 718
3. By Azido
BENZOFURAZAN OXIDE,
IV, 75
2-NITROCARBAZOLE, **V,** 829
4. By Cyano
o-TOLUNITRILE, **I,** 514
p-TOLUNITRILE, **I,** 514
5. By Halogen
(a)By Br
m-BROMOBENZALDE-
HYDE, **II,** 132
o-BROMOTOLUENE, **I,** 135
p-BROMOTOLUENE, **I,** 136
o-CHLOROBROMOBEN-
ZENE, **III,** 185

(b) By Cl
m-CHLOROBENZALDE-
HYDE, **II**, 130
1-CHLORO-2,6-DINITRO-
BENZENE, **IV**, 160
m-CHLORONITROBEN-
ZENE, **I**, 162
2-CHLOROPYRIMIDINE,
IV, 182
o-CHLOROTOLUENE, **I**, 170
p-CHLOROTOLUENE, **I**, 170
(c) By F
1-BROMO-2-FLUOROBEN-
ZENE, **V**, 133
4,4'-DIFLUOROBIPHENYL,
II, 188
FLUOROBENZENE, **II**, 295
p-FLUOROBENZOIC ACID,
II, 299
(d) By I
IODOBENZENE, **II**, 351
o-IODOBROMOBENZENE,
V, 1120
p-IODOPHENOL, **II**, 355
1,2,3-TRIODO-5-NITROBEN-
ZENE, **II**, 604
6. By Hydrogen
m-BROMOTOLUENE, **I**, 133
3,3'-DIMETHOXYBI-
PHENYL, **III**, 295
3,3'-DIMETHYLBIPHENYL,
III, 295
m-IODOBENZOIC ACID, **II**,
353
sym.-TRIBROMOBENZENE,
II, 592
2,4,6-TRIBROMOBENZOIC
ACID, **IV**, 947
7. By Hydroxyl
3-BROMO-4-HYDROXY-
TOLUENE, **III**, 130
m-HYDROXYBENZALDE-
HYDE, **III**, 453, 564
3-METHOXY-5-NITRO-
PHENOL, **I**, 405
m-NITROPHENOL, **I**, 404

3-NITRO-4,6-XYLENOL, **I**,
405
2,4,6-TRIPHENYLPHENOL,
V, 1130
8. By Nitro
o-DINITROBENZENE, **II**, 226
p-DINITROBENZENE, **II**, 225
1,4-DINITRONAPHTHAL-
ENE, **III**, 341
m-NITROTOLUENE, **I**, 415
9. By Selenium
DIPHENYL SELENIDE, **II**,
238
DI-*m*-TOLYL SELENIDE, **II**,
239
DI-*o*-TOLYL SELENIDE, **II**,
239
DI-*p*-TOLYL SELENIDE, **II**,
239
10. By Thiol
m-THIOCRESOL, **III**, 809
11. Miscellaneous
BIPHENYLENE, **V**, 54
DIPHENIC ACID, **I**, 222
MERCURY DI-*β*-NAPH-
THYL, **II**, 381
β-NAPHTHYLMERCURIC
CHLORIDE, **II**, 432
PHENYLARSONIC ACID, **II**,
494
THIOSALICYCLIC ACID, **II**,
580
dl-4,4',6,6'-TETRACHLORO-
DIPHENIC ACID, **IV**, 872
DIAZO TRANSFER
t-BUTYL DIAZOACETATE, **V**, 179
ELECTROLYTIC REACTIONS
2,7-DIMETHYL-2,7-DINI-
TROÖCTANE, **V**, 445
DIMETHYL OCTADECANE-
DIOATE, **V**, 463
ELIMINATION (See also Dehalogenation;
Dehydrogenation; Dehydrohal-
ogenation.) The group elimi-
nated is given in brackets.
3(5)-AMINOPYRAZOLE(ArSO$_2$Na),
V, 39

p-CHLOROPHENYL ISOTHIO-
CYANATE(Zn(SCH₂·
CO₂NH₄)₂), **V, 223**
CROTYL DIAZOACETATE-
(ArSO₂H), **V, 258**
trans-CYCLOOCTENE((CH₃)₃N), **V,
315**
1,3,5-HEXATRIENE(C₆H₅·
CH₂N(CH₃)₂), **V, 608**
D-2-OXO-7,7-DIMETHYL-1-VI-
NYLBICYCLO-
(2.2.1)HEPTANE(SO₂), **V, 877**
ENAMINE OR IMINE FORMATION (See
also Grignard and Related Reactions).
1-MORPHOLINO-1-CYCLOHEX-
ENE, **V, 808**
2-PHENYLINDAZOLE, **V, 941**
EPOXIDATION
2-*t*-BUTYL-3-PHENYLOXAZIR-
ANE, **V, 191**
DIHYDROXYCYCLOPENTENE,
V, 414
2,7-DIMETHYLOXEPIN, **V, 467**
ISOPHORONE OXIDE, **IV, 552**
trans-STILBENE OXIDE, **IV, 860**
TETRACYANOETHYLENE OX-
IDE, **V, 1007**
ESTERIFICATION (See also Acylation),
n-AMYL BORATE, **II, 107**
n-AMYL PHOSPHATE, **II, 110**
t-BUTYL ACETOACETATE, **V,
155**
n-BUTYL BORATE, **II, 106**
n-BUTYL *n*-BUTYRATE, **I, 138**
t-BUTYL CARBAMATE, **V, 162**
t-BUTYL CARBAZATE, **V, 166**
t-BUTYL CYANOACETATE, **V,
171**
n-BUTYL NITRITE, **II, 108**
n-BUTYL PHOSPHATE, **II, 109**
sec.-BUTYL PHOSPHATE, **II, 110**
n-BUTYL SULFATE, **II, 111**
n-BUTYL SULFITE, **II, 112**
n-BUTYL *p*-TOLUENESULFON-
ATE, **I, 145**

α-CHLOROPHENYLACETIC
ACID, **IV, 169**
γ-CHLOROPROPYL ACETATE,
III, 203
CROTYL DIAZOACETATE, **V,
258**
DI-*t*-BUTYL MALONATE, **IV, 261**
DIETHYL *cis*-Δ⁴-TETRAHYDRO-
PHTHALATE, **IV, 304**
DIMETHYL ACETYLENEDI-
CARBOXYLATE, **IV, 329**
DIMETHYL 3-METHYLENECY-
CLOBUTANE-1,2-DICAR-
BOXYLATE, **V, 459**
n-DODECYL *p*-TOLUENESUL-
FONATE, **III, 366**
ETHYL ACETONEDICARBOXY-
LATE, **I, 237**
ETHYL ADIPATE, **II, 264**
ETHYL *p*-AMINOBENZOATE, **II,
300**
ETHYL BENZOYLFORMATE, **I,
241**
ETHYL BROMOACETATE, **III,
381**
ETHYL β-BROMOPROPIONATE,
I, 246
ETHYL *t*-BUTYL MALONATE,
IV, 417
ETHYL CHLOROACETATE, **II,
263**
ETHYL CYANOACETATE, **I, 254**
ETHYL α-CYANO-β-PHENYLA-
CRYLATE, **I, 451**
ETHYL ETHOXYACETATE, **II,
261**
ETHYL HYDROGEN ADIPATE,
II, 277
ETHYL HYDROGEN SEBA-
CATE, **II, 276**
ETHYL OXALATE, **I, 261**
ETHYL PHENYLACETATE, **I, 270**
ETHYL γ-PHENYLBUTYRATE,
II, 196
ETHYL PIMELATE, **II, 536**

ETHYL 2-PYRIDYLACETATE, **III**, 413
ETHYL SEBACATE, **II**, 277
ETHYL *n*-TRIDECYLATE, **II**, 292
2-FURYLMETHYL ACETATE, **I**, 285
6-HYDROXYNICOTINIC ACID, **IV**, 532
ISOPROPYL LACTATE, **II**, 365
METHYL BENZOYLFORMATE, **I**, 244
3-METHYLHEPTANOIC ACID, **V**, 762
METHYL HOMOVERATRATE, **II**, 333
METHYL HYDROGEN HENDE-CANEDIOATE, **IV**, 635
METHYL NITRATE, **II**, 412
METHYL OXALATE, **I**, 264, **II**, 414
METHYL PYRUVATE, **III**, 610
METHYL *p*-TOLUENESULFON-ATE, **I**, 146
1-METHYL-3-*p*-TOLYLTRI-AZENE, **V**, 797
2-NORBORNANONE, **V**, 852
NORTRICYCLANOL, **V**, 863
sec-OCTYL HYDROGEN PHTHALATE, **I**, 418
2,3,4,5,6-PENTA-*O*-ACETYL-D-GLUCONIC ACID and 2,3,-4,5,6-PENTA-*O*-ACETYL-D-GLUCONYL CHLORIDE, **V**, 887
n-PROPYL PHOSPHATE, **II**, 110
n-PROPYL SULFATE, **II**, 112
ETHERIFICATION
CHOLESTANYL METHYL ETHER, **V**, 245
3-ETHOXY-2-CYCLOHEXEN-ONE, **V**, 539
(+) and (−)-α-(2,4,5,7-TETRANI-TRO-9-FLUORENYLIDE-NEAMINOOXY) PRO-PIONIC ACID, **V**, 1031

EXCHANGE REACTIONS (See also Acy-lation).
A. ESTER-ACID
ACRYLIC ACID, **III**, 33
VINYL LAURATE, **IV**, 977
B. ESTER-ALCOHOL
METHYL MYRISTATE, **III**, 605
METHYL PALMITATE, **III**, 605
NORTRICYCLANOL, **V**, 863
C. ESTER-AMINE
BENZOYLACETANILIDE, **III**, 108
LACTAMIDE, **III**, 516
MANDELAMIDE, **III**, 536
SALICYL-*o*-TOLÚIDE, **III**, 765
D. ESTER-ESTER
ALLYL LACTATE, **III**, 46
n-BUTYL ACRYLATE, **III**, 146
FORMYLATION REACTIONS
2-*n*-BUTYL-2-METHYLCYCLO-HEXANONE, **V**, 187
2-CHLORO-1-FORMYL-1-CY-CLOHEXENE, **V**, 215
p-DIMETHYLAMINOBENZAL-DEHYDE, **IV**, 331
N-ETHYL-*p*-CHLOROANILINE, **IV**, 420
INDOLE-3-ALDEHYDE, **IV**, 539
MESITALDEHYDE, **V**, 49
2-PYRROLEALDEHYDE, **IV**, 831
SYRINGIC ALDEHYDE, **IV**, 866
2-THENALDEHYDE, **IV**, 915
3-THENALDEHYDE, **IV**, 918
FREUND REACTION
CYCLOPROPYLBENZENE, **V**, 328
FRIEDEL-CRAFTS REACTION
β-(3-ACENAPHTHOYL)-PROPIONIC ACID, **III**, 6
ACETO-*p*-CYMEME, **II**, 3
ACETOPHENONE, **I**, 111
2-ACETOTHIENONE, **II**, 8; **III**, 14
9-ACETYLANTHRACENE, **IV**, 8
2-ACETYLFLUORENE, **III**, 23
2-AMINOBENZOPHENONE, **IV**, 34
BENZOPHENONE, **I**, 95

β-BENZOYLACRYLIC ACID, **III,** 109

γ-BENZOYLBUTYRIC ACID, **II,** 82

β-BENZOYLPROPIONIC ACID, **II,** 81

3-BENZOYLPYRIDINE, **IV,** 88

4,4′-BIS(DIMETHYLAMINO)-BENZIL, **V,** 111

p-BROMOACETOPHENONE, **I,** 109

p-CHLOROACETOPHENONE, **I,** 111

p-CHLOROACETYLACETANI-LIDE, **III,** 183

m-CYMENE, **V,** 332

DESOXYBENZOIN **II,** 156

1,4-DIBENZOYLBUTANE, **II,** 169

trans-DIBENZOYLETHYLENE, **III,** 248

α,α-DIPHENYLACETONE, **III,** 343

DIPHENYLACETONITRILE, **III,** 347

DIPHENYL SULFIDE, **II,** 242

DURENE, **II,** 248

HEXAMETHYLBENZENE, **II,** 248

MESITALDEHYDE, **III,** 549

MESITOIC ACID, **V,** 706

p-METHOXYACETOPHENONE, **I,** 111

p-METHYLACETOPHENONE, **I,** 111

1-METHYL-3-ETHYLOXINDOLE, **IV,** 620

NEOPHYL CHLORIDE, **IV,** 702

PENTAMETHYLBENZENE, **II,** 248

PHENYLDICHLOROPHOS-PHINE, **IV,** 784

PHENYL THIENYL KETONE, **II,** 520

RESACETOPHENONE, **III,** 761

α-TETRALONE, **II,** 569; **IV,** 898

2α-THIOHOMOPHTHALIMIDE, **V,** 1051

p-TOLUYL-o-BENZOIC ACID, **I,** 517

TRIPHENYLCHLOROME-THANE, **III,** 842

TRIPHENYLMETHANE, **I,** 548

α,β,β-TRIPHENYLPROPIONIC ACID, **IV,** 960

TRIPHENYLSELENONIUM CHLORIDE, **II,** 241

GRIGNARD REACTIONS

9-ACETYLPHENANTHRENE, **III,** 26

n-AMYLBENZENE, **II,** 47

o-ANISALDEHYDE, **V,** 46

BENZENEBORONIC ANHY-DRIDE, **IV,** 68

BENZOIC ACID, **I,** 363

3-BENZYL-3-METHYLPENTAN-OIC ACID, **IV,** 93

BIALLYL, **III,** 121

2-BROMODECENE-1, **I,** 187

CHLORODIISOPROPYLPHOS-PHINE, **V,** 211

m-CHLOROPHENYLMETHYL-CARBINOL, **III,** 200

CYCLOHEXANECARBOXYLIC ACID, **I,** 364

3-CYCLOHEXYL-2-BROMOPRO-PENE, **I,** 186

CYCLOHEXYLCARBINOL, **I,** 188

CYCLOHEXYL METHYL KE-TONE, **V,** 775

3,12-DIACETOXY-bis nor-CHO-LANYLDIPHENYLETHY-LENE, **III,** 237

DI-n-BUTYL CARBONOL, **II,** 179

DI-n-BUTYLDIVINYLTIN, **IV,** 258

2,4-DIMETHYL-5-CARBETHOXY-PYRROLE, **II,** 198

4,6-DIMETHYL-1-HEPTEN-4-OL, **V,** 452

DIPHENYL-p-BROMOPHENYL-PHOSPHINE, **V,** 496

DIPHENYL KETIMINE, **V,** 520

1,1-DIPHENYLETHYLENE, **I,** 226

2,3-DIPHENYLINDONE, **III,** 353
ETHYL α-NAPHTHOATE, **II,** 282
n-HEXALDEHYDE, **II,** 323
1,3,5-HEXATRIENE, **V,** 608
n-HEXYL ALCOHOL, **I,** 306
2-HYDROXYTHIOPHENE, **V,** 642
ISODURENE, **II,** 360
MESITOIC ACID, **III,** 553
ω-METHOXYACETOPHENONE,
     III, 562
6-METHOXY-2-NAPHTHOL, **V,**
     918
10-METHYL-10,9-BORAZARO-
     PHENANTHRENE, **V,** 727
2-METHYL-2,5-DECANEDIOL,
     IV, 601
N-METHYL-1,2-DIPHENYL-
     ETHYLAMINE, **IV,** 605
3-METHYLHEPTANOIC ACID, **V,**
     762
METHYL ISOPROPYL CARBI-
     NOL, **II,** 406
METHYL 4-KETO-7-METHYLOC-
     TANOATE, **III,** 601
dl-METHYLETHYLACETIC
     ACID, **I,** 361
METHYL 2-THIENYL SULFIDE,
     IV. 667
α-NAPHTHOIC ACID, **II,** 425
PENTACHLOROBENZOIC ACID,
     V, 890
PENTANE, **II,** 478
3-PENTEN-2-OL, **III,** 696
PHENANTHRENE-9-ALDE-
     HYDE, **III,** 701
4-PHENYL-2-BROMOBUTENE-1,
     I, 187
trans-1-PHENYL-1,3-BUTADIENE,
     IV, 771
PHENYL *t*-BUTYL ETHER
     (Method I), **V,** 924
1-PHENYLNAPHTHALENE, **III,**
     729
1-PHENYL-1-PENTEN-4-YN-3-OL,
     IV, 792
n-PROPYLBENZENE, **I,** 471

SELENOPHENOL, **III,** 771
TETRAETHYLTIN, **IV,** 881
TETRAMETHYLBIPHOSPHINE
     DISULFIDE, **V,** 1016
TETRAPHENYLARSONIUM
     CHLORIDE HYDRO-
     CHLORIDE, **IV,** 910
TETROLIC ACID, **V,** 1043
1-*p*-TOLYLCYCLOPROPANOL, **V,**
     1058
TRI-*n*-AMYL CARBINOL, **II,** 603
TRI-*n*-BUTYL CARBINOL, **II,** 603
TRIETHYL CARBINOL, **II,** 602
TRI-*n*-HEPTYL CARBINOL, **II,**
     603
TRIMETHYLACETIC ACID, **I,**
     524
TRIPHENYLCARBINOL, **III,** 839
TRIPHENYLETHYLENE, **II,** 606
TRIPHENYLSTIBINE, **I,** 550
TRI-*n*-PROPYL CARBINOL, **II,**
     603
TRI-*p*-TOLYLSTIBINE, **I,** 551
n-VALERIC ACID, **I,** 363
HALOFORM REACTION
     β-NAPHTHOIC ACID, **II,** 428
HALOGENATION
     A. BROMINATION
          ε-BENZOYLAMINO-α-BROMO-
               CAPROIC ACID, **II,** 74
          BROMAL, **II,** 87
          N-BROMOACETAMIDE, **IV,** 104
          3-BROMO-4-ACETAMINOTOL-
               UENE, **I,** 111
          BROMOACETONE, **II,** 88
          3-BROMOACETOPHENONE, **V,**
               117
          p-BROMOBENZALDEHYDE, **II,**
               89
          α-BROMO-*n*-CAPROIC ACID, **I,**
               115; **II,** 95
          4-BROMO-2-HEPTENE, **IV,** 108
          α-BROMOISOCAPROIC ACID, **II,**
               95
          α-BROMOISOVALERIC ACID, **II,**
               93

p-BROMOMANDELIC ACID, **IV**, 110
BROMOMESITYLENE, **II**, 95
2-BROMO-3-METHYLBENZOIC ACID, **IV**, 114
α-BROMO-β-METHYLVALERIC ACID, **II**, 95
α-BROMONAPHTHALENE, **I**, 121
6-BROMO-2-NAPHTHOL, **III**, 132
m-BROMONITROBENZENE, **I**, 123
p-BROMOPHENACYL BROMIDE, **I**, 127
9-BROMOPHENANTHRENE, **III**, 134
o-BROMOPHENOL, **II**, 97
p-BROMOPHENOL, **I**, 128
α-BROMO-α-PHENYLACTONE, **III**, 343
α-BROMO-α-PHENYLACETONI-TRILE, **III**, 347
3-BROMOPHTHALIDE, **V**, 145
3-BROMOPYRENE, **V**, 147
4-BROMO-*o*-XYLENE, **III**, 138
γ-CROTONOLACTONE, **V**, 255
CYANOGEN BROMIDE, **II**, 150
CYCLOPROPYLBENZENE, **V**, 328
2,3-DIAMINOPYRIDINE, **V**, 346
DIBENZOYLDIBROMOME-THANE, **II**, 244
DIBROMOACETONITRILE, **IV**, 254
2,6-DIBROMOANILINE, **III**, 262
9,10-DIBROMOANTHRACENE, **I**, 207
4,4'-DIBROMOBIPHENYL, **IV**, 256
2,6-DIBROMO-4-NITROPHENOL, **II**, 173
2,4-DIHYDROXY-5-BROMOBEN-ZOIC ACID, **II**, 100
2,7-DIMETHYLOXEPIN, **V**, 467
DIPHENYLCYCLOPROPENONE, **V**, 514
trans-2-DODECENOIC ACID, **IV**, 398
ETHYL BROMOACETATE, **III**, 381
ETHYL BROMOMALONATE, **I**, 245
dl-ISOLEUCINE, **III**, 495
dl-LEUCINE, **III**, 523
1-METHYLAMINO-4-BROMOAN-THRAQUINONE, **III**, 575
METHYL 5-BROMOVALERATE, **III**, 578
trans-2-METHYL-2-DODECENOIC ACID, **IV**, 608
2-METHYLENEDODECANOIC ACID, **IV**, 616
MUCOBROMIC ACID, **III**, 621; **IV**, 688
MUCONIC ACID, **III**, 623
o-NITROBENZALDEHYDE, **V**, 825
p-NITROBENZYL BROMIDE, **II**, 443
PHENACYL BROMIDE, **II**, 480
dl-PHENYLALANINE, **III**, 705
PHENYLBROMOETHYNE, **V**, 921
o-PHTHALALDEHYDE, **IV**, 807
PHTHALALDEHYDIC ACID, **III**, 737
dl-SERINE, **III**, 774
TEREPHTHALALDEHYDE, **III**, 788
3-THENYL BROMIDE, **IV**, 921
dl-THREONINE, **III**, 813
sym.-TRIBROMOBENZENE, **II**, 592
2,4,6-TRIBROMOBENZOIC ACID, **IV**, 947
TRIMETHYLETHYLENE DI-BROMIDE, **II**, 409
dl-VALINE, **III**, 848
o-XYLYLENE DIBROMIDE, **IV**, 984
B. CHLORINATION
t-BUTYL HYPOCHLORITE, **IV**, 125; **V**, 184
9-CHLOROANTHRACENE, **V**, 206
α-CHLOROANTHRAQUINONE, **II**, 128

p-CHLOROBENZALDEHYDE, **II**, 133

o-CHLOROBENZOYL CHLORIDE, **I**, 155

2-CHLOROCYCLOHEXANONE, **III**, 188

2-CHLORO-1-FORMYL-1-CYCLOHEXENE, **V**, 215

2-CHLORO-2-METHYLCYCLOHEXANONE, **IV**, 162

p-CHLOROPHENOXYMETHYL CHLORIDE, **V**, 221

2-CHLOROTHIIRANE 1,1-DIOXIDE, **V**, 231

2,6-DICHLOROANILINE, **III**, 262

2,6-DICHLOROPHENOL, **III**, 267

DIETHYLTHIOCARBAMYL CHLORIDE, **IV**, 307

2,4-DINITROPHENYLSULFUR CHLORIDE, **II**, 456

DIPHENYLCARBODIIMIDE (Method II), **V**, 504

3-HYDROXYQUINOLINE, **V**, 635

MANDELIC ACID, **III**, 538

3-METHYLCOUMARONE, **IV**, 590

o-NITROBENZENESULFONYL CHLORIDE, **II**, 471

o-NITRO-*p*-CHLOROPHENYLSULFUR CHLORIDE, **II**, 456

o-NITROPHENYLSULFUR CHLORIDE, **II**, 455

PHENACYLAMINE HYDROCHLORIDE, **V**, 909

dl-4,4',6,6'-TETRACHLORODIPHENIC ACID, **IV**, 872

p-TOLUENESULFENYL CHLORIDE, **IV**, 934

C. IODINATION

CYANOGEN IODIDE, **IV**, 207

2,6-DIIODO-*p*-NITROANILINE, **II**, 196

2-HYDROXY-3,5-DIIODOBENZOIC ACID, **II**, 343

4-HYDROXY-3,5-DIIODOBENZOIC ACID, **II**, 344

p-IODOANILINE, **II**, 347

5-IODOANTHRANILIC ACID, **II**, 349

N-IODOSUCCINIMIDE, **V**, 663

2-IODOTHIOPHENE, **II**, 357, **IV**, 545

4-IODOVERATROLE, **IV**, 547

TETRAIODOPHTHALIC ANHYDRIDE, **III**, 796

HOESCH REACTION

PHLOROACETOPHENONE, **II**, 522

HOFMANN DECOMPOSITION (See Elimination)

HUNSDIECKER REACTION

BROMOCYCLOPROPANE, **V**, 126

HYDROLYSIS. The subheadings illustrate the types of compounds hydrolyzed. The reagent used to effect hydrolysis is included in brackets after the name of the preparation.

A. ACID ANHYDRIDE

CITRACONIC ACID, **II**, 141

ITACONIC ACID, **II**, 368

MESACONIC ACID, **II**, 382

B. ACYL PEROXIDE

PERBENZOIC ACID [NaOCH$_3$], **I**, 431

C. AMIDE

ε-AMINOCAPROIC ACID, **IV**, 39

ATROLACTIC ACID, **IV**, 58

3-BROMO-4-AMINOTOLUENE [HCl], **I**, 111

t-BUTYLAMINE, **III**, 154

DIAZOMETHANE, **V**, 351

1,3-DIHYDROISOINDOLE, **V**, 406

N-ETHYL-*p*-CHLOROANILINE, **IV**, 420

GLUTARIC ACID, **IV**, 496

HOMOPHTHALIC ACID, **V**, 612

dl-LYSINE DIHYDROCHLORIDE [HCl], **II**, 375

dl-METHIONINE [HCl], **II**, 385

1-METHYLISOQUINOLINE, **IV**, 641

2-NITRO-4-METHOXYANILINE,
III, 661
dl-β-PHENYLALANINE [HCl], II,
491
α-PHENYLETHYLAMINE [HCl],
II, 503
β-VALEROLACTAM, V, 673
D. AMINE
1-NITRO-2-NAPHTHOL [NaOH],
II, 451
PHENYLPYRUVIC ACID [HCl],
II, 519
PHLOROGLUCINOL [HCl], I, 455
E. AZLACTONE
α-ACETAMINOCINNAMIC
ACID, II, 1
α-BENZOYLAMINOCINNAMIC
ACID [NaOH], II, 491
3,4-DIMETHOXYPHENYLPYRU-
VIC ACID [NaOH], II, 335
dl-β-PHENYLALANINE, II, 489
F. CARBOHYDRATE
l-ARABINOSE [H₂SO₄], I, 67
CELLOBIOSE [NaOCH₃], II, 122
α-CELLOBIOSE OCTAACETATE
[H₂SO₄—(CH₃CO)₂O], II, 124
LEVULINIC ACID [HCl], I, 335
G. DIHALIDE
exo-cis-BICYCLO(3.3.0)OCTANE-2-
CARBOXYLIC ACID, V, 93
p-BROMOBENZALDEHYDE, II,
89
p-BROMOMANDELIC ACID, IV,
110
p-CHLOROBENZALDEHYDE
[H₂SO₄], II, 133
ETHYL CHLOROFLUOROACE-
TATE, IV, 423
o-PHTHALALDEHYDE, IV, 807
TEREPHTHALALDEHYDE, III,
788
H. ESTER
2-BENZYLCYCLOPENTANONE,
V, 76
3-BENZYL-3-METHYLPENTAN-
OIC ACID, IV, 93

p-BROMOBENZALDEHYDE
[H₂SO₄], II, 442
α-BROMO-n-CAPROIC ACID
[KOH], II, 95
α-BROMOISOCAPROIC ACID
[KOH], II, 95
α-BROMOISOVALERIC ACID
[KOH], II, 93
α-BROMO-β-METHYLVALERIC
ACID [KOH], II, 95
n-CAPROIC ACID [KOH], II, 417,
474
CHELIDONIC ACID [HCl], II, 126
2,6-DICHLOROPHENOL, III, 267
2,5-DIMETHYLMANDELIC
ACID, III, 326
2,6-DIMETHYLPYRIDINE [KOH],
II, 214
2,4-DIMETHYLPYRROLE [KOH],
II, 217
ERUCIC ACID [KOH], II, 258
ETHYL BENZOYLACETATE
[NH₄OH], II, 266
ETHYL t-BUTYL MALONATE,
IV, 417
GLUTARIC ACID [HCl], I, 291
HOMOVERATRIC ACID [NaOH],
II, 333
6-HYDROXYNICOTINIC ACID,
IV, 532
IDOSOBEZENE, V, 658
dl-ISOLEUCINE, III, 495
α-KETOGLUTARIC ACID, III, 510
LINOLEIC ACID, III, 526
LINOLENIC ACID, III, 531
METHYL n-AMYL KETONE,
[NaOH], I, 351
o-METHYLBENZYL ALCOHOL,
IV, 582
METHYL n-BUTYL KETONE
[NaOH], I, 352
METHYL iso-BUTYL KETONE
[NaOH], I, 352
METHYL CYCLOPROPYL KE-
TONE, IV, 597

trans-2-METHYL-2-DODECENOIC
ACID, **IV**, 608

2-METHYLENEDODECANOIC
ACID, **IV**, 616

3-METHYL-2-FUROIC ACID, **IV**,
628

β-METHYLGLUTARIC ANHY-
DRIDE, **IV**, 630

METHYLGLYOXAL-ω-PHENYL-
HYDRAZONE, **IV**, 633

4-METHYLHEXANONE-2
[NaOH], **I**, 352

METHYL HYDROGEN HENDE-
CANEDIOATE, **IV**, 635

3-METHYLPENTANOIC ACID
[KOH], **II**, 416

6-METHYLURACIL, [NAOH], **II**,
422

MYRISTIC ACID [NaOH], **I**, 379

o-NITROACETOPHENONE, **IV**,
708

p-NITROBENZALDEHYDE
[H₂SO₄], **II**, 441

m-NITROBENZOIC ACID
[NaOH], **I**, 391

p-NITROBENZYL ALCOHOL, **III**,
652

d- and *l*-OCTANOL-2- [NaOH], **I**,
418

PELARGONIC ACID, [KOH], **II**,
474

dl-PHENYLALANINE, **III**, 705

PHENYLSUCCINIC ACID [HCl],
I, 451

1-PHENYL-2-THIOBIURET, **V**, 966

PIMELIC ACID [NaOH], **II**, 531

(+)- and (−)-α-(2,4,5,7-TETRANI-
TRO-9-FLUORENYLIDE-
NEAMINOOXY)-PRO-
PIONIC ACID, **V**, 1031

TRICARBALLYLIC ACID [HCl],
I, 523

1,2,5-TRIHYDROXYPENTANE,
III, 833

I. ETHER
CATECHOL [HBr], **I**, 150

PHENYL α-METHOXYSTYRYL
KETONE [HCl], **I**, 205

J. IMIDE
ACID AMMONIUM α-SULFO-
BENZOATE [HCl], **I**, 14

γ-AMINOBUTYRIC ACID
[H₂SO₄], **II**, 25

dl-ASPARTIC ACID, **IV**, 55

γ-DI-*n*-BUTYLAMINOPROPY-
LAMINE, **III**, 256

β-ETHYL-β-METHYLGLUTARIC
ACID, **IV**, 441

β-HYDROXYPHENYLPYRUVIC
ACID, **V**, 627

β-METHYL-β-PHENYLGLU-
TARIC ACID, **IV**, 664

1-METHYLTHIOL-3-PHTHALAM-
IDOPROPANE-3,3-DICAR-
BOXYLIC ACID [NaOH], **II**,
385

4-NITROPHTHALIC ACID
[NaOH], **II**, 457

K. IMINE
ETHYL α-PHENYLACETOACE-
TATE [H₂SO₄], **II**, 284

PHLOROACETOPHENONE, **II**,
522

o-TOLUALDEHYDE, **III**, 818

L. LACTAM
ε-AMINOCAPROIC ACID [HCl],
II, 28

ε-BENZOYLAMINOCAPROIC
ACID [H₂SO₄], **II**, 76

M. NITRILE
β-ALANINE, **III**, 34

dl-ALANINE [HCl], **I**, 21

γ-AMINOBUTYRIC ACID
[H₂SO₄], **II**, 25

α-AMINODIETHYLACETIC
ACID, **III**, 66

dl-α-AMINOPHENYLACETIC
ACID, **III**, 84

dl-α-AMINO-α-PHENYLPRO-
PIONIC ACID, **III**, 88

ATROLACTIC ACID, **IV**, 58

BENZOYLFORMIC ACID, **III**, 114

3-BENZYL-3-METHYLPENTAN-
OIC ACID, **IV**, 93
2-BROMO-3-METHYLBENZOIC
ACID, **IV**, 414
β-BROMOPROPIONIC ACID
[HBr], **I**, 131
3-CHLORO-2,2,3-TRIFLUORO-
PROPIONIC ACID, **V**, 239
DIALLYLAMINE [H₂SO₄], **I**, 201
DI-n-BUTYLAMINE [H₂SO₄], **I**,
202
β-ETHYL-β-METHYLGLUTARIC
ACID, **IV**, 441
α-ETHYL-α-METHYLSUCCINIC
ACID, **V**, 572
ETHYL n-TRIDECYLATE [KOH],
II, 292
GLUTARIC ACID [HCl], **I**, 289;
IV, 496
GLYCINE [Ba(OH)₂], **I**, 298
GLYCINE ETHYL ESTER HY-
DROCHLORIDE [HCl], **II**,
310
D-GULONIC-γ-LACTONE, **IV**, 506
β-HYDROXYPROPIONIC ACID
[NaOH], **I**, 321
MALONIC ACID [NaOH], **II**, 376
MANDELIC ACID [HCl], **I**, 336
MESITYLACETIC ACID, **III**, 557
METHYL BENZYL KETONE
[H₂SO₄], **II**, 391
β-METHYLGLUTARIC ACID, **III**,
591
β-METHYL-β-PHENYLGLU-
TARIC ACID, **IV**, 664
p-NITROPHENYLACETIC ACID
[H₂SO₄], **I**, 406
PHENYLACETAMIDE, **IV**, 760
PHENYLACETIC ACID [H₂SO₄],
I, 436
α-PHENYLGLUTARIC ANHY-
DRIDE, **IV**, 790
PHENYLNITROMETHANE
[NaOH], **II**, 512
PHENYLSUCCINIC ACID, **IV**,
804; [HCl], **I**, 451

o-TOLUAMIDE [NaOH—H₂O₂], **II**,
586
o-TOLUIC ACID [H₂SO₄], **II**, 588
p-TOLUIC ACID [H₂SO₄], **II**, 589
VERATRIC AMIDE [NaOH—
H₂O₂], **II**, 44
VINYLACETIC ACID, **III**, 851
N. OXIME
HYDROXYLAMINE HYDRO-
CHLORIDE [HCl], **I**, 318
O. PROTEIN
d-ARGININE HYDROCHLORIDE
[HCl], **II**, 49
l-CYSTINE [HCl], **I**, 194
d-GLUTAMIC ACID [HCl], **I**, 286
l-HISTIDINE MONOHYDRO-
CHLORIDE [HCl], **II**, 330
l-TYROSINE [Pancreatin], **II**, 612
l-TRYPTOPHANE [Pancreatin], **II**,
612
P. SCHIFF BASE
d-ARGININE HYDROCHLORIDE
[HCl], **II**, 50
2,4-DINITROBENZALDEHYDE
[HCl], **II**, 223
GLYCINE [H₂SO₄], **I**, 298
GLYCINE ETHYL ESTER HY-
DROCHLORIDE [HCl], **II**,
310
β-NAPHTHOL PHENYLAMINO-
METHANE [HCl], **I**, 381
Q. SULFONIC ACID
p-CRESOL [KOH], **I**, 175
o-NITROANILINE [H₂SO₄], **I**, 388
R. MISCELLANEOUS
ACETONYLACETONE [H₂SO₄],
II, 219
ACETYLBENZOYL, **III**, 20
ALLYLAMINE [HCl], **II**, 24
2-AMINOBENZOPHENONE, **IV**,
34
BICYCLO(2.2.1)HEPTEN-7-ONE,
V, 91
BICYCLO(2.2.0) PENTANE, **V**, 96
α-BROMOHEPTALDEHYDE, **III**,
127

o-BROMOPHENOL [H₂SO₄], II, 97
2-CYCLOHEXANONE, V, 294
sym.-DIMETHYLHYDRAZINE
DIHYDROCHLORIDE
[HCl], II, 208
2,6-DINITROANILINE, IV, 364
DOCOSANEDIOIC ACID, V, 533
ETHYL BENZOYLACETATE, IV,
415
2-FURFURYL MERCAPTAN, IV,
491
dl-GLYCERALDEHYDE [H₂SO₄],
II, 305
d-GLUCOSAMINE HYDRO-
CHLORIDE, III, 430
GLUTARIC ACID, IV, 496
5-HYDROXYPENTANAL, III, 470
6-KETOHENDECANEDIOIC
ACID, IV, 555
LAURONE, IV, 560
D-MANNOSE, III, 541
N-METHYL-3,4-DIHYDROXY-
PHENYLALANINE, III, 586
3-METHYL-1,5-PENTANEDIOL,
IV, 660
o-NITROBENZALDEHYDE, V,
825
ORTHANILIC ACID [Na₂CO₃], II,
471
γ-PHENYLALLYLSUCCINIC
ACID, IV, 766
PHENYLPROPARGYL ALDE-
HYDE, III, 731
PSEUDOPELLETIERINE, IV, 816
PYOCYANINE, III, 753
dl-SERINE, III, 774
β-TETRALONE, IV, 903
2-THIOPHENEALDEHYDE, III,
811
JAPP-KLINGEMANN REACTION
METHYLGLYOXAL-ω-PHENYL-
HYDRAZONE, IV, 633
KOLBE ELECTROLYSIS
DIMETHYL OCTADECANE-
DIOATE, V, 463
2,7-DIMETHYL-2,7-DINI-
TROÖCTANE, V, 445

KNORR PYRROLE SYNTHESIS
2,4-DIMETHYL-3,5-DICARBETH-
OXYPYRROLE, II, 202
KRYPTOPYRROLE, III, 513
LEUCKART REACTION
α-p-BROMOPHENYLETHYLA-
MINE, II, 505
α-p-CHLOROPHENYLETHYLA-
MINE, II, 505
α-(β-NAPHTHYL)-ETHYLAMINE,
II, 505
α-PHENYLETHYLAMINE, II, 503
α-p-XENYLETHYLAMINE, II, 505
α-p-TOLYLETHYLAMINE, II, 505
MANNICH REACTION
1-DIETHYLAMINO-3-BUTA-
NONE, IV, 281
N,N-DIMETHYLAMINOME-
THYLFERROCENE METH-
IODIDE, V, 434
β-DIMETHYLAMINOPROPIO-
PHENONE HYDRO-
CHLORIDE, III, 305
5-METHYLFURFURYLDIME-
THYLAMINE, IV, 626
MEERWEIN REACTION
1(p-NITROPHENYL)-1,3-BUTA-
DIENE, IV, 727
MERCURATION
o-CHLOROMERCURIPHENOL, I,
161
METALATION REACTIONS (See also
Grignard and Related Reac-
tions).
CYCLOPENTANECARBOXAL-
DEHYDE, V, 320
DI-n-BUTYLDIVINYLTIN, IV, 258
4,6-DIMETHYL-1-HEPTEN-4-OL,
V, 452
ETHYL 2-PYRIDYLACETATE,
III, 413
FERROCENE, IV, 473
PHENYL(TRICHLOROMETHYL)-
MERCURY, V, 969
1-(α-PYRIDYL)-2-PROPANOL, III,
757
RUTHENOCENE, V, 1001

TETRAETHYLTIN, **IV**, 881

TRIPHENYLALUMINUM, **V**, 1116

UNSOLVATED *n*-BUTYLMAG-
NESIUM CHLORIDE, **V**,
1141

MICHAEL REACTION

1,6-DIOXO-8*a*-METHYL-1,2,3,4,6,-
7,8*a*-OCTAHYDRONA-
PHTHALENE, **V**, 486

α-ETHYL-*α*-METHYLSUCCINIC
ACID, **V**, 572

Δ¹⁽⁹⁾-OCTALONE-2, **V**, 869

NENCKI REACTION

GALLACETOPHENONE, **II**, 304

NITRATION

2-AMINO-3-NITROTOLUENE, **IV**,
42

2,3-DIAMINOPYRIDINE, **V**, 346

2,6-DINITROANILINE, **IV**, 364

3,5-DINITROBENZOIC ACID, **III**,
337

DINITRODURENE, **II**, 254

3,5-DINITRO-*o*-TOLUNITRILE, **V**,
480

ETHYL *m*-NITROBENZOATE, **I**,
373

METHYL *m*-NITROBENZOATE,
I, 372

3-METHYL-4-NITROPYRIDINE-1-
OXIDE, **IV**, 654

m-NITROACETOPHENONE, **II**,
434

1-NITRO-2-ACETYLAMINONA-
PHTHALENE, **II**, 438

9-NITROANTHRACENE, **IV**, 711

NITROANTHRONE, **I**, 390

NITROBARBITURIC ACID, **II**,
440

m-NITROBENZALDEHYDE, **III**,
644

p-NITROBENZYL CYANIDE, **I**,
396

o-NITROCINNAMALDEHYDE,
IV, 722

2-NITRO-*p*-CYMENE, **III**, 653

m-NITRODIMETHYLANILINE,
III, 658

2-NITROFLUORENE, **II**, 447

NITROGUANIDINE, **I**, 399

NITROMESITYLENE, **II**, 449

2-NITRO-4-METHOXYANILINE,
III, 661

N-NITROMORPHOLINE, **V**, 839

3-NITROPHTHALIC ACID, **I**, 408

4-NITROPHTHALIMIDE, **II**, 459

2-NITROTHIOPHENE, **II**, 466

NITROUREA, **I**, 417

6-NITROVERATRALDEHYDE,
IV, 735

2,4,5,7-TETRANITROFLUOREN-
ONE, **V**, 1029

2,4,5-TRIAMINONITROBEN-
ZENE, **V**, 1067

2,4,7-TRINITROFLUORENONE,
III, 837

NITROSATION

AMINOMALONONITRILE *p*-TO-
LUENESULFONATE, **V**, 32

BIACETYL MONOXIME, **II**, 205

t-BUTYL AZIDOFORMATE, **V**,
157

ω-CHLOROISONITROSOACETO-
PHENONE, **III**, 191

CUPFERRON, **I**, 177

DIAMINOURACIL HYDRO-
CHLORIDE, **IV**, 247

DIAZOMETHANE, **III**, 244

DIETHYL ACETAMIDOMALON-
ATE, **V**, 373

2,4-DIMETHYL-3,5-DICARBETH-
OXYPYRROLE, **II**, 202

N-ETHYL-*m*-TOLUIDINE, **II**, 290

2-HYDROXY-3-METHYLISOCAR-
BOSTYRIL, **V**, 623

INDAZOLE, **V**, 650

p-NITROSODIETHYLANILINE,
II, 224

NITROSODIMETHYLAMINE, **II**,
211

p-NITROSODIMETHYLANILINE
HYDROCHLORIDE, **II**, 223

N-NITROSOMETHYLANILINE,
II, 460

NITROSOMETHYLUREA, **II**, 461

NITROSOMETHYLURETHANE,
II, 464
NITROSO-β-NAPHTHOL, I, 411
NITROSOTHYMOL, I, 511
NORTRICYCLANOL, V, 863
2-PHENYLCYCLOHEPTANONE,
IV, 780
2-PHENYLETHYL BENZOATE,
V, 336
3-PHENYLSYDNONE, V, 962
p-TOLYLSULFONYLMETHYL-
NITROSAMIDE, IV, 943
OXIDATION. Only those reactions are
included under this heading
which by common usage, structural change, and necessary reagents permit reasonably certain classification. Other headings such as "Dehydrogenation" and "Epoxidation"
should also be consulted. The
subheadings indicate the type of
oxidation, and the oxidizing
agent is included in brackets
after the name of the preparation.
A. CH$_3$ → CHO
p-BROMOBENZALDEHYDE
[CrO$_3$], II, 442
p-NITROBENZALDEHYDE
[CrO$_3$], II, 441
PHENYLGLYOXAL [SeO$_2$], II, 509
B. CH$_3$ → CO$_2$H
o-CHLOROBENZOIC ACID
[KMnO$_4$], II, 135
p-CHLOROMERCURIBENZOIC
ACID [KMnO$_4$], I, 159
2-HYDROXYISOPHTHALIC
ACID [PbO$_2$], V, 617
2,3-NAPHTHALENEDICARBOX-
YLIC ACID [H$_2$CrO$_4$], V, 810
p-NITROBENZOIC ACID
[Na$_2$Cr$_2$O$_7$], I, 392
PICOLINIC ACID HYDROCHLO-
RIDE [KMnO$_4$], III, 740
o-TOLUIC ACID [HNO$_3$], III, 820

2,4,6-TRINITROBENZOIC ACID
[Na$_2$Cr$_2$O$_7$], I, 543
C. CH$_2$ → CHOH
ACENAPHTHENOL-7,
[Pb(CH$_3$CO$_2$)$_4$], III, 3
D. CH$_2$ → CHOR
3-BENZOYLOXYCYCLOHEXENE
[C$_6$H$_5$CO$_3$C(CH$_3$)$_3$], V, 70
7-t-BUTOXYNORBORNADIENE
[C$_6$H$_5$CO$_3$C(CH$_3$)$_3$], V, 151
DIETHYL(O-BENZOYL)ETHYL-
TARTRONATE [(C$_6$H$_5$CO$_2$)$_2$],
V, 379
E. CH$_2$ → CO
ACENAPHTHENEQUINONE
[Na$_2$Cr$_2$O$_7$], III, 1
ALLOXAN MONOHYDRATE
[CrO$_3$], IV, 23
Δ^4-CHOLESTEN-3,6-DIONE
[Na$_2$Cr$_2$O$_7$], IV, 189
1,2-CYCLOHEXANEDIONE
DIOXIME [SeO$_2$], IV, 229
ETHYL OXOMALONATE [N$_2$O$_3$],
I, 266
GLYOXAL BISULFITE [H$_2$SeO$_3$],
III, 438
METHYL p-ACETYLBENZOATE
[Cr$_2$O$_3$ + O$_2$], IV, 579
o-NITROBENZALDEHYDE
[p-(CH$_3$)$_2$NC$_6$H$_4$NO], V, 825,
[CrO$_3$], III, 641
o- and p-NITROBENZALDI-
ACETATE [CrO$_3$], IV, 713
PROPIOLALDEHYDE [CrO$_3$], IV,
813
F. CH$_2$OH → CHO
PROPIONALDEHYDE [K$_2$Cr$_2$O$_7$],
II, 541
G. CH$_2$OH → CO$_2$H
n-BUTYL n-BUTYRATE
[Na$_2$Cr$_2$O$_7$] I, 138
β-CHLOROPROPIONIC ACID
[HNO$_3$], I, 168
H. CHOH → CO
BENZIL [CuSO$_4$ + pyridine], I, 87
BENZOYLFORMIC ACID
[KMnO$_4$], I, 244

CHOLANE-24-AL [(C$_6$H$_{11}$N)$_2$C; (CH$_3$)$_2$SO], **V**, 242

CHOLESTANONE [CrO$_3$], **II**, 139

Δ^4-CHOLESTEN-3,6-DIONE [Na$_2$Cr$_2$O$_7$], **IV**, 189

CHOLESTENONE [CH$_3$COCH$_3$], **III**, 207

Δ^4-CHOLESTEN-3-ONE [cyclohexanone], **IV**, 192

Δ^5-CHOLESTEN-3-ONE [Na$_2$Cr$_2$O$_7$], **IV**, 195

CYCLOÖCTANONE [H$_2$CrO$_4$], **V**, 310

2-CYCLOPENTENE-1,4-DIONE [H$_2$CrO$_4$], **V**, 324

α,γ-DICHLOROACETONE [Na$_2$Cr$_2$O$_7$], **I**, 211

ETHYL BENZOYLFORMATE [KMnO$_4$], **I**, 241

ETHYL PYRUVATE [KMnO$_4$], **IV**, 467

l-MENTHONE [NaCr$_2$O$_7$], **I**, 340

METHYL BENZOYLFORMATE [KMnO$_4$], **I**, 244

1-METHYL-2-PYRIDONE [K$_3$Fe(CN)$_6$], **II**, 419

2-NORBORNANONE [H$_2$CrO$_4$], **V**, 852

NORTRICYCLANONE [H$_2$CrO$_4$], **V**, 866

SEBACIL [Cu(O$_2$CCH$_3$)$_2$], **IV**, 838

I. CHO → CO$_2$H

β-CHLOROPROPIONIC ACID [HNO$_3$], **I**, 166

2-FUROIC ACID [CuO, Ag$_2$O, O$_2$], **IV**, 493

GLUTARIC ACID [HNO$_3$] **IV**, 496

n-HEPTANOIC ACID [KMnO$_4$], **II**, 315

PIPERONYLIC ACID [KMnO$_4$], **II**, 538

3-THENOIC ACID [Ag$_2$O], **IV**, 919

VANILLIC ACID [Ag$_2$O], **IV**, 972; [KOH], **IV**, 974

J. COCH$_3$ → CO$_2$H

3β-ACETOXYETIENIC ACID [NaOBr], **V**, 8

β,β-DIMETHYLACRYLIC ACID [KOCl], **III**, 302

FLUORENONE-2-CARBOXYLIC ACID [Na$_2$Cr$_2$O$_7$], **III**, 420

β-NAPHTHOIC ACID [NaOCl], **II**, 428

TEREPHTHALIC ACID [HNO$_3$, KMnO$_4$], **III**, 791

K. C=C → CO$_2$H

AZELAIC ACID [KMnO$_4$], **II**, 53

L. CH$_2$I → CHO

OCTANAL [((CH$_3$)$_3$NO], **V**, 872

M. CO$_2$H → CO$_3$H

MONOPERPHTHALIC ACID [H$_2$O$_2$], **V**, 805

PEROXYBENZOIC ACID [H$_2$O$_2$], **V**, 904

N. R$_3$N → R$_3$NO

N,N-DIMETHYLCYCLOHEXYL-METHYLAMINE OXIDE [H$_2$O$_2$], **IV**, 612

NICOTINAMIDE-1-OXIDE [H$_2$O$_2$], **IV**, 704

PYRIDINE-*N*-OXIDE [CH$_3$CO$_3$H], **IV**, 828

O. NH$_2$ → NO$_2$

2,6-DICHLORONITROBENZENE [CF$_3$CO$_3$H], **V**, 367

4-NITRO-2,2,4-TRIMETHYLPENTANE [KMnO$_4$], **V**, 845

P. NH—NH → N=N

BICYCLO(2.1.0) PENTANE [CuCl$_2$], **V**, 96

t-BUTYL AZODIFORMATE [(CH$_3$CO)$_2$NBr], **V**, 160

3,3-PENTAMETHYLENEDIAZIRINE [Ag$_2$O], **V**, 897

Q. Miscellaneous

δ-ACETYL-*n*-VALERIC ACID [CrO$_3$], **IV**, 19

ADIPIC ACID [HNO$_3$], **I**, 18

ALLANTOIN [KMnO$_4$], **II**, 21

ALLOXAN MONOHYDRATE [HNO$_3$], **III**, 37; [CrO$_3$], **III**, 39

ALLOXANTIN DIHYDRATE
[KClO$_3$], **III,** 42

p-AMINOBENZALDEHYDE [Na$_2$S$_x$,
NaOH], **IV,** 31

p-AMINOPHENYL DISULFIDE
[H$_2$O$_2$], **III,** 86

ANTHRAQUINONE [NaClO$_3$·
V$_2$O$_5$], **II,** 554

BENZILIC ACID [NaBrO$_3$ +
NaOH], **I,** 89

BENZOFURAZAN OXIDE
[NaOCl], **IV,** 74

BENZOYL DISULFIDE [I$_2$], **III,**
116

n-BUTYL GLYOXYLATE
[Pb(OCOCH$_3$)$_4$], **IV,** 124

β-(*o*-CARBOXYPHENYL)-PRO-
PIONIC ACID [CH$_3$CO$_3$H],
IV, 136

CHLORO-*p*-BENZOQUINONE
[Na$_2$Cr$_2$O$_7$], **IV,** 148

2-CHLOROTHIIRANE 1,1-DIOX-
IDE [Cl$_2$], **V,** 231

trans-1,2-CYCLOHEXANEDIOL
[H$_2$O$_2$], **III,** 217

1,1'-azo-*bis*-1-CYCLOHEXANENI-
TRILE [Br$_2$], **IV,** 66

CYSTEIC ACID MONOHY-
DRATE [Br$_2$], **III,** 226

nor-DESOXYCHOLIC ACID
[CrO$_3$], **III,** 234

4,4'-DIAMINOAZOBENZENE
[NaBO$_3$/H$_3$BO$_3$], **V,** 341

2,6-DIBROMOQUINONE-4-CHLO-
ROIMIDE [NaOCl], **II,** 175

3,5-DICARBETHOXY-2,6-DIME-
THYLPYRIDINE [HNO$_3$],
II, 215

2,2-DIFLUOROSUCCINIC ACID
[KMnO$_4$], **V,** 393

β,β-DIMETHYLGLUTARIC ACID
[NaOCl], **IV,** 345

2,5-DINITROBENZOIC ACID
[K$_2$Cr$_2$O$_7$], **III,** 334

p,p'-DINITROBIBENZYL [O$_2$], **IV,**
367

DIPHENYLACETYLENE [HgO],
IV, 377

DIPHENYLDIACETYLENE
[Cu(OCOCH$_3$)$_2$], **V,** 517

DIPHENYLDIAZOMETHANE
[HgO], **III,** 351

DIPHENYLKETENE [HgO], **III,**
356

DUROQUINONE [FeCl$_3$], **II,** 254

ETHYL AZODICARBOXYLATE
[HNO$_3$], **IV,** 411

2-FLUOROHEPTANOIC ACID
[HNO$_3$], **V,** 580

FUMARIC ACID [NaClO$_3$], **II,** 302

GLUTARIC ACID [HNO$_3$], **I,** 290

dl-GLYCERALDEHYDE ACETAL
[KMnO$_4$], **II,** 307

2,3,4,5,6,6-HEXAMETHYL-2,4-CY-
CLOHEXADIEN-1-ONE
[CF$_3$CO$_3$H], **V,** 598

HOMOPHTHALIC ACID
[K$_2$Cr$_2$O$_7$], **III,** 449

HOMOVERATRIC ACID [H$_2$O$_2$],
II, 335

HYDRAZINE SULFATE [NaOCl],
I, 309

α-HYDRINDONE [CrO$_3$], **II,** 336

2-HYDROXYTHIOPHENE
[(C$_6$H$_5$CO$_2$OC(CH$_3$)$_3$], **V,** 642

2-INDANONE [HCO$_3$H], **V,** 647

IODOSOBENZENE DIACETATE
[CH$_3$CO$_3$H], **V,** 660

IODOXYBENZENE [CH$_3$CO$_3$H],
V, 665; [NaOCl], **III,** 486

DL-KETOPINIC ACID [KMnO$_4$], **V,**
689

METHANESULFINYL CHLOR-
IDE [Cl$_2$], **V,** 709

6-METHOXY-2-NAPHTHOL
[H$_2$O$_2$], **V,** 918

METHYL BENZENESULFINATE
[Pb(OCOCH$_3$)$_4$], **V,** 723

METHYL PHENYL SULFOXIDE
[NaIO$_4$], **V,** 791

MONOPERPHTHALIC ACID
[H$_2$O$_2$], **III,** 619

1,2-NAPHTHOQUINONE [FeCl₃],
  II, 430
1,4-NAPHTHOQUINONE
  [K₂Cr₂O₇], **I**, 383; [CrO₃], **IV**,
  698
1,2-NAPHTHOQUINONE-4-SUL-
  FONIC ACID SALTS
  [HNO₃], **III**, 633
NEOPENTYL ALCOHOL [H₂O₂],
  V, 818
NICOTINIC ACID [HNO₃], **I**, 385
o-NITROBENZENESULFONYL
  CHLORIDE [HNO₃—HCl],
  II, 471
p-NITROBENZOYL PEROXIDE
  [Na₂O₂], **III**, 649
NITROSOBENZENE [Na₂Cr₂O₇],
  III, 668
2-NITROSO-5-NITROTOLUENE
  [K₂S₂O₈], **III**, 334
PHENYL-t-BUTYL ETHER
  [C₆H₅CO₃C(CH₃)₃], **V**, 924
PHENYLGLYOXAL [internal], **V**,
  937
PHENYLSULFUR TRIFLUO-
  RIDE [AgF₂], **V**, 959
PHTHALALDEHYDIC ACID
  [KMnO₄], **II**, 523
2,3-PYRAZINEDICARBOXYLIC
  ACID [KMNO₄], **IV**, 824
PYROGALLOL 1-MONOMETHYL
  ETHER [H₂O₂], **III**, 759
PYROMELLITIC ACID [H₂SO₄],
  II, 551
QUINONE [Na₂Cr₂O₇], **I**, 482;
  [NaClO₃·V₂O₅], **II**, 553
STYRENE OXIDE [C₆H₅COO₂H],
  I, 494
TETRALIN HYDROPEROXIDE
  [O₂]; **IV**, 895
TETRAPHENYLARSONIUM
  CHLORIDE HYDRO-
  CHLORIDE [H₂O₂], **IV**, 910
THYMOQUINONE [HNO₂], **I**, 511
p-TOLUIC ACID [HNO₃], **III**, 822
TRIMETHYLACETIC ACID

[NaOBr], **I**, 524
2,4,6-TRIPHENYLPHENOXY
  [K₃(Fe(CN)₂)], **V**, 1130
TROPYLIUM FLUOBORATE
  [PCl₅], **V**, 1138
OXIMATION
  INDOLE-3-CARBONITRILE, **V**,
  656
  (+)- and)−)-α-(2,4,5,7-TETRANI-
  TRO-9-FLUORENYLIDE-
  NEAMINOOXY)-PRO-
  PIONIC ACID, **V**, 1031
OZONIŻATION
  DIPHENALDEHYDE, **V**, 489
  DIPHENALDEHYDIC ACID, **V**,
  493
  5-FORMYL-4-PHENANTHROIC
  ACID, **IV**, 484
PHOTOCHEMICAL REACTIONS
  cis- and trans-1,2-DIVINYLCYCLO-
  BUTANE, **V**, 528
  CYCLOHEXYLIDENECYCLO-
  HEXANE, **V**, 297
  9-PHENYLPHENANTHRENE, **V**,
  952
PRINS REACTION
  4-PHENYL-m-DIOXANE, **IV**, 786
PYROLYSIS (See also Decarboxylation;
  Dehydration)
  ACRYLIC ACID, **III**, 30
  1,3-BUTADIENE, **II**, 102
  EHYL 2-KETOHEXAHYDRO-
  BENZOATE, **II**, 532
  ETHYL METHYLMALONATE, **II**,
  279
  ETHYL PHENYLMALONATE, **II**,
  288
  KETENE, **I**, 330
  METHYL BENZYL KETONE, **II**,
  389
  SUCCINIMIDE, **II**, 562
RACEMIZATION
  dl-TARTARIC ACID, **I**, 497
REARRANGEMENTS
  1-ACETYLCYCLOHEXENE, **III**,
  22

ACETYL METHYLUREA, **II,** 462
ADAMANTANE, **V,** 16
β-ALANINE, **II,** 19
ALLANTOIN, **II,** 21
2-ALLYLCYCLOHEXANONE, **V,** 25
4-AMINOVERATROLE, **II,** 44
BENZILIC ACID, **I,** 89
β-BENZOPINACOLONE, **II,** 73
ε-BENZOYLAMINOCAPROIC ACID, **II,** 76
CHLORO-*p*-BENZOQUINONE, **IV,** 148
Δ⁴-CHOLESTEN-3-ONE, **IV,** 192
CITRACONIC ANHYDRIDE, **II,** 140
7-CYANOHEPTANAL, **V,** 266
CYCLOBUTYLAMINE, **V,** 273
1,2-CYCLONONADIENE, **V,** 306
CYCLOPENTANECARBOXAL-DEHYDE, **V,** 320
2-CYCLOPENTENONE, **V,** 326
m-CYMENE, **V,** 332
n,n-DIETHYL-1,2,2-TRICHLO-ROVINYLAMINE, **V,** 387
2,5-DIHYDROXYACETOPHEN-ONE, **III,** 280
DIMETHYLKETENE β-LAC-TONE DIMER, **V,** 456
DIPHENYLACETALDEHYDE, **IV,** 375
o-EUGENOL, **III,** 418
FLAVONE, **IV,** 478
2,3,4,5,6,6-HEXAMETHYL-2,4-CY-CLOHEXADIEN-1-ONE, **V,** 598
p-HYDROXYBENZOIC ACID, **II,** 341
2-HYDROXY-3-METHYLISOCAR-BOSTYRIL, **V,** 623
2-INDANONE, **V,** 647
2-KETOHEXAMETHYLENI-MINE, **II,** 371
MESACONIC ACID, **II,** 382
2-METHYLBENZYLDIMETHY-LAMINE, **IV,** 585

METHYL CYCLOPENTANECAR-BOXYLATE, **IV,** 594
METHYL-*n*-HEXYLCARBINOL, **I,** 366
METHYL ISOPROPYL KETONE, **II,** 408
METHYL 3-METHYL-2-FURO-ATE, **IV,** 649
1-METHYL-2-PYRIDONE, **II,** 419
2,6-NAPHTHALENEDICARBOX-YLIC ACID, **V,** 813
PHENACYLAMINE HYDRO-CHLORIDE, **V,** 909
4-PHENYL-5-ANILINO-1,2,3-TRIAZOLE, **IV,** 380
2-PHENYLETHYL BENZOATE, **V,** 336
PHENYLUREA, **I,** 453
unsym. o-PHTHALYL CHLORIDE, **II,** 528
PINACOLONE, **I,** 462
o- AND *p*-PROPIOPHENOL, **II,** 543
2,4,4-TRIMETHYLCYCLOPEN-TANONE, **IV,** 957
XANTHONE, **I,** 552

REDUCTION. The subheadings illustrate types of reductions. The reducing agent is included in brackets. The reduction reaction is frequently only one step in the preparation of the compound indexed. (See also; *Replacement Reactions* (C) *Halogen by Hydrogen.*)

A. C=C → CH—CH

N-ACETYLHEXAHYDROPHEN-YLALANINE [Catalytic, Pt], **II,** 493
N-ACETYLPHENYLALANINE [Catalytic, Pt], **II,** 493
ADAMANTANE [H₂, PtO₂], **V,** 16
BENZYLACETOPHENONE [Catalytic, Pt], **I,** 101
BICYCLO(2.1.0) PENTANE [H₂Pd], **V,** 96

β-(o-CARBOXY-
PHENYL)PROPIONIC
ACID [Ni—Al, NaOH], **IV**,
136
CYCLODECANONE [H₂, Pd], **V**,
277
cis-CYCLODODECENE [H₂NNH₂,
O₂·CuSO₄], **V**, 281
DIETHYL CIS-HEXAHYDRO-
PHTHALATE [Pt or Pd, H₂],
IV, 304
DIETHYL METHYLENEMALON-
ATE [Ni, H₂], **IV**, 298
DIETHYL SUCCINATE [CrSO₄],
V, 993
9,10-DIHYDROANTHRACENE
[Na; C₂H₅OH], **V**, 398
1,4-DIHYDROBENZOIC ACID
[Na, NH₃], **V**, 400
DIHYDROCHOLESTEROL [C-
atalytic, Pt], **II**, 191
9,10-DIHYDROPHENANTHRENE
[CuCr₂O₄, H₂], **IV**, 313
2,7-DIMETHYLOXEPIN [Na,
NH₃], **V**, 467
ETHYL α-ACETYL-β-(2,3-DIME-
THOXYPHENYL)PRO
PIONATE [Pd, H₂], **IV**, 408
ETHYL n-BUTYLCYANOACE-
TATE [Pd, H₂], **III**, 385
β-FURYLPROPIONIC ACID
[Electrolytic], **I**, 313
HEXAHYDROGALLIC ACID
AND HEXAHYDROGAL-
LIC ACID TRIACETATE
[H₂, Rh], **V**, 591
HYDROCINNAMIC ACID
[Electrolytic], **I**, 311
3-ISOQUINUCLIDONE [H₂, Rh],
V, 670
2-METHYL-1,3-CYCLOHEXANE-
DIONE [H₂, Ni], **V**, 743
N-METHYL-3,4-DIHYDROXY-
PHENYLALANINE [Hg—
Na], **III**, 586

dl-β-PHENYLALANINE [P + HI]
[Catalytic, Pt], **II**, 489, 491
γ-n-PROPYLBUTYROLACTONE
[Ni—Al, NaOH], **III**, 742
3-QUINUCLIDONE HYDRO-
CHLORIDE [H₂, Pd], **V**, 989
SUCCINIC ACID [Catalytic, Pt], **I**,
64
TETRAHYDROFURAN [Catalytic,
Pd], **II**, 566
β-(TETRAHYDRO-
FURYL)PROPIONIC ACID
[Ni—Al, NaOH], **III**, 742
ar-TETRAHYDRO-α-NAPHTHOL
[Pd, H₂], **IV**, 887
ac-TETRAHYDRO-β-NAPHTHYL-
AMINE [Na + C₅H₁₁OH], **I**,
499
TETRAHYDROPYRAN [Ni, H₂], **III**,
794
β-TETRALONE [Na, C₂H₅OH], **IV**,
903
B. COH → CH
DEOXYANISOIN [Sn, HCl], **V**, 339
DIPHENYLACETIC ACID [P + I₂],
I, 224
C. CHO → CH₂OH
BENZYL ALCOHOL [CH₂O—
KOH], **II**, 591
n-HEPTYL ALCOHOL [Fe +
CH₃CO₂H], **I**, 304
PIPERONYL ALCOHOL [CH₂O—
KOH], **II**, 591
p-TOLYL CARBINOL [CH₂O—
KOH], **II**, 590
TRICHLOROETHYL ALCOHOL
[Al(OC₂H₅)₃], **II**, 598
D. C=O → CHOH
ALLOXANTIN DIHYDRATE
[H₂S], **IV**, 25
BENZOHYDROL [Zn + NaOH], **I**,
90
trans-4-t-BUTYLCYCLOHEX-
ANOL [LiAlH₄], **V**, 175
1,2-CYCLODECANEDIOL
[CuCr₂O₄, H₂], **IV**, 216

2-CYCLOHEXANONE [LiAlH$_4$], **V**, 294

2-HEPTANOL [Na + C$_2$H$_5$OH], **II**, 317

HEXAHYDROXYBENZENE [SnCl$_2$; HCl], **V**, 595

18,20-LACTONE of 3β-ACETOXY-20β-HYDROXY-5-PREG-NENE-18-OIC ACID [LiAlH(OC$_4$H$_9$)$_3$], **V**, 692

3-METHYL-1,5-PENTANEDIOL [Ni, H$_2$], **IV**, 660

l-PROPYLENE GLYCOL [Reductase], **II**, 545

XANTHYDROL [Na·Hg], **I**, 554

E. C=O → CH$_2$

ANTHRONE [Sn + HCl], **I**, 60

CREOSOL [Zn, HCl], **IV**, 203

DOCOSANEDIOIC ACID [H$_2$NNH$_2$], **V**, 533

HENDECANEDIOIC ACID [H$_2$NNH$_2$, KOH], **IV**, 510

o-n-HEPTYLPHENOL [Zn, HCl], **III**, 444

2-METHYLCYCLOPENTANE-1,3-DIONE [H$_2$NNHCONH$_2$], **V**, 747

γ-PHENYLBUTYRIC ACID [Zn + HCl], **II**, 499

PHTHALIDE [Zn—Cu + NaOH], **II**, 526

F. CO$_2$R → CH$_2$OH

CETYL ALCOHOL [Na + C$_2$H$_5$OH], **II**, 374

DECAMETHYLENE GLYCOL [Na + C$_2$H$_5$OH], **II**, 154

HEPTAMETHYLENE GLYCOL [Na + C$_2$H$_5$OH], **II**, 155

HEXAMETHYLENE GLYCOL [Catalytic, Copper Chromite], **II**, 325

LAURYL ALCOHOL [Na + C$_2$H$_5$OH], **II**, 372

MYRISTYL ALCOHOL [Na + C$_2$H$_5$OH], **II**, 374

NONAMETHYLENE GLYCOL [Na + C$_2$H$_5$OH], **II**, 155

OCTADECAMETHYLENE GLY-COL [Na + C$_2$H$_5$OH], **II**, 155

OLEYL ALCOHOL [Na, C$_2$H$_5$OH], **III**, 671; [Na + C$_4$H$_9$OH], **II**, 468

2-(1-PYRROLIDYL)PROPANOL [LiAlH$_4$], **IV**, 834

TETRADECAMETHYLENE GLY-COL [Na + C$_2$H$_5$OH], **II**, 155

TRIDECAMETHYLENE GLYCOL [Na + C$_2$H$_5$OH], **II**, 155

UNDECAMETHYLENE GLYCOL [Na + C$_2$H$_5$OH], **II**, 155

UNDECYLENYL ALCOHOL [Na + C$_2$H$_5$OH], **II**, 374

G. CONR$_2$ → CH$_2$NR$_2$

N,N-DIMETHYLCYCLOHEXYL-METHYLAMINE [LiAlH$_4$], **IV**, 339

2,2-DIMETHYLPYRROLIDINE [LiAlH$_4$], **IV**, 354

LAURYLMETHYLAMINE [LiAlH$_4$], **IV**, 564

H. CN → CH$_2$NH$_2$

DECAMETHYLENEDIAMINE [Ni, H$_2$], **III**, 229

2,4-DIPHENYLPYRROLE [Ni, H$_2$], **III**, 358

β-PHENYLETHYLAMINE [Ni, H$_2$], **III**, 720

I. C=NOH → CH$_2$NH$_2$

DIETHYL ACETAMIDOMALO-NATE [Zn; CH$_3$CO$_2$H], **V**, 373

DIETHYL AMINOMALONATE HYDROCHLORIDE [H$_2$, Pd], **V**, 376

n-BUTYLAMINE [Na + C$_2$H$_5$OH], **II**, 319

sec-BUTYLAMINE [Na + C$_2$H$_5$OH], **II**, 319

CYCLOHEXYLAMINE [Na + C$_2$H$_5$OH], **II**, 319

n-HEPTYLAMINE [Na + C$_2$H$_5$OH], **II**, 318

J. NO$_2$ → NHOH
β-PHENYLHYDROXYLAMINE
[Zn + NH$_4$Cl], **I**, 445

K. NO$_2$ → NH$_2$
o-AMINOBENZALDEHYDE
[FeSO$_4$], **III**, 56
p-AMINOBENZALDEHYDE [Na$_2$S$_x$,
NaOH], **IV**, 31
m-AMINOBENZALDEHYDE DI-
METHYLACETAL [Ni, H$_2$],
III, 59
2-AMINO-p-CYMENE [Ni, H$_2$], **III**,
63
5-AMINO-2,3-DIHYDRO-1,4-
PHTHALAZINEDIONE
[Na$_2$S$_2$O$_4$], **III**, 69
2-AMINOFLUORENE [H$_2$NNH$_2$;
Pd], **V**, 30; [Zn + H$_2$O], **II**, 448
AMINOGUANIDINE BICAR-
BONATE [Zn], **III**, 73
2-AMINO-4-NITROPHENOL
[(NH$_4$)$_2$S], **III**, 82
p-AMINOPHENYLACETIC ACID
[(NH$_4$)$_2$S], **I**, 52
p-AMINOPHENYL DISULFIDE
[Na$_2$S], **III**, 86
m-BROMOBENZALDEHYDE
[SnBr$_2$ + HBr], **II**, 132
m-CHLOROBENZALDEHYDE
[SnCl$_2$ + HCl], **II**, 130
CYCLOHEPTANONE [Ni, H$_2$], **IV**,
221
4,4'-DIAMINODIPHENYLSUL-
FONE [SnCl$_2$], **III**, 239
DIAMINODURENE CHLORO-
STANNATE [SnCl$_2$ + HCl],
II, 255
1,2-DIAMINO-4-NITROBENZENE
[(NH$_4$)$_2$S], **III**, 242
2,3-DIAMINOPYRIDINE [Fe;
HCl], **V**, 346
2,4-DIAMINOTOLUENE [Fe +
HCl], **II**, 160
2,6-DIBROMO-4-AMINOPHENOL
CHLOROSTANNATE [Sn +
HCl], **II**, 175

5,5-DIMETHYL-2-PYRROLIDONE
[Ni, H$_2$], **IV**, 357
ETHYL p-AMINOBENZOATE
[Catalytic, Pt], **I**, 240
m-HYDROXYBENZALDEHYDE
[SnCl$_2$], **III**, 453
2-NITROCARBAZOLE [H$_2$, Pd], **V**,
829
ORTHANILIC ACID [Fe +
CH$_3$CO$_2$H], **II**, 471
o-PHENYLENEDIAMINE [Zn +
NaOH], **II**, 501
PHLOROGLUCINOL [Sn + HCl],
I, 455
SEMICARBAZIDE SULFATE
[Electrolytic] **I**, 485
2,4,5-TRIAMINONITROBENZENE
[Na$_2$S$_2$], **V**, 1067
2,4,6-TRIPHENYLPHENOXYL
[H$_2$, Ni], **V**, 1130
URAMIL [Sn + HCl], **II**, 617

L. NO$_2$ → N=N
AZOBENZENE [Zn], **III**, 103

M. NO → NH$_3$
1,2-AMINONAPHTHOL HY-
DROCHLORIDE [Na$_2$S$_2$O$_4$],
II, 33
2,4-DIMETHYL-3,5-DICARBETH-
OXYPYRROLE [Zn +
CH$_3$CO$_2$H], **II**, 202
unsym.-DIMETHYLHYDRAZINE
HYDROCHLORIDE [Zn +
CH$_3$CO$_2$H], **II**, 211
α-METHYL-α-PHENYLHYDRA-
ZINE [Zn + CH$_3$CO$_2$H], **II**,
418

N. N=N → NH$_2$
1,2-AMINONAPHTHOL HY-
DROCHLORIDE [Na$_2$S$_2$O$_4$],
II, 35
1,4-AMINONAPHTHOL HY-
DROCHLORIDE [Na$_2$S$_2$O$_4$],
II, 39

O. C—X → CH
BENZENE [Mg; (CH$_3$)$_2$CHOH], **V**,
998

2,3-DIAMINOPYRIDINE [H$_2$, Pd], **V**, 346

7,7-DIMETHOXYBICY-CLO(2.2.1)HEPTENE [Na; (CH$_3$)$_3$COH], **V**, 424

INDAZOLE [HI, P], **III**, 475

LEPIDINE [Pd, H$_2$], **III**, 519

MESITALDEHYDE [Pd, H$_2$], **III**, 551

β-NAPHTHALDEHYDE [Pd, H$_2$], **III**, 627

o-TOLUALDEHYDE [SnCl$_2$], **III**, 818

TRIPHENYLMETHANE [(C$_2$H$_5$)$_2$O], **I**, 548

P. MISCELLANEOUS

p-ACETAMINOBENZENESUL-FINIC ACID [Na$_2$SO$_3$], **I**, 7

ALLOXANTIN DIHYDRATE [H$_2$S], **III**, 42

o-AMINOBENZYL ALCOHOL [Elect.], **III**, 60

AMINOMALONONITRILE p-TOL-UENESULFONATE [Al(Hg)], **V**, 32

1,4-AMINONAPHTHOL HY-DROCHLORIDE [Na$_2$S$_2$O$_4$], **I**, 49

dl-β-AMINO-β-PHENYLPRO-PIONIC ACID [NH$_2$OH], **III**, 91

AMINOTHYMOL [NH$_4$SH], **I**, 512

ARSENOACETIC ACID [NaH$_2$PO$_2$], **I**, 74

AZOXYBENZENE [Na$_3$AsO$_3$][Dextrose], **II**, 57

BENZOPINACOL [CH$_3$CHOHCH$_3$], **II**, 71

BENZOYL PIPERIDINE [Na + C$_2$H$_5$OH], **I**, 99

t-BUTYLAMINE [Ni, H$_2$], **III**, 148

CHLORO-p-BENZOQUINONE [elect.], **IV**, 148

CHLORONICOTINONITRILE [PCl$_5$], **IV**, 166

CYCLODECANONE [Zn, HCl], **IV**, 218

2-CYCLOHEXYLOXYETHANOL [LiAlH$_4$], **V**, 303

2,5-DIAMINO-3,4-DICYANO-THIOPHENE [H$_2$S], **IV**, 243

DIAMINOURACIL HYDRO-CHLORIDE [Na$_2$S$_2$O$_4$], **IV**, 247

DI-t-BUTYL NITROXIDE [Na], **V**, 355

2,2-DICHLOROETHANOL [LiAlH$_4$], **IV**, 271

DIETHYL MERCAPTOACETAL [Na, NH$_3$], **IV**, 295

N,N-DIETHYL-1,2,2-TRICHLO-ROVINYLAMINE [(C$_4$H$_9$)$_3$P], **V**, 387

DIHYDRORESORCINOL [Ni, H$_2$], **III**, 278

1,5-DIMETHYL-2-PYROLLIDONE [Ni, H$_2$], **III**, 328

1,1-DIPHENYLCYCLOPROPANE [LiAlH$_4$], **V**, 509

ETHYL p-DIMETHYLAMINO-PHENYLACETATE [H$_2$, Pd], **V**, 552

N-ETHYL-m-TOLUIDINE [SnCl$_2$ + HCl], **II**, 290

GLYCINE t-BUTYL ESTER [H$_2$, Pd], **V**, 586

HEMIMELLITENE [Hg—Na], **IV**, 508

o-METHOXYPHENYLACETONE [Fe, HCl], **IV**, 573

N-METHYL-2,3-DIMETHOXY-BENZYLAMINE [Ni, H$_2$], **IV**, 603

METHYLENE BROMIDE [Na$_3$AsO$_3$], **I**, 357

METHYLENE IODIDE [Na$_3$AsO$_3$], **I**, 358

β-NAPHTHALDEHYDE [SnCl$_2$], **III**, 626

1,5-NAPHTHALENEDITHIOL [Zn, H$_2$SO$_4$], **IV**, 695

m-NITROPHENYL DISULFIDE [HI], **V**, 843

NITROBENZENE [Zn], **III**, 668

PALLADIUM CATALYST FOR PARTIAL REDUCTION OF ACETYLENES [Pd(Pb)], **V**, 887

1,5-PENTANEDIOL [CuCr$_2$O$_4$, H$_2$], **III**, 693

α-PHENYLETHYLAMINE [Ni, H$_2$], **III**, 717

PHENYLHYDRAZINE [Na$_2$SO$_3$], **I**, 442

3-PHENYL-1-PROPANOL [Na, C$_4$H$_9$OH], **IV**, 798

PIMELIC ACID [Na + iso-C$_5$H$_{11}$OH], **II**, 535

PINCOL HYDRATE [Mg], **I**, 459

γ-n-PROPYLBUTYROLACTONE [Ni—Al, NaOH], **III**, 742

SODIUM p-TOLUENESULFI-NATE [Zn], **I**, 492

trans-STILBENE [Zn—Hg], **III**, 786

ar-TETRAHYDRO-α-NAPHTHOL [Li, C$_2$H$_5$OH], **IV**, 887

2,2,6,6-TETRAMETHYLOLCY-CLOHEXANOL [HCO$_2$H], **IV**, 907

THIOPHENOL [Zn + H$_2$SO$_4$], **I**, 504

THIOPHOSGENE [Sn + HCl], **I**, 509

THIOSALICYCLIC ACID [Zn + CH$_3$CO$_2$H], **II**, 580

m-TOLYLBENZYLAMINE [Ni, H$_2$], **III**, 827

REFORMATSKY REACTION

4-ETHYL-2-METHYL-2-OCTEN-OIC ACID, **IV**, 444

ETHYL β-PHENYL-β-HYDROXY-PROPIONATE, **III**, 408

REIMER–TIEMANN REACTION

2-HYDROXY-1-NAPHTHALDE-HYDE, **III**, 463

REISSERT REACTION

1-METHYLISOQUINOLINE, **IV**, 641

REMOVAL OF HX. The reagent used is included in brackets after the name of the preparation.

ACETYLENEDICARBOXYLIC ACID [KOH], **II**, 10

ACROLEIN ACETAL [KOH], **II**, 17

CYCLOHEXENE OXIDE [NaOH], **I**, 185

3-CYCLOHEXYLPROPINE [NaNH$_2$], **I**, 191

DECINE-1 [NaNH$_2$], **I**, 192

DIBENZOYLMETHANE [NaOCH$_3$], **I**, 205

2,3-DIBROMOPROPENE [NaOH], **I**, 209

EPIBROMOHYDRIN [Ca(OH)$_2$], **II**, 257

EPICHLOROHYDRIN [Ca(OH)$_2$] **II**, 256; [NaOH], **I**, 233

PHENYLACETYLENE [KOH], **I**, 438

4-PHENYLBUTINE-1 [NaNH$_2$], **I**, 192

PHENYLPROPIOLIC ACID [K-OH], **II**, 515

REPLACEMENT REACTIONS. (See also Diazotization; Reduction; Grignard Reaction). The subheadings illustrate types of replacement reactions, and X represents Cl, Br, or I. Frequently the replacement reaction is an intermediate step in the synthesis of the compound indexed.

A. Halogen

1. By CN

ALLYL CYANIDE, **I**, 46

BENZOYL CYANIDE, **III**, 112

BENZYL CYANIDE, **I**, 107

γ-CHLOROBUTYRONITRILE, **I**, 156

9-CYANOPHENANTHRENE, **III**, 212

β-ETHOXYPROPIONITRILE, **III**, 372

ETHYL CYANOACETATE, **I**, 254

ETHYLENE CYANOHYDRIN, **I**, 256

ETHYL *n*-TRIDECYLATE, **II**, 292

3-HYDROXYGLUTARONI-TRILE, **V**, 614

MESITYLACETIC ACID, **III**, 557

p-METHOXYPHENYLACE-TONITRILE, **IV**, 576

α-NAPHTHONITRILE, **III**, 631

MALONIC ACID, **II**, 376

SODIUM CYANOACETATE, **I**, 181

TRIMETHYLENE CYANIDE, **I**, 536

2. by H

n-HEXADECANE, **II**, 320

METHYLENE BROMIDE, **I**, 357

METHYLENE IODIDE, **I**, 358

5-METHYLFURFURAL, **II**, 393

THIOPHOSGENE, **I**, 509

3. by Metals (See also Grignard Reaction)

DI-*n*-BUTYL ZINC, **II**, 187

DIETHYL ZINC, **II**, 184

DIISOAMYL ZINC, **II**, 187

DIPHENYLMERCURY, **I**, 228

DI-*n*-PROPYL ZINC, **II**, 187

TRIPHENYLMETHYL SO-DIUM, **II**, 607

4. by NH$_2$; NHR; NR$_2$

dl-ALANINE, **I**, 23

AMINOACETAL, **III**, 50

9-AMINOACRIDINE, **III**, 53

2-AMINOBENZOPHENONE, **IV**, 34

α-AMINO-*n*-CAPROIC ACID, **I**, 48

dl-ε-BENZOYLLYSINE, **II**, 374

2-BROMOALLYLAMINE, **V**, 121

N-(2-BROMOALLYL)ETHYLA-MINE, **V**, 124

β-DI-*n*-BUTYLAMINOETHYL-AMINE, **III**, 254

β-DIETHYLAMINOETHYL AL-COHOL, **II**, 183

3,4-DIMETHYLANILINE, **III**, 307

2,4-DINITROANILINE, **II**, 221

2,6-DINITROANILINE, **IV**, 364

2,4-DINITROPHENYLHYDRA-ZINE, **II**, 228

ETHYL *N*-METHYLCARBA-MATE, **II**, 278

GLYCINE, **I**, 300

HIPPURIC ACID, **II**, 328

ISOBUTYRAMIDE, **III**, 490

dl-ISOLEUCINE, **III**, 495

dl-LEUCINE, **III**, 523

METHYLIMINODIACETIC ACID, **II**, 397

METHYL SEBACAMATE, **III**, 613

dl-PHENYLALANINE, **III**, 705

N-PHENYLANTHRANILIC ACID, **II**, 15

dl-SERINE, **III**, 774

TAURINE, **II**, 563

dl-THREONINE, **III**, 813

2,4,5-TRIAMINONITROBEN-ZENE, **V**, 1067

TRIPHENYLAMINE, **I**, 544

dl-VALINE, **III**, 848

5. by NO$_2$

NITROMETHANE, **I**, 401

6. by OH

2-BUTYN-1-OL, **IV**, 128

β-*d*-GLUCOSE-2,3,4,6-TETRA-ACETATE, **III**, 434

3-HYDROXYPYRENE, **V**, 632

PHTHALALDEHYDIC ACID, **III**, 737

TEREPHTHALALDEHYDE, **III**, 788

7. by OR

p-ARSONOPHENOXYACETIC ACID, **I**, 75

DIBENZOYLMETHANE, **I**, 205

ETHYL ORTHOFORMATE, **I**, 258

α-GLYCERYL PHENYL ETHER, **I**, 296

β-PHENOXYETHYL BRO-

MIDE, I, 436
γ-PHENOXYPROPYL BRO-
MIDE, I, 435
8. by SO₃H
SODIUM 2-BROMOETHANE-
SULFONATE, II, 558
TAURINE, II, 564
9. by various groups
ACETOL, II, 5
γ-AMINOBUTYRIC ACID, II, 25
ARSONOACETIC ACID, I, 73
BENZOPHENONE, I, 95
β-BROMOETHYLPHTHALIM-
IDE, I, 119
DIALLYLCYANAMIDE, I, 203
DI-n-BUTYLCYANAMIDE, I,
204
DI-o-NITROPHENYL DISUL-
FIDE, I, 220
DIPHENYLMETHANE, II, 232
DIPHENYLTRIKETONE, II, 244
ETHOXYACETIC ACID, II, 260
ETHYL BENZOYLACETOACE-
TATE, II, 266
ETHYL BENZOYLDIMETHY-
LACETATE, II, 268
ETHYL n-BUTYLACETOACE-
TATE, I, 248
ETHYL n-BUTYLMALON-
ATE,I, 250
ETHYL ISOBUTYRYLISOBU-
TYRATE, II, 270
ETHYL 1-METHYLTHIOL-3-
PHTHALIMIDOPROPANE-
3,3-DICARBOXYLATE, II,
384
ETHYL PHTHALIMIDOMALON-
ATE, I, 271
α-GLYCERYL PHENYL
ETHER, I, 296
β-HYDROXYETHYL METHYL
SULFIDE, II, 345
ISOPROPYL THIOCYANATE,
II, 366
2-PHENYLPYRIDINE, II, 517
SODIUM p-ARSONO-N-PHEN-

YLGLYCINAMIDE, I, 488
THIOBENZOPHENONE, II, 573
β-THIODIGLYCOL, II, 576
TRICARBETHOXYMETHANE,
II, 594
TRICARBOMETHOXYME-
THANE, II, 596
10. by X
BENZOYL FLUORIDE, V, 66
2,4-DINITROIODOBENZENE,
V, 478
B. HYDROXYL
1. By Br
ALLYL BROMIDE, I, 27
iso-AMYL BROMIDE, I, 27
β-BROMOETHYLAMINE HY-
DROBROMIDE, II, 91
β-BROMOETHYLPHTHALI-
MIDE, IV, 106
2-BROMONAPHTHALENE, V,
142
β-BROMOPROPIONIC ACID, I,
131
n-BUTYL BROMIDE, I, 28, 37
sec-BUTYL BROMIDE, I, 38; II,
359
t-BUTYL BROMIDE, I, 38
CINNAMYL BROMIDE, V, 249
CYCLOHEXYL BROMIDE, II,
247
DECAMETHYLENE BROMIDE,
III, 227
β-DIBUTYLAMINOETHYL
BROMIDE HYDROBRO-
MIDE, II, 92
β-DIETHYLAMINOETHYL
BROMIDE HYDROCHLO-
RIDE, II, 92
β-DIMETHYLAMINOETHYL
BROMIDE HYDROBRO-
MIDE, II, 92
β-DIPROPYLAMINOETHYL
BROMIDE HYDROBRO-
MIDE, II, 92
n-DODECYL BROMIDE, I, 29;
II, 246

β-ETHOXYETHYL BROMIDE, III, 370
ETHYL BROMIDE, I, 29, 36
GLYCEROL α,γ-DIBROMOHY-
DRIN, II, 308
n-HEPTYL BROMIDE, II, 247
1,3,5-HEXATRIENE, V, 608
ISOBUTYL BROMIDE, II, 358
ISOPROPYL BROMIDE, II, 359
MONOBROMOPENTAERY-
THRITOL, IV, 681
OCTADECYL BROMIDE, II, 247
n-OCTYL BROMIDE, I, 30
PENTAERYTHRITYL BRO-
MIDE, II, 476
iso-PROPYL BROMIDE, I, 37
n-PROPYL BROMIDE, I, 37; II,
359
TETRADECYL BROMIDE, II,
247
TETRAHYDROFURFURYL
BROMIDE, III, 793
TRIMETHYLENE BROMIDE, I,
30
TRIMETHYLENE CHLORO-
BROMIDE, I, 157
2. by Cl
2-ACETAMIDO-3,4,6-tri-O-ACE-
TYL-2-DEOXY-α-D-GLUCO-
PYRANOSYL CHLORIDE, V,
1
ACETYLMANDELYL CHLO-
RIDE, I, 13
2-AMINOBENZOPHENONE, IV,
34
3-BENZOYLPYRIDINE, IV, 88
n-BUTYL CHLORIDE, I, 142
sec-BUTYL CHLORIDE, I, 143
t-BUTYL CHLORIDE, I, 144
t-BUTYL CYANOACETATE, V,
171
n-BUTYRYL CHLORIDE, I, 147
dl-10-CAMPHORSULFONYL
CHLORIDE, V, 196
β-CARBOMETHOXYPRO-
PIONYL CHLORIDE, III,
169

N-CHLOROBETAINYL CHLO-
RIDE, IV, 154
β-CHLOROETHYL METHYL
SULFIDE, II, 136
2-CHLOROLEPIDINE, III, 194
α-CHLOROPHENYLACETIC
ACID, IV, 169
CROTYL DIAZOACETATE, V,
258
DESYL CHLORIDE, II, 159
Di-t-BUTYL MALONATE, IV,
261
4,7-DICHLOROQUINOLINE,
III, 272
1,6-DIIODOHEXANE, IV, 323
β-DIMETHYLAMINOETHYL
CHLORIDE HYDRO-
CHLORIDE, IV, 333
n,n-DIMETHYLCYCLOHEXYL-
METHYLAMINE, IV, 339
GLYCEROL α,γ-DICHLORO-
HYDRIN, I, 292
GLYCEROL α-MONOCHLO-
ROHYDRIN, I, 294
HEXAMETHYLENE CHLO-
ROHYDRIN, III, 446
INDAZOLE, III, 475
ISOBUTYRAMIDE, III, 490
ITACONYL CHLORIDE, IV, 554
l-MENTHOXYACETYL CHLO-
RIDE, III, 547
MESITOYL CHLORIDE, III, 555
METHANESULFONYL CHLOR-
IDE, IV, 571
p-METHOXYPHENYLACETONI-
TRILE, IV, 576
1-METHYL-3-ETHYLOXINDOLE,
IV, 620
MUCONIC ACID, III, 623
NAPHTHALENE-1,5-DISUL-
FONYL CHLORIDE, IV, 693
m-NITROBENZAZIDE, IV, 715
p-NITROBENZOYL CHLOR-
IDE, I, 394
OLEOYL CHLORIDE, IV, 739
2,3,4,5,6-PENTA-O-ACETYL-D-
GLUCONIC ACID and 2,3,-

4,5,6-PENTA-*O*-ACETYL-D-
GLUCONYL CHLORIDE,
V, 887
4-PENTEN-1-OL, **III**, 698
p-PHENYLAZOBENZOYL
CHLORIDE, **III**, 712
PHENYL CINNAMATE, **III**, 714
n-PROPYL CHLORIDE, **I**, 143
β-STYRENESULFONYL CHLO-
RIDE, **IV**, 846
p-TOLUENESULFINYL CHLO-
RIDE, **IV**, 937
TRIMETHYLENE CHLOROHY-
DRIN, **I**, 533
TRIPHENYLCHLOROME-
THANE, **III**, 841
3. by I
n-AMYL IODIDE, **II**, 402
n-BUTYL IODIDE, **II**, 402
ETHYL IODIDE, **II**, 402
n-HEXADECYL IODIDE, **II**, 322
ISOAMYL IODIDE, **II**, 402
ISOBUTYL IODIDE, **II**, 402
METHYL IODIDE, **II**, 399
n-PROPYL IODIDE, **II**, 402
C. MISCELLANEOUS
3-AMINO-2-NAPHTHOIC ACID,
III, 78
3-AMINOPYRIDINE, **IV**, 45
BENZOYLCHOLINE CHLORIDE,
IV, 84
BENZOYL DISULFIDE, **III**, 116
BENZYLTRIMETHYLAMMO-
NIUM ETHOXIDE, **IV**, 98
BROMOCYCLOPROPANE, **V**, 126
2-BROMO-3-METHYLBENZOIC
ACID, **IV**, 114
2-BROMO-3-NITROBENZOIC
ACID, **I**, 125
2-BROMOPYRIDINE, **III**, 136
t-BUTYL CARBAZATE, **V**, 166
CARBOBENZOXY CHLORIDE,
III, 167
CHLOROMETHYLPHOSPHONO-
THIOIC DICHLORIDE, **V**,
218
CYCLOHEXENE SULFIDE, **IV**,

232
4,4'-DICHLORODIBUTYL
ETHER, **IV**, 266
DICHLOROMETHYL METHYL
ETHER, **V**, 365
1,2-DI-1-(1-CYANO)CYCLO-
HEXYLHYDRAZINE,
IV, 274
DICYCLOPROPYL KETONE, **IV**,
278
DIETHYL MERCAPTOACETAL,
IV, 295
α,α-DIFLUOROTOLUENE, **V**, 396
2,5-DIHYDRO-2,5-DIMETHOXY-
FURAN, **V**, 403
3,5-DIHYDROXYBENZOIC
ACID, **III**, 288
1,4-DIIODOBUTANE, **IV**, 321
2,6-DIMETHOXYBENZONI-
TRILE, **III**, 293
α-*N*,*N*-DIMETHYLAMINOPHEN-
YLACETONITRILE, **V**, 437
2-(DIMETHYLAMINO)PYRIMI-
DINE, **IV**, 336
1,4-DINITROBUTANE, **IV**, 368
N,*N*'-DIPHENYLBENZAMIDINE,
IV, 383
DIPHENYLCARBODIIMIDE, **V**,
501
ETHYL ISOCYANIDE, **IV**, 438
ETHYL α-NITROBUTYRATE, **IV**,
454
ETHYL ORTHOCARBONATE, **IV**,
457
FERROCENYLACETONITRILE,
V, 578
2-FLUOROHEPTANOIC ACID, **V**,
580
FORMAMIDINE ACETATE, **V**,
582
FUMARYL CHLORIDE, **III**, 422
GLYCINE *t*-BUTYL ESTER, **V**,
586
n-HEXYL FLUORIDE, **IV**, 525
4-HYDROXY-1-BUTANESUL-
FONIC ACID SULTONE,
IV, 529

HYDROXYMETHYLFERRO-
CENE, **V**, 621
2-HYDROXY-1,4-NAPHTHOQUI-
NONE, **III**, 465
p-IODOBENZOIC ACID, **I**, 325
2-IODO-3-NITROBENZOIC ACID,
I, 126
o-IODOPHENOL, **I**, 326
1-METHYLAMINONTHRAQUI-
NONE, **III**, 573
o-METHYLBENZYL ALCOHOL,
IV, 582
METHYL CYCLOPROPYL KE-
TONE, **IV**, 597
METHYL IODIDE, **II**, 404
2-METHYLMERCAPTO-*N*-
METHYL-Δ^2-PYRROLINE,
V, 780
N-METHYL-1-NAPHTHYLCY-
ANAMIDE, **III**, 608
METHYL *P*-TOLYL SULFONE,
IV, 674
N-β-NAPHTHYLPIPERIDINE, **V**,
816
m-NITROBENZAZIDE, **IV**, 715
p-NITROBENZYL ACETATE, **III**,
650
1-NITROÖCTANE, **IV**, 724
p-NITROPHENYL SULFIDE, **III**,
667
PENTAERYTHRITYL IODIDE,
II, 477
PENTAERYTHRITYL TETRA-
BROMIDE, **IV**, 753
sym. o-PHTHALYL CHLORIDE,
II, 528
PUTRESCINE DIHYDROCHLO-
RIDE, **IV**, 819
THIOBENZOIC ACID, **IV**, 924
THIOBENZOYLTHIOGLYCOLIC
ACID, **V**, 1046
THIOLACETIC ACID, **IV**, 928
TRICHLOROMETHYLPHOS-
PHONYL DICHLORIDE,
IV, 950
1,1,1-TRIFLUOROHEPTANE, **V**,
1082

RESOLUTION
d- and *l*-OCTANOL-2, **I**, 418
d- and *l*-α-PHENYLETHYLA-
MINE, **II**, 506
$R(+)$ - and $S(-)$-αPHENYLE-
THYLAMINE, **V**, 932
RING CONTRACTION REACTIONS
PHENYLCYCLOPROPANE, **V**,
929
RING EXPANSION REACTIONS
CYCLODECANONE, **V**, 277
CYCLOHEPTANONE, **IV**, 221
1,3-DIHYDRO-2,3,7-TRIMETHYL-
2*H*-AZEPIN-2-ONE, **V**, 408
2,7-DIMETHYLOXEPIN, **V**, 467
3-HYDROXYQUINOLINE, **V**, 635
2-PHENYLCYCLOHEPTANONE,
IV, 780
4,6,8-TRIMETHYLAZULENE, **V**,
1088
RITTER REACTION
N-BENZYLACRYLAMIDE, **V**, 73
α,α-DIMETHYL-β-PHENYLE-
THYLAMINE, **V**, 471
SCHMIDT REACTION
CYCLOBUTYLAMINE, **V**, 273
SKRAUP REACTION
6-METHOXY-8-NITROQUINO-
LINE, **III**, 568
QUINOLINE, **I**, 478
SOMMELET REACTION
ISOPHTHALALDEHYDE, **V**, 668
1-NAPHTHALDEHYDE, **IV**, 690
STRECKER SYNTHESIS
α-AMINODIETHYLACETIC
ACID, **III**, 66
α-AMINOISOBUTYRIC ACID, **II**,
29
dl-α-AMINOPHENYLACETIC
ACID, **III**, 84
dl-α-AMINO-α-PHENYLPRO-
PIONIC ACID, **III**, 88
SULFONATION
p-ACETAMINOBENZENESUL-
FONYL CHLORIDE, **I**, 8
BARIUM 2-PHENANTHRENE-
SULFONATE, **II**, 482

BENZENESULFONYL CHLO-
RIDE, **I,** 85
o-BROMOPHENOL, **II,** 97
DL-10-CAMPHORSULFONIC
ACID, **V,** 194
3,5-DIHYDROXYBENZOIC
ACID, **III,** 288
2,6-DINITROANILINE, **IV,** 364
POTASSIUM ANTHRAQUI-
NONE-α-SULFONATE, **II,**
539
POTASSIUM 3-PHENANTHRE-
NESULFONATE, **II,** 482
4-PYRIDINESULFONIC ACID, **V,**
977
SODIUM β-STYRENESULFON-
ATE, **IV,** 846
α-SULFOPALMITIC ACID, **IV,**
862
o-TOLUIDINESULFONIC ACID,
III, 824
**THERMAL DECOMPOSITION REAC-
TIONS**
1-CHLORO-1,4,4-TRIFLUOROBU-
TADIENE, **V,** 235
trans-CYCLOÖCTENE, **V,** 315
1,1'-DICYANO-1,1'-BICYCLO-
HEXYL, **IV,** 273
DIETHYL METHYLENEMALON-
ATE, **IV,** 298

DIHYDROXCYCLOPENTENE, **V,**
414
DIPHENYLCARBODIIMIDE
(METHOD II), **V,** 504
ETHYLENE SULFIDE, **V,** 562
HEXAPHENYLBENZENE, **V,** 604
2-HYDROXYTHIOPHENE, **V,** 642
KETENE, **V,** 679
METHYL BUTADIENOATE, **V,**
734
METHYLENECYCLOHEXANE
AND *N,N*-DIMETHYLHY-
DROXYLAMINE HY-
DROCHLORIDE, **IV,** 612
α-NAPHTHYL ISOTHIOCYAN-
ATE, **IV,** 700
1,4-PENTADIENE, **IV,** 746
PUTRESCINE DIHYDROCHLO-
RIDE, **IV,** 819
STEARONE, **IV,** 854
2-VINYLTHIOPHENE, **IV,** 980
THIOCYANATION
p-THIOCYANODIMETHYLANI-
LINE; **II,** 574
ULLMAN REACTION
2,2'-DINITROBIPHENYL, **III,** 339
WITTIG REACTION (See Condensation)
YLIDE FORMATION (See Condensation)
VINYL TRIPHENYLPHOSPHON-
IUM BROMIDE, **V,** 1145

TYPE OF COMPOUND INDEX

Preparations are listed by functional groups or by ring systems. Phenyl, ethylenic, and acetylenic groups are not considered as substituents unless otherwise stated. Where possible, compounds are listed according to the group introduced. For example, *m*-bromonitrobenzene, prepared by the bromination of nitrobenzene, is listed under halogen derivatives and not under nitro compounds. *p*-Acetaminobenzenesulfonyl chloride is listed under acid chlorides and not under amides or sulfur compounds. Salts are included with the corresponding acids and bases.

ACETALS (See Esters; Ethers)
  ACETAL, **I,** 1
  ACETONE DIBUTYL ACETAL,
    V, 5
  ACROLEIN ACETAL, **II,** 17; **IV,**
    21
  AMINOACETAL, **III,** 50
  m-AMINOBENZALDEHYDE
    DIMETHYLACETAL, **III,** 59
  BROMOACETAL, **III,** 123
  β-CHLOROPROPIONALDE-
    HYDE ACETAL, **II,** 137
  CROTONALDEHYDE DI-
    ETHYL ACETAL, **IV,** 22
  CYCLOHEXANONE DIALLYL
    ACETAL, **V,** 292
  DICYANOKETENE ETHYL-
    ENE ACETAL, **IV,** 276
  DIETHYL MERCAPTOACE-
    TAL, **IV,** 295
  7,7-DIMETHOXYBICY-
    CLO(2.2.1)HEPTENE, **V,** 424

β-ETHOXYPROPIONALDE-
  HYDE ACETAL, **III,** 371
ETHYL DIETHOXYACETATE,
  IV, 427
ETHYL ORTHOFORMATE, **I,**
  258
KETENE DIETHYLACETAL,
  III, 506
KETENE DI(2-METHOXY-
  ETHYL) ACETAL, **V,** 684
β-KETOISOÖCTALDEHYDE
  DIMETHYL ACETAL, **IV,**
  558
MONOBENZALPENTAERY-
  THRITOL, **IV,** 679
m-NITROBENZALDEHYDE DI-
  METHYLACETAL, **III,** 644
4-PHENYL-*m*-DIOXANE, **IV,** 786
PHENYLPROPARGYLALDE-
  HYDE DIETHYL ACETAL,
  IV, 801
1,1,1′,1′-TETRAETHOXYETHYL-

POLYSULFIDE, **IV**, 295
TIGLYLALDEHYDE DIETHYL
ACETAL, **IV**, 22
ACID AMIDES (See Amides).
ACID ANHYDRIDES
BENZENEBORONIC ANHY-
DRIDE, **IV**, 68
BENZOIC ANHYDRIDE, **I**, 91
BENZOIC-CARBONIC ANHY-
DRIDE, **IV**, 286
n-CAPROIC ANHYDRIDE, **III**,
164
p-CHLOROBENZOIC ANHY-
DRIDE, **III**, 29
CITRACONIC ANYDRIDE, **II**,
140
DIACETYL-*d*-TARTARIC AN-
HYDRIDE, **IV**, 242
3,4-DIHYDRO-1,2-NAPH-
THALIC ANHYDRIDE, **II**,
194
HEPTOIC ANHYDRIDE, **III**, 28
HOMOPHTHALIC ANHY-
DRIDE, **III**, 450
ITACONIC ANHYDRIDE, **II**,
368
β-METHYLGLUTARIC ANHY-
DRIDE, **IV**, 630
1,2-NAPHTHALIC ANHY-
DRIDE, **II**, 423
NICOTINIC ANHYDRIDE, **V**,
822
3-NITROPHTHALIC ANHY-
DRIDE, **I**, 410
α-PHENYLGLUTARIC ANHY-
DRIDE, **IV**, 790
SUCCINIC ANHYDRIDE, **II**,
560
o-SULFOBENZOIC ANHY-
DRIDE, **I**, 495
cis-Δ⁴-TETRAHYDRO-
PHTHALIC ANHYDRIDE,
IV, 890
TETRAIODOPHTHALIC AN-
HYDRIDE, **III**, 796
TETRAPHENYLPHTHALIC

ANHYDRIDE, **III**, 807
p-TOLUENESULFONIC ANHY-
DRIDE, **IV**, 940
ACID HALIDES
ACETYLMANDELYL CHLOR-
IDE, **I**, 13
p-ACETAMINOBENZENESUL-
FONYL CHLORIDE, **I**, 8
BENZENESULFINYL FLUOR-
IDE, **V**, 396
BENZENESULFONYL CHLOR-
IDE, **I**, 84
BENZOYL FLUORIDE, **V**, 66
α-BROMOISOBUTYRYL BROM-
IDE, **IV**, 348
n-BUTYRYL CHLORIDE, **I**, 147
DL-10-CAMPHORSULFONYL
CHLORIDE, **V**, 196
CARBOBENZOXY CHLORIDE,
III, 167
β-CARBOMETHOXYPRO-
PIONYL CHLORIDE, **III**,
169
δ-CARBOMETHOXYVALERYL
CHLORIDE, **IV**, 555
o-CHLOROBENZOYL CHLOR-
IDE, **I**, 155
N-CHLOROBETAINYL CHLOR-
IDE, **IV**, 154
CHLOROMETHYLPHOSPHONO-
THIOIC DICHLORIDE, **V**, 218
CHLOROSULFONYL ISO-
CYANATE, **V**, 226
2,4-DI-*t*-AMYLPHENOXYACE-
TYL CHLORIDE, **IV**, 742
DIETHYLTHIOCARBAMYL
CHLORIDE, **IV**, 307
2,4-DINITROBENZENESUL-
FENYL CHLORIDE, **V**, 474
ISOBUTYRYL CHLORIDE, **III**,
490
β-ISOVALEROLACTAM-*N*-
SULFONYL CHLORIDE, **V**,
673
ITACONYL CHLORIDE, **IV**, 554
MALONYL DICHLORIDE, **IV**,
263

l-MENTHOXYACETYL CHLOR-
IDE, **III,** 547

MESITOYL CHLORIDE, **III,** 555

METHANESULFINYL CHLOR-
IDE, **V,** 709

METHANESULFONYL CHLOR-
IDE, **IV,** 571

NAPHTHALENE-1,5-DISUL-
FONYL CHLORIDE, **IV,** 693

o-NITROBENZENESULFONYL
CHLORIDE, **II,** 471

m-NITROBENZOYL CHLOR-
IDE, **IV,** 715

OLEOYL CHLORIDE, **IV,** 739

PALMITOYL CHLORIDE, **IV,**
742

2,3,4,5,6-PENTA-*O*-ACETYL-D-
GLUCONYL CHLORIDE,
V, 887

p-PHENYLAZOBENZOYL
CHLORIDE, **III,** 712

PHENYLDICHLOROPHOS-
PHINE, **IV,** 784

sym. AND *unsym.* *o*-PHTHALYL
CHLORIDE, **II,** 528

RICINOLEOYL CHLORIDE, **IV,**
742

β-STYRENESULFONYL CHLO-
RIDE, **IV,** 846

STYRYLPHOSPHONIC DI-
CHLORIDE, **V,** 1005

THIOPHOSGENE, **I,** 506

p-TOLUENESULFENYL
CHLORIDE, **IV,** 934

p-TOLUENESULFINYL CHLOR-
IDE, **IV,** 937

TRICHLOROMETHYLPHOS-
PHONYL DICHLORIDE,
IV, 950

2,2,4-TRIMETHYL-3-OXOVAL-
ERYL CHLORIDE, **V,** 1103

ACIDS

A. UNSUBSTITUTED ACIDS

1. Monobasic

ACRYLIC ACID, **III,** 30

1-ADAMANTANECARBOX-
YLIC ACID, **V,** 20

BENZOIC ACID, **I,** 363

3-BENZYL-3-METHYLPEN-
TANOIC ACID, **IV,** 93

exo-cis-BICYCLO(3.3.0)OCTANE-
2-CARBOXYLIC ACID, **V,**
93

n-CAPROIC ACID, **II,** 417, 475

CYCLOBUTANECARBOXYLIC
ACID, **III,** 213

CYCLOHEXANECARBOXYLIC
ACID, **I,** 364

CYCLOPROPANECARBOX-
YLIC ACID, **III,** 221

1,4-DIHYDROBENZOIC ACID,
V, 400

β,β-DIMETHYLACRYLIC
ACID, **III,** 302

DIPHENYLACETIC ACID, **I,**
224

α,β-DIPHENYLPROPIONIC
ACID, **V,** 526

trans-2-DODECENOIC ACID,
IV, 398

ERUCIC ACID, **II,** 258

3-ETHYL-3-METHYLHEXAN-
OIC ACID, **IV,** 97

4-ETHYL-2-METHYL-2-OCTEN-
OIC ACID, **IV,** 444

9-FLUORENECARBOXYLIC
ACID, **IV,** 482

β-FURYLPROPIONIC ACID, **I,**
313

n-HEPTANOIC ACID, **II,** 315

HYDROCINNAMIC ACID, **I,**
311

HYDROGEN CYANIDE, **I,** 314

INDOLE-3-ACETIC ACID, **V,**
654

LEVOPIMARIC ACID, **V,** 699

LINOLEIC ACID, **III,** 526

LINOLENIC ACID, **III,** 531

MESITOIC ACID, **III,** 553; **V,** 706

MESITYLACETIC ACID, **III,**
557

1-METHYLCYCLOHEXANE-
    CARBOXYLIC ACID, **V**, 739
2-METHYLDODECANOIC
    ACID, **IV**, 618
trans-2-METHYL-2-DODECEN-
    OIC ACID, **IV**, 608
dl-METHYLETHYLACETIC
    ACID, **I**, 361
3-METHYLHEPTANOIC ACID,
    V,762
2-METHYLENEDODECANOIC
    ACID, **IV**, 616
3-METHYLPENTANOIC ACID,
    II, 416
3-METHYL-3-PHENYLPEN-
    TANOIC ACID, **IV**, 97
MYRISTIC ACID, **I**, 379
α-NAPHTHOIC ACID, **II**, 425
β-NAPHTHOIC ACID, **II**, 428
PELARGONIC ACID, **II**, 474
PERBENZOIC ACID, **I**, 431
PHENYLACETIC ACID, **I**, 436
γ-PHENYLBUTYRIC ACID, **II**,
    499
α-PHENYLCINNAMIC ACID,
    IV, 777
PHENYLPROPIOLIC ACID, **II**,
    515
SORBIC ACID, **III**, 783
STEAROLIC ACID, **III**, 785; **IV**,
    851
TETROLIC ACID, **V**, 1043
o-TOLUIC ACID, **II**, 588, **III**, 820
p-TOLUIC ACID, **II**, 589, **III**, 822
TRIMETHYLACETIC ACID, **I**,
    524
α,β,β-TRIPHENYLPROPIONIC
    ACID, **IV**, 960
10-UNDECYNOIC ACID, **IV**, 969
n-VALERIC ACID, **I**, 363
VINYLACETIC ACID, **III**, 851
2. Polybasic
ACETYLENEDICARBOXYLIC
    ACID, **II**, 10
ACONITIC ACID, **II**, 12
ADIPIC ACID, **I**, 18

AZELAIC ACID, **II**, 53
CALCIUM MALONATE, **II**, 376
o-CARBOXYCINNAMIC ACID,
    IV, 136
β-(*o*-CARBOXY-
    PHENYL)PROPIONIC
    ACID, **IV**, 136
CITRACONIC ACID, **II**, 140
1,1-CYCLOBUTANEDICAR-
    BOXYLIC ACID, **III**, 213
β,β-DIMETHYLGLUTARIC
    ACID, **IV**, 345
DIPHENIC ACID, **I**, 222
DOCOSANEDIOIC ACID, **V**,
    533
β-ETHYL-β-METHYLGLU-
    TARIC ACID, **IV**, 441
α-ETHYL-α-METHYLSUC-
    CINIC ACID, **V**, 572
FUMARIC ACID, **II**, 302
GLUTARIC ACID, **I**, 289; **IV**, 496
HENDECANEDIOIC ACID, **IV**,
    510
HOMOPHTHALIC ACID, **III**,
    449, **V**, 612
ITACONIC ACID, **II**, 368
MALONIC ACID, **II**, 376
MESACONIC ACID, **II**, 382
β-METHYLGLUTARIC ACID,
    III, 591
β-METHYL-β-PHENYLGLU-
    TARIC ACID, **IV**, 664
METHYLSUCCINIC ACID, **III**,
    615
MUCONIC ACID, **III**, 623
2,3-NAPHTHALENEDICAR-
    BOXYLIC ACID, **V**, 810
2,6-NAPHTHALENEDICAR-
    BOXYLIC ACID, **V**, 813
OXALIC ACID (ANHYDROUS),
    I, 421
γ-PHENYLALLYLSUCCINIC
    ACID, **IV**, 766
PHENYLSUCCINIC ACID, **I**,
    451; **IV**, 804
PIMELIC ACID, **II**, 531

PYROMELLITIC ACID, **II**, 551
3,4-SECO-Δ⁵-CHOLESTEN-3,4-
DIOIC ACID, **IV**, 191
TEREPHTHALIC ACID, **III**, 791
TRICARBALLYLIC ACID, **I**,
523
(B) *ACIDS, SUBSTITUTED*
1. **Amino Acids**
p-ACETAMINOBENZENESUL-
FINIC ACID, **I**, 7
α-ACETAMINOCINNAMIC
ACID, **II**, 1
N-ACETYLHEXAHYDRO-
PHENYLALANINE, **II**, 493
N-ACETYLPHENYLALANINE,
II, 493
β-ALANINE, **II**, 19; **III**, 34
dl-ALANINE, **I**, 21
γ-AMINOBUTYRIC ACID, **II**, 25
ε-AMINOCAPROIC ACID, **II**,
28; **IV**, 39
α-AMINODIETHYLACETIC
ACID, **III**, 66
α-AMINOISOBUTYRIC ACID,
II, 29
3-AMINO-2-NAPHTHOIC
ACID, **III**, 78
dl-α-AMINOPHENYLACETIC
ACID, **III**, 84
dl-α-AMINO-α-PHENYLPRO-
PIONIC ACID, **III**, 88
dl-β-AMINO-β-PHENYLPRO-
PIONIC ACID, **III**, 91
3-AMINO-2,4,6-TRIBROMOBEN-
ZOIC ACID, **IV**, 947
d-ARGININE HYDROCHLOR-
IDE, **II**, 49
ARSANILIC ACID, **I**, 70
DL-ASPARTIC ACID, **IV**, 55
ε-BENZOYLAMINOCAPROIC
ACID, **II**, 76
α-BENZOYLAMINOCINNAM-
IC ACID, **II**, 491
CYSTEIC ACID MONOHY-
DRATE, **III**, 226
l-CYSTINE, **I**, 194

3,5-DICHLORO-2-AMINOBEN-
ZOIC ACID, **IV**, 872
GUANIDOACETIC ACID, **III**,
440
d-GLUTAMIC ACID, **I**, 286
GLYCINE, **I**, 298
HIPPURIC ACID, **II**, 328
l-HISTIDINE MONOHYDRO-
CHLORIDE, **II**, 330
6-HYDROXYNICOTINIC
ACID, **IV**, 532
dl-ISOLEUCINE, **III**, 495
dl-LEUCINE, **III**, 523
dl-METHIONINE, **II**, 384
N-METHYL-3,4-DIHYDROXY-
PHENYLALANINE, **III**, 586
METHYLIMINODIACETIC
ACID, **II**, 397
NICOTINIC ACID, **I**, 385
dl-PHENYLALANINE, **III**, 705
dl-β-PHENYLALANINE, **II**, 489
N-PHENYLANTHRANILIC
ACID, **II**, 15
N-PHTHALYL-L-β-PHENYL-
ALANINE, **V**, 973
dl-SERINE, **III**, 774
TAURINE, **II**, 563
dl-THREONINE, **III**, 813
l-TRYPTOPHANE, **II**, 612
l-TYROSINE, **II**, 612
dl-VALINE, **III**, 848
2. **Halogen Acids**
3-AMINO-2,4,6-TRIBROMOBEN-
ZOIC ACID, **IV**, 947
ε-BENZOYLAMINO-α-BROMO-
CAPROIC ACID, **II**, 74
BROMOACETIC ACID, **III**, 381
α-BROMOISOCAPROIC ACID,
III, 523
α-BROMOISOVALERIC ACID,
III, 848
p-BROMOMANDELIC ACID,
IV, 110
2-BROMO-3-METHYLBENZOIC
ACID, **IV**, 114
2-BROMO-3-NITROBENZOIC

ACID, **I,** 125

β-BROMOPROPIONIC ACID, **I,** 131

o-CHLOROBENZOIC ACID, **II,** 135

p-CHLOROMANDELIC ACID, **IV,** 112

α-CHLOROPHENYLACETIC ACID, **IV,** 169

β-CHLOROPROPIONIC ACID, **I,** 166

3-CHLORO-2,2,3-TRIFLUORO-PROPIONIC ACID, **V,** 239

DIBROMOHYDROCINNAMIC ACID, **IV,** 961

DICHLOROACETIC ACID, **II,** 181

3,5-DICHLORO-2-AMINOBEN-ZOIC ACID, **IV,** 872

DI-(p-CHLORO-PHENYL)ACETIC ACID, **III,** 270

2,2-DIFLUOROSUCCINIC ACID, **V,** 393

2-FLUOROHEPTANOIC ACID, **V,** 580

m-IODOBENZOIC ACID, **II,** 353

p-IODOMANDELIC ACID, **IV,** 112

2-IODO-3-NITROBENZOIC ACID, **I,** 126

MUCOBROMIC ACID, **III,** 621; **IV,** 688

PENTACHLOROBENZOIC ACID, **V,** 890

dl-4,4′,6,6′-TETRACHLORODI-PHENIC ACID, **IV,** 872

2,4,6-TRIBROMOBENZOIC ACID, **IV,** 947

3. Hydroxy Acids

ATROLACTIC ACID, **IV,** 58

BENZILIC ACID, **I,** 89

nor-DESOXYCHOLIC ACID, **III,** 234

5,5-DIHYDROXYBARBITURIC ACID, **IV,** 23

2,5-DIHYDROXY-p-BENZENE-DIACETIC ACID, **III,** 286

3,5-DIHYDROXYBENZOIC ACID, **III,** 288

9,10-DIHYDROXYSTEARIC ACID, **IV,** 317

2,5-DIMETHYLMANDELIC ACID, **III,** 326

HEXAHYDROGALLIC ACID, **V,** 591

p-HYDROXYBENZOIC ACID, **II,** 341

2-HYDROXYISOPHTHALIC ACID, **V,** 617

6-HYDROXYNICOTINIC ACID, **IV,** 532

p-HYDROXYPHENYLPYRUV-IC ACID, **V,** 627

β-HYDROXYPROPIONIC ACID, **I,** 321

MANDELIC ACID, **I,** 336; **III,** 538

PROTOCATECHUIC ACID, **III,** 745

β-RESORCYLIC ACID, **II,** 557

SODIUM 2-PHENYL-2-HYDROX-YETHANE-1-SULFON-ATE, **IV,** 850

dl-TARTARIC ACID, **I,** 497

VANILLIC ACID, **IV,** 972

4. Keto Acids

β-(1-ACENAPHTHOYL)-PROPIONIC ACID, **III,** 9

β-(3-ACENAPHTHOYL)-PROPIONIC ACID, **III,** 6

δ-ACETYL-n-VALERIC ACID, **IV,** 19

β-BENZOYLACRYLIC ACID, **III,** 109

γ-BENZOYLBUTYRIC ACID, **II,** 82

BENZOYLFORMIC ACID, **I,** 244; **III,** 114

β-BENZOYLPROPIONIC ACID, **II,** 81

FLUORENONE-2-CARBOX-

YLIC ACID, **III,** 420

p-HYDROXYPHENYLPYRUV-
   IC ACID, **V,** 627

α-KETOGLUTARIC ACID, **III,**
   510; **V,** 687

6-KETOHENDECANEDIOIC
   ACID, **IV,** 555

DL-KETOPINIC ACID, **V,** 689

LEVULINIC ACID, **I,** 335

γ-OXOCAPRIC ACID, **IV,** 432

PHENYLPYRUVIC ACID, **II,**
   519

p-TOLUYL-o-BENZOIC ACID,
   I, 517

5. Nitro Acids

2,5-DINITROBENZOIC ACID,
   III, 334

3,5-DINITROBENZOIC ACID,
   III, 337

m-NITROBENZOIC ACID, **I,** 391

p-NITROBENZOIC ACID, **I,** 392

m-NITROCINNAMIC ACID, **I,**
   398; **IV,** 731

p-NITROPHENYLACETIC
   ACID, **I,** 406

trans-o-NITRO-α-PHENYLCIN-
   NAMIC ACID, **IV,** 730

3-NITROPHTHALIC ACID, **I,**
   408

4-NITROPHTHALIC ACID, **II,**
   457

(+)- and (−)-α-(2,4,5,7-TETRANI-
   TRO-9-FLUORENYLIDE-
   NEAMINOOXY)PRO-
   PIONIC ACID, **V,** 1031

2,4,6-TRINITROBENZOIC
   ACID, **I,** 543

6. Miscellaneous

ABIETIC ACID, **IV,** 1

3β-ACETOXYETIENIC ACID,
   V, 8

ACID AMMONIUM o-SULFO-
   BENZOATE, **I,** 14

AMMONIUM SALT OF AURIN
   TRICARBOXYLIC ACID, **I,**
   54

p-ARSONOPHENOXYACETIC
   ACID, **I,** 75

BENZENEDIAZONIUM-2-CAR-
   BOXYLATE, **V,** 54

β-CARBETHOXY-γ,γ-DIPHEN-
   YLVINYLACETIC ACID,
   IV, 132

CARBOBENZOXYGLYCINE,
   III, 168

o-CARBOXYPHENYLACETO-
   NITRILE, **III,** 174

p-CHLOROMERCURIBENZOIC
   ACID, **I,** 159

COUMALIC ACID, **IV,** 201

COUMARILIC ACID, **III,** 209

ω-CYANOPELARGONIC ACID,
   III, 768

α-CYANO-β-PHENYLACRYLIC
   ACID, **I,** 181

CYCLOHEXYLIDENECYANO-
   ACETIC ACID, **IV,** 234

2,3-DIMETHOXYCINNAMIC
   ACID, **IV,** 327

6,7-DIMETHOXY-3,4-DIHY-
   DRO-2-NAPHTHOIC ACID,
   III, 300

3,4-DIMETHOXYPHENYL-
   PYRUVIC ACID, **II,** 335

DIPHENALDEHYDIC ACID, **V,**
   493

ETHOXYACETIC ACID, **II,** 260

5-FORMYL-4-PHENANTHROIC
   ACID, **IV,** 484

2-FURANCARBOXYLIC ACID,
   I, 276

2-FUROIC ACID, **IV,** 493

FURYLACRYLIC ACID, **III,** 425

HEXAHYDROGALLIC ACID
   AND HEXAHYDROGAL-
   LIC ACID TRIACETATE,
   V, 591

HOMOVERATRIC ACID, **II,** 333

o-HYDRAZINOBENZOIC ACID
   HYDROCHLORIDE, **III,** 475

ISODEHYDROACETIC ACID,
   IV, 549

l-MENTHOXYACETIC ACID, **III**, 544

3-METHYLCOUMARILIC ACID, **IV**, 591

3-METHYL-2-FUROIC ACID, **IV**, 628

2-METHYL-5,6-PYRAZINEDI-CARBOXYLIC ACID, **IV**, 827

MONOPERPHTHALIC ACID, **III**, 619; **V**, 805

sec.-OCTYL HYDROGEN PHTHALATE, **I**, 418

PARABANIC ACID, **IV**, 744

2,3,4,5,6-PENTA-*O*-ACETYL-D-GLUCONIC ACID, **V**, 887

p-PHENYLAZOBENZOIC ACID, **III**, 711

PHTHALALDEHYDIC ACID, **II**, 523

α-PHTHALIMIDO-*o*-TOLUIC ACID, **IV**, 810

PIPERONYLIC ACID, **II**, 538

2,3-PYRAZINEDICARBOXYLIC ACID, **IV**, 824

SODIUM *p*-ARSONO-*N*-PHEN-YLGLYCINAMIDE, **I**, 488

SODIUM *p*-HYDROXYPHENYL-ARSONATE, **I**, 490

SODIUM *p*-TOLUENESULFI-NATE, **I**, 492

α-SULFOPALMITIC ACID, **IV**, 862

β-(TETRAHYDROFURYL)-PROPIONIC ACID, **III**, 742

(+)- and (−)-α-(2,4,5,7-TETRANI-TRO-9-FLUORENYLIDE-NEAMINOOXY)/PRO-PIONIC ACID, **V**, 1031

3-THENOIC ACID, **IV**, 919

THIOBENZOIC ACID, **IV**, 924

THIOBENZOYLTHIOGLY-COLIC ACID, **V**, 1046

THIOLACETIC ACID, **IV**, 928

p-TOLUENESULFONYLAN-THRANILIC ACID, **IV**, 34

TRITHIOCARBODIGLYCOLIC ACID, **IV**, 967

ALCOHOLS

(A) *UNSUBSTITUTED ALCOHOLS*

1. **Primary**

ALLYL ALCOHOL, **I**, 42

BENZYL ALCOHOL, **II**, 591

2-BUTYN-1-OL, **IV**, 128

CETYL ALCOHOL, **II**, 374

CYCLOHEXYLCARBINOL, **I**, 188

DECAMETHYLENE GLYCOL, **II**, 154

2,3-DIMETHYLBENZYL ALCO-HOL, **IV**, 584

HEPTAMETHYLENE GLYCOL, **II**, 155

n-HEPTYL ALCOHOL, **I**, 304

HEXAMETHYLENE GLYCOL, **II**, 325

n-HEXYL ALCOHOL, **I**, 306

LAURYL ALCOHOL, **II**, 372

o-METHYLBENZYL ALCO-HOL, **IV**, 582

3-METHYL-1,5-PENTANEDIOL, **IV**, 660

MYRISTYL ALCOHOL, **II**, 374

NEOPENTYL ALCOHOL, **V**, 818

NONAMETHYLENE GLYCOL, **II**, 155

OCTADECAMETHYLENE GLY-COL, **II**, 155

OLEYL ALCOHOL, **II**, 468; **III**, 671

PENTAERYTHRITOL, **I**, 425

1,5-PENTANEDIOL, **III**, 693

4-PENTEN-1-OL, **III**, 698

4-PENTYN-1-OL, **IV**, 755

3-PHENYL-1-PROPANOL, **IV**, 798

TETRADECAMETHYLENE GLYCOL, **II**, 155

p-TOLYL CARBINOL, **II**, 590

TRIDECAMETHYLENE GLY-COL, **II**, 155

UNDECAMETHYLENE GLY-

COL, **II**, 155
UNDECYLENYL ALCOHOL,
 II, 374
2. **Primary-Secondary**
BENZOHYDROL, **I**, 90
METHYL-*n*-HEXYLCARBINOL,
 I, 366
d- and *l*-OCTANOL-2, **I**, 418
l-PROPYLENE GLYCOL, **II**, 545
1,2,5-TRIHYDROXYPENTANE,
 III, 833
XANTHYDROL, **I**, 554
3. **Secondary**
ACENAPHTHENOL-7, **III**, 3
trans-4-*t*-BUTYLCYCLOHEX-
 ANOL, **V**, 175
CHOLESTEROL, **IV**, 195
1,2-CYCLODECANEDIOL, **IV**,
 216
trans-1,2-CYCLOHEXANEDIOL,
 III, 217
DI-*n*-BUTYL CARBINOL, **II**, 179
DIHYDROCHOLESTEROL, **II**,
 191
DIHYDROXYCYCLOPEN-
 TENE, **V**, 414
2-HEPTANOL, **II**, 317
METHYL ISOPROPYL CARBI-
 NOL, **II**, 406
trans-METHYLSTYRYLCARBI-
 NOL, **IV**, 773
exo/endo-7-NORCARANOL, **V**,
 859
NORTRICYCLANOL, **V**, 863
3-PENTEN-2-OL, **III**, 696
1-PHENYL-1-PENTEN-4-YN-3-
 OL, **IV**, 792
4. **Tertiary**
1-ACETYLCYCLOHEXANOL, **IV**,
 13
BENZOPINACOL, **II**, 71
DIMETHYLETHYNYLCARBI-
 NOL, **III**, 320
4,6-DIMETHYL-1-HEPTEN-4-OL,
 V, 452
1-ETHYNYLCYCLOHEXANOL,

III, 416
1,1'-ETHYNYLENE-*bis*-CYCLO-
 HEXANOL, **IV**, 471
PINACOL HYDRATE, **I**, 459
α,α,α',α'-TETRAMETHYLTE-
 TRAMETHYLENE GLYCOL,
 V, 1026
1-*p*-TOLYLCYCLOPROPANOL,
 V, 1058
TRI-*n*-AMYL CARBINOL, **II**, 603
TRIBIPHENYLCARBINOL, **III**,
 831
TRI-*n*-BUTYL CARBINOL, **II**, 603
TRIETHYL CARBINOL, **II**, 602
TRI-*n*-HEPTYL CARBINOL, **II**,
 603
TRIPHENYLCARBINOL, **III**, 839
TRI-*n*-PROPYL CARBINOL, **II**,
 603
(B) *SUBSTITUTED ALCOHOLS. See
 also Acids, B, 3; Amides, B, 1;
 Esters (B) 2*
1. **Halogen Alcohols**
2-BROMOETHANOL, **I**, 117
2-CHLOROCYCLOHEXANOL, **I**,
 158
trans-2-CHLOROCYCLOPEN-
 TANOL, **IV**, 157
m-CHLOROPHENYLMETHYL-
 CARBINOL, **III**, 200
CHOLESTEROL DIBROMIDE,
 IV, 195
2,2-DICHLOROETHANOL, **IV**,
 271
HEXAMETHYLENE CHLOR-
 OHYDRIN, **III**, 446
MONOBROMOPENTAERYTHRI-
 TOL, **IV**, 681
TRICHLOROETHYL ALCOHOL,
 II, 598
2. **Keto Alcohols**
ACETOL, **II**, 5
BENZOIN, **I**, 94
BUTYROIN, **II**, 114
DIACETONE ALCOHOL, **I**, 199
ISOBUTYROIN, **II**, 115

PIVALOIN, **II**, 115
PROPIONOIN, **II**, 115
SEBACOIN, **IV**, 840
3. **Miscellaneous**
o-AMINOBENZYL ALCOHOL,
   III, 60
1-(AMINOMETHYL)-
   CYCLOHEXANOL, **IV**, 224
2-CYCLOHEXYLOXYETHANOL,
   V, 303
DIETHYL BIS(HYDROXY-
   METHYL)MALONATE, **V**,
   381
2-FURYLCARBINOL, **I**, 276
dl-GLYCERALDEHYDE DI-
   ETHYL ACETAL, **II**, 307
HEXAHYDROGALLIC ACID, **V**,
   591
3-HYDROXYGLUTARONITRILE,
   V, 614
HYDROXYMETHYLFERRO-
   CENE, **V**, 621
5-HYDROXYPENTANAL, **III**, 470
p-HYDROXYPHENYLPYRUVIC
   ACID, **V**, 627
2-ISOPROPYLAMINOETHAN-
   OL, **III**, 501
dl-ISOPROPYLIDENEGLY-
   CEROL, **III**, 502
MANDELIC ACID, **I**, 336
2-METHYL-2,5-DECANEDIOL,
   IV, 601
MONOBENZALPENTAERY-
   THRITOL, **IV**, 679
p-NITROBENZYL ALCOHOL,
   III, 652
2-NITROETHANOL, **V**, 833
PIPERONYL ALCOHOL, **II**, 591
1-(α-PYRIDYL)-2-PROPANOL, **III**,
   757
2-(1-PYRROLIDYL)PROPANOL,
   IV, 834
dl-TARTARIC ACID, **I**, 497
2,2,6,6-TETRAMETHYLOLCY-
   CLOHEXANOL, **IV**, 907

ALDEHYDES
A.1.   ALIPHATIC ALDEHYDES—
   UNSUBSTITUTED
ACROLEIN, **I**, 15
CHOLANE-24-AL, **V**, 242
CYCLOPENTANECARBOXAL-
   DEHYDE, **V**, 320
n-HEXALDEHYDE, **II**, 323
OCTANAL, **V**, 872
PROPIOLALDEHYDE, **IV**, 813
PROPIONALDEHYDE, **II**, 541
A.2.   ALIPHATIC, SUBSTITUTED
   ALDEHYDES
α-BROMOHEPTALDEHYDE, **III**,
   127
BUTYRCHLORAL, **IV**, 130
n-BUTYL GLYOXYLATE, **IV**, 124
α-CHLOROCROTONALDEHYDE,
   IV, 131
2-CHLORO-1-FORMYL-1-CY-
   CLOHEXENE, **V**, 215
7-CYANOHEPTANAL, **V**, 266
dl-GLYCERALDEHYDE, **II**, 305
5-HYDROXYPENTANAL, **III**, 470
PHENYLGLYOXAL, **II**, 509; **V**,
   937
PHENYLPROPARGYL ALDE-
   HYDE, **III**, 731
α-PHENYLPROPIONALDE-
   HYDE, **III**, 733
B.1.   AROMATIC ALDEHYDES
9-ANTHRALDEHYDE, **III**, 98
DIPHENYLACETALDEHYDE,
   IV, 375
MESITALDEHYDE, **III**, 549
1-NAPHTHALDEHYDE, **IV**, 690
β-NAPHTHALDHYDE, **III**, 626
PHENANTHRENE-9-ALDE-
   HYDE, **III**, 701
o-TOLUALDEHYDE, **IV**, 932; **III**,
   818
p-TOLUALDEHYDE, **II**, 583
B.2.   AROMATIC, SUBSTITUTED
   ALDEHYDES
p-ACETAMIDOBENZALDE-

HYDE, IV, 32
o-AMINOBENZALDEHYDE, III,
    56
p-AMINOBENZALDHYDE, IV, 31
o-ANISALDEHYDE, V, 46
p-BROMOBENZALDEHYDE, II,
    89, 442
2-BROMO-4-METHYLBENZAL-
    DEHYDE, V, 139
p-CHLOROBENZALDEHYDE, II,
    133
p-DIMETHYLAMINOBENZAL-
    DEHYDE, I, 214; IV, 331
2,4-DINITROBENZALDEHYDE,
    II, 223
DIPHENALDEHYDIC ACID, V,
    493
2-ETHOXY-1-NAPHTHALDE-
    HYDE, III, 98
5-FORMYL-4-PHENANTHROIC
    ACID, IV, 484
m-HYDROXYBENZALDEHYDE,
    III, 453
2-HYDROXY-1-NAPHTHALDE-
    HYDE, III, 463
MESITALDEHYDE, V, 49
m-METHOXYBENZALDEHYDE,
    III, 564
o-NITROBENZALDEHYDE, III,
    641; V, 825
p-NITROBENZALDEHYDE, II,
    441
o-NITROCINNAMALDEHYDE,
    IV, 722
6-NITROVERATRALDEHYDE,
    IV, 735
PHTHALALDEHYDIC ACID, III,
    737
SYRINGIC ALDEHYDE, IV, 866
C. DIALDEHYDES
DIPHENALDEHYDE, V, 489
GLYOXAL BISULFITE, III, 438
ISOPHTHALALDEHYDE, V, 668
β-METHYLGLUTARALDE-
    HYDE, IV, 661

o-PHTHALALDEHYDE, IV, 807
SODIUM NITROMALONALDE-
    HYDE MONOHYDRATE, IV,
    844
TEREPHTHALALDEHYDE, III,
    788
D. HETEROCYCLIC ALDEHYDES
FURFURAL, I, 280
INDOLE-3-ALDEHYDE, IV, 539
2-PYRROLEALDEHYDE, IV, 831
2-THENALDEHYDE, IV, 915
3-THENALDEHYDE, IV, 918
2-THIOPHENEALDEHYDE, III,
    811
ALUMINUM COMPOUNDS (See also
    OrganoMetallic Compounds)
    ALUMINUM t-BUTOXIDE, III, 48
AMIDES
A. UNSUBSTITUTED on NITROGEN
    ACETAMIDE, I, 3
    CHLOROACETAMIDE, I, 153
    CYANOACETAMIDE, I, 179
    α,α-DICHLOROACETAMIDE, III,
        260
    2-ETHYLHEXANAMIDE, IV, 436
    FUMARAMIDE, IV, 486
    n-HEPTAMIDE, IV, 513
    ISOBUTYRAMIDE, III, 490
    LACTAMIDE, III, 516
    MANDELAMIDE, III, 536
    METHACRYLAMIDE, III, 560
    NICOTINAMIDE-1-OXIDE, IV,
        704
    PHENYLACETAMIDE, IV, 760
    o-TOLUAMIDE, II, 586
    VERATRIC AMIDE, II, 44
B. Substituted on nitrogen (See also
        Heterocyclic Compounds)
    3-ACETAMIDO-2-BUTANONE,
        IV, 5
    2-ACETAMIDO-3,4,6-TRI-O-ACE-
        TYL-2-DEOXY-a-D-GLUCO-
        PYRANOSYL CHLORIDE, V,
        1
    p-ACETOTOLUIDE, I, 111

ACETYLGLYCINE, II, 11
BENZANILIDE, I, 82
BENZOYLACETANILIDE, III,
    108; IV, 80
BENZOYL-2-METHOXY-4-NI-
    TROACETANILIDE, IV, 82
BENZOYL PIPERIDINE, I, 99
N-BENZYLACRYLAMIDE, V, 73
N-BROMOACETAMIDE, IV, 104
CARBOBENZOXYGLYCINE, III,
    168
p-CHLOROACETYLACETANIL-
    IDE, III, 183
DIETHYL ACETAMIDOMAL-
    ONATE, V, 373
N,N-DIMETHYLCYCLOHEXA-
    NECARBOXAMIDE, IV, 339
N-ETHYL-p-CHLOROFORMANI-
    LIDE, IV, 420
ISONITROSOACETANILIDE, I,
    327
N-METHYLFORMANILIDE, III,
    590
2-NITRO-4-METHOXYACETANI-
    LIDE, III, 661
SALICYL-o-TOLUIDE, III, 765
SODIUM p-ARSONO-N-PHENYL-
    GLYCINAMIDE, I, 488
C. SULFONAMIDES
p-TOLUENESULFONYLAN-
    THRANILIC ACID, IV, 34
p-TOLYLSULFONYLMETHYL-
    NITROSAMIDE, IV, 943
D. MISCELLANEOUS
2-BENZOYL-1-CYANO-1-
    METHYL-1,2-DIHYDROISO-
    QUINOLINE, IV, 642
1-CYANO-2-BENZOYL-1,2-DI-
    HYDROISOQUINOLINE, IV,
    641
2,4-DITHIOBIURET, IV, 504
GUANYLTHIOUREA, IV, 502
HEXAHYDRO-1,3,5-TRIPRO-
    PIONYL-s-TRIAZINE, IV, 518
HEXAMETHYLPHOSPHORUS
    TRIAMIDE, V, 602

ISATIN, I, 327
3-ISOQUINUCLIDONE, V, 670
AMIDINES AND GUANIDINES
ACETAMIDINE HYDRO-
    CHLORIDE, I, 5
BENZAMIDINE HYDRO-
    CHLORIDE, I, 6
CREATININE, I, 172
N-N'-DIPHENYLBENZAMI-
    DINE, IV, 383
FORMAMIDINE ACETATE, V,
    582
GUANIDINE NITRATE, I, 302
NITROGUANIDINE, I, 399
N-PHENYLBENZAMIDINE, IV,
    769
AMINE OXIDES
BENZOFURAZAN OXIDE, IV, 74
3-METHYL-4-NITROPYRIDINE-1-
    OXIDE, IV, 654
3-METHYLPYRIDINE-1-OXIDE,
    IV, 655
NICOTINAMIDE-1-OXIDE, IV,
    704
PYRIDINE-N-OXIDE, IV, 828
AMINES (See also Acids, B, 1. (Amino-
    acids); Heterocyclic Compounds)
A. ALIPHATIC AMINES
ALLYLAMINE, II, 24
AMINOACETAL, III, 50
AMINOACETONE SEMICARBA-
    ZONE HYDROCHLOR-
    IDE, V, 27
AMINOMALONONITRILE p-TO-
    LUENESULFONATE, V, 32
1-AMINO-1-METHYLCYCLO-
    HEXANE, V, 35
1-(AMINOMETHYL)CYCLO-
    HEXANOL, IV, 224
β-AMINOPROPIONITRILE, III, 93
2-BROMOALLYLAMINE, V, 121
N-(2-BROMOALLYL)ETHYL-
    AMINE, V, 124
α-p-BROMOPHENYLETHYL-
    AMINE, II, 505
n-BUTYLAMINE, II, 319

sec.-BUTYLAMINE, **II,** 319

t-BUTYLAMINE, **III,** 148

t-BUTYLAMINE HYDROCHLOR-
IDE, **III,** 153

CARBOXYMETHOXYLAMINE
HEMIHYDROCHLORIDE,
III, 172

α-p-CHLOROPHENYLETHYL-
AMINE, **II,** 505

bis-(*β*-CYANOETHYL)AMINE,
III, 93

CYCLOBUTYLAMINE, **V,** 273

CYCLOHEXYLAMINE, **II,** 319

DIACETONAMINE HYDROGEN
OXALATE, **I,** 196

DIALLYLAMINE, **I,** 201

DI-*n*-BUTYLAMINE, **I,** 202

DI-*β*-CARBETHOXYETHYLME-
THYLAMINE, **III,** 258

DIETHYLAMINOACETONI-
TRILE, **III,** 275

1-DIETHYLAMINO-3-BUTAN-
ONE, **IV,** 281

β-DIETHYLAMINOETHYL AL-
COHOL, **II,** 183

DIETHYL AMINOMALONATE
HYDROCHLORIDE, **V,** 376

N,N-DIETHYL-1,2,2-TRICHLOR-
OVINYLAMINE, **V,** 387

β-DIMETHYLAMINOETHYL
CHLORIDE HYDROCHLOR-
IDE, **IV,** 333

α-N,N-DIMETHYLAMINOPHEN-
YLACETONITRILE, **V,** 437

β-DIMETHYLAMINOPROPIO-
PHENONE HYDROCHLOR-
IDE, **III,** 305

2,3-DIMETHYLBENZYLDIME-
THYLAMINE, **IV,** 587

N,N-DIMETHYLCYCLOHEXYL-
METHYLAMINE, **IV,** 339

α,α-DIMETHYL-*β*-PHEN-
ETHYLAMINE, **V,** 471

1-ETHYL-3-(3-DIMETHYL-
AMINO)PROPYLCARBO-
DIIMIDE HYDROCHLOR-

IDE, **V,** 555

GLYCINE-*t*-BUTYL ESTER, **V,**
586

n-HEPTYLAMINE, **II,** 318

2-ISOPROPYLAMINOETHANOL,
III, 501

LAURYLMETHYLAMINE, **IV,**
564

METHYLAMINE HYDROCH-
LORIDE, **I,** 347

2-METHYLBENZYLDIMETHYL-
AMINE, **IV,** 585

N-METHYLBUTYLAMINE, **V,**
736

N-METHYL-2,3-DIMETHOXY-
BENZYLAMINE, **IV,** 603

N-METHYL-1,2-DIPHENYL-
ETHYLAMINE,
IV, 605

N-METHYLETHYLAMINE, **V,**
758

5-METHYLFURFURYLDIME-
THYLAMINE, **IV,** 626

METHYLIMINODIACETIC ACID,
II, 397

β-NAPHTHOL PHENYLAMIN-
OMETHANE, **I,** 381

α-(*β*-NAPHTHYL)-ETHYL-
AMINE, **II,** 505

PHENACYLAMINE HY-
DROCHLORIDE, **V,** 909

α-PHENYLETHYLAMINE, **III,**
717

β-PHENYLETHYLAMINE, **III,**
720

dl-α-PHENYLETHYLAMINE, **II,**
503

d- and *l-α*-PHENYLETHYL-
AMINE, **II,** 506

R(+)- and *S*(−)-*α*-PHENYL-
ETHYLAMINE, **V,** 932

β-PHENYLETHYLDIMETHYL-
AMINE, **III,** 723

PSEUDOPELLETIERINE, **IV,** 816

ac-TETRAHYDRO-*β*-NAPH-
THYLAMINE, **I,** 499

α-p-TOLYLETHYLAMINE, **II**, 505
TRIMETHYLAMINE, **I**, 528
TRIMETHYLAMINE HYDROCH-
LORIDE, **I**, 531
α-p-XENYLETHYLAMINE, **II**,
505
B. *AROMATIC AMINES*
o-AMINOBENZALDEHYDE, **III**,
56
p-AMINOBENZALDEHYDE, **IV**,
31
m-AMINOBENZALDEHYDE DI-
METHYLACETAL, **III**, 59
2-AMINOBENZOPHENONE, **IV**,
34
o-AMINOBENZYL ALCOHOL,
III, 60
2-AMINO-p-CYMENE, **III**, 63
2-AMINOFLUORENE, **II**, 448; **V**,
30
1,2-AMINONAPHTHOL HY-
DROCHLORIDE, **II**, 33
1,4-AMINONAPHTHOL HY-
DROCHLORIDE, **I**, 49; **II**, 39
2-AMINO-4-NITROPHENOL, **III**,
82
2-AMINO-3-NITROTOLUENE, **IV**,
42
p-AMINOPHENYL DISULFIDE,
III, 86
p-AMINOTETRAPHENYLME-
THANE, **IV**, 47
AMINOTHYMOL, **I**, 512
4-AMINOVERATROLE, **II**, 44
N-BENZYLANILINE, **I**, 102; **IV**,
92
3-(o-CHLOROANILINO)-
PROPIONITRILE, **IV**, 146
N-2-CYANOETHYLANILINE, **IV**,
205
DIAMINODURENE CHLORO-
STANNATE, **II**, 255
2,4-DIAMINOTOLUENE, **II**, 160
2,6-DIBROMO-4-AMINOPHENOL
CHLOROSTANNATE, **II**, 175
2,6-DIBROMOANILINE, **III**, 262

2,6-DICHLOROANILINE, **III**, 262
β,β-DICHLORO-p-DIMETHY-
LAMINOSTYRENE, **V**, 361
p-DIMETHYLAMINOBENZAL-
DEHYDE, **IV**, 331
6-(DIMETHYLAMINO)-
FULVENE, **V**, 431
3,4-DIMETHYLANILINE, **III**, 307
2,4-DINITROANILINE, **II**, 221
2,6-DINITROANILINE, **IV**, 364
ETHYL p-AMINOBENZOATE, **I**,
240
N-ETHYL-p-CHLOROANILINE,
IV, 420
ETHYL p-DIMETHYLAMINO-
PHENYLACETATE, **V**, 552
N-ETHYL-m-TOLUIDINE, **II**, 290
1-METHYLAMINOANTHRAQUI-
NONE, **III**, 573
4'-METHYL-2-AMINOBENZO-
PHENONE, **IV**, 38
1-METHYLAMINO-4-BROM-
OANTHRAQUINONE, **III**,
575
o-NITROANILINE, **I**, 388
m-NITRODIMETHYLANILINE,
III, 658
2-NITRO-4-METHOXYANILINE,
III, 661
4-NITRO-1-NAPHTHYLAMINE,
III, 664
N-PHENYLCARBAZOLE, **I**, 547
o-PHENYLENEDIAMINE, **II**, 501
SODIUM p-ARSONO-N-PHENYL-
GLYCINAMIDE, **I**, 488
o-TOLUIDINESULFONIC ACID,
III, 824
m-TOLYLBENZYLAMINE, **III**,
827
p-TRICYANOVINYL-N,N-DIME-
THYLANILINE, **IV**, 953
m-TRIFLUOROMETHYL-N,N-DI-
METHYLANILINE, **V**, 1085
TRIPHENYLAMINE, **I**, 544
C. *HETEROCYCLIC AMINES*
9-AMINOACRIDINE, **III**, 53

2-AMINO-4-ANILINO-6-(CHLOR-
OMETHYL)-*s*-TRIAZINE, **IV,**
29
5-AMINO-2,3-DIHYDRO-1,4-
PHTHALAZINEDIONE, **III,**
69
4-AMINO-2,6-DIMETHYLPY-
RIMIDINE, **III,** 71
2-AMINO-6-METHYLBENZO-
THIAZOLE, **III,** 76
3(5)-AMINOPYRAZOLE, **V,** 39
3-AMINOPYRIDINE, **IV,** 45
1-AMINOPYRIDINIUM IODIDE,
V, 43
3-AMINO-1*H*-1,2,4-TRIAZOLE,
III, 95
4-AMINO-4*H*-1,2,4-TRIAZOLE,
III, 96
BENZOGUANAMINE, **IV,** 78
2-BENZYLAMINOPYRIDINE, **IV,**
91
2-CHLOROMETHYL-4,6-DI-
AMINO-*s*-TRIAZINE, **IV,** 30
2,5-DIAMINO-3,4-DICYANO-
THIOPHENE, **IV,** 243
2,4-DIAMINO-6-HYDROXYPYR-
IMIDINE, **IV,** 245
2,4-DIAMINO-6-PHENYL-*s*-TRIA-
ZINE, **IV,** 78
DIAMINOURACIL HYDROCH-
LORIDE, **IV,** 247
2-(DIMETHYLAMINO)-
PYRIMIDINE, **IV,** 336
1,4-DIPHENYL-5-AMINO-1,2,3-
TRIAZOLE, **IV,** 380
N-ETHYLPIPERIDINE, **V,** 575
2-MERCAPTO-4-AMINO-5-CARB-
ETHOXYPYRIMIDINE, **IV,**
566
N-METHYLAMINOPYRIMIDINE,
IV, 336
4-PHENYL-5-ANILINO-1,2,3-
TRIAZOLE, **IV,** 380
URAMIL, **II,** 617
D. *DIAMINES*
4,4'-BIS(DIMETHYLAMINO)-
BENZIL, **V,** 111
DECAMETHYLENEDIAMINE,
III, 229
4,4'-DIAMINOAZOBENZENE, **V,**
341
4,4'-DIAMINODIPHENYL SUL-
FONE, **III,** 239
DIAMINOMALEONITRILE, **V,**
344
1,2-DIAMINO-4-NITROBEN-
ZENE, **III,** 242
2,3-DIAMINOPYRIDINE, **V,** 346
N,N'-DIBENZYL-*p*-PHENYL-
ENEDIAMINE, **IV,** 92
β-DI-*n*-BUTYLAMINOETHYL-
AMINE, **III,** 254
γ-DI-*n*-BUTYLAMINOPROPYL-
AMINE, **III,** 256
N,N'-DIETHYLBENZIDINE, **IV,**
283
PUTRESCINE DIHYDRO-
CHLORIDE, **IV,** 819
TETRAMETHYL-*p*-PHENYL-
ENEDIAMINE, **V,** 1018
2,4,5-TRIAMINONITROBEN-
ZENE, **V,** 1067
AMINO ACIDS (See *Acids*; B.1)
**ARSENIC AND ANTIMONY COM-
POUNDS**
ARSANILIC ACID, **I,** 70
ARSENOACETIC ACID, **I,** 74
ARSONOACETIC ACID, **I,** 73
p-ARSONOPHENOXYACETIC
ACID, **I,** 75
p-NITROPHENYLARSONIC
ACID, **III,** 665
SODIUM *p*-ARSONO-*N*-PHENYL-
GLYCINAMIDE, **I,** 488
SODIUM *p*-HYDROXYPHENY-
LARSONATE, **I,** 490
TETRAPHENYLARSONIUM
CHLORIDE HYDRO-
CHLORIDE, **IV,** 910
TRIPHENYLARSINE, **IV,** 910
TRIPHENYLARSINE OXIDE, **IV,**
911

TRIPHENYLSTIBINE, **I**, 550
TRI-*p*-TOLYLSTIBINE, **I**, 551
AZIDES
    t-BUTYL AZIDOFORMATE, **V**, 157
    m-NITROBENZAZIDE, **IV**, 715
    o-NITROPHENYLAZIDE, **IV**, 75
    PHENYL AZIDE, **III**, 710
AZO COMPOUNDS
    AZOBENZENE, **III**, 103
    1,1'-AZO-*bis*-1-CYCLOHEXA-NENITRILE, **IV**, 66
    2,2'-AZO-*bis*-ISOBUTYRONI-TRILE, **IV**, 67
    AZO DYE FROM *o*-TOLIDINE AND CHICAGO ACID, **II**, 145
    t-BUTYL AZODIFORMATE, **V**, 160
    4,4'-DIAMINOAZOBENZENE, **V**, 341
    DIAZOAMINOBENZENE, **II**, 163
    ETHYL AZODICARBOXYLATE, **III**, 375; **IV**, 411
    METHYL AZODICARBOXY-LATE, **IV**, 414
    METHYL RED, **I**, 374
    p-PHENYLAZOBENZOIC ACID, **III**, 711
    p-PHENYLAZOBENZOYL CHLORIDE, **III**, 712
AZOXY COMPOUNDS
    AZOXYBENZENE, **II**, 57
BORON COMPOUNDS
    BENZENEBORONIC ANHY-DRIDE, **IV**, 68
    10-METHYL-10,9-BORAZARO-PHENANTHRENE, **V**, 727
    TRIETHYLOXONIUM FLUO-BORATE, **V**, 1080
    TRIMETHYLOXONIUM FLUO-BORATE, **V**, 1096
    2,4,6-TRIMETHYLPYRYLIUM TETRAFLUOROBORATE, **V**, 1112
    2,4,6-TRIPHENYLPYRYLIUM TETRAFLUOROBORATE, **V**, 1135

TROPYLIUM FLUOBORATE, **V**, 1138
CARBAMATES
    t-BUTYL CARBAMATE, **V**, 162
CARBODIIMIDES
    DIPHENYLCARBODIIMIDE (Method I), **V**, 501
    DIPHENYLCARBODIIMIDE (Method II), **V**, 504
    1-ETHYL-3-(3-DIMETHYLA-MINO)PROPYLCARBODI-IMIDE HYDROCHLORIDE, **V**, 555
    1-ETHYL-3-(3-DIMETHYLA-MINO)PROPYLCARBODI-IMIDE METHIODIDE, **V**, 555
CARBOHYDRATES AND DERIVATIVES
    2-ACETMIDO-3,4,6-TRI-*O*-ACE-TYL-2-DEOXY-*a*-D-GLUCO-PYRANOSYL CHLORIDE, **V**, 1
    ACETOBROMOGLUCOSE, **III**, 11
    l-ARABINOSE, **I**, 67
    D-ARABINOSE, **III**, 101
    CELLOBIOSE, **II**, 122
    α-CELLOBIOSE OCTAACETATE, **II**, 124
    β-GENTIOBIOSE OCTAACE-TATE, **III**, 428
    d-GLUCOSAMINE HYDRO-CHLORIDE, **III**, 430
    β-*d*-GLUCOSE-1,2,3,4-TETRA-ACETATE, **III**, 432
    β-*d*-GLUCOSE-2,3,4,6-TETRA-ACETATE, **III**, 434
    D-GULONIC-γ-LACTONE, **IV**, 506
    D-MANNOSE, **III**, 541
    α-METHYL *d*-GLUCOSIDE, **I**, 364
    α-METHYL MANNOSIDE, **I**, 371
    PENTAACETYL *d*-GLUCONO-NITRILE, **III**, 690
    2,3,4,5,6-PENTA-*O*-ACETYL-D-GLUCONIC ACID, **V**, 887
    2,3,4,5,6-PENTA-*O*-ACETYL-D-GLUCONYL CHLORIDE, **V**, 887

2,3,4,6-TETRAMETHYL-d-GLU-
COSE, **III,** 800

CYANAMIDES
o-CHLOROPHENYLCYAN-
AMIDE, **IV,** 172
DIALLYLCYANAMIDE, **I,** 203
DI-n-BUTYLCYANAMIDE, **I,** 204
N-METHYL-1-NAPHTHYLCY-
ANAMIDE, **III,** 608
α-NAPHTHYCYANAMIDE, **IV,**
174

CYANIDES (see Nitriles).

DIAZO AND DIAZONIUM COMPOUNDS
BENZENEDIAZONIUM-2-CAR-
BOXYLATE, **V,** 54
t-BUTYL DIAZOACETATE, **V,**
179
CROTYL DIAZOACETATE, **V,**
258
DIAZOMETHANE, **II,** 165; **III,** 244;
IV, 250; **V,** 351
DIPHENYLDIAZOMETHANE,
III, 351
ETHYL DIAZOACETATE, **III,**
392; **IV,** 424
PHENYLBENZOYLDIAZO-
METHANE, **II,** 496

DIAZOIUM SALTS
4,4'-BIPHENYLENE-bis-DI-
AZONIUM FLUOBORATE,
II, 188
p-CARBETHOXYBENZENEDI-
AZONIUM FLUOBORATE,
II, 299
β-NAPHTHALENEDIAZONIUM
CHLORIDE–MERCURIC
CHLORIDE COMPOUND, **II,**
432
o-NITROBENZENEDIAZO-
NIUM FLUOBORATE, **II,** 226
p-NITROBENZENEDIAZONIUM
FLUOBORATE, **II,** 225

ENAMINES
1-MORPHOLINO-1-CYCLO-
HEXENE, **V,** 808

ESTERS OF INORGANIC ACIDS
n-AMYL BORATE, **II,** 107
n-AMYL PHOSPHATE, **II,** 110
n-BUTYL BORATE, **II,** 106
t-BUTYL HYPOCHLORITE, **IV,**
125
n-BUTYL NITRITE, **II,** 108
n-BUTYL PHOSPHATE, **II,** 109
sec.-BUTYL PHOSPHATE, **II,** 110
n-BUTYL SULFATE, **II,** 111
n-BUTYL SULFITE, **II,** 112
ETHYL NITRITE, **II,** 204
ISOPROPYL THIOCYANATE, **II,**
366
METHYL NITRATE, **II,** 412
METHYL NITRITE, **II,** 363
n-PROPYL PHOSPHATE, **II,** 110
n-PROPYL SULFATE, **II,** 112
TRIETHYL PHOSPHITE, **IV,** 955
TRIISOPROPYL PHOSPHITE, **IV,**
956

**ESTERS OF ORGANIC ACIDS (See also
Lactones)**
A. *OF UNSUBSTITUTED MONO-
BASIC ACIDS*
ACENAPHTHENOL ACETATE,
III, 3
2-ACETAMIDO-3,4,6-TRI-O-ACE-
TYL-2-DEOXY-a-D-GLUCO-
PYRANOSYL CHLORIDE, **V,**
1
3β-ACETOXYETIENIC ACID, **V,**
8
3-BENZOYLOXYCYCLOHEX-
ENE, **V,** 70
BENZYL BENZOATE, **I,** 104
t-BUTYL ACETATE, **III,** 141; **IV,**
263
n-BUTYL ACRYLATE, **III,** 146
n-BUTYL n-BUTYRATE, **I,** 138
n-BUTYL CARBAMATE, **I,** 140
n-BUTYL OLEATE, **II,** 469
4-CHLOROBUTYL BENZOATE,
III, 187
2-CHLOROETHYL BENZOATE,
IV, 84

γ-CHLOROPROPYL ACETATE, **III**, 203

CHOLESTERYL ACETATE, **II**, 193

2,3-DIMETHYLBENZYL ACETATE, **IV**, 584

ETHYL CINNAMATE, **I**, 252

ETHYL CYCLOHEXYLIDENE-ACETATE, **V**, 547

ETHYL 2,4-DIPHENYLBUTANOATE, **V**, 559

ETHYL INDOLE-2-CARBOXYLATE, **V**, 567

ETHYL α-NAPHTHOATE, **II**, 282

ETHYL PHENYLACETATE, **I**, 270

ETHYL γ-PHENYLBUTYRATE, **II**, 196

ETHYL *n*-TRIDECYLATE, **II**, 292

FURFURAL DIACETATE, **IV**, 489

2-FURYLMETHYL ACETATE, **I**, 285

HEXAHYDROGALLIC ACID TRIACETATE, **V**, 591

HYDROXYHYDROQUINONE TRIACETATE, **I**, 317

2-IODOETHYL BENZOATE, **IV**, 84

18,20-LACTONE OF 3β-ACETOXY-20β-HYDROXY-5-PREGNENE-18-OIC ACID, **V**, 692

1-(2-METHOXYCARBONYL-PHENYL)-PYRROLE, **V**, 716

o-METHYLBENZYL ACETATE, **IV**, 582

METHYL CYCLOPENTANECARBOXYLATE, **IV**, 594

METHYL HOMOVERATRATE, **II**, 333

METHYL MYRISTATE, **III**, 605

METHYL PALMITATE, **III**, 605

o- AND *p*-NITROBENZALDIACETATE, **IV**, 713

p-NITROBENZYL ACETATE, **III**, 650

1,5-PENTANEDIOL DIACETATE, **IV**, 748

2,3,4,5,6-PENTA-*O*-ACETYL-D-GLUCONIC ACID, **V**, 887

2,3,4,5,6-PENTA-*O*-ACETYL-D-GLUCONYL CHLORIDE, **V**, 887

PHENYL CINNAMATE, **III**, 714

2-PHENYLETHYL BENZOATE, **V**, 336

VINYL LAURATE, **IV**, 977

TRIMYRISTIN, **I**, 538

B. OF SUBSTITUTED MONOBASIC ACIDS

  1. Cyano Esters

  t-BUTYL CYANOACETATE, **V**, 171

  ETHYL β-BROMOPROPIONATE, **I**, 246

  ETHYL *n*-BUTYLCYANOACETATE, **III**, 385

  ETHYL *sec*-BUTYLIDENECYANOACETATE, **IV**, 94

  ETHYL CYANOACETATE, **I**, 254

  ETHYL α-CYANO-β-PHENYLACRYLATE, **I**, 451

  ETHYL (1-ETHYLPROPENYL)-METHYLCYANOACETATE, **III**, 397

  ETHYL (1-ETHYLPROPYLIDENE)-CYANOACETATE, **III**, 399

  ETHYL PHENYLCYANOACETATE, **IV**, 461

  ETHYL β-PHENYL-β-CYANOPROPIONATE, **IV**, 804

  ETHYL PHENYLCYANOPYRUVATE, **II**, 287

  ETHYL (1-PHENYLETHYLIDENE)-CYANOACETATE, **IV**, 463

  METHYL ω-CYANOPELARGONATE, **III**, 584

  α-PHENYL-α-CARBETHOXYGLUTARONITRILE, **IV**, 776

  2. Halogen Esters

  p-BROMOBENZALDIACETATE, **II**, 442

t-BUTYL BROMOACETATE, **IV**, 263

sec-BUTYL α-BROMOPROPION-
ATE, **IV**, 122

t-BUTYL CHLOROACETATE, **IV**, 263

t-BUTYL α-CHLOROPROPIO-
NATE, **IV**, 263

ETHYL BROMOACETATE, **III**, 381

ETHYL γ-BROMOBUTYRATE, **V**, 545

ETHYL CHLOROACETATE, **II**, 263

ETHYL α-CHLOROACETOACE-
TATE, **IV**, 592

ETHYL CHLOROFLUORO-
ACETATE, **IV**, 423

ETHYL α-CHLOROPHENYL-
ACETATE, **IV**, 169

METHYL β-BROMOPROPIO-
NATE, **III**, 576

METHYL 5-BROMOVALERATE, **III**, 578

VINYL CHLOROACETATE, **III**, 853

3. **Hydroxy Esters**

ALLYL LACTATE, **III**, 46

p-CHLOROPHENYL SALICY-
LATE, **IV**, 178

ETHYL-4-ETHYL-2-METHYL-3-
HYDROXYOCTANOATE, **IV**, 447

ETHYL β-HYDROXY-β,β-DI-
PHENYLPROPIONATE, **V**, 564

ETHYL MANDELATE, **IV**, 169

ETHYL β-PHENYL-β-HYDROX-
YPROPIONATE, **III**, 408

ISOPROPYL LACTATE, **II**, 365

4. **Keto Esters**

2-*p*-ACETYLPHENYLHYDRO-
QUINONE DIACETATE, **IV**, 16

BENZOIN ACETATE, **II**, 69

o-BENZOYLOXYACETOPHEN-
ONE, **IV**, 478

t-BUTYL ACETOACETATE, **V**, 155

t-BUTYL *o*-BENZOYLBEN-
ZOATE, **IV**, 263

sec-BUTYL α-*n*-CAPROYLPRO-
PIONATE, **IV**, 120

2-CARBETHOXYCYCLOOCTAN-
ONE, **V**, 198

2-CARBETHOXYCYCLOPEN-
TANONE, **II**, 116

3,5-DIMETHYL-4-CARBETHOXY-
2-CYCLOHEXEN-1-ONE, **III**, 317

ETHYL ACETOACETATE, **I**, 235

ETHYL α-ACETOGLUTARATE, **II**, 263

ETHYL ACETONEDIOXALATE, **II**, 126

ETHYL ACETOPYRUVATE, **I**, 238

ETHYL α-ACETYL-β-(2,3-DI-
METHOXYPHENYL)-
ACRYLATE, **IV**, 408

ETHYL α-ACETYL-β-(3,4-DI-
METHOXYPHENYL)-
ACRYLATE, **IV**, 410

ETHYL α-ACETYL-β-(2,3-DI-
METHOXYPHENYL)-
PROPIONATE, **IV**, 408

ETHYL α-ACETYL-β-(3,4-DI-
METHOXYPHENYL)-
PROPIONATE, **IV**, 410

ETHYL BENZOYLACETATE, **II**, 266; **III**, 379; **IV**, 415

ETHYL BENZOYLACETOACE-
TATE, **II**, 266

ETHYL BENZOYLFORMATE, **I**, 241

ETHYL *n*-BUTYLACETOACE-
TATE, **I**, 248

ETHYL DIACETYLACETATE, **III**, 390

ETHYL ISOBUTYRYLISOBU-
TYRATE, **II**, 270

ETHYL α-ISOPROPYLACETO-
ACETATE, **III**, 405

ETHYL 2-KETOCYCLOHEXYL-

GLYOXALATE, **II,** 531

ETHYL 2-KETOHEXAHYDRO-
BENZOATE, **II,** 532

ETHYL α-PHENYLACETOACE-
TATE, **II,** 284

ETHYL PYRUVATE, **IV,** 467

METHYL p-ACETYLBENZOATE,
IV, 579

METHYL BENZOYLFORM-
ATE, **I,** 244

METHYL-4-KETO-7-METHYL-
OCTANOATE, **III,** 601

METHYL PYRUVATE, **III,** 610

5. Nitrile Esters (See Cyano Esters)

6. Miscellaneous

ACETYLMANDELIC ACID, **I,** 12

ACETYLMANDELYL CHLO-
RIDE, **I,** 13

BENZYL CARBAMATE, **III,** 168

t-BUTYL AZODIFORMATE, **V,**
160

t-BUTYL AZIDOFORMATE, **V,**
157

t-BUTYL CARBAMATE, **V,** 162

t-BUTYL CARBAZATE, **V,** 166

t-BUTYL DIAZOACETATE, **V,**
179

n-BUTYL GLYOXYLATE, **IV,** 124

3-CARBETHOXYCOUMARIN, **III,**
165

CROTYL DIAZOACETATE, **V,**
258

DI-β-CARBETHOXYETHYLME-
THYLAMINE, **III,** 258

N,N'-DICARBETHOXYPUTRES-
CINE, **IV,** 822

ETHYL p-AMINOBENZOATE, **II,**
300

ETHYL β-ANILINOCROTO-
NATE, **III,** 374

ETHYL N-BENZYLCARBA-
MATE, **IV,** 780

ETHYL DIAZOACETATE, **III,**
392; **IV,** 424

ETHYL DIETHOXYACETATE,
IV, 427

ETHYL 6,7-DIMETHOXY-3-
METHYLINDENE-2-CAR-
BOXYLATE, **V,** 550

ETHYL p-DIMETHYLAMINO-
PHENYLACETATE, **V,** 552

ETHYL ETHOXYACETATE, **II,**
261

ETHYL HYDRAZINECARBOXY-
LATE, **III,** 404

ETHYL ISODEHYDROACE-
TATE, **IV,** 549

ETHYL N-METHYLCARBAM-
ATE, **II,** 278

ETHYL 3-METHYLCOUMARIL-
ATE, **IV,** 590

ETHYL α-NITROBUTYRATE, **IV,**
454

ETHYL N-NITROSO-N-BENZYL-
CARBAMATE, **IV,** 780

ETHYL β,β-PENTAMETHYL-
ENEGLYCIDATE, **IV,** 459

ETHYL N-PHENYLFORMIMID-
ATE, **IV,** 464

ETHYL 2-PYRIDYLACETATE,
III, 413

ETHYL α-(1-PYRROLI-
DYL)PROPIONATE, **IV,** 466

GLYCINE t-BUTYL ESTER, **V,**
586

GLYCINE ETHYL ESTER HY-
DROCHLORIDE, **II,** 310

HYDROQUINONE DIACETATE,
III, 452

2-MERCAPTO-4-AMINO-5-CAR-
BETHOXYPYRIMIDINE, **IV,**
566

METHYL COUMALATE, **IV,** 532

METHYL 5,5-DIMETHOXY-3-
METHYL-2,3-EPOXYPEN-
TANOATE, **IV,** 649

METHYL p-ETHYLBENZOATE,
IV, 580

METHYL 3-METHYL-2-FURO-
ATE, **IV** 649

METHYL γ-METHYL-γ-NITROV-
ALERATE, **IV,** 652

METHYL β-THIODIPROPIO-
NATE, **IV**, 669
p-NITROBENZALDIACETATE,
II, 441
PHENYLMETHYLGLYCIDIC ES-
TER, **III**, 727
7. **SULFONIC ACID ESTERS**
n-BUTYL p-TOLUENESULFON-
ATE, **I**, 145
METHYL p-TOLUENESULFON-
ATE, **I**, 146
C. *OF DICARBOXYLIC ACIDS*
β-CARBETHOXY-γ,γ-DIPHEN-
YLVINYLACETIC ACID, **IV**,
132
β-CARBOMETHOXYPROPIONYL
CHLORIDE, **III**, 169
CETYLMALONIC ESTER, **IV**, 141
DI-t-BUTYL β,β-DIMETHYL-
GLUTARATE, **IV**, 263
DI-t-BUTYL GLUTARATE, **IV**,
263
DI-t-BUTYL MALONATE, **IV**, 261
DI-t-BUTYL SUCCINATE, **IV**, 263
DIETHYL ACETAMIDOMALON-
ATE, **V**, 373
DIETHYL ACETYLENEDICAR-
BOXYLATE, **IV**, 330
DIETHYL(0-BENZOYL)-
ETHYLTARTRONATE, **V**,
379
DIETHYL BENZOYLMALON-
ATE, **IV**, 285
DIETHYL BIS(HYDROXY-
METHYL)-MALONATE, **V**,
381
DIETHYL 1,1-CYCLOBUTANE-
DICARBOXYLATE, **IV**, 288
DIETHYL CYCLOHEXANONE-
2,6-DICARBOXYLATE, **V**,
439
DIETHYL Δ²-CYCLOPENTEN-
YLMALONATE, **IV**, 291
DIETHYL ETHYLIDENEMA-
LONATE, **IV**, 293
DIETHYL ETHYLPHOSPHON-

ATE, **IV**, 326
DIETHYL cis-HEXAHYDRO-
PHTHALATE, **IV**, 304
DIETHYL β-KETOPIMELATE, **V**,
384
DIETHYL METHYLENEMA-
LONATE, **IV**, 298
DIETHYL METHYLPHOSPHON-
ATE, **IV**, 326
DIETHYL γ-OXOPIMELATE, **IV**,
302
DIETHYL SUCCINATE, **V**, 993
DIETHYL cis-Δ⁴-TETRAHYDRO-
PHTHALATE, **IV**, 304
DIISOPROPYL ETHYLPHOS-
PHONATE, **IV**, 326
DIISOPROPYL METHYLPHOS-
PHONATE, **IV**, 325
DIMETHYL ACETYLENEDI-
CARBOXYLATE, **IV**, 329
DIMETHYL cis-HEXAHYDRO-
PHTHALATE, **IV**, 306
DIMETHYL 3-METHYLENECY-
CLOBUTANE-1,2-DICAR-
BOXYLATE, **V**, 459
DIMETHYL OCTADECANE-
DIOATE, **V**, 463
DIMETHYL cis-Δ⁴-TETRAHY-
DROPHTHALATE, **IV**, 306
DIPHENYL SUCCINATE, **IV**, 390
ETHOXYMAGNESIUM-
MALONIC ESTER, **IV**, 285
ETHYL ACETONEDICARBOXY-
LATE, **I**, 237
ETHYL ACETOSUCCINATE, **II**,
262
ETHYL ADIPATE, **II**, 264
ETHYL AZODICARBOXYL-
ATE, **III**, 375; **IV**, 411
ETHYL BENZALMALONATE,
III, 377
ETHYL n-BUTYLMALONATE, **I**,
250
ETHYL sec.-BUTYLMALON-
ATE, **II**, 417
ETHYL t-BUTYL MALONATE,
IV, 417

ETHYL ENANTHYLSUCCIN-
ATE, **IV,** 430
ETHYL ETHOXALYLPROPION-
ATE, **II,** 272
ETHYL ETHOXYMETHYL-
ENEMALONATE, **III,** 395
ETHYL 1,16-HEXADECANEDI-
CARBOXYLATE, **III,** 401
ETHYL HYDRAZODICARBOXY-
LATE, **III,** 375; **IV,** 411
ETHYL HYDROGEN ADIP-
ATE, **II,** 277
ETHYL HYDROGEN SEBAC-
ATE, **II,** 276
ETHYL ISOPROPYLMALON-
ATE, **II,** 94
ETHYL METHYLMALONATE,
II, 279
ETHYL ORTHOFORMATE, **I,** 258
ETHYL OXALATE, **I,** 261
ETHYL PHENYLMALONATE,
II, 288
ETHYL PHTHALIMIDOMALON-
ATE, **I,** 271
ETHYL PIMELATE, **II,** 536
ETHYL SEBACATE, **II,** 277
METHYL AZODICARBOXY-
LATE, **IV,** 414
METHYL BUTADIENOATE, **V,**
734
METHYL HYDRAZODICARBOX-
YLATE, **IV,** 413
METHYL HYDROGEN HENDE-
CANEDIOATE, **IV,** 635
METHYL OXALATE, **I,** 264
METHYL OXALATE, **II,** 414
METHYL SEBACAMATE, **III,** 613
METHYL β-THIODIPROPION-
ATE, **IV,** 669
sec.-OCTYL HYDROGEN
PHTHALATE, **I,** 418
o-PHENYLENE CARBONATE,
IV, 788
D. *OF TRICARBOXYLIC ACIDS*
ETHYL *N*-TRICARBOXYLATE,
III, 415

TRICARBETHOXYMETHANE,
II, 594
TRICARBOMETHOXYME-
THANE, **II,** 596
TRIETHYL α-PHTHALIMIDOE-
THANE-α,α,β-TRICARBOXY-
LATE, **IV,** 55
E. *OF TETRACARBOXYLIC ACIDS*
ETHYL ETHYLENETETRACAR-
BOXYLATE, **II,** 273
ETHYL ORTHOCARBONATE,
IV, 457
ETHYL PROPANE-1,1,2,3-
TETRACARBOXYLATE, **I,**
272
ETHYL PROPANE-1,1,3,3-
TETRACARBOXYLATE, **I,**
290
**ETHERS (See also Heterocyclic com-
pounds.)**
α-ALLYL-β-BROMOETHYL
ETHYL ETHER, **IV,** 750
o-ANISALDEHYDE, **V,** 46
ANISOLE, **I,** 58
p-ARSONOPHENOXYACETIC
ACID, **I,** 75
BENZHYDRYL β-CHLORO-
ETHYL ETHER, **IV,** 72
BISCHLOROMETHYL ETHER,
IV, 101
o-*n*-BUTOXYNITROBENZENE,
III, 140
7-*t*-BUTOXYNORBORNADIENE,
V, 151
α-CHLOROETHYL ETHYL
ETHER, **IV,** 748
p-CHLOROPHENOXYMETHYL
CHLORIDE, **V,** 221
2-CHLORO-1,1,2-TRIFLUORO-
ETHYL ETHYL ETHER, **IV,**
184
CHOLESTANYL METHYL
ETHER, **V,** 245
CREOSOL, **IV,** 203
2-CYCLOHEXYLOXYETHANOL,
V, 303

DEOXYANISOIN, **V**, 339
α,β-DIBROMOETHYL ETHYL
  ETHER, **IV**. 749
4,4'-DICHLORODIBUTYL
  ETHER, **IV**, 266
DICHLOROMETHYL METHYL
  ETHER, **V**, 365
DIETHYL AMINOMALONATE
  HYDROCHLORIDE, **V**, 376
2,5-DIHYDRO-2,5-DIMETHOXY-
  FURAN, **V**, 403
2,6-DIMETHOXYBENZONI-
  TRILE, **III**, 293
3,3'-DIMETHOXYBIPHENYL, **III**,
  295
2,3-DIMETHOXYCINNAMIC
  ACID, **IV**, 327
6,7-DIMETHOXY-3,4-DIHYDRO-
  2-NAPHTHOIC ACID, **III**, 300
trans-4,4'-DIMETHOXYSTIL-
  BENE, **V**, 428
3,5-DINITROANISOLE, **I**, 219
ETHOXYACETYLENE, **IV**, 404
3-ETHOXY-2-CYCLOHEXE-
  NONE, **V**, 539
β-ETHOXYETHYL BROMIDE,
  III, 370
2-ETHOXY-1-NAPHTHALDE-
  HYDE, **III**, 98
p-ETHOXYPHENYLUREA, **IV**, 52
β-ETHOXYPROPIONITRILE, **III**,
  372
ETHYL α-ACETYL-β-(2,3-DI-
  METHOXYPHENYL)-
  ACRYLATE, **IV**, 408
ETHYL α-ACETYL-β-(3,4-DI-
  METHOXYPHENYL)-
  ACRYLATE, **IV**, 410
ETHYL α-ACETYL-β-(2,3-DI-
  METHOXYPHENYL)-
  PROPIONATE, **IV**, 408
ETHYL α-ACETYL-β-(3,4-DI-
  METHOXYPHENYL)-
  PROPIONATE, **IV**, 410
ETHYL 6,7-DIMETHOXY-3-

METHYLINDENE-2-CAR-
  BOXYLATE, **V**, 550
α-GLYCERYL PHENYL ETHER,
  I, 296
4-IODOVERATROLE, **IV**, 547
l-MENTHOXYACETIC ACID, **III**,
  544
l-MENTHOXYACETYL CHLOR-
  IDE, **III**, 547
METHOXYACETONITRILE, **II**,
  387
ω-METHOXYACETOPHE-
  NONE, **III**, 562
METHOXYACETYLENE, **IV**, 406
m-METHOXYBENZALDEHYDE,
  III, 564
2-METHOXYDIPHENYL ETHER,
  III, 566
6-METHOXY-8-NITROQUINO-
  LINE, **III**, 568
o-METHOXYPHENYLACETONE,
  IV, 573
p-METHOXYPHENYLACETONI-
  TRILE, **IV**, 576
1-(*o*-METHOXYPHENYL)-2-NI-
  TRO-1-PROPENE, **IV**, 573
1-(*p*-METHOXYPHENYL)-5-
  PHENYL-1,3,5-PENTANE-
  TRIONE, **V**, 718
2-(*p*-METHOXYPHENYL)-6-
  PHENYL-4-PYRONE, **V**, 721
p-METHOXYPHENYLUREA, **IV**,
  53
O-METHYLCAPROLACTIM, **IV**,
  588
N-METHYL-2,3-DIMETHOXY-
  BENZYLAMINE, **IV**, 603
METHYL 5,5-DIMETHOXY-3-
  METHYL-2,3-EPOXYPEN-
  TANOATE, **IV**, 649
METHYLISOUREA HYDROCH-
  LORIDE, **IV**, 645
METHYL β-NAPHTHYL ETHER,
  I, 59
MONOCHLOROMETHYL
  ETHER, **I**, 377

o-NITRODIPHENYL ETHER, **II**, 446

p-NITRODIPHENYL ETHER, **II**, 445

2-NITRO-4-METHOXYACET-ANILIDE, **III**, 661

2-NITRO-4-METHOXYANIL-INE, **III**, 661

6-NITROVERATRALDEHYDE, **IV**, 735

PHENOLS: 6-METHOXY-2-NA-PHTHOL, **V**, 918

β-PHENOXYETHYL BROMIDE, **I**, 436

γ-PHENOXYPROPYL BROMIDE, **I**, 435

PHENYL *t*-BUTYL ETHER(METHOD I), **V**, 924

PHENYL *t*-BUTYL ETHER(METHOD II), **V**, 926

PYROGALLOL 1-MONO-METHYL ETHER, **III**, 759

QUINACETOPHENONE DI-METHYL ETHER, **IV**, 837

QUINACETOPHENONE MONO-METHYL ETHER, **IV**, 836

SYRINGIC ALDEHYDE, **IV**, 866

2,3,4,6-TETRAMETHYL-*d*-GLU-COSE, **III**, 800

TRIMETHYLGALLIC ACID, **I**, 537

VANILLIC ACID, **IV**, 972

VERATRALDEHYDE, **II**, 619

FREE RADICALS

DI-*t*-BUTYL NITROXIDE, **V**, 355

2,4,6-TRIPHENYLPHENOXYL, **V**, 1130

GUANIDINES (See also Amidines)

AMINOGUANIDINE BICARBON-ATE, **III**, 73

GUANIDOACETIC ACID, **III**, 440

HALOGENATED COMPOUNDS

A. *ALIPHATIC HYDROCARBON HALIDES*

1. **Mono Bromides**

ALLYL BROMIDE, **I**, 27

iso-AMYL BOMIDE, **I**, 27

2-BROMODECENE-1, **I**, 187

4-BROMO-2-HEPTENE, **IV**, 108

n-BUTYL BROMIDE, **I**, 28, 37

sec.-BUTYL BROMIDE, **I**, 38; **II**, 359

t-BUTYL BROMIDE, **I**, 38

CYCLOHEXYL BROMIDE, **II**, 247

3-CYCLOHEXYL-2-BROMOPRO-PENE, **I**, 186

n-DODECYL BROMIDE, **I**, 29; **II**, 246

ETHYL BROMIDE, **I**, 29, 36

n-HEPTYL BROMIDE, **II**, 247

ISOBUTYL BROMIDE, **II**, 358

ISOPROPYL BROMIDE, **II**, 359

OCTADECYL BROMIDE, **II**, 247

n-OCTYL BROMIDE, **I**, 30

4-PHENYL-2-BROMOBUTENE-1, **I**, 187

ISOPROPYL BROMIDE, **I**, 37

n-PROPYL BROMIDE, **I**, 37; **II**, 359

TETRADECYL BROMIDE, **II**, 247

2. **Mono Chlorides**

n-BUTYL CHLORIDE, **I**, 142

sec.-BUTYL CHLORIDE, **I**, 143

t-BUTYL CHLORIDE, **I**, 144

NEOPHYL CHLORIDE, **IV**, 702

n-PROPYL CHLORIDE, **I**, 143

3. **Mono Fluorides**

n-HEXYL FLUORIDE, **IV**, 525

4. **Mono Iodides**

n-AMYL IODIDE, **II**, 402

n-BUTYL IODIDE, **II**, 402; **IV**, 322

t-BUTYL IODIDE, **IV**, 324

ETHYL IODIDE, **II**, 402

n-HEXADECYL IODIDE, **II**, 322

ISOAMYL IODIDE, **II**, 402

ISOBUTYL IODIDE, **II**, 402; **IV**, 324

ISOPROPYL IODIDE, **IV**, 322

METHYL IODIDE, **II**, 399

n-PROPYL IODIDE, **II**, 402; **IV**, 324

5. Polyhalides

1-BROMO-2-FLUOROHEPT-
ANE, **V**, 136

1-CHLORO-1,4,4-TRIFLUOROBU-
TADIENE, **V**, 235

DECAMETHYLENE BROMIDE,
III, 227

1,2-DIBROMOCYCLOHEXANE,
II, 171

2,3-DIBROMOPROPENE, **I**, 209

1,4-DIIODOBUTANE, **IV**, 321

1,6-DIIODOHEXANE, **IV**, 323

unsym.-HEPTACHLOROPRO-
PANE, **II**, 312

METHYLENE BROMIDE, **I**, 357

METHYLENE IODIDE, **I**, 358

PENTAERYTHRITYL BROM-
IDE, **II**, 476

PENTAERYTHRITYL IODIDE,
II, 477

PENTAERYTHRITYL TETRA-
BROMIDE, **IV**, 753

PENTAMETHYLENE BROM-
IDE, **I**, 428; **III**, 692

1,2,3-TRIBROMOPROPANE, **I**, 521

1,1,3-TRICHLORO-n-NONANE,
V, 1076

1,1,1-TRIFLUOROHEPTANE, **V**,
1082

TRIMETHYLENE BROMIDE, **I**,
30

TRIMETHYLENE CHLORO-
BROMIDE, **I**, 157

TRIMETHYLETHYLENE DI-
BROMIDE, **II**, 419

**B. AROMATIC HYDROCARBON
HALIDES**

1. Mono Bromides

p-BROMOBIPHENYL, **I**, 113

p-BROMODIPHENYLMETHANE,
V, 130

BROMOMESITYLENE, **II**, 95

α-BROMONAPHTHALENE, **I**, 121

2-BROMONAPHTHALENE, **V**,
142

9-BROMOPHENANTHRENE, **III**,
134

o-BROMOTOLUENE, **I**, 135

m-BROMOTOLUENE, **I**, 133

p-BROMOTOLUENE, **I**, 136

4-BROMO-o-XYLENE, **III**, 138

2. Mono Chlorides

o-CHLOROTOLUENE, **I**, 170

p-CHLOROTOLUENE, **I**, 170

3. Mono Fluorides

FLUOROBENZENE, **II**, 295

4. Mono Iodides

IODOBENZENE, **I**, 323; **II**, 351

5. Polyhalides

BENZOPHENONE DICHLOR-
IDE, **II**, 573

1-BROMO-2-FLUOROBENZENE,
V, 133

o-CHLOROBROMOBENZENE,
III, 185

9,10-DIBROMOANTHRACENE, **I**,
207

4,4'-DIBROMOBIBENZYL, **IV**,
257

4,4'-DIBROMOBIPHENYL, **IV**,
256

3,4-DICHLOROBIPHENYL, **V**, 51

2,6-DICHLORONITROBENZENE,
V, 367

4,4'-DIFLUOROBIPHENYL, **II**,
188

sym.-TRIBROMOBENZENE, **II**,
592

**C. HALOGENATED POLYFUNC-
TIONAL COMPOUNDS** (See
also entries under other types
of compounds; for example,
Acids, Acid halides, Amides,
Ethers, Esters, Ketones, and so
on).

1. Bromo Compounds

ACETOBROMOGLUCOSE, **III**, 11

α-ALLYL-β-BROMOETHYL
ETHYL ETHER, **IV**, 750

ALLYLMAGNESIUM BROMIDE,
IV, 749

BENZALACETOPHENONE DI-

BROMIDE, **I**, 205

BROMAL, **II**, 87

BROMOACETAL, **III**, 123, 506

N-BROMOACETAMIDE, **IV**, 104

3-BROMO-4-ACETAMINO-TOLUENE, **I**, 111

BROMOACETONE, **II**, 88

3-BROMOACETOPHENONE, **V**, 117

2-BROMOALLYLAMINE, **V**, 121

N-(2-BROMOALLYL)ETHYL-AMINE, **V**, 124

3-BROMO-4-AMINOTOLUENE, **I**, 111

m-BROMOBENZALDEHYDE, **II**, 132

α-BROMO-*n*-CAPROIC ACID, **I**, 115; **II**, 95

BROMOCYCLOPROPANE, **V**, 126

2-BROMOETHANOL, **I**, 117

β-BROMOETHYLAMINE HY-DROBROMIDE, **II**, 91

β-BROMOETHYLPHTHAL-IMIDE, **I**, 119; **IV**, 106

N-BROMOGLUTARIMIDE, **IV**, 498

α-BROMOHEPTALDEHYDE, **III**, 127

3-BROMO-4-HYDROXY-TOLUENE, **III**, 130

α-BROMOISOCAPROIC ACID, **II**, 95

α-BROMOISOVALERIC ACID, **II**, 93

2-BROMO-4-METHYLBENZAL-DEHYDE, **V**, 139

α-BROMO-β-METHYLVALERIC ACID, **II**, 95

6-BROMO-2-NAPHTHOL, **III**, 132

m-BROMONITROBENZENE, **I**, 123

2-BROMO-3-NITROBENZOIC ACID, **I**, 125

2-BROMO-4-NITROTOLUENE, **IV**, 114

p-BROMOPHENACYL BROM-IDE, **I**, 127

o-BROMOPHENOL, **II**, 97

p-BROMOPHENOL, **I**, 128

p-BROMOPHENYLUREA, **IV**, 49

3-BROMOPHTHALIDE, **V**, 145

β-BROMOPROPIONIC ACID, **I**, 131

3-BROMOPYRENE, **V**, 147

2-BROMOPYRIDINE, **III**, 136

4-BROMORESORCINOL, **II**, 100

3-BROMOTHIOPHENE, **V**, 149

2-CHLOROCYCLOHEXANOL, **I**, 158

CINNAMYL BROMIDE, **V**, 249

COUMARIN DIBROMIDE, **III**, 209

DIBENZOYLDIBROMOME-THANE, **II**, 244

DIBROMOACETONITRILE, **IV**, 254

2,6-DIBROMOANILINE, **III**, 262

α,β-DIBROMOETHYL ETHYL ETHER, **IV**, 749

2,6-DIBROMO-4-NITROPHENOL, **II**, 173

α,β-DIBROMOSUCCINIC ACID, **II**, 177

β-DIBUTYLAMINOETHYL BROM-IDE HYDROBROMIDE, **II**, 92

β-DIETHYLAMINOETHYL BROM-IDE HYDROBROMIDE, **II**, 92

2,4-DIHYDROXY-5-BROMOBEN-ZOIC ACID, **II**, 100

β-DIMETHYLAMINOETHYL BROMIDE HYDROBROM-IDE, **II**, 92

DIPHENYL-*p*-BROMOPHENYL-PHOSPHINE, **V**, 496

β-DIPROPYLAMINOETHYL BROMIDE HYDROBROM-IDE, **II**, 92

β-ETHOXYETHYL BROMIDE, **III**, 370

ETHYL γ-BROMOBUTYRATE, **V**, 545

ETHYL BROMOMALONATE, I,
245
ETHYL α,β-DIBROMO-β-PHEN-
YLPROPIONATE, II, 270
ETHYNYLMAGNESIUM BROM-
IDE, IV, 792
GLYCEROL α,γ-DIBROMOHY-
DRIN, II, 308
1-METHYLAMINO-4-BROMO-
ANTHRAQUINONE, III, 575
MONOBROMOPENTAERYTHRI-
TOL, IV, 681
p-NITROBENZYL BROMIDE, II,
443
PHENACYL BROMIDE, II, 480
β-PHENOXYETHYL BROM-
IDE, I, 436
γ-PHENOXYPROPYL BROM-
IDE, I, 435
PHENYLBROMOETHYNE, V,
921
α,α,α',α'-TETRABROMO-o-XY-
LENE, IV, 807
3-THENYL BROMIDE, IV, 921
VINYL TRIPHENYLPHOSPHON-
IUM BROMIDE, V, 1145
o-XYLYLENE DIBROMIDE, IV,
984
2. Chloro Compounds
2-ACETAMIDO-3,4,6-tri-O-ACE-
TYL-2-DEOXY-α-D-GLUCO-
PYRANOSYL CHLORIDE, V,
1
α-ACETYL-δ-CHLORO-γ-VALER-
OLACTONE, IV, 10
2-AMINO-4-ANILINO-6-(CHLOR-
OMETHYL)-s-TRIAZINE, IV,
29
BENZHYDRYL β-CHLORO-
ETHYL ETHER, IV, 72
BENZOYLCHOLINE CHLOR-
IDE, IV, 84
BISCHLOROMETHYL ETHER,
IV, 101
t-BUTYL HYPOCHLORITE, V,
184

BUTYRCHLORAL, IV, 130
CHLOROACETONITRILE, IV,
144
p-CHLOROACETYLACETANIL-
IDE, III, 260
α-CHLOROACETYL ISOCYAN-
ATE, V, 204
9-CHLOROACRIDINE, III, 54
3-(o-CHLOROANILINO)PRO-
PIONITRILE, IV, 146
9-CHLOROANTHRACENE, V,
206
α-CHLOROANTHRAQUINONE,
II, 128
m-CHLOROBENZALDEHYDE,
II, 130
p-CHLOROBENZOIC ANHY-
DRIDE, III, 29
CHLORO-p-BENZOQUINONE,
IV, 148
N-CHLOROBETAINYL CHLOR-
IDE, IV, 154
α-CHLOROCROTONALDEHYDE,
IV, 130
N-CHLOROCYCLOHEXYL-
IDENEIMINE, V, 208
trans-2-CHLOROCYCLOPEN-
TANOL, IV, 157
3-CHLOROCYCLOPENTENE, IV,
239
CHLORODIISOPROPYLPHOS-
PHINE, V, 211
1-CHLORO-2,6-DINITROBEN-
ZENE, IV, 160
α-CHLOROETHYL ETHYL
ETHER, IV, 748
β-CHLOROETHYL METHYL
SULFIDE, II, 136
2-CHLORO-1-FORMYL-1-CY-
CLOHEXENE, V, 215
α-CHLOROHYDRINDENE, II, 336
3-CHLOROINDAZOLE, III, 476
ω-CHLOROISONITROSOACETO-
PHENONE, III, 191
2-CHLOROLEPIDINE, III, 194

2-CHLOROMETHYL-4,6-DI-
AMINE-*s*-TRIAZINE, **IV**, 30
1-CHLOROMETHYLNAPH-
THALENE, **III**, 195
CHLOROMETHYLPHOSPHONO-
THIOIC DICHLORIDE,
V, 218
2-CHLOROMETHYLTHIO-
PHENE, **III**, 197
2-CHLORONICOTINONITRILE,
IV, 166
m-CHLORONITROBENZENE, **I**,
162
p-CHLOROPHENOXYMETHYL
CHLORIDE, **V**, 221
o-CHLOROPHENYLCYAN-
AMIDE, **IV**, 172
p-CHLOROPHENYL ISOTHIO-
CYANATE, **V**, 223
m-CHLOROPHENYLMETHYL-
CARBINOL, **III**, 200
α-(4-CHLOROPHENYL)-γ-PHEN-
YLACETOACETONITRILE,
IV, 174
o-CHLOROPHENYLTHIOUREA,
IV, 180
β-CHLOROPROPIONIC ACID, **I**,
166
2-CHLOROPYRIMIDINE, **IV**, 182
m-CHLOROSTYRENE, **III**, 204
CHLOROSULFONYL ISOCYAN-
ATE, **V**, 226
2-CHLOROTHIIRANE 1,1-DIOX-
IDE, **V**, 231
3-CHLOROTOLUQUINONE, **IV**,
152
1-CHLORO-1,4,4-TRIFLUOROBU-
TADIENE, **V**, 235
2-CHLORO-1,1,2-TRIFLUORO-
ETHYL ETHYL ETHER, **IV**,
184
3-CHLORO-2,2,3-TRIFLUORO-
PROPIONIC ACID, **V**, 239
CHOLESTEROL DIBROMIDE,
IV, 195
DESYL CHLORIDE, **II**, 159

α,α-DICHLOROACETAMIDE, **III**,
260
α,γ-DICHLOROACETONE, **I**, 211
DICHLOROACETONITRILE, **IV**,
255
2,6-DICHLOROANILINE, **III**, 262
4,4'-DICHLORODIBUTYL
ETHER, **IV**, 266
1,1-DICHLORO-2,2-DIFLUOR-
OETHYLENE, **IV**, 268
β,β-DICHLORO-*p*-DIMETHYL-
AMINOSTYRENE, **V**, 361
2,2'-DICHLORO-α,α'-EPOXIBI-
BENZYL, **V**, 358
2,2-DICHLOROETHANOL, **IV**,
271
DICHLOROMETHYLENETRI-
PHENYLPHOSPHORANE, **V**,
361
DICHLOROMETHYL METHYL
ETHER, **V**, 365
2,6-DICHLOROPHENOL, **III**, 267
4,7-DICHLOROQUINOLINE, **III**,
272
2,5-DICHLOROQUINONE, **IV**, 152
3,4-DICHLORO-1,2,3,4-TETRA-
METHYLCYCLOBUTENE,
V, 370
N,*N*-DIETHYL-1,2,2-TRICHLOR-
OVINYLAMINE, **V**, 387
β-DIMETHYLAMINOETHYL
CHLORIDE HYDRO-
CHLORIDE, **IV**, 333
N-ETHYL-*p*-CHLOROANILINE,
IV, 420
N-ETHYL-*p*-CHLOROFORM-
ANILIDE, **IV**, 420
GLYCEROL α,γ-DICHLORO-
HYDRIN, **I**, 292
GLYCEROL α-MONOCHLOR-
OHYDRIN, **I**, 294
HEXAMETHYLENE CHLOR-
OHYDRIN, **III**, 446
2-HYDROXY-5-NITROBENZYL
CHLORIDE, **III**, 468
MONOCHLOROMETHYL

ETHER, **I,** 377

1-(1-NITROPHENYL)-4-CHLORO-
2-BUTENE, **IV,** 727

2-OXA-7,7-DICHLORONORCAR-
ANE, **V,** 874

PENTACHLOROBENZOIC ACID,
V, 890

1,2,3,4,5-PENTACHLORO-5-
ETHYLCYCLOPENTA-
DIENE, **V,** 893

PERCHLOROFULVALENE, **V,**
901

PHENYLDICHLOROPHOS-
PHINE, **IV,** 784

PHENYL(TRICHLORO-
METHYL)-MERCURY, **V,** 969

TETRAMETHYLENE CHLOR-
OHYDRIN, **II,** 571

TETRAPHENYLARSONIUM
CHLORIDE HYDRO-
CHLORIDE, **IV,** 910

α,α,α-TRICHLOROACETANIL-
IDE, **V,** 1074

TRICHLOROMETHYLPHOS-
PHONYL DICHLORIDE, **IV,**
950

TRIMETHYLENE CHLOROHY-
DRIN, **I,** 533

2,4,6-TRIMETHYLPYRYLIUM
PERCHLORATE, **V,** 1106

TRIPHENYLCHLOROME-
THANE, **III,** 841

3. Fluoro Compounds

2-CHLORO-1,1,2-TRIFLUORO-
ETHYL ETHYL ETHER,
IV, 184

1,1-DICHLORO-2,2-DIFLUOROE-
THYLENE, **IV,** 268

β,β-DIFLUOROSTYRENE, **V,** 390

2,2-DIFLUOROSUCCINIC ACID,
V, 393

α,α-DIFLUOROTOLUENE, **V,** 396

p-FLUOROBENZOIC ACID, **II,**
299

2-FLUOROHEPTANOIC ACID, **V,**
580

2-PHENYLPERFLUOROPRO-
PENE, **V,** 949

PHENYLSULFUR TRIFLUO-
RIDE, **V,** 959

TRIETHYLOXONIUM FLUO-
BORATE, **V,** 1080

m-TRIFLUOROMETHYL-N,N-DI-
METHYLANILINE, **V,** 1085

TRIMETHYLOXONIUM FLUO-
BORATE, **V,** 1096

2,4,6-TRIMETHYLPYRYLIUM
TETRAFLUOROBORATE, **V,**
1112

2,4,6-TRIMETHYLPYRYLIUM
TRIFLUOROMETHANESUL-
FONATE, **V,** 1114

2,4,6-TRIPHENYLPYRYLIUM
TETRAFLUOROBORATE, **V,**
1135

TROPYLIUM FLUOBORATE, **V,**
1138

4. Iodo Compounds

BENZOYLCHOLINE IODIDE, **IV,**
84

BENZYLTRIMETHYLAMMON-
IUM IODIDE, **IV,** 585

CYANOGEN IODIDE, **IV,** 207

2,6-DIIODO-p-NITROANILINE,
II, 196

N,N-DIMETHYLAMINOMETH-
YLFERROCENE METHIOD-
IDE, **V,** 434

2,4-DINITROIODOBENZENE, **V,**
478

DIPHENYLIODONIUM IODIDE,
III, 355

1-ETHYL-3-(3-DIMETHYL-
AMINO)PROPYLCARBODI-
IMIDE METHIODIDE, **V,** 555

2-HYDROXY-3,5-DIIODOBEN-
ZOIC ACID, **II,** 343

4-HYDROXY-3,5-DIIODOBEN-
ZOIC ACID, **II,** 344

p-IODOANILINE, **II,** 347

5-IODOANTHRANILIC ACID, **II,**
349

IODOBENZENE DICHLORIDE, III, 482
p-IODOBENZOIC ACID, I, 325
IODOCYCLOHEXANE, IV, 324, 543
2-IODO-3-NITROBENZOIC ACID, I, 126
o-IODOPHENOL, I, 326
p-IODOPHENOL, II, 355
IODOSOBENZENE, III, 483; V, 658
IODOSOBENZENE DIACETATE, V, 660
N-IODOSUCCINIMIDE, V, 663
2-IODOTHIOPHENE, II, 357; IV, 545
4-IODOVERATROLE, IV, 547
IODOXYBENZENE, III, 485; V, 665
TETRAIODOPHTHALIC ANHYDRIDE, III, 796
1,2,3-TRIIODO-5-NITROBENZENE, II, 604
HETEROCYCLIC COMPOUNDS. These are classified according to the size of the *hetero*-ring, number and nature of the heteroatoms. The system follows the one in *Collective Volume* III, IV, and V; the entries from *Collective Volumes* I and II have been reclassified.
A. *THREE-MEMBERED, ONE NITROGEN*
cis-2-BENZYL-3-PHENYLAZIRIDINE, V, 83
N-ETHYLALLENIMINE, V, 541
ETHYLENIMINE, IV, 433
B. *THREE-MEMBERED, TWO NITROGEN*
3,3-PENTAMETHYLENEDIAZIRINE, V, 897
C. *THREE-MEMBERED, ONE NITROGEN, ONE OXYGEN*
2-t-BUTYL-3-PHENYLOXAZIRANE, V, 191

D. *THREE-MEMBERED, ONE OXYGEN*
CYCLOHEXENE OXIDE, I, 185
2,2'-DICHLORO-$\mu\alpha'$-EPOXYBIBENZYL, V, 358
EPIBROMOHYDRIN, II, 257
EPICHLOROHYDRIN, I, 233
EPICHLOROHYDRIN, II, 256
ETHYL β,β-PENTAMETHYLENEGLYCIDATE, IV, 459
ISOPHORONE OXIDE, IV, 552
METHYL 5,5-DIMETHOXY-3-METHYL-2,3-EPOXYPENTANOATE, IV, 649
METHYLENECYCLOHEXANE OXIDE, V, 755
PHENYLMETHYLGLYCIDIC ESTER, III, 727
trans-STILBENE OXIDE, IV, 860
STYRENE OXIDE, I, 494
TETRACYANOETHYLENE OXIDE, V, 1007
E. *THREE-MEMBERED, ONE SULFUR*
2-CHLOROTHIIRANE 1,1-DIOXIDE, V, 231
CYCLOHEXENE SULFIDE, IV, 232
ETHYLENE SULFIDE, V, 562
F. *FOUR-MEMBERED, ONE NITROGEN*
β-ISOVALEROLACTAM, V, 673
β-ISOVALEROLACTAM-N-SULFONYL CHLORIDE, V, 673
G. *FOUR-MEMBERED, ONE OXYGEN*
KETENE DIMER, III, 508
TRIMETHYLENE OXIDE, III, 835
H. *FIVE-MEMBERED, ONE NITROGEN*
3-ACETYLOXINDOLE, V, 12
1,2-BENZO-3,4-DIHYDROCARBAZOLE, IV, 885
β-BROMOETHYLPHTHALIMIDE, I, 119
1-n-BUTYLPYRROLIDINE, III, 159

1,3-DIHYDROISOINDOLE, V, 406
2,4-DIMETHYL-5-CARBETHOX-
YPYRROLE, II, 198
2,4-DIMETHYL-3,5-DICARBETH-
OXYPYRROLE, II, 202
2,4-DIMETHYLPYRROLE, II, 217
2,5-DIMETHYLPYRROLE, II, 219
2,2-DIMETHYLPYRROLIDINE,
IV, 354
1,5-DIMETHYL-2-PYRROLI-
DONE, III, 328
5,5-DIMETHYL-2-PYRROLI-
DONE, IV, 357
2,4-DIPHENYLPYRROLE, III, 358
ETHYL INDOLE-2-CARBOXYL-
ATE, V, 567
ETHYL α-(1-PYRROLIDYL)-
PROPIONATE, IV, 466
INDOLE, III, 479
INDOLE-3-ACETIC ACID, V, 654
INDOLE-3-ALDEHYDE, IV, 539
INDOLE-3-CARBONITRILE, V,
656
ISATIN, I, 327
KRYPTOPYRROLE, III, 513
1-(2-METHOXYCARBONYL-
PHENYL)PYRROLE, V, 716
1-METHYL-3-ETHYLOXINDOLE,
IV, 620
1-METHYLINDOLE, V, 769
2-METHYLINDOLE, III, 597
2-METHYLMERCAPTO-N-
METHYL-Δ²-PYRROLINE, V,
780
3-METHYLOXINDOLE, IV, 657
2-NITROCARBAZOLE, V, 829
2-PHENYLINDOLE, III, 725
N-PHENYLMALEIMIDE, V, 944
2-PHENYL-3-n-PROPYLISOX-
AZOLIDINE-4,5-cis-DICAR-
BOXYLIC ACID N-PHENYL-
IMIDE, V, 957
PHTHALIMIDE, I, 457
PYRROLE, I, 473
2-PYRROLEALDEHYDE, IV, 831

2-(1-PYRROLIDYL)PROPANOL,
IV, 834
1,2,3,4-TETRAHYDROCARBA-
ZOLE, IV, 884
2,3,4,5-TETRAMETHYLPYR-
ROLE, V, 1022
2-(p-TOLYLSULFONYL)-
DIHYDROISOINDOLE, V,
1064
I. FIVE-MEMBERED, TWO NITRO-
GEN
ALLANTOIN, II, 21
3(5)-AMINOPYRAZOLE, V, 39
BENZIMIDAZOLE, II, 65
BIS(1,3-DIPHENYLIMIDAZOL-
IDINYLIDENE-2), V, 115
1,1'-CARBONYLDIIMIDAZOLE,
V, 201
3-CHLOROINDAZOLE, III, 476
CREATININE, I, 172
5,5-DIMETHYLHYDANTOIN, III,
323
3,5-DIMETHYLPYRAZOLE, IV,
351
4,5-DIPHENYLGLYOXALONE,
II, 231
ETHYLENE THIOUREA, III, 394
4(5)-HYDROXYMETHYLIMID-
AZOLE HYDROCHLOR-
IDE, III, 460
IMIDAZOLE, III, 471
INDAZOLE, III, 475; IV, 536; V,
650
INDAZOLONE, III, 476
2-MERCAPTOBENZIMIDAZOLE,
IV, 569
2-METHYLBENZIMIDAZOLE, II,
66
5-NITROINDAZOLE, III, 660
PARABANIC ACID, IV, 744
1-PHENYL-3-AMINO-5-PYRAZ-
OLONE, III, 708
2-PHENYLINDAZOLE, V, 941
4,5,6,7-TETRAHYDROINDAZ-
OLE, IV, 537
J. FIVE-MEMBERED, THREE NI-
TROGEN

3-AMINO-1H-1,2,4-TRIAZOLE, **III,** 95
4-AMINO-4H-1,2,4-TRIAZOLE, **III,** 96
1,2,3-BENZOTRIAZOLE, **III,** 106
1,4-DIPHENYL-5-AMINO-1,2,3-TRIAZOLE, **IV,** 380
4-PHENYL-5-ANILINO-1,2,3-TRIAZOLE, **IV,** 380
1,2,4-TRIAZOLE, **V,** 1070
K. *FIVE-MEMBERED, ONE NITRO-GEN, ONE OXYGEN*
AZLACTONE OF α-ACETAMINOCINNAMIC ACID, **II,** 3
AZLACTONE OF α-BENZOYLAMINOCINNAMIC ACID, **II,** 490
AZLACTONE OF α-BENZOYLAMINO-β-(3,4-DIMETHOXYPHENYL) ACRYLIC ACID, **II,** 55
2-MERCAPTOBENZOXAZOLE, **IV,** 570
2-PHENYL-5-OXAZOLONE, **V,** 946
2-PHENYL-3-n-PROPYLISOXAZOLIDINE-4,5-*cis*-DICARBOXYLIC ACID N-PHENYL-IMIDE, **V,** 957
2,3,5-TRIPHENYLISOXAZOLIDINE, **V,** 1124
L. *FIVE-MEMBERED, TWO NITRO-GEN, ONE OXYGEN*
BENZOFURAZAN OXIDE, **IV,** 74
DIMETHYLFURAZAN, **IV,** 342
3-PHENYLSYDNONE, **V,** 962
M. *FIVE-MEMBERED, ONE NITRO-GEN, ONE SULFUR*
N-(p-ACETYLAMINOPHENYL)RHODANINE, **IV,** 6
2-AMINO-6-METHYLBENZOTHIAZOLE, **III,** 76
2-AMINO-4-METHYLTHIAZOLE, **II,** 31
2,4-DIMETHYLTHIAZOLE, **III,** 332

1-METHYL-2-IMINO-β-NAPHTHOTHIAZOLINE, **III,** 595
PSEUDOTHIOHYDANTOIN, **III,** 751
RHODANINE, **III,** 763
N. *FIVE-MEMBERED, TWO NI-TROGEN, ONE SULFUR*
1,2,3-BENZOTHIADIAZOLE 1,1-DIOXIDE, **V,** 61
O. *FIVE-MEMBERED, ONE OXY-GEN*
BENZALPHTHALIDE, **II,** 61
α-BENZYLIDENE-γ-PHENYL-Δ^{β,γ}-BUTENOLIDE, **V,** 80
COUMARILIC ACID, **III,** 209
COUMARONE, **V,** 251
γ-CROTONOLACTONE, **V,** 255
2,5-DIHYDRO-2,5-DIMETHOXYFURAN, **V,** 403
5,5-DIMETHYL-2-n-PENTYLTETRAHYDROFURAN, **IV,** 350
ETHYL 3-METHYLCOUMARILATE, **IV,** 590
FURAN, **I,** 274
2-FURANCARBOXYLIC ACID, **I,** 276
2-FUROIC ACID, **IV,** 493
FURFURAL, **I,** 280
2-FURFURALACETONE, **I,** 283
FURFURAL DIACETATE, **IV,** 489
2-FURFURYL MERCAPTAN, **IV,** 491
FURYLACRYLIC ACID, **III,** 425
3-(2-FURYL)ACRYLONITRILE, **V,** 585
2-FURYLCARBINOL, **I,** 276
2-FURYLMETHYL ACETATE, **I,** 285
β-FURYLPROPIONIC ACID, **I,** 313
3-HYDROXYTETRAHYDROFURAN, **IV,** 534
3-METHYLCOUMARILIC ACID, **IV,** 591
3-METHYLCOUMARONE, **IV,** 590
3-METHYLFURAN, **IV,** 628

5-METHYLFURFURAL, **II,** 393
5-METHYLFURFURYLDIME-
THYLAMINE, **IV,** 626
3-METHYL-2-FUROIC ACID, **IV,**
628
METHYL 3-METHYL-2-FUR-
OATE, **IV,** 649
PHTHALIDE, **II,** 526
γ-*n*-PROPYLBUTYROLACT-
ONE, **III,** 742
TETRAHYDROFURAN, **II,** 566
TETRAHYDROFURFURYL BRO-
MIDE, **III,** 793
β-(TETRAHYDROFURYL)-
PROPIONIC ACID, **III,** 742
2,2,5,5-TETRAMETHYLTETRAH-
YDRO-3-KETOFURAN, **V,**
1024

P. FIVE-MEMBERED, ONE SULFUR
2-ACETOTHIENONE, **III,** 14
3-BROMOTHIOPHENE, **V,** 149
2-CHLOROMETHYLTHIO-
PHENE, **III,** 197
2,5-DIAMINO-3,4-DICYANO-
THIOPHENE, **IV,** 243
2-HYDROXYTHIOPHENE, **V,** 642
2-IODOTHIOPHENE, **IV,** 545
ISOPRENE CYCLIC SULFONE,
III, 499
METHYL 2-THIENYL SULFIDE,
IV, 667
3-METHYLTHIOPHENE, **IV,** 671
TETRAHYDROTHIOPHENE, **IV,**
892
2-THENALDEHYDE, **IV,** 915
3-THENALDEHYDE, **IV,** 918
3-THENOIC ACID, **IV,** 919
3-THENYL BROMIDE, **IV,** 921
THIOPHENE, **II,** 578
2-THIOPHENEALDEHYDE, **III,**
811
2-VINYLTHIOPHENE, **IV,** 980
*Q. SIX-MEMBERED, ONE NITRO-
GEN*
ACRIDONE, **II,** 15
9-AMINOACRIDINE, **III,** 53
3-AMINOPYRIDINE, **IV,** 45

1-AMINOPYRIDINIUM IODIDE,
V, 43
2-BENZOYL-1-CYANO-1-
METHYL-1,2-DIHYDROISO-
QUINOLINE, **IV,** 642
BENZOYL PIPERIDINE, **I,** 99
3-BENZOYLPYRIDINE, **IV,** 88
4-BENZOYLPYRIDINE, **IV,** 89
2-BENZYLAMINOPYRIDINE, **IV,**
91
2,2'-BIPYRIDINE, **V,** 102
2-BROMOPYRIDINE, **III,** 136
9-CHLOROACRIDINE, **III,** 54
2-CHLOROLEPIDINE, **III,** 194
2-CHLORONICOTINONITRILE,
IV, 166
1-CYANO-2-BENZOYL-1,2-DI-
HYDROISOQUINOLINE, **IV,**
641
2-CYANO-6-METHYLPYRIDINE,
V, 269
3-CYANO-6-METHYL-2(1)-PYRI-
DONE, **IV,** 210
2,3-DIAMINOPYRIDINE, **V,** 346
3,5-DICARBETHOXY-2,6-DIME-
THYLPYRIDINE, **II,** 216
4,7-DICHLOROQUINOLINE, **III,**
272
1,4-DIHYDRO-3,5-DICARBETH-
OXY-2,6-DIMETHYLPYRI-
DINE, **II,** 216
2,6-DIMETHYLPYRIDINE, **II,** 214
2,4-DIMETHYLQUINOLINE, **III,**
329
5-ETHYL-2-METHYLPYRIDINE,
IV, 451
4-ETHYLPYRIDINE, **III,** 410
ETHYL 2-PYRIDYLACETATE,
III, 413
2-HYDROXYCINCHONINIC
ACID, **III,** 456
2-HYDROXY-3-METHYLISOCAR-
BOSTYRIL, **V,** 623
6-HYDROXYNICOTINIC ACID,
IV, 532
3-HYDROXYQUINOLINE, **V,** 635
3-ISOQUINUCLIDONE, **V,** 670

JULOLIDINE, **III**, 504
LEPIDINE, **III**, 519
6-METHOXY-8-NITROQUINOL-
    INE, **III**, 568
4-METHYLCARBOSTYRIL, **III**,
    580
2-METHYL-4-HYDROXY-
    QUINOLINE, **III**, 593
1-METHYLISOQUINOLINE, **IV**,
    641
3-METHYL-4-NITROPYRIDINE-1-
    OXIDE, **IV**, 654
3-METHYLPYRIDINE-1-OXIDE,
    IV, 655
1-METHYL-2-PYRIDONE, **II**, 419
QUINOLINE, **I**, 478
N-β-NAPHTHYLPIPERIDINE, **V**,
    816
NICOTINAMIDE-1-OXIDE, **IV**,
    704
NICOTINIC ACID, **I**, 385
NICOTINIC ACID ANHYDRIDE,
    V, 822
NICOTINONITRILE, **IV**, 706
1-PHENYLPIPERIDINE, **IV**, 795
2-PHENYLPYRIDINE, **II**, 517
PICOLINIC ACID HYDRO-
    CHLORIDE, **III**, 740
PYRIDINE-1-OXIDE, **IV**, 828
4-PYRIDINESULFONIC ACID, **V**,
    977
1-(α-PYRIDYL)-2-PROPANOL, **III**,
    757
3-QUINUCLIDONE HYDROCH-
    LORIDE, **V**, 989
2*a*-THIOHOMOPHTHALIMIDE,
    V, 1051
*R. SIX-MEMBERED, TWO NITRO-
    GEN*
ALLOXAN MONOHYDRATE, **III**,
    37; **IV**, 23
ALLOXANTIN DIHYDRATE, **III**,
    42; **IV**, 25
5-AMINO-2,3-DIHYDRO-1,4-
    PHTHALAZINEDIONE, **III**,
    69

4-AMINO-2,6-DIMETHYLPYRIM-
    IDINE, **III**, 71
BARBITURIC ACID, **II**, 60
BENZALBARBITURIC ACID, **III**,
    39
BENZOYLENE UREA, **II**, 79
1-BENZYLPIPERAZINE, **V**, 88
2-CHLOROPYRIMIDINE, **IV**, 182
2,4-DIAMINO-6-HYDROXYPYR-
    IMIDINE, **IV**, 245
DIAMINOURACIL HYDRO-
    CHLORIDE, **IV**, 247
2-(DIMETHYLAMINO)PYR-
    IMIDINE, **IV**, 336
3-*n*-HEPTYL-5-CYANOCYTO-
    SINE, **IV**, 515
2-MERCAPTO-4-AMINO-5-CAR-
    BETHOXYPYRIMIDINE,
    IV, 566
2-MERCAPTO-4-HYDROXY-5-CY-
    ANOPYRIMIDINE, **IV**, 566
2-MERCAPTOPYRIMIDINE, **V**,
    703
N-METHYLAMINOPYR-
    IMIDINE, **IV**, 336
4-METHYL-6-HYDROXYPYRIMI-
    DINE, **IV**, 638
2-METHYL-5,6-PYRAZINEDI-
    CARBOXYLIC ACID, **IV**, 827
4-METHYLPYRIMIDINE, **V**, 794
6-METHYLURACIL, **II**, 422
5-NITRO-2,3-DIHYDRO-1,4-
    PHTHALAZINEDIONE, **III**,
    656
PYOCYANINE, **III**, 753
2,3-PYRAZINEDICARBOXYLIC
    ACID, **IV**, 824
QUINOXALINE, **IV**, 824
2-THIO-6-METHYLURACIL, **IV**,
    638
*S. SIX-MEMBERED, THREE NITRO-
    GEN*
2-AMINO-4-ANILINO-6-(CHLOR-
    OMETHYL)-*s*-TRIAZINE, **IV**,
    29
BENZOGUANAMINE, **IV**, 78
2-CHLOROMETHYL-4,6-DI-

AMINO-*s*-TRIAZINE, **IV,** 30
2,4-DIAMINO-6-PHENYL-*s*-TRIA-
ZINE, **IV,** 78
HEXAHYDRO-1,3,5-TRIPRO-
PIONYL-*s*-TRIAZINE, **IV,** 518
T. *SIX-MEMBERED, ONE NITRO-
GEN, ONE OXYGEN*
ISATOIC ANHYDRIDE, **III,** 488
1-MORPHOLINO-1-CYCLO-
HEXENE, **V,** 808
N-NITROMORPHOLINE, **V,** 839
U. *SIX-MEMBERED, ONE OXYGEN*
3-CARBETHOXYCOUMARIN, **III,**
165
CHELIDONIC ACID, **II,** 126
COUMALIC ACID, **IV,** 201
COUMARIN DIBROMIDE, **III,**
209
DEHYDROACETIC ACID, **III,** 231
3,4-DIHYDRO-2-METHOXY-4-
METHYL-2*H*-PYRAN, **IV,** 311
2,3-DIHYDROPYRAN, **III,** 276
4,6-DIMETHYLCOUMALIN, **IV,**
337
2,6-DIMETHYL-3,5-DIPHENYL-
4*H*-PYRAN-4-ONE, **V,** 450
2-ETHYLCHROMONE, **III,** 387
ETHYL ISODEHYDROACET-
ATE, **IV,** 549
FLAVONE, **IV,** 478
ISODEHYDROACETIC ACID, **IV,**
549
2-(*p*-METHOXYPHENYL)-6-
PHENYL-4-PYRONE, **V,** 721
METHYL COUMALATE, **IV,** 532
4-METHYLCOUMARIN, **III,** 581
4-METHYLESCULETIN, **I,** 360
2-OXA-,7-DICHLORONORCAR-
ANE, **V,** 874
α-PYRONE, **V,** 982
TETRAHYDROPYRAN, **III,** 794
2,4,6-TRIMETHYLPYRYLIUM
PERCHLORATE, **V,** 1106
2,4,6-TRIMETHYLPYRYLIUM
TETRAFLUOROBORATE, **V,**
1112
2,4,6-TRIMETHYLPYRYLIUM

TRIFLUOROMETHANESUL-
FONATE, **V,** 1114
2,4,6-TRIPHENYLPYRYLIUM
TETRAFLUOROBORATE, **V,**
1135
XANTHONE, **I,** 552
XANTHYDROL, **I,** 554
V. *SEVEN-MEMBERED, ONE
NITROGEN*
1,3-DIHYDRO-3,5,7-TRIMETH-
YL-2*H*-AZEPIN-2-ONE, **V,**
408
O-ETHYLCAPROLACTIM, **IV,**
589
2-KETOHEXAMETHYLEN-
IMINE, **II,** 371
O-METHYLCAPROLACTIM, **IV,**
588
W. *SEVEN-MEMBERED, ONE OXY-
GEN*
2,7-DIMETHYLOXEPIN, **V,** 467
X. *MISCELLANEOUS*
ANHYDRO-*o*-HYDROXYMERCU-
RIBENZOIC ACID, **I,** 57
ANHYDRO-2-HYDROXYMER-
CURI-3-NITROBENZOIC
ACID, **I,** 56
p-DITHIANE, **IV,** 396
HEMIN, **III,** 443
10-METHYL-10,9-BORAZARO-
PHENANTHRENE, **V,** 727
PHENOXTHIN, **II,** 485
o-SULFOBENZOIC ANHY-
DRIDE, **I,** 495
sym.-TRITHIANE, **II,** 610
HYDRAZINES AND DERIVATIVES
AMINOACETONE SEMICARBA-
ZONE HYDRO-
CHLORIDE, **V,** 27
BENZALAZINE, **II,** 395
BENZIL HYDRAZONE, **II,** 497
BETAINE HYDRAZIDE HY-
DROCHLORIDE, **II,** 85
t-BUTYL CARBAZATE, **V,** 166
sym.-DIBENZOYLDIMETHYL-
HYDRAZINE, **II,** 209
1,2-DI-1-(1-CYANO)CYCLO-

HEXYLHYDRAZINE, **IV**, 274
1,2 DI-2-(2-CYANO)PROPYL-
HYDRAZINE, **IV**, 275
unsym.-DIMETHYLHYDRAZINE,
II, 213
sym.-DIMETHYLHYDRAZINE
DIHYDROCHLORIDE, **II**, 208
unsym.-DIMETHYLHYDRAZINE
HYDROCHLORIDE, **II**, 211
2,4-DINITROPHENYLHYDRA-
ZINE, **II**, 228
ETHYL HYDRAZODICAR-
BOXYLATE, **IV**, 411
HYDRAZINE SULFATE, **I**, 309
METHYLHYDRAZINE SUL-
FATE, **II**, 395
METHYL HYDRAZODICARBOX-
YLATE, **IV**, 413
α-METHYL-α-PHENYLHYDRA-
ZINE, **II**, 418
PHENYLHYDRAZINE, **I**, 442
4-PHENYLSEMICARBAZIDE, **I**,
450
β-PROPIONYLPHENYLHYDRA-
ZINE, **IV**, 657
SEMICARBAZIDE SULFATE, **I**,
485
p-TOLUENESULFONYLHYDRA-
ZIDE, **V**, 1055
**HYDROCARBONS (Classified according to
the customary dominant func-
tion)**
A. *ALIPHATIC*
1. **Paraffins**
n-HEXADECANE, **II**, 320
PENTANE, **II**, 478
2. **Olefins, Dienes, Trienes**
ALLENE, **V**, 22
BIALLYL, **III**, 121
1,3-BUTADIENE, **II**, 102
2,3-DIMETHYL-1,3-BUTADIENE,
III, 312
1,3,5-HEXATRIENE, **V**, 608
1,4-PENTADIENE, **IV**, 746
2-PENTENE, **I**, 430

3. **Acetylenes**
n-AMYLACETYLENE, **IV**, 119
n-BUTYLACETYLENE, **IV**, 117
3-CYCLOHEXYLPROPINE, **I**, 191
DECINE-1, **I**, 192
n-HEXYLACETYLENE, **IV**, 119
ISOAMYLACETYLENE, **IV**, 119
MONOVINYLACETYLENE, **IV**,
683
n-PROPYLACETYLENE, **IV**, 119
B. *ALICYCLIC*
1. **Saturated**
ADAMANTANE, **V**, 16
BICYCLO(2.1.0)PENTANE, **V**, 96
NORCARANE, **V**, 855
2. **Unsaturated**
cis-CYCLODODECENE, **V**, 281
1,3-CYCLOHEXADIENE, **V**, 285
CYCLOHEXENE, **II**, 152; **I**, 183
CYCLOHEXYLIDENECY-
CLOHEXANE, **V**, 297
1,2-CYCLONONADIENE, **V**, 306
trans-CYCLOOCTENE, **V**, 315
CYCLOPENTADIENE, **IV**, 238
CYCLOPENTENE, **II**, 153
cis- and *trans*-1,2-DIVINYLCY-
CLOBUTANE, **V**, 528
METHYLENECYCLOHEXANE,
IV, 612 **V**, 751
NORBORNYLENE, **IV**, 738
4,6,8-TRIMETHYLAZULENE,
V, 1088
C. *AROMATIC*
n-AMYLBENZENE, **II**, 47
BENZENE, **V**, 998
BIPHENYLENE, **V**, 54
n-BUTYLBENZENE, **III**, 157
CYCLOHEXYLBENZENE, **II**,
151
CYCLOPROPYLBENZENE, **V**,
328, 929
m-CYMENE, **V**, 332
9,10-DIHYDROANTHRACENE,
V, 398
9,10-DIHYDROPHENAN-
THRENE, **IV**, 313

DIMESITYLMETHANE, V, 422
3,3'-DIMETHYLBIPHENYL, III, 295
1,1-DIPHENYLCYCLOPRO-PANE, V, 509
DIPHENYLMETHANE, II, 232
1,1-DIPHENYLPENTANE, V, 523
1,1-DI-*p*-TOLYLETHANE, I, 229
DURENE, II, 248
HEMIMELLITENE, IV, 508
HEXAMETHYLBENZENE, II, 248; IV, 520
HEXAPHENYLBENZENE, V, 604
ISODURENE, II, 360
MESITYLENE, I, 341
9-METHYLFLUORENE, IV, 623
1-METHYL-3-PHENYLIN-DANE, IV, 665
(2.2)PARACYCLOPHANE, V, 883
PENTAMETHYLBENZENE, II, 248
PHENANTHRENE, IV, 313
PHENYLCYCLOPROPANE, V, 328, 929
1-PHENYLNAPHTHALENE, III, 729
9-PHENYLPHENANTHRENE, V, 952
n-PROPYLBENZENE, I, 471
p-QUINQUEPHENYL, V, 985
1,2,3,4-TETRAPHENYLNAPH-THALENE, V, 1037
1,1,3-TRIMETHYL-3-PHENYL-INDANE, IV, 666
TRIPHENYLENE, V, 1120
TRIPHENYLMETHANE, I, 548
TRIPTYCENE, IV, 964
D. *ARYL SUBSTITUTED OLEFINS, DIENES, ACETYLENES*
DIPHENYLACETYLENE, III, 350; IV, 377
1,4-DIPHENYLBUTADIENE, II, 229

1,4-DIPHENYL-1,3-BUTA-DIENE, V, 499
DIPHENYLDIACETYLENE, V, 517
1,1-DIPHENYLETHYLENE, I, 226
PHENYLACETYLENE, I, 438; IV, 763
trans-1-PHENYL-1,3-BUTA-DIENE, IV, 771
4-PHENYLBUTINE-1, I, 192
PHENYLETHYLENE, I, 440
cis-STILBENE, IV, 857
trans-STILBENE, III, 786
TETRAPHENYLETHYLENE, IV, 914
TRIPHENYLETHYLENE, II, 606
HYDROXYLAMINES AND DERIVA-TIVES
ACETOXIME, I, 318
BENZOHYDROXAMIC ACID, II, 67
BENZOPHENONE OXIME, II, 70
BIACETYL MONOXIME, II, 205
CUPFERRON, I, 177
CYCLOHEXANONE OXIME, II, 76, 314
DIMETHYLGLYOXIME, II, 204
HEPTALDOXIME, II, 313
HYDROXYLAMINE HY-DROCHLORIDE, I, 318
2-HYDROXY-3-METHYLISO-CARBOSTYRIL, V, 623
ISONITROSOACETANILIDE, I, 327
ISONITROSOPROPIOPHEN-ONE, II, 363
β-PHENYLHYDROXYLAMINE, I, 445
POTASSIUM BENZOHYDROX-AMATE, II, 67
VERATRALDOXIME, II, 622
IMIDES
BENZYLPHTHALMIDE, II, 83
β-BROMOETHYLPHTHALIM-

IDE, **I**, 119; **IV**, 106
N-BROMOGLUTARIMIDE, **IV**, 498
t-BUTYLPHTHALIMIDE, **III**, 152
2,6-DIBROMOQUINONE-4-CHLOROIMIDE, **II**, 175
α,α'-DICYANO-β-ETHYL-β-METHYLGLUTARIMIDE, **IV**, 441
GLUTARIMIDE, **IV**, 496
N-IODOSUCCINIMIDE, **V**, 663
β-METHYL-β-PHENYL-α,α'-DICYANOGLUTARIMIDE, **IV**, 662
N-PHENYLMALEIMIDE, **V**, 944
2-PHENYL-3-*n*-PROPYLISOXAZOLIDINE-4,5-*cis*-DICARBOXYLIC ACID *N*-PHENYLIMIDE, **V**, 957
PHTHALIMIDE, **I**, 457
α-PHTHALIMIDO-*o*-TOLUIC ACID, **IV**, 810
N-PHTHALYL-L-β-PHENYLALANINE, **V**, 973
SUCCINIMIDE, **II**, 562
2α-THIOHOMOPHTHALIMIDE, **V**, 1051
TRIETHYL α-PHTHALIMIDO-ETHANE-α,α,β-TRICARBOXYLATE, **IV**, 55
IMINES (See Amidines and Guanidines; Heterocyclic Compounds)
BENZALANILINE, **I**, 80
BENZYLIDENEARGININE, **II**, 49
N-CHLOROCYCLOHEXYLIDENEIMINE, **V**, 208
DIPHENYL KETIMINE, **V**, 520
DIPHENYLMETHANE IMINE HYDROCHLORIDE, **II**, 234
METHYLENEAMINOACETONITRILE, **I**, 355
ISOCYANATES
α-CHLOROACETYL ISOCYANATE, **V**, 204
CHLOROSULFONYL ISOCYANATE, **V**, 226
HEXAMETHYLENE DIISOCYANATE, **IV**, 521
p-NITROPHENYL ISOCYANATE, **II**, 453
UNDECYL ISOCYANATE, **III**, 846
ISONITRILES
CYCLOHEXYL ISOCYANIDE, **V**, 300
ETHYL ISOCYANIDE, **IV**, 438
METHYL ISOCYANIDE, **V**, 772
o-TOLYL ISOCYANIDE, **V**, 1060
ISOTHIOCYANATES
α-NAPHTHYL ISOTHIOCYANATE, **IV**, 700
KETALS
dl-ISOPROPYLIDENEGLYCEROL, **III**, 502
KETENES
DIMETHYLKETENE, **IV**, 348
DIPHENYLKETENE, **III**, 356
KETENE, **I**, 330; **V**, 679
LACTONES (See Heterocyclic Compounds)
α-ACETYL-δ-CHLORO-γ-VALEROLACTONE, **IV**, 10
α-BENZYLIDENE-γ-PHENYL-Δ^{β,γ}-BUTENOLIDE, **V**, 80
3-BROMOPHTHALIDE, **V**, 145
γ-CAPRILACTONE, **IV**, 432
COUMALIC ACID, **IV**, 201
γ-CROTONOLACTONE, **V**, 255
4,6-DIMETHYLCOUMALIN, **IV**, 337
DIMETHYLKETENE β-LACTONE DIMER, **V**, 456
4-ETHYL-4-HYDROXY-2-METHYLOCTANOIC ACID γ-LACTONE, **IV**, 447
ETHYL ISODEHYDROACETATE, **IV**, 549
D-GULONIC-γ-LACTONE, **IV**, 506
ISODEHYDROACETIC ACID, **IV**, 549
18,20-LACTONE of 3β-ACETOXY-20β-HYDROXY-5-PREGNENE-18-OIC ACID, **V**, 692
METHYL COUMALATE, **IV**, 532

β-METHYL-δ-VALEROLACT-
ONE, **IV**, 677
KETONES
A. MONOKETONES
  1. Unsubstituted
  ACETO-*p*-CYMENE, **II**, 3
  ACETOPHENONE, **I**, 111
  9-ACETYLANTHRACENE, **IV**, 8
  1-ACETYLCYCLOHEXENE, **III**,
  22
  2-ACETYLFLUORENE, **III**, 23
  9-ACETYLPHENANTHRENE,
  III, 26
  2-ALLYLCYCLOHEXANONE,
  III, 44; **V**, 25
  ANTHRONE, **I**, 60
  BENZANTHRONE, **II**, 62
  BENZOPHENONE, **I**, 95
  β-BENZOPINACOLONE, **II**, 73
  BENZYLACETOPHENONE, **I**,
  101
  2-BENZYLCYCLOPENTAN-
  ONE, **V**, 76
  BICYCLO(2.2.1)HEPTEN-7-ONE,
  V, 91
  2-*n*-BUTYL-2-METHYLCY-
  CLOHEXANONE, **V**, 187
  CHOLESTANONE, **II**, 139
  CHOLESTENONE, **III**, 207
  Δ⁴-CHOLESTEN-3-ONE, **IV**, 192
  Δ⁵-CHOLESTEN-3-ONE, **IV**, 195
  CYCLODECANONE, **IV**, 218; **V**,
  277
  CYCLOHEPTANONE, **IV**, 221
  2-CYCLOHEXENONE, **V**, 294
  CYCLOHEXYL METHYL KE-
  TONE, **V**, 775
  CYCLOÖCTANONE, **V**, 310
  CYCLOPENTANONE, **I**, 192
  2-CYCLOPENTENONE, **V**, 326
  DESOXYBENZOIN, **II**, 156
  DICYCLOPROPYL KETONE,
  IV, 278
  3,5-DIMETHYL-2-CYCLO-
  HEXEN-1-ONE, **III**, 317
  DI-(2-METHYLCYCLOPRO-
  PYL)KETONE, **IV**, 280

α,α-DIPHENYLACETONE, **III**,
343
DIPHENYLCYCLOPROPEN-
ONE, **V**, 514
2,3-DIPHENYLINDONE, **III**, 353
β,β-DIPHENYLPROPIOPHEN-
ONE, **II**, 236
DYPNONE, **III**, 367
4-HEPTANONE, **V**, 589
2,3,4,5,6,6-HEXAMETHYL-2,4-
CYCLOHEXADIEN-1-ONE,
V, 598
α-HYDRINDONE, **II**, 336
2-INDANONE, **V**, 647
LAURONE, **IV**, 560
l-MENTHONE, **I**, 340
p-METHYLACETOPHENONE,
I, 111
METHYL *n*-AMYL KETONE, **I**,
351
METHYL BENZYL KETONE,
II, 389
METHYL *n*-BUTYL KETONE, **I**,
352
2-METHYL-2-CYCLOHEXEN-
ONE, **IV**, 162
METHYL CYCLOPROPYL KE-
TONE, **IV**, 597
4-METHYLHEXANONE-2, **I**, 352
5-METHYL-5-HEXEN-2-ONE, **V**,
767
METHYL ISOBUTYL KETONE,
I, 352
METHYL ISOPROPYL KE-
TONE, **II**, 408
2-NORBORNANONE, **V**, 852
NORTRICYCLANONE, **V**, 866
Δ¹⁽⁹⁾-OCTALONE-2, **V**, 869
D-2-OXO-7,7-DIMETHYL-1-VI-
NYLBICYCLO(2.2.1-HEP-
TANE, **V**, 877
2-PHENYLCYCLOHEPTAN-
ONE, **IV**, 780
PINACOLONE, **I**, 462
PSEUDOIONONE, **III**, 747
STEARONE, **IV**, 854
α-TETRALONE, **II**, 569; **III**, 798;
IV, 898

β-TETRALONE, **IV**, 903
TETRAPHENYLCYCLOPEN-
TADIENONE, **III**, 806
2,6,6-TRIMETHYL-2,4-CY-
CLOHEXADIENONE, **V**,
1092
2,4,4-TRIMETHYLCYCLOPEN-
TANONE, **IV**, 957

2. Halogen Ketones
BENZALACETONE DIBROM-
IDE, **III**, 105
BENZYL CHLOROMETHYL
KETONE, **III**, 119
3-BROMOACETOPHENONE, **V**,
117
p-BROMOACETOPHENONE, **I**,
109
α-BROMOBENZALACETONE,
III, 125
p-CHLOROACETOPHENONE,
I, 111
2-CHLOROCYCLOHEXAN-
ONE, **III**, 188
2-CHLORO-2-METHYLCY-
CLOHEXANONE, **IV**, 162
5-CHLORO-2-PENTANONE, **IV**,
597
1-(p-CHLOROPHENYL)-3-
PHENYL-2-PROPANONE,
IV, 176
β-CHLOROVINYL ISOAMYL
KETONE, **IV**, 186
5α,6β-DIBROMOCHOLESTAN-
3-ONE, **IV**, 197
α,γ-DICHLOROACETONE, **I**,
211
DICHLOROACETOPHENONE,
III, 538
1,7-DICHLORO-4-HEPTANONE,
IV, 279
p,α,α-TRIBROMOACETOPHEN-
ONE, **IV**, 110

3. Phenol Ketones
2-p-ACETYLPHENYLHYDRO-
QUINONE, **IV**, 15

2,5-DIHYDROXYACETOPHEN-
ONE, **III**, 280
2,6-DIHYDROXYACETOPHEN-
ONE, **III**, 281
GALLACETOPHENONE, **II**, 304
o-HYDROXYDIBENZOYLME-
THANE, **IV**, 479
PHLOROACETOPHENONE, **II**,
522
o- AND p-PROPIOPHENOL, **II**,
543
QUINACETOPHENONE MON-
OMETHYL ETHER, **IV**, 836
RESACETOPHENONE, **III**, 761

4. Miscellaneous, Substituted
3-ACETAMIDO-2-BUTANONE,
IV, 5
ACETOACETANILIDE, **III**, 10
ACETONEDICARBOXYLIC
ACID, **I**, 10
2-ACETOTHIENONE, **II**, 8; **III**,
14
α-ACETYL-δ-CHLORO-γ-VAL-
EROLACTONE, **IV**, 10
1-ACETYLCYCLOHEXANOL,
IV, 13
3-ACETYLOXINDOLE, **V**, 12
2-p-ACETYLPHENYLHYDRO-
QUINONE DIACETATE,
IV, 17
2-p-ACETYLPHENYLQUIN-
ONE, **IV**, 16
2-AMINOBENZOPHENONE, **IV**,
34
ANISALACETONE, **I**, 78
BENZALACETONE, **I**, 77
BENZALACETOPHENONE, **I**,
78
BENZALPINACOLONE, **I**, 81
BENZOIN, **I**, 94
BENZOYLACETANILIDE, **III**,
108; **IV**, 80
BENZOYL-2-METHOXY-4-NI-
TROACETANILIDE, **IV**, 82
o-BENZOYLOXYACETOPHEN-
ONE, **IV**, 478

3-BENZOYLPYRIDINE, **IV**, 88
4-BENZOYLPYRIDINE, **IV**, 89
t-BUTYL ACETOACETATE, **V**, 155
DL-10-CAMPHORSULFONIC-ACID, **V**, 194
DL-10-CAMPHORSULFONYL CHLORIDE, **V**, 196
2-CARBETHOXYCYCLOOCT-ANONE, **V**, 198
p-CHLOROACETYLACETANIL-IDE, **III**, 183
ω-CHLOROISONITROSOACET-OPHENONE, **III**, 191
DIACETONE ALCOHOL, **I**, 199
DIBENZALACETONE, **II**, 167
1-DIETHYLAMINO-3-BUTAN-ONE, **IV**, 281
DIETHYL BENZOYLMALON-ATE, **IV**, 285
DIETHYL β-KETOPIMELATE, **V**, 384
DIETHYL γ-OXOPIMELATE, **IV**, 302
DIHYDRORESORCINOL, **III**, 278
p-DIMETHYLAMINOBENZO-PHENONE, **I**, 217
β-DIMETHYLAMINOPROPIO-PHENONE HYDRO-CHLORIDE, **III**, 305
3,5-DIMETHYL-4-CARBETHOXY-2-CYCLOHEXEN-1-ONE, **III**, 317
DIMETHYL CYCLOHEXAN-ONE-2,6-DICARBOXYL-ATE, **V**, 439
3-ETHOXY-2-CYCLOHEXEN-ONE, **V**, 539
ETHYL ACETOACETATE, **I**, 235
ETHYL ACETOPYRUVATE, **I**, 238
ETHYL BENZOYLFORMATE, **I**, 241
ETHYL *n*-BUTYLACETOACET-ATE, **I**, 248
2-ETHYL CHROMONE, **III**, 387

ETHYL ENANTHYLSUCCIN-ATE, **IV**, 430
ETHYL OXOMALONATE, **I**, 266
FLAVONE, **IV**, 478
2-FURFURALACETONE, **I**, 283
2-HYDROXYMETHYLENECY-CLOHEXANONE, **IV**, 536
p-HYDROXYPHENYLPYRUV-IC ACID, **V**, 627
ISOPHORONE OXIDE, **IV**, 552
β-KETOISOÖCTALDEHYDE DI-METHYL ACETAL, **IV**, 558
DL-KETOPINIC ACID, **V**, 689
LEVULINIC ACID, **I**, 335
MESITYL OXIDE, **I**, 345
ω-METHOXYACETOPHENONE, **III**, 562
p-METHOXYACETOPHENONE, **I**, 111
o-METHOXYPHENYLACET-ONE, **IV**, 573
METHYL *p*-ACETYLBENZO-ATE, **IV**, 579
4'-METHYL-2-AMINOBENZO-PHENONE, **IV**, 38
METHYL BENZOYLFORM-ATE, **I**, 244
o-NITROACETOPHENONE, **IV**, 708
NITROANTHRONE, **I**, 390
PHENACYLAMINE HYDROCH-LORIDE, **V**, 909
α-PHENYLACETOACETONI-TRILE, **II**, 487
PHENYLGLYOXAL, **V**, 937
PHENYL 4-PYRIDYL KETONE, **IV**, 89
PHENYL THIENYL KETONE, **II**, 520
PSEUDOPELLETIERINE, **IV**, 816
PYRUVIC ACID, **I**, 475
QUINACETOPHENONE MONO-METHYL ETHER, **IV**, 836
3-QUINUCLIDONE HYDROCH-LORIDE, **V**, 989
SEBACOIN, **IV**, 840

2,2,5,5-TETRAMETHYLTETRA-
HYDRO-3-KETOFURAN, **V,**
1024
2,4,5,7-TETRANITROFLUOREN-
ONE, **V,** 1029
THIOBENZOPHENONE, **IV,** 927
p-TOLUYL-*o*-BENZOIC ACID, **I,**
517
2,2,4-TRIMETHYL-3-OXOVAL-
ERYL CHLORIDE, **V,** 1103
2,4,7-TRINITROFLUORENONE,
III, 837
XANTHONE, **I,** 552
B. *DIKETONES*
ACETONYLACETONE, **II,** 219
ACETYLACETONE, **III,** 16
ACETYLBENZOYL, **III,** 20
BENZIL, **I,** 87
4,4'-BIS(DIMETHYLAMINO)-
BENZIL, **V,** 111
Δ⁴-CHOLESTEN-3,6-DIONE, **IV,**
189
1,2-CYCLOHEXANEDIONE, **IV,**
229
1,4-CYCLOHEXANEDIONE, **V,**
288
2-CYCLOPENTENE-1,4-DIONE,
V, 324
1,4-DIBENZOYLBUTANE,**II,** 169
trans-DIBENZOYLETHYLENE,
III, 248
DIBENZOYLMETHANE, **I,** 205;
III, 251
DIISOVALERYLMETHANE, **III,**
291
5,5-DIMETHYL-1,3-CYCLO-
HEXANEDIONE, **II,** 200
1,6-DIOXO-8*a*-METHYL-1,2,3,4,6,-
7,8*a*-OCTAHYDRONAPH-
THALENE, **V,** 486
o-HYDROXYDIBENZOYLME-
THANE, **IV,** 479
a-KETOGLUTARIC ACID, **V,** 687
2-METHYL-1,3-CYCLOHEX-
ANEDIONE, **V,** 743

2-METHYLCYCLOPENTANE-1,3-
DIONE, **V,** 747
3-METHYLPENTANE-2,4-DIONE,
V, 785
2,4-NONANEDIONE, **V,** 848
SEBACIL, **IV,** 838
C. *TRIKETONES*
DIPHENYL TRIKETONE, **II,** 244
DIPHENYL TRIKETONE HY-
DRATE, **II,** 244
1-(*p*-METHOXYPHENYL)-5-
PHENYL-1,3,5-PENTANE-
TRIONE, **V,** 718
1,3,5-TRIACETYLBENZENE, **III,**
829
D. *TETRAKETONES*
TETRAACETYLETHANE, **IV,** 869
MERCURY COMPOUNDS
ANHYDRO-*o*-HYDROXYMERCU-
RIBENZOIC ACID, **I,** 57
ANHYDRO-2-HYDROXYMER-
CURI-3-NITROBENZOIC
ACID, **I,** 56
p-CHLOROMERCURIBENZOIC
ACID, **I,** 159
o-CHLOROMERCURIPHENOL, **I,**
161
DIPHENYLMERCURY, **I,** 228
DI-*p*-TOLYLMERCURY, **I,** 231
MERCURY DI-*β*-NAPHTHYL, **II,**
381
β-NAPHTHYLMERCURIC
CHLORIDE, **II,** 432
p-TOLYMERCURIC CHLORIDE,
I, 519
NITRILES (See also Acids, B. 2; Esters, B.
4)
A. *UNSUBSTITUTED*
ALLYL CYANIDE, **I,** 46; **III,** 852
AZELANITRILE, **IV,** 62
BENZYL CYANIDE, **I,** 107
3-BENZYL-3-METHYLPENT-
ANENITRILE, **IV,** 95
1-CYANOBENZOCYCLOBUT-
ENE, **V,** 263

2-CYANO-6-METHYLPYRIDINE,
V, 269

9-CYANOPHENANTHRENE, III,
212

1-CYCLOHEXENYLACETONI-
TRILE, IV, 234

α-CYCLOHEXYLPHENYLACE-
TONITRILE, III, 219

CYCLOPROPYL CYANIDE, III,
223

DIAMINOMALEONITRILE, V,
344

1,1'-DICYANO-1,1'-BICYCLO-
HEXYL, IV, 273

DIPHENYLACETONITRILE, III,
347

2-ETHYLHEXANONITRILE, IV,
436

3-ETHYL-3-METHYLHEXANEN-
ITRILE, IV, 97

FUMARONITRILE, IV, 486

3-(2-FURYL)ACRYLONITRILE,
V, 585

HYDROGEN CYANIDE, I, 314

INDOLE-3-CARBONITRILE, V,
656

ISOBUTYRONITRILE, III, 493

MALONONITRILE, II, 379; III,
535

3-METHYL-3-PHENYLPENT-
ANENITRILE, IV, 97

α-NAPHTHONITRILE, III, 631

PALMITONITRILE, IV, 437

α-PHENYLCINNAMONITRILE,
III, 715

SEBACONITRILE, III, 768

TETRACYANOETHYLENE, IV,
877

TETRACYANOETHYLENE OX-
IDE, V, 1007

TETRAMETHYLAMMONIUM 1,-
1,2,3,3-PENTACYANOPRO-
PENIDE, V, 1013

o-TOLUNITRILE, I, 514

p-TOLUNITRILE, I, 514

TRIMETHYLENE CYANIDE, I,
536

B. SUBSTITUTED NITRILES

ACETONE CYANOHYDRIN, II, 7

AMINOMALONONITRILE p-TO-
LUENESULFONATE, V, 32

β-AMINOPROPIONITRILE, III, 93

1,1'-AZO-bis-1-CYCLOHEX-
ANENITRILE, IV, 66

2,2'-AZO-bis-ISOBUTYRONI-
TRILE, IV, 67

BENZOYL CYANIDE, III, 112

2-BENZOYL-1-CYANO-1-
METHYL-1,2-DIHYDROISO-
QUINOLINE, IV, 642

t-BUTYL CYANOACETATE, V,
171

o-CARBOXYPHENYLACETONI-
TRILE, III, 174

CHLOROACETONITRILE, IV,
144

3-(o-CHLOROANILINO)PRO-
PIONITRILE, IV, 146

γ-CHLOROBUTYRONITRILE, I,
156

2-CHLORONICOTINONITRILE,
IV, 166

α-(4-CHLOROPHENYL)-γ-PHEN-
YLACETOACETONITRILE,
IV, 174

CYANAMIDE, IV, 645

1-CYANO-2-BENZOYL-1,2-DI-
HYDROISOQUINOLINE,
IV,641

bis-(β-CYANOETHYL)AMINE,
III,93

N-2-CYANOETHYLANILINE, IV,
205

CYANOGEN IODIDE, IV, 207

7-CYANOHEPTANAL, V, 266

3-CYANO-6-METHYL-2(1)-PYRI-
DONE, IV, 210

1-CYANO-3-α-NAPHTHYL-
UREA, IV, 215

ω-CYANOPELARGONIC ACID,
II, 768

α-CYANO-β-PHENYLACRYLIC
ACID, **I,** 181
1-CYANO-3-PHENYLUREA, **IV,**
213
CYCLOHEXYLIDENECY-
ANOACETIC ACID, **IV,** 234
2,5-DIAMINO-3,4-DICYANO-
THIOPHENE, **IV,** 243
DIAMINOMALEONITRILE, **V,**
344
DIBROMOACETONITRILE, **IV,**
254
DICHLOROACETONITRILE, **IV,**
255
1,2-DI-1-(1-CYANO)CYCLO-
HEXYLHYDRAZINE, **IV,** 274
α,α'-DICYANO-β-ETHYL-β-
METHYLGLUTARIMIDE, **IV,**
441
DICYANOKETENE ETHYLENE
ACETAL, **IV,** 276
1,2-DI-2-(2-CYANO)PRO-
PYLHYDRAZINE, **IV,** 275
DIETHYLAMINOACETONI-
TRILE, **III,** 275
2,6-DIMETHOXYBENZONI-
TRILE, **III,** 293
α-N,N-DIMETHYLAMINOPHEN-
YLACETONITRILE, **V,** 437
3,5-DINITRO-o-TOLUNITRILE,
V, 480
α,β-DIPHENYLCINNAMONI-
TRILE, **IV,** 387
2,3-DIPHENYLSUCCINONI-
TRILE, **IV,** 392
β-ETHOXYPROPIONITRILE, **III,**
372
ETHYL α,β-DICYANO-β-PHEN-
YLPROPIONATE, **I,** 452
ETHYL CYANOACETATE, **I,** 254
ETHYLENE CYANOHYDRIN, **I,**
256
FERROCENYLACETONITRILE,
V, 578

GLYCOLONITRILE, **III,** 436
3-n-HEPTYL-5-CYANOCYTO-
SINE, **IV,** 515
3-n-HEPTYLUREIDOMETHYL-
ENEMALONONITRILE, **IV,**
515
β-(2-HYDROXYETHYLMER-
CAPTO)-PROPIONITRILE,
III, 458
3-HYDROXYGLUTARONITRILE,
V, 614
2-MERCAPTO-4-HYDROXY-5-CY-
ANOPYRIMIDINE, **IV,** 566
METHOXYACETONITRILE, **II,**
387
p-METHOXYPHENYLACETONI-
TRILE, **IV,** 576
METHYLENEAMINOACETO-
NITRILE, **I,** 355
β-METHYL-β-PHENYL-α,α'-DI-
CYANOGLUTARIMIDE, **IV,**
662
NICOTINONITRILE, **IV,** 706
p-NITROBENZONITRILE, **III,** 646
PENTAACETYL-d-GLUCONONI-
TRILE, **III,** 690
α-PHENYL-β-BENZOYLPRO-
PIONITRILE, **II,** 498
SODIUM CYANOACETATE, **I,**
181
TETRAMETHYLSUCCINONI-
TRILE, **IV,** 273
p-TRICYANOVINYL-N,N-DIME-
THYLANILINE, **IV,** 953
α,α,β-TRIPHENYLPROPIONI-
TRILE, **IV,** 962
VERATRONITRILE, **II,** 622
NITRO COMPOUNDS (See also Acids, B.6)
A. *UNSUBSTITUTED*
2,7-DIMETHYL-2,7-DINI-
TROÖCTANE, **V,** 445
o-DINITROBENZENE, **II,** 226
p-DINITROBENZENE, **II,** 225
p,p'-DINITROBIBENZYL, **IV,** 367
2,2'-DINITROBIPHENYL, **III,** 339

1,4-DINITROBUTANE, **IV,** 368
2,3-DINITRO-2-BUTENE, **IV,** 374
DINITRODURENE, **II,** 254
2-NITROFLUORENE, **II,** 447
1,6-DINITROHEXANE, **IV,** 370
3,4-DINITRO-3-HEXENE, **IV,** 372
1,4-DINITRONAPHTHALENE,
    III, 341
1,5-DINITROPENTANE, **IV,** 370
1,3-DINITROPROPANE, **IV,** 370
9-NITROANTHRACENE, **IV,** 711
m-NITROBIPHENYL, **IV,** 718
2-NITROCARBAZOLE, **V,** 829
2-NITRO-*p*-CYMENE, **III,** 653
NITROMESITYLENE, **II,** 449
NITROMETHANE, **I,** 401
1-NITROÖCTANE, **IV,** 724
NITROSTYRENE, **I,** 413
m-NITROSTYRENE, **IV,** 731
m-NITROTOLUENE, **I,** 415
4-NITRO-2,2,4-TRIMETHYLPEN-
    TANE, **V,** 845
PHENYLNITROMETHANE, **II,**
    512
TETRANITROMETHANE, **III,** 803
1,3,5-TRINITROBENZENE, **I,** 541
2,4,6-TRIPHENYLNITROBEN-
    ZENE, **V,** 1128
B. *SUBSTITUTED*
  1. **Nitro Amines**
    2-AMINO-4-NITROPHENOL, **III,**
      82
    2-AMINO-3-NITROTOLUENE, **IV,**
      42
    2-AMINO-5-NITROTOLUENE, **IV,**
      44
    1,2-DIAMINO-4-NITROBEN-
      ZENE, **III,** 242
    2,6-DINITROANILINE, **IV,** 364
    o-NITROANILINE, **I,** 388
    m-NITRODIMETHYLANILINE,
      III, 658
    2-NITRO-4-METHOXYANIL-
      INE, **III,** 661
    N-NITROMORPHOLINE, **V,** 839

4-NITRO-1-NAPHTHYLAMINE,
    III, 664
  2. **Miscellaneous**
    BENZOYL-2-METHOXY-4-NI-
      TROACETANILIDE, **IV,** 82
    2-BROMO-4-NITROTOLUENE,
      IV, 114
    o-(*n*-BUTOXY)NITRO-
      BENZENE, **III,** 140
    1-CHLORO-2,6-DINITROBEN-
      ZENE, **IV,** 160
    2,6-DICHLORONITROBENZENE,
      V, 367
    2,4-DINITROBENZENESUL-
      FENYL CHLORIDE, **V,** 474
    2,4-DINITROIODOBENZENE, **V,**
      478
    3,5-DINITRO-*o*-TOLUNITRILE,
      V, 480
    ETHYL *m*-NITROBENZOATE, **I,**
      373
    ETHYL α-NITROBUTYRATE, **IV,**
      454
    2-HYDROXY-5-NITROBENZYL
      CHLORIDE, **III,** 468
    6-METHOXY-8-NITROQUINO-
      LINE, **III,** 568
    1-(*o*-METHOXYPHENYL)-2-NI-
      TRO-1-PROPENE, **IV,** 573
    METHYL γ-METHYL-γ-NITROV-
      ALERATE, **IV,** 652
    METHYL *m*-NITROBENZOATE,
      I, 372
    3-METHYL-4-NITROPYRIDINE-1-
      OXIDE, **IV,** 654
    o-NITROACETOPHENONE, **IV,**
      708
    m-NITROACETOPHENONE, **II,**
      434
    1-NITRO-2-ACETYLAMINO-
      NAPHTHALENE, **II,** 438
    NITROANTHRONE, **I,** 390
    NITROBARBITURIC ACID, **II,**
      440
    o-NITROBENZALDEHYDE, **III,**
      641, **V,** 825

m-NITROBENZALDEHYDE DI-
METHYLACETAL, **III,** 644
o- AND p-NITROBENZALDIACE-
TATE, **IV,** 713
m-NITROBENZAZIDE, **IV,** 715
p-NITROBENZONITRILE, **III,** 646
m-NITROBENZOYL CHLOR-
IDE, **IV,** 715
p-NITROBENZOYL PEROXIDE,
III, 649
p-NITROBENZYL ACETATE, **III,**
650
p-NITROBENZYL ALCOHOL,
III, 652
o-NITROCINNAMALDEHYDE,
IV, 722
5-NITRO-2,3-DIHYDRO-1,4-
PHTHALAZINEDIONE, **III,**
656
2-NITROETHANOL, **V,** 833
5-NITROINDAZOLE, **III,** 660
2-NITRO-4-METHOXYACETANI-
LIDE, **III,** 661
p-NITROPHENYLARSONIC
ACID, **III,** 665
o-NITROPHENYLAZIDE, **IV,** 75
1-(p-NITROPHENYL)-1,3-BUTA-
DIENE, **IV,** 727
1-(p-NITROPHENYL)-4-
CHLORO-2-BUTENE, **IV,** 727
1-(m-NITROPHENYL)-3,3-DIME-
THYLTRIAZINE, **IV,** 718
m-NITROPHENYL DISULFIDE,
V, 843
p-NITROPHENYL SULFIDE, **III,**
667
4-NITROPHTHALIMIDE, **II,** 459
2-NITROTHIOPHENE, **II,** 466
6-NITROVERATRALDEHYDE,
IV, 735
SODIUM NITROMALONALDE-
HYDE MONOHYDRATE, **IV,**
844
2,4,5,7-TETRANITROFLUOREN-
ONE, **V,** 1029

(+)- and (−)-α-(2,4,5,7-TETRANI-
TRO-9-FLUORENYLIDENE-
AMINOOXY)PRO-
PIONIC ACID, **V,** 1031
TRIMETHYLOXONIUM 2,4,6-
TRINITROBENZENESUL-
FONATE, **V,** 1099
2,4,7-TRINITROFLUOREN-
ONE,**III,** 837
NITROSO COMPOUNDS
CUPFERRON, **I,** 177
ETHYL N-NITROSO-N-BENZYL-
CARBAMATE, **IV,** 780
NITROSOBENZENE, **III,** 668
p-NITROSODIETHYLANILINE,
II, 224
NITROSODIMETHYLAMINE, **II,**
211
p-NITROSODIMETHYLANILINE
HYDROCHLORIDE, **II,** 223
N-NITROSOMETHYLANILINE,
II, 460
NITROSOMETHYLUREA, **II,** 461
NITROSOMETHYLURETHANE,
II, 464
NITROSO-β-NAPHTHOL, **I,** 411
NITROSOTHYMOL, **I,** 511
p-TOLYLSULFONYLMETHYL-
NITROSAMIDE, **IV,** 943
ORGANO-METALLIC COMPOUNDS
ALLYLMAGNESIUM BROM-
IDE, **IV,** 749
DI-n-BUTYLDIVINYLTIN, **IV,**
258
N,N-DIMETHYLAMINOME-
THYLFERROCENE METH-
IODIDE, **V,** 434
ETHYNYLMAGNESIUM BRO-
MIDE, **IV,** 792
FERROCENE, **IV,** 473
FERROCENYLACETONITRILE,
V, 578
HYDROXYMETHYLFERRO-
CENE, **V,** 621
PHENYL(TRICHLOROMETH-
YL)MERCURY, **V,** 969

RUTHENOCENE, V, 1001
TETRAETHYLTIN, IV, 881
TRIPHENYLALUMINUM, V, 1116
UNSOLVATED n-BUTYLMAG-
  NESIUM CHLORIDE, V, 1141
OXIDES (See Ethers; Heterocyclic compounds)
OXIMES
  1,2-CYCLOHEXANEDIONE
    DIOXIME, IV, 229
  (+)- and (−)-α-(2,4,5,7-TETRANI-
    TRO-9-FLUORENYLI-
    DENEAMINOOXY)PRO-
    PIONIC ACID, V, 1031
OXONIUM COMPOUNDS
  TRIETHYLOXONIUM FLUO-
    BORATE, V, 1080
  TRIMETHYLOXONIUM FLUO-
    BORATE, V, 1096
  TRIMETHYLOXONIUM 2,4,6-
    TRINITROBENZENESUL-
    FONATE, V, 1099
  2,4,6-TRIMETHYLPYRYLIUM
    TETRAFLUOROBORATE, V, 1112
  2,4,6-TRIMETHYLPYRYLIUM
    TRIFLUOROMETHANESUL-
    FONATE, V, 1114
  2,4,6-TRIMETHYLPYRYLIUM
    PERCHLORATE, V, 1106
  2,4,6-TRIPHENYLPYRYLIUM
    TETRAFLUOROBORATE, V, 1135
PEROXY ACIDS
  MONOPERPHTHALIC ACID, V, 805
  PEROXYBENZOIC ACID, V, 904
PHENOLS (See also Acids, B.4; Ketones, A.3)
  2-AMINO-4-NITROPHENOL, III, 82
  AMMONIUM SALT OF AURIN
    TRICARBOXYLIC ACID, I, 54
  3-BROMO-4-HYDROXYTOL-

UENE, III, 130
  6-BROMO-2-NAPHTHOL, III, 132
  CATECHOL, I, 149
  p-CHLOROPHENYL SALICYL-
    ATE, IV, 178
  CREOSOL, IV, 203
  p-CRESOL, I, 175
  2,6-DICHLOROPHENOL, III, 267
  2,5-DIHYDROXY-p-BENZENEDI-
    ACETIC ACID, III, 286
  3,5-DIHYDROXYBENZOIC ACID, III, 288
  3,3′-DIHYDROXYBIPHENYL, V, 412
  o-EUGENOL, III, 418
  o-n-HEPTYLPHENOL, III, 444
  HEXAHYDROXYBENZENE, V, 595
  m-HYDROXYBENZALDEHYDE, III, 453
  2-HYDROXYISOPHTHALIC
    ACID, V, 617
  2-HYDROXY-1-NAPHTHALDE-
    HYDE, III, 463
  2-HYDROXY-1,4-NAPHTHOQUI-
    NONE, III, 465
  2-HYDROXY-5-NITROBENZYL
    CHLORIDE, III, 468
  3-HYDROXYPYRENE, V, 632
  3-HYDROXYQUINOLINE, V, 635
  2-HYDROXYTHIOPHENE, V, 642
  6-METHOXY-2-NAPHTHOL, V, 918
  3-METHOXY-5-NITROPHENOL, I, 405
  4-METHYLESCULETIN, I, 360
  β-NAPHTHOL PHENYLAMIN-
    OMETHANE, I, 381
PHENOLS (See also Acids, B.4; Ketones, A.3)
  NAPHTHORESORCINOL, III, 637
  1-NITRO-2-NAPHTHOL, II, 451
  m-NITROPHENOL, I, 404
  3-NITRO-4,6-XYLENOL, I, 405
  PHLOROGLUCINOL, I, 455
  PROTOCATECHUALDEHYDE, II, 549

PROTOCATECHUIC ACID, III, 745
PYROGALLOL 1-MONOME-THYL ETHER, III, 759
QUINIZARIN, I, 476
SALICYL-o-TOLUIDE, III, 765
SYRINGIC ALDEHYDE, IV, 866
ar-TETRAHYDRO-α-NAPHTHOL, IV, 887
TETRAHYDROXYQUINONE, V, 1011

PHOSPHORUS COMPOUNDS
CHLORODIISOPROPYLPHOS-PHINE, V, 211
CHLOROMETHYLPHOSPHONO-THIOIC DICHLORIDE, V, 218
DICHLOROMETHYLENETRI-PHENYLPHOSPHORANE, V, 361
DIETHYL ETHYLPHOSPHON-ATE, IV, 326
DIETHYL METHYLPHOSPHON-ATE, IV, 326
DIISOPROPYL ETHYLPHOS-PHONATE, IV, 326
DIISOPROPYL METHYLPHOS-PHONATE, IV, 325
DIPHENYL-p-BROMOPHENYL-PHOSPHINE, V, 496
HEXAMETHYLPHOSPHORUS TRIAMIDE, V, 602
3-METHYL-1-PHENYLPHOS-PHOLENE OXIDE, V, 787
PHENYLDICHLOROPHOS-PHINE, IV, 784
STYRYLPHOSPHONIC DICHLO-RIDE, V, 1005
TETRAMETHYLBIPHOSPHINE DISULFIDE, V, 1016
TRICHLOROMETHYLPHOS-PHONYL DICHLORIDE, IV, 950
TRIETHYL PHOSPHITE, IV, 955
VINYL TRIPHENYLPHOSPHON-IUM BROMIDE, V, 1145

PROTEINS
CASEIN, II, 120
QUARTERNARY AMMONIUM COM-POUNDS
1-AMINOPYRIDINIUM IODIDE, V, 43
BENZOYLCHOLINE CHLOR-IDE, IV, 84
BENZOYLCHOLINE IODIDE, IV, 84
BENZYLTRIMETHYLAMMON-IUM ETHOXIDE, IV, 98
BENZYLTRIMETHYLAMMON-IUM IODIDE, IV, 585
N-CHLOROBETAINYL CHLO-RIDE, IV, 154
N,N-DIMETHYLAMINOME-THYLFERROCENE METH-IODIDE, V, 434
1-ETHYL-3-(3-DIMETHYLA-MINO)PROPYLCARBODI-IMIDE METHIODIDE, V, 555
TETRAHYDROXYQUINONE, V, 1011
TETRAMETHYLAMMONIUM 1,-1,2,3,3-PENTACYANOPRO-PENIDE, V, 1013
VINYL TRIPHENYLPHOSPHON-IUM BROMIDE, V, 1145

QUINONES
ACENAPHTHENEQUINONE, III, 1
2-p-ACETYLPHENYLQUINONE, IV, 16
ANTHRAQUINONE, II, 554
p-BENZOQUINONE, IV, 152
CHLORO-p-BENZOQUINONE, IV, 148
3-CHLOROTOLUQUINONE, IV, 152
2,5-DICHLOROQUINONE, IV, 152
2,3-DIMETHYLANTHRAQUI-NONE, III, 310
DUROQUINONE, II, 254
2-HYDROXY-1,4-NAPHTHOQUI-NONE, III, 465

METHOXYQUINONE, **IV,** 153
1-METHYLAMINOANTHRAQUI-
NONE, **III,** 573
1-METHYLAMINO-4-BROMOAN-
THRAQUINONE, **III,** 575
β-METHYLANTHRAQUINONE,
I, 353
1,2-NAPHTHOQUINONE, **II,** 430
1,4-NAPHTHOQUINONE, **I,** 383;
IV, 698
1,2-NAPHTHOQUINONE-4-SUL-
FONIC ACID, SALTS OF **III,**
633
p-NITROBENZYL CYANIDE, **I,**
396
NITROGUANIDINE, **I,** 399
3-NITROPHTHALIC ACID, **I,** 408
NITROUREA, **I,** 417
PHENANTHRENEQUINONE, **IV,**
757
QUINIZARIN, **I,** 476
QUINONE, **I,** 482; **II,** 553
THYMOQUINONE, **I,** 511
SALTS (See Acids; Amines)
SCHIFF BASES (See Imines)
SELENIUM COMPOUNDS
N,N-DIMETHYLSELENO-
UREA, **IV,** 359
DIPHENYL SELENIDE, **II,** 238
DIPHENYLSELENIUM DICHLO-
RIDE, **II.** 240
DI-m-TOLYL SELENIDE, **II,** 239
DI-o-TOLYL SELENIDE, **II,** 239
DI-p-TOLYL SELENIDE, **II,** 239
SELENOPHENOL, **III,** 771
TRIPHENYLSELENONIUM
CHLORIDE, **II,** 241
SODIUM COMPOUNDS
ETHYL SODIUM PHTHALIM-
IDOMALONATE, **II,** 384
SODIUM PHENYL-aci-NITRO-
METHANE, **II,** 512
TRIPHENYLMETHYL-
SODIUM, **II,** 607
SULFONIC ACIDS (See Acids)
SULFONYL HALIDES (SEE Acid Halides)

**SULFUR COMPOUNDS (See also Hetero-
cyclic compounds)**
A. *DISULFIDES*
p-AMINOPHENYL DISULF-
IDE, **III,** 86
BENZOYL DISULFIDE, **III,** 116
l-CYSTINE, **I,** 194
DIBENZOHYDRYL DISULF-
IDE, **II,** 574
DI-o-NITROPHENYL DISULF-
IDE, **I,** 220
METHYL 2-THIENYL SULF-
IDE, **IV,** 667
m-NITROPHENYL DISULF-
IDE, **V,** 843
TETRAMETHYLBIPHOSPHINE
DISULFIDE, **V,** 1016
B. *MERCAPTANS*
1,2-BIS(n-BUTYLTHIO)BEN-
ZENE, **V,** 107
DIETHYL MERCAPTOACET-
AL, **IV,** 295
1,2-DIMERCAPTOBENZENE, **V,**
419
n-DODECYL MERCAPTAN, **III,**
363
ETHANEDITHIOL, **IV,** 401
2-FURFURYL MERCAPTAN, **IV,**
491
2-MERCAPTO-4-AMINO-5-CAR-
BETHOXYPYRIMIDINE, **IV,**
566
2-MERCAPTOBENZIMIDAZOLE,
IV, 569
2-MERCAPTOBENZOXAZOLE,
IV, 570
2-MERCAPTO-4-HYDROXY-5-CY-
ANOPYRIMIDINE, **IV,** 566
2-THIO-6-METHYLURACIL, **IV,**
638
C. *SULFIDES*
DIPHENYL SULFIDE, **II,** 242
ETHYL 1-METHYLTHIOL-3-
PHTHALIMIDOPROPANE-
3,3-DICARBOXYLATE, **II,**
384

β-(2-HYDROXYETHYLMER-
CAPTO)-PROPIONITRILE,
III, 458

β-HYDROXYETHYL METHYL
SULFIDE, **II,** 345

dl-METHIONINE, **II,** 384

2-METHYLMERCAPTO-N-
METHYL-Δ²-PYRROLINE, **V,**
780

METHYL β-THIODIPROPION-
ATE, **IV,** 669

p-NITROPHENYL SULFIDE, **III,**
667

n-PROPYL SULFIDE, **II,** 547

1,1,1′,1′-TETRAETHOXYETHYL
POLYSULFIDE, **IV,** 295

β-THIODIGLYCOL, **II,** 576

D. *SULFONES*

4,4′-DIAMINODIPHENYLSULF-
ONE, **III,** 239

ISOPRENE CYCLIC SULFONE,
III, 499

METHYL p-TOLYL SULFONE,
IV, 674

E. *SULFONIC ACIDS AND DERIV-
ATIVES*

ACID AMMONIUM o-SULFO-
BENZOATE, **I,** 14

1-AMINO-2-NAPHTHOL-4-SUL-
FONIC ACID, **II,** 42

BARIUM 2-PHENANTHRENE-
SULFONATE, **II,** 482

n-BUTYL p-TOLUENESULF-
ONATE, **I,** 145

CYSTEIC ACID MONOHY-
DRATE, **III,** 226

n-DODECYL p-TOLUENESULF-
ONATE, **III,** 366

4-HYDROXY-1-BUTANESULF-
ONIC ACID SULTONE, **IV,**
529

METHYL p-TOLUENESULFON-
ATE, **I,** 146

1,2-NAPHTHOQUINONE-4-SUL-
FONIC ACID, SALTS, **III,** 633

POTASSIUM ANTHRAQUIN-

ONE-α-SULFONATE, **II,** 539

POTASSIUM 3-PHENAN-
THRENESULFONATE, **II,**
482

4-PYRIDINESULFONIC ACID, **V,**
977

SODIUM 2-BROMOETHANE-
SULFONATE, **II,** 558

SODIUM 2-PHENYL-2-HYDROX-
YETHANE-1-SULFONATE,
IV, 850

SODIUM β-STYRENESULFON-
ATE, **IV,** 846

o-SULFOBENZOIC ANHY-
DRIDE, **I,** 495

α-SULFOPALMITIC ACID, **IV,**
862

TAURINE, **II,** 563

p-TOLUENESULFONIC ANHY-
DRIDE, **IV,** 940

p-TOLUENESULFONYLAN-
THRANILIC ACID, **IV,** 34

p-TOLUENESULFONYLHYDRA-
ZIDE, **V,** 1055

o-TOLUIDINESULFONIC ACID,
III, 824

2-(p-TOLYLSULFONYL)DIHY-
DROISOINDOLE, **V,** 1064

p-TOLYLSULFONYLMETHYL-
NITROSAMIDE, **IV,** 943

TRIMETHYLOXONIUM 2,4,6-
TRINITROBENZENESUL-
FONATE, **V,** 1099

2,4,6-TRIMETHYLPYRYLIUM
TRIFLUOROMETHANESUL-
FONATE, **V,** 1114

F. *SULFONYL, SULFINYL, SUL-
FENYL, or SULFUR HA-
LIDES*

p-ACETAMINOBENZENESUL-
FONYL CHLORIDE, **I,** 8

BENZENESULFINYL FLUO-
RIDE, **V,** 396

BENZENESULFONYL CHLOR-
IDE, **I,** 84

DL-10-CAMPHORSULFONIC

ACID, **V**, 194

DL-10-CAMPHORSULFONYL CHLORIDE, **V**, 196

CHLOROSULFONYL ISOCYANATE, **V**, 226

2,4-DINITROPHENYLSULFUR CHLORIDE, **II**, 456

β-ISOVALEROLACTAM-N-SULFONYL CHLORIDE, **V**, 673

METHANESULFINYL CHLORIDE, **V**, 709

METHANESULFONYL CHLORIDE, **IV**, 571

NAPHTHALENE-1,5-DISULFONYL CHLORIDE, **IV**, 693

o-NITRO-p-CHLOROPHENYLSULFUR CHLORIDE, **II**, 456

o-NITROPHENYLSULFUR CHLORIDE, **II**, 455

β-STYRENESULFONYL CHLORIDE, **IV**, 846

p-TOLUENESULFENYL CHLORIDE, **IV**, 934

p-TOLUENESULFINYL CHLORIDE, **IV**, 937

G. *THIOPHENOLS*

2-MERCAPTOPYRIMIDINE, **V**, 703

1,5-NAPHTHALENEDITHIOL, **IV**, 695

m-THIOCRESOL, **III**, 809

THIOPHENOL, **I**, 504

THIOSALICYLIC ACID, **II**, 580

H. *THIOUREA DERIVATIVES*

o-CHLOROPHENYLTHIOUREA, **IV**, 180

2,6-DIMETHYLPHENYLTHIOUREA, **V**, 802

2,4-DITHIOBIURET, **IV**, 504

DITHIZONE, **III**, 360

ETHYLENE THIOUREA, **III**, 394

GUANYLTHIOUREA, **IV**, 502

S-METHYL ISOTHIOUREA SULFATE, **II**, 411

1-METHYL-1-(1-NAPHTHYL)-2-THIOUREA, **III**, 609

METHYLTHIOUREA, **III**, 617

1-PHENYL-2-THIOBIURET, **V**, 966

α-PHENYLTHIOUREA, **III**, 735

I. *MISCELLANEOUS*

p-ACETAMINOBENZENESULFINIC ACID, **I**, 7

m-BROMOPHENYL ISOTHIOCYANATE, **I**, 448

p-BROMOPHENYL ISOTHIOCYANATE, **I**, 448

CHLOROMETHYLPHOSPHONOTHIOIC DICHLORIDE, **V**, 218

p-CHLOROPHENYL ISOTHIOCYANATE, **I**, 165; **V**, 223

DIETHYLTHIOCARBAMYL CHLORIDE, **IV**, 307

p-DIMETHYLAMINOPHENYL ISOTHIOCYANATE, **I**, 448

2,4-DINITROBENZENESULFENYL CHLORIDE, **V**, 474

p-IODOPHENYL ISOTHIOCYANATE, **I**, 448

p-ISOPROPYLPHENYL ISOTHIOCYANATE, **I**, 448

METHYL BENZENESULFINATE, **V**, 723

METHYL ISOTHIOCYANATE, **III**, 599

METHYL PHENYL SULFOXIDE, **V**, 791

2-METHYL-4-iso-PROPYLPHENYL ISOTHIOCYANATE, **I**, 448

MORLAND SALT, **II**, 555

α-NAPHTHYL ISOTHIOCYANATE, **IV**, 700

β-NAPHTHYL ISOTHIOCYANATE, **I**, 448

PHENYL ISOTHIOCYANATE, **I**, 447

PHENYLSULFUR TRIFLUORIDE, **V**, 959

REINECKE SALT, **II**, 555

SODIUM p-TOLUENESULFINATE, **I**, 492

THIOBENZOIC ACID, **IV**, 924
THIOBENZOPHENONE, **II**, 573;
  IV, 927
THIOBENZOYLTHIOGLYCOLIC
  ACID, **V**, 1046
THIOCARBONYL PERCHLOR-
  IDE, **I**, 506
p-THIOCYANODIMETHYL-
  ANILINE, **II**, 574
2*a*-THIOHOMOPHTHALIMIDE,
  V, 1051
THIOLACETIC ACID, **IV**, 928
THIOPHOSGENE, **I**, 509
m-(TRIFLUOROMETHYL)-
  PHENYL ISOTHIOCYAN-
  ATE, **I**, 448
TRITHIOCARBODIGLYCOLIC
  ACID, **IV**, 967

TRIAZENES
  1-METHYL-3-*p*-TOLYLTRI-
    AZENE, **V**, 797

TROPYLIUM COMPOUNDS
  TROPYLIUM FLUOBORATE, **V**,
    1138

UREA DERIVATIVES
  ACETYL METHYLUREA, **II**, 462
  p-BROMOPHENYLUREA, **IV**, 49
  t-BUTYLUREA, **III**, 151
  1,1'-CARBONYLDIIMIDAZOLE,
    V, 201
  1-CYANO-3-*α*-NAPHTHYL-
    UREA, **IV**, 215
  1-CYANO-3-PHENYLUREA, **IV**,
    213
  CYCLOHEXYLUREA, **V**, 801
  DIAMINOBIURET, **III**, 404
  N,N-DIMETHYLSELENO-
    UREA, **IV**, 359

asym-DIMETHYLUREA, **IV**, 361
sym.-DIPHENYLUREA, **I**, 453
p-ETHOXYPHENYLUREA, **IV**, 52
N-*n*-HEPTYLUREA, **IV**, 515
3-*n*-HEPTYLUREIDOMETHYL-
  ENEMALONONITRILE, **IV**,
  515
HYDROXYUREA, **V**, 645
p-METHOXYPHENYLUREA, **IV**,
  53
METHYLISOUREA HYDRO-
  CHLORIDE, **IV**, 645
NITROUREA, **I**, 417
4-PHENYLSEMICARBAZIDE, **I**,
  450
PHENYLUREA, **I**, 453
SEMICARBAZIDE SULFATE, **I**,
  485

ZINC COMPOUNDS
  DI-*n*-BUTYL ZINC, **II**, 187
  DIETHYL ZINC, **II**, 184
  DIISOAMYL ZINC, **II**, 187
  DI-*n*-PROPYL ZINC, **II**, 187

UNCLASSIFIED
  ADIPYL HYDRAZIDE, **IV**, 819
  N-BENZYLIDENEMETHYL-
    AMINE, **IV**, 605
  N-*N*-DIMETHYLHYDROXYL-
    AMINE HYDROCHLOR-
    IDE, **IV**, 612
  METHYLGLYOXAL-*ω*-PHENYL-
    HYDRAZONE, **IV**, 633
  1-(*m*-NITROPHENYL)-3,3-DI-
    METHYLTRIAZENE, **IV**, 718
  PALLADIUM CATALYST FOR
    PARTIAL REDUCTION OF
    ACETYLENES, **V**, 880
  TETRALIN HYDROPEROXIDE,
    IV, 895

FORMULA INDEX

All preparations listed in the Contents are recorded in this index. The system of indexing is that used by *Chemical Abstracts*. The essential principles involved are as follows: (1) The arrangement of symbols in formulas is alphabetical except that in carbon compounds C always comes first, followed immediately by H if hydrogen is also present: (2) The arrangement of formulas is also alphabetical except that the number of atoms of any specific kind influences the order of compounds, for example, all formulas with one carbon atom precede those with two carbon atoms; thus, CH_2I_2, CH_3NO_2, CH_5N, C_2H_2O: (3) The arrangement of entries under any heading is strictly alphabetical according to the names of the isomers: (4) Inorganic salts of organic acids and inorganic addition compounds of organic compounds are listed under the formulas of the compounds from which they are derived.

The names in this index are the common Title Names of the preparations. To obtain the *Chemical Abstracts Index* Name, Index No. 1, the Cumulative Contents Index should be consulted.

CBrN CYANOGEN BROMIDE, II, 150
CClNO₂S CHLOROSULFONYL ISO-
    CYANATE, V, 226
CCl₂S THIOPHOSGENE, I, 509
CCl₄S THIOCARBONYL PERCHLORIDE,
    I, 506
CCl₅OP TRICHLOROMETHYLPHOS-
    PHONYL DICHLORIDE, IV, 950
CIN CYANOGEN IODIDE, IV, 207
CN₄O₈ TETRANITROMETHANE, III, 803
CHN HYDROGEN CYANIDE (ANHY-
    DROUS), I, 314
CH₂Br₂ METHYLENE BROMIDE, I, 357
CH₂Cl₃PS CHLOROMETHYLPHOS-

PHONOTHIOIC DICHLORIDE, V,
    218
CH₂I₂ METHYLENE IODIDE, I, 358
CH₂N₂, DIAZOMETHANE, II, 165; III,
    244; IV,250; V, 351
CH₃ClOS METHANESULFINYL CHLOR-
    IDE, V, 709
CH₃ClO₂S METHANESULFONYL CHLOR-
    IDE, IV, 571
CH₃I METHYL IODIDE, II, 399
CH₃NO₂ METHYL NITRITE, II, 363
    NITROMETHANE, I, 401
CH₃NO₃ METHYL NITRATE, II, 412
CH₃N₃O₃ NITROUREA, I, 417

CH$_4$N$_2$O$_2$ HYDROXYUREA, V, 645

CH$_4$N$_4$O$_2$ NITROGUANIDINE, I, 399

CH$_4$S METHYL MERCAPTAN, II, 345

CH$_5$N METHYLAMINE HYDROCHLO-
RIDE, I, 347

CH$_5$N$_3$ GUANIDINE NITRATE, I, 302

CH$_5$N$_3$O SEMICARBAZIDE SULFATE, I,
485

CH$_6$N$_2$ METHYLHYDRAZINE SULFATE,
II, 395

CH$_6$N$_4$ AMINOGUANIDINE BICARBON-
ATE, III, 73

C$_2$Cl$_2$F$_2$ 1,1-DICHLORO-2,2-DIFLUORO-
ETHYLENE, IV, 268

C$_2$HBr$_2$N DIBROMOACETONITRILE, IV,
254

C$_2$HBr$_3$O BROMAL, II, 87

C$_2$H$_2$ClN CHLOROACETONITRILE, IV,
144

C$_2$H$_2$Cl$_2$O$_2$ DICHLOROACETIC ACID, II,
181

C$_2$H$_2$O KETENE, I, 330; V, 679

C$_2$H$_2$O$_2$ GLYOXAL, III, 438

C$_2$H$_2$O$_4$ OXALIC ACID (ANHYDROUS), I,
421

C$_2$H$_3$BrO$_2$ BROMOACETIC ACID, III, 381

C$_2$H$_3$ClO$_2$S 2-CHLOROTHIIRANE 1,1-
DIOXIDE, V, 231

C$_2$H$_3$Cl$_2$NO α,α-DICHLOROACETAMIDE,
III, 260

C$_2$H$_3$Cl$_3$O TRICHLOROETHYL ALCO-
HOL, II, 598

C$_2$H$_3$N METHYL ISOCYANIDE, V, 772

C$_2$H$_3$NO GLYCOLONITRILE, III, 436

C$_2$H$_3$NS METHYL ISOTHIOCYANATE,
III, 599

C$_2$H$_3$N$_3$ 1,2,4-TRIAZOLE, V, 1070

C$_2$H$_4$BrNO N-BROMOACETAMIDE, IV,
104

C$_2$H$_4$ClNO CHLOROACETAMIDE, I, 153

C$_2$H$_4$Cl$_2$O BISCHLOROMETHYL ETHER,
IV, 101

2,2-DICHLOROETHANOL, IV, 271

DICHLOROMETHYL METHYL
ETHER, V, 365

C$_2$H$_4$N$_4$ 3-AMINO-1H-1,2,4-TRIAZOLE,

III, 95

4-AMINO-4H-1,2,4-TRIAZOLE, III,
96

C$_2$H$_4$OS THIOLACETIC ACID, IV, 928

C$_2$H$_4$S ETHYLENE SULFIDE, V, 562

C$_2$H$_5$AsO$_5$ ARSONOACETIC ACID, I, 73

C$_2$H$_5$Br ETHYL BROMIDE, I, 29, 36

C$_2$H$_5$BrO 2-BROMOETHANOL, I, 117

C$_2$H$_5$BrO$_3$S SODIUM 2-BROMOETHANE-
SULFONATE, II, 558

C$_2$H$_5$ClO MONOCHLOROMETHYL
ETHER, I, 377

C$_2$H$_5$I ETHYL IODIDE, II, 402

C$_2$H$_5$N ETHYLENIMINE, IV, 433

C$_2$H$_5$NO ACETAMIDE, I, 3

C$_2$H$_5$NO$_2$ GLYCINE, I, 298

C$_2$H$_5$NO$_3$ CARBOXYMETHOXYLAMINE
HEMIHYDROCHLORIDE, III, 172

C$_2$H$_5$NO$_3$ 2-NITROETHANOL, V, 833

C$_2$H$_5$N$_3$ 3(5)-AMINOPYRAZOLE, V, 39

C$_2$H$_5$N$_3$O$_2$ NITROSOMETHYLUREA, II,
461

C$_2$H$_6$BrN β-BROMOETHYLAMINE HY-
DROBROMIDE, II, 91

C$_2$H$_6$N$_2$ ACETAMIDINE HYDROCHLOR-
IDE, I, 5

C$_2$H$_6$N$_2$O METHYLISOUREA HYDRO-
CHLORIDE, IV, 645

C$_2$H$_6$N$_2$O NITROSODIMETHYLAMINE,
II, 211

C$_2$H$_6$N$_2$ S-METHYL ISOTHIOUREA SUL-
FATE, II, 411

METHYLTHIOUREA, III, 617

C$_2$H$_6$N$_4$S GUANYLTHIOUREA, IV, 502

C$_2$H$_6$S$_2$ ETHANEDITHIOL, IV, 401

C$_2$H$_7$NO N,N-DIMETHYLHYDROXY-
LAMINE HYDROCHLORIDE, IV,
612

C$_2$H$_7$NO$_3$S TAURINE, II, 563

C$_2$H$_7$N$_5$O$_2$ DIAMINOBIURET, III, 404

C$_2$H$_8$N$_2$ sym.-DIMETHYLHYDRAZINE
DIHYDROCHLORIDE, II, 208

unsym.-DIMETHYLHYDRAZINE, II,
213

unsym.-DIMETHYLHYDRAZINE HY-
DROCHLORIDE, II, 212

C₃HCl₇ *unsym.*-HEPTACHLOROPRO-
PANE, II, 312

C₃H₂ClF₃O₂ 3-CHLORO-2,2,3-TRIFLUOR-
OPROPIONIC ACID, V, 239

C₃H₂ClNO₂ α-CHLOROACETYL ISO-
CYANATE, V, 204

C₃H₂N₂ MALONONITRILE, II, 379; III,
535

C₃H₂N₂O₃ PARABANIC ACID, IV, 744

C₃H₂O PROPIOLALDEHYDE, IV, 813

C₃H₃NOS₂ RHODANINE, III, 763

C₃H₃NO₂ SODIUM CYANOACETATE, I,
181

C₃H₃NO₄ SODIUM NITROMALONALDE-
HYDE MONOHYDRATE, IV, 844

C₃H₄ ALLENE, V, 22

C₃H₄Br₂ 2,3-DIBROMOPROPENE, I, 209

C₃H₄Cl₂O α,γ-DICHLOROACETONE, I,
211

C₃H₄NOS PSEUDOTHIOHYDANTOIN,
III, 751

C₃H₄N₂ IMIDAZOLE, III, 471
METHYLENEAMINOACETONI-
TRILE, I, 355

C₃H₄N₂O CYANOACETAMIDE, I, 179

C₃H₄O ACROLEIN, I, 15

C₃H₄O₂ ACRYLIC ACID, III, 30

C₃H₄O₃ PYRUVIC ACID, I, 475

C₃H₄O₄ CALCIUM MALONATE, II, 376
MALONIC ACID, II, 376

C₃H₅Br ALLYL BROMIDE, I, 27
BROMOCYCLOPROPANE, V, 126

C₃H₅BrO BROMOACETONE, II, 88
EPIBROMOHYDRIN, II, 257

C₃H₅BrO₂ β-BROMOPROPIONIC ACID,
I, 131

C₃H₅Br₃ 1,2,3-TRIBROMOPROPANE, I,
521

C₃H₅ClO EPICHLOROHYDRIN, I, 233; II,
256

C₃H₅ClO₂ β-CHLOROPROPIONIC ACID,
I,166

C₃H₅N ETHYL ISOCYANIDE, IV, 438

C₃H₅NO ETHYLENE CYANOHYDRIN, I,
256
METHOXYACETONITRILE, II, 387

C₃H₆BrCl TRIMETHYLENE CHLORO-
BROMIDE, I, 156

C₃H₆BrN 2-BROMOALLYLAMINE, V, 121

C₃H₆Br₂ TRIMETHYLENE BROMIDE, I,
30

C₃H₆Br₂O GLYCEROL α,γ-DIBROMOHY-
DRIN, II, 308

C₃H₆Cl₂O GLYCEROL α,γ-DICHLORO-
HYDRIN, I, 292

C₃H₆N₂ β-AMINOPROPIONITRILE, III,
93

C₃H₆N₂S ETHYLENE THIOUREA, III, 394

C₃H₆O ALLYL ALCOHOL, I, 42
PROPIONALDEHYDE, II, 541
TRIMETHYLENE OXIDE, III, 835

C₃H₆O₂ ACETOL, II, 5

C₃H₆O₃ *dl*-GLYCERALDEHYDE, II, 305
β-HYDROXYPROPIONIC ACID, I,
321

C₃H₆S₃ *sym.*-TRITHIANE, II, 610

C₃H₇Br ISOPROPYL BROMIDE, I, 37; II,
359
n-PROPYL BROMIDE, I, 37; II, 359

C₃H₇Cl *n*-PROPYL CHLORIDE, I, 143

C₃H₇ClO TRIMETHYLENE CHLOROHY-
DRIN, I, 533

C₃H₇ClO₂ GLYCEROL α-MONOCHLO-
ROHYDRIN, I, 294

C₃H₇ClS β-CHLOROETHYL METHYL
SULFIDE, II, 136

C₃H₇I *n*-PROPYL IODIDE, II, 402

C₃H₇N ALLYLAMINE, II, 24

C₃H₇NO ACETOXIME, I, 318

C₃H₇NO₂ β-ALANINE, II, 19; III, 34
dl-ALANINE, I, 21
LACTAMIDE, III, 516

C₃H₇NO₃ *dl*-SERINE, III, 774

C₃H₇NO₅S CYSTEIC ACID MONOHY-
DRATE, III, 226

C₃H₇N₃O₂ GUANIDOACETIC ACID, III,
440

C₃H₈N₂O *asym.*-DIMETHYLUREA, IV, 361

C₃H₈N₂O₂ ETHYL HYDRAZINECARBOX-
YLATE, III, 404
FORMAMIDINE ACETATE, V, 582

C₃H₈N₂Se N,N-DIMETHYLSELENOUREA,
IV, 359

C_3H_8OS β-HYDROXYETHYL METHYL SULFIDE, II, 345

$C_3H_8O_2$ *l*-PROPYLENE GLYCOL, II, 545

$C_3H_9BF_4O$ TRIMETHYLOXONIUM FLUOBORATE, V, 1096

C_3H_9N *N*-METHYLETHYLAMINE, V, 758
TRIMETHYLAMINE, I, 528
TRIMETHYLAMINE HYDROCHLORIDE, I, 531

$C_4H_2Br_2O_3$ MUCOBROMIC ACID, III, 621; IV, 688

$C_4H_2ClF_2$ 1-CHLORO-1,4,4-TRIFLUOROBUTADIENE, V, 235

$C_4H_2F_4O_4$ 2,2-DIFLUOROSUCCINIC ACID, V, 393

$C_4H_2N_2$ FUMARONITRILE, IV, 486

$C_4H_2N_2O_4$ ALLOXAN MONOHYDRATE, III, 37; IV, 23

$C_4H_2O_4$ ACETYLENEDICARBOXYLIC ACID, II, 10

C_4H_3BrS 3-BROMOTHIOPHENE, V, 149

$C_4H_3ClN_2$ 2-CHLOROPYRIMIDINE, IV, 182

C_4H_3IS 2-IODOTHIOPHENE, II, 357; IV,545

$C_4H_3NO_2S$ 2-NITROTHIOPHENE, II, 466

$C_4H_3N_3O_5$ NITROBARBITURIC ACID, II, 440

C_4H_4 MONOVINYLACETYLENE, IV, 683

$C_4H_4Br_2O_4$ α,β-DIBROMOSUCCINIC ACID, II, 177

$C_4H_4INO_2$ *N*-IODOSUCCINIMIDE, V, 663

$C_4H_4N_2O_3$ BARBITURIC ACID, II, 60

$C_4H_4N_2S$ 2-MERCAPTOPYRIMIDINE, V, 703

$C_4H_4N_4$ DIAMINOMALEONITRILE, V, 344

C_4H_4O FURAN, I, 274

C_4H_4OS 2-HYDROXYTHIOPHENE, V, 642

$C_4H_4O_2$ γ-CROTONLACTONE, V, 255
KETENE DIMER, III, 508
TETROLIC ACID, V, 1043

$C_4H_4O_3$ SUCCINIC ANHYDRIDE, II, 560

$C_4H_4O_4$ FUMARIC ACID, II, 302

C_4H_4S THIOPHENE, II, 578

$C_4H_5ClO_2$ VINYL CHLOROACETATE, III, 853

$C_4H_5Cl_3O$ BUTYRCHLORAL, IV, 130

C_4H_5N ALLYL CYANIDE, I, 46; III, 852
CYCLOPROPYL CYANIDE, III, 223
PYRROLE, I, 473

$C_4H_5NO_2$ SUCCINIMIDE, II, 562

$C_4H_5N_3O_3$ URAMIL, II, 617

C_4H_6 1,3-BUTADIENE, II, 102

$C_4H_6As_2O_4$ ARSENOACETIC ACID, I, 74

$C_4H_6ClFO_2$ ETHYL CHLOROFLUOROACETATE, IV, 423

$C_4H_6ClF_3O$ 2-CHLORO-1,1,2-TRIFLUOROETHYL ETHYL ETHER, IV, 184

C_4H_6ClN γ-CHLOROBUTYRONITRILE, I, 156

$C_4H_6N_2$ 4(5)-HYDROXYMETHYLIMIDAZOLE HYDROCHLORIDE, III, 460

$C_4H_6N_2O$ DIMETHYLFURAZAN, IV, 342

$C_4H_6N_2O_2$ ETHYL DIAZOACETATE, III, 392; IV, 424

$C_4H_6N_2S$ 2-AMINO-4-METHYLTHIAZOLE, II, 31

$C_4H_6N_4O$ 2,4-DIAMINO-6-HYDROXYPYRIMIDINE, IV, 245

$C_4H_6N_4O_2$ DIAMINOURACIL HYDROCHLORIDE, IV, 247

$C_4H_6N_4O_3$ ALLANTOIN, II, 21

C_4H_6O 2-BUTYN-1-OL, IV, 128
DIMETHYLKETENE, IV, 348
ETHOXYACETYLENE, IV, 404

$C_4H_6O_2$ CYCLOPROPANECARBOXYLIC ACID, III, 221
VINYLACETIC ACID, III, 851

$C_4H_6O_3$ METHYL PYRUVATE, III, 610

$C_4H_6O_4$ METHYL OXALATE, I, 264; II, 414

$C_4H_6O_6$ *dl*-TARTARIC ACID, I, 497

$C_4H_7BrO_2$ ETHYL BROMOACETATE, III, 381
METHYL β-BROMOPROPIONATE, III, 576

C_4H_7ClO *n*-BUTYRYL CHLORIDE, I, 147
ISOBUTYRYL CHLORIDE, III, 490

$C_4H_7ClO_2$ ETHYL CHLOROACETATE, II, 263

C₄H₇N ISOBUTYRONITRILE, III, 493

C₄H₇NO ACETONE CYANOHYDRIN, II, 7

METHACRYLAMIDE, III, 560

C₄H₇NO₂ BIACETYL MONOXIME, II, 205

C₄H₇NO₃ ACETYLGLYCINE, II, 11

C₄H₇NO₄ dl-ASPARTIC ACID, IV, 55

C₄H₇NS ISOPROPYL THIOCYANATE, II, 366

C₄H₇N₃O CREATININE, I, 172

C₄H₈I₂ 1,4-DIIODOBUTANE, IV, 321

C₄H₈N₂O₂ ACETYL METHYLUREA, II, 462

DIMETHYL GLYOXIME, II, 204

SODIUM DIMETHYLGLYOXIMATE, II, 206

C₄H₈N₂O₃ NITROSOMETHYLURE-THANE, II, 464

N-NITROMORPHOLINE, V, 839

C₄H₈N₂O₄ 1,4-DINITROBUTANE, IV, 368

C₄H₈O TETRAHYDROFURAN, II, 566

C₄H₈O₂ 3-HYDROXYTETRAHYDRO-FURAN, IV, 534

C₄H₈O₃ ETHOXYACETIC ACID, II, 260

C₄H₈O₃S 4-HYDROXY-1-BUTANESUL-FONIC ACID SULTONE, IV, 529

C₄H₈S TETRAHYDROTHIOPHENE, IV, 892

C₄H₈S₂ p-DITHIANE, IV, 396

C₄H₉Br, n-BUTYL BROMIDE, I, 28, 37

sec-BUTYL BROMIDE, I, 38; II, 359

t-BUTYL BROMIDE, I, 38

ISOBUTYL BROMIDE, II, 358

C₄H₉BrO β-ETHOXYETHYL BROMIDE, III, 370

C₄H₉Cl n-BUTYL CHLORIDE, I, 142

sec.-BUTYL CHLORIDE, I, 143

t-BUTYL CHLORIDE, I, 144

C₄H₉ClMg UNSOLVATED n-BUTYLMAG-NESIUM CHLORIDE, V, 1141

C₄H₉ClO t-BUTYL HYPOCHLORITE, IV, 125; V, 184

TETRAMTHYLENE CHLOROHY-DRIN, II, 571

C₄H₉I n-BUTYL IODIDE, II, 402

ISOBUTYL IODIDE, II, 402

C₄H₉N CYCLOBUTYLAMINE, V, 273

2,2-DIMETHYLETHYLENIMINE, III, 148

C₄H₉NO ISOBUTYRAMIDE, III, 490

C₄H₉NO₂ γ-AMINOBUTYRIC ACID, II, 25

α-AMINOISOBUTYRIC ACID, II, 29

n-BUTYL NITRITE, II, 108

ETHYL N-METHYLCARBAMATE, II, 278

GLYCINE ETHYL ESTER HY-DROCHLORIDE, II, 310

C₄H₉NO₃ dl-THREONINE, III, 813

C₄H₁₀BrN β-DIMETHYLAMINOETHYL BROMIDE HYDROBROMIDE, II, 92

C₄H₁₀ClN β-DIMETHYLAMINOETHYL CHLORIDE HYDROCHLORIDE, IV, 333

C₄H₁₀CrN₇S₄ REINECKE SALT, II, 555

C₄H₁₀O₂S β-THIODIGLYCOL, II, 576

C₄H₁₀Zn DIETHYL ZINC, II, 184

C₄H₁₁ClN₄O AMINOACETONE SEMI-CARBAZONE HYDROCHLORIDE, V, 27

C₄H₁₁N n-BUTYLAMINE, II, 319

sec.-BUTYLAMINE, II, 319

t-BUTYLAMINE, III, 148

t-BUTYLAMINE HYDROCHLORIDE, III, 153

C₄H₁₂N₂ PUTRESCINE DIHYDROCHLOR-IDE, IV, 819

C₄H₁₂P₂S₂ TETRAMETHYLBIPHOSPHINE DISULFIDE, V, 1016

C₅H₃N₃OS 2-MERCAPTO-4-HYDROXY-5-CYANOPYRIMIDINE, IV, 566

C₅H₄Cl₂O₂ ITACONYL CHLORIDE, IV, 554

C₅H₄OS 2-THENALDEHYDE, IV, 915

3-THENALEHYDE, IV, 918

2-THIOPHENEALDEHYDE, III, 811

C₅H₄O₂ 2-CYCLOPENTENE-1,4-DIONE, V, 324

FURFURAL, I, 280

α-PYRONE, V, 982

C₅H₄O₂S 3-THENOIC ACID, IV, 919

C₅H₄O₃ CITRACONIC ANHYDRIDE, II, 140

2-FURANCARBOXYLIC ACID, I, 276
2-FUROIC ACID, IV, 493
ITACONIC ANHYDRIDE, II, 368
C₅H₄O₆ α-KETOGLUTARIC ACID, V, 687
C₅H₅BrS 3-THENYL BROMIDE, IV, 921
C₅H₅ClS 2-CHLOROMETHYLTHIO-
PHENE, III, 197
C₅H₅NO PYRIDINE-N-OXIDE, IV, 828
2-PYRROLEALDEHYDE, IV, 831
C₅H₅NO₃S 4-PYRIDINESULFONIC ACID,
V, 977
C₅H₆ CYCLOPENTADIENE, IV, 238
C₅H₆N₂ 3-AMINOPYRIDINE, IV, 45
3-HYDROXYGLUTARONITRILE, V,
614
4-METHYLPYRIMIDINE, V, 794
TRIMETHYLENE CYANIDE, I, 536
C₅H₆N₂O 4-METHYL-6-HYDROXYPYR-
IMIDINE, IV, 638
C₅H₆N₂O₂ 6-METHYLURACIL, II, 422
C₅H₆O 2-CYCLOPENTENONE, V, 326
3-METHYLFURAN, IV, 628
C₅H₆OS 2-FURFURYL MERCAPTAN, IV,
491
C₅H₆O₂ 2-FURYLCARBINOL, I, 276
C₅H₆O₄ CITRACONIC ACID, II, 141
ITACONIC ACID, II, 368
MESACONIC ACID, II, 382
C₅H₆O₄S₃ TRITHIOCARBODIGLYCOLIC
ACID, IV, 967
C₅H₆O₅ ACETONEDICARBOXYLIC
ACID, I, 10
α-KETOGLUTARIC ACID, III, 510
C₅H₆S 3-METHYLTHIOPHENE, IV, 671
C₅H₆S₂ METHYL 2-THIENYL SULFIDE,
IV, 667
C₅H₇Cl 3-CHLOROCYCLOPENTENE, IV,
238
C₅H₇ClO₃ β-CARBOMETHOXYPRO-
PIONYL CHLORIDE, III, 169
C₅H₇IN₂ 1-AMINOPYRIDINIUM IODIDE,
V, 43
C₅H₇NO₂ ETHYL CYANOACETATE, I,
254
GLUTARIMIDE, IV, 496

C₅H₇NS 2,4-DIMETHYLTHIAZOLE, III,
332
C₅H₇N₃ 2,3-DIAMINOPYRIDINE, V, 346
C₅H₈ BICYCLO(2.1.0)PENTANE, V, 96
1,4-PENTADIENE, IV, 746
C₅H₈Br₄ PENTAERYTHRITYL BROM-
IDE, II, 476; IV, 753
C₅H₈ClNO₃S β-ISOVALEROLACTAM-N-
SULFONYL CHLORIDE, V, 673
C₅H₈I₄ PENTAERYTHRITYL IODIDE, II,
477
C₅H₈N₂ 3,5-DIMETHYLPYRAZOLE, IV,
351
C₅H₈N₂O₂ 5,5-DIMETHYLHYDANTOIN,
III, 323
2,3-DIHYDROPYRAN, III, 276
C₅H₈O CYCLOPENTANONE, I, 192
DIMETHYLETHYNYLCARBINOL,
III, 320
METHYL CYCLOPROPYL KETONE,
IV, 597
4-PENTYN-1-OL, IV, 755
C₅H₈O₂ ACETYLACETONE, III, 16
CYCLOBUTANECARBOXYLIC
ACID, III, 213
DIHYDROXYCYCLOPENTENE, V,
414
β,β-DIMETHYLACRYLIC ACID, III,
302
C₅H₈O₂S ISOPRENE CYCLIC SULFONE,
III, 499
C₅H₈O₃ ETHYL PYRUVATE, IV, 467
LEVULINIC ACID, I, 335
C₅H₈O₄ GLUTARIC ACID, I, 289; IV, 496
METHYLSUCCINIC ACID, III, 615
C₅H₉BrO TETRAHYDROFURFURYL
BROMIDE, III, 793
C₅H₉BrO₂ α-BROMOISOVALERIC ACID,
II, 93; III, 848
ETHYL β-BROMOPROPIONATE, I,
246
C₅H₉ClO trans-2-CHLOROCYCLOPEN-
TANOL, IV, 157
C₅H₉N N-ETHYLALLENIMINE, V, 541
C₅H₉NO β-ETHOXYPROPIONITRILE,
III, 372

β-ISOVALEROLACTAM, V, 673
C₅H₉NOS β-(2-HYDROXYETHYLMER-
    CAPTO)PROPIONITRILE, III, 458
C₅H₉NO₄ d-GLUTAMIC ACID, I, 286
    METHYLIMINODIACETIC ACID, II,
    397
C₅H₉N₃ 4-AMINO-2,6-DIMETHYLPYRIM-
    IDINE, III, 71
C₅H₉N₃O₂ t-BUTYL AZIDOFORMATE, V,
    157
C₅H₁₀ 2-PENTENE, I, 430
C₅H₁₀BrN N-(2-BROMOALLYL)ETHYL-
    AMINE, V, 124
C₅H₁₀Br₂ PENTAMETHYLENE BROM-
    IDE, I, 428; III, 692
    TRIMETHYLETHYLENE DIBROM-
    IDE, II, 409
C₅H₁₀ClNS DIETHYLTHIOCARBAMYL
    CHLORIDE, IV, 307
C₅H₁₀O METHYL ISOPROPYL KETONE,
    II, 408
    3-PENTEN-2-OL, III, 696
    4-PENTEN-1-OL, III, 698
    TETRAHYDROPYRAN, III, 794
C₅H₁₀O₂ 5-HYDROXYPENTANAL, III, 470
    dl-METHYLETHYLACETIC ACID, I,
    361
    TRIMETHYLACETIC ACID, I, 524
    n-VALERIC ACID, I, 363
C₅H₁₀O₅ D-ARABINOSE, III, 101
    l-ARABINOSE, I, 67
C₅H₁₁Br iso-AMYL BROMIDE, I, 27
C₅H₁₁BrO₃ MONOBROMOPENTAERY-
    THRITOL, IV, 681
C₅H₁₁Cl₂NO N-CHLOROBETAINYL
    CHLORIDE, IV, 154
C₅H₁₁I n-AMYL IODIDE, II, 402
    ISOAMYL IODIDE, II, 402
C₅H₁₁NO₂ n-BUTYL CARBAMATE, I, 140
    t-BUTYL CARBAMATE, V, 162
    dl-VALINE, III, 848
C₅H₁₁NO₂S dl-METHIONINE, II, 384
C₅H₁₂ n-PENTANE, II, 478
C₅H₁₂CrN₉S₄ MORLAND SALT, II, 555
C₅H₁₂N₂O t-BUTYLUREA, III, 151
C₅H₁₂N₂O₂ t-BUTYL CARBAZATE, V, 166

C₅H₁₂O METHYL ISOPROPYL CARBI-
    NOL, II, 406
    NEOPENTYL ALCOHOL, V, 818
C₅H₁₂O₂ 1,5-PENTANEDIOL, III, 693
C₅H₁₂O₃ 1,2,5-TRIHYDROXYPENTANE,
    III, 833
C₅H₁₂O₄ PENTAERYTHRITOL, I, 425
C₅H₁₃N N-METHYLBUTYLAMINE, V, 736
C₅H₁₃NO 2-ISOPROPYLAMINO-
    ETHANOL, III, 501
C₅H₁₄ClN₃O BETAINE HYDRAZIDE HY-
    DROCHLORIDE, II, 85
C₆H₂Br₂ClNO 2,6-DIBROQUINONE-4-
    CHLOROIMIDE, II, 175
C₆H₂I₃NO₂ 1,2,3-TRIIODO-5-NITROBEN-
    ZENE, II, 604
C₆H₃Br₂NO₃ 2,6-DIBROMO-4-NITRO-
    PHENOL, II, 173
C₆H₃Br₃ sym-TRIBROMOBENZENE, II,
    592
C₆H₃ClNO₂S o-NITRO-p-CHLOROPHEN-
    YLSULFUR CHLORIDE, II, 456
C₆H₃ClN₂ 2-CHLORONICOTINONI-
    TRILE, IV, 166
C₆H₃ClN₂O₄ 1-CHLORO-2,6-DINITRO-
    BENZENE, IV, 160
C₆H₃ClN₂O₄S 2.4-DINTROBENZENESUL-
    FENYL CHLORIDE, V, 474
    2,4-DINITROPHENYLSULFUR
    CHLORIDE, II, 456
C₆H₃ClO₂ CHLORO-p-BENZOQUINONE,
    IV, 148
C₆H₃Cl₂NO₂ 2,6-DICHLORONITROBEN-
    ZENE, V, 367
C₆H₃IN₂O₄ 2,4-DINITROIODOBENZENE,
    V, 478
C₆H₃N₃O₆ 1,3,5-TRINITROBENZENE, I,
    541
C₆H₄BF₄N₃O₂ o-NITROBENZENEDIAZO-
    NIUM FLUOBORATE, II, 226
    p-NITROBENZENEDIAZONIUM
    FLUOBORATE, II, 225
C₆H₄BrCl o-CHLOROBROMOBENZENE,
    III, 185
C₆H₄BrF 1-BROMO-2-FLUOROBENZENE,
    V, 133

C_6H_4BrN 2-BROMOPYRIDINE, III, 136

$C_6H_4BrNO_2$ m-BROMONITROBENZENE, I, 123

$C_6H_4ClNO_2$ m-CHLORONITROBENZENE, I, 162

$C_6H_4ClNO_2S$ o-NITROPHENYLSULFUR CHLORIDE, II, 455

$C_6H_4ClNO_4S$ o-NITROBENZENESUL- FONYL CHLORIDE, II, 471

$C_6H_4Cl_2O$ 2,6-DICHLOROPHENOL, III, 267

$C_6H_4I_2N_2O_2$ 2,6-DIIODO-p-NITROANI- LINE, II, 196

$C_6H_4N_2$ NICOTINONITRILE, IV, 706

$C_6H_4N_2O_2$ BENZOFURAZAN OXIDE, IV, 74

DICYANOKETENE ETHYLENE ACETAL, IV, 276

$C_6H_4N_2O_2S$ 1,2,3-BENZOTHIADIAZOLE 1,1-DIOXIDE, V, 61

$C_6H_4N_2O_4$ o-DINITROBENZENE, II, 226

p-DINITROBENZENE, II, 225

2,3-PYRAZINEDICARBOXYLIC ACID, IV, 824

$C_6H_4N_4S$ 2,5-DIAMINO-3,4-DICYANO- THIOPHENE, IV, 243

$C_6H_4O_2$ QUINONE, I, 482; II, 553

$C_6H_4O_4$ COUMALIC ACID, IV, 201

$C_6H_4O_6$ TETRAHYDROXYQUINONE, V, 1011

C_6H_5BrO o-BROMOPHENOL, II, 97

p-BROMOPHENOL, I, 128

$C_6H_5BrO_2$ 4-BROMORESORCINOL, II, 100

$C_6H_5Br_2N$ 2,6-DIBROMOANILINE, III, 262

$C_6H_5Br_2NO$ 2,6-DIBROMO-4-AMINO- PHENOL CHLOROSTANNATE, II, 175

C_6H_5ClHgO o-CHLOROMERCURI- PHENOL, I, 161

$C_6H_5ClO_2S$ BENZENESULFONYL CHLO- RIDE, I, 84

$C_6H_5Cl_2I$ IODOBENZENE DICHLORIDE, III, 482

$C_6H_5Cl_2N$ 2,6-DICHLOROANILINE, III, 262

$C_6H_5Cl_2P$ PHENYLDICHLOROPHOS- PHINE, IV, 784

C_6H_5F FLUOROBENZENE, II, 295

C_6H_5FOS BENZYLSULFINYL FLUOR- IDE, V, 396

$C_6H_5F_3S$ PHENYLSULFUR TRIFLUOR- IDE, V, 959

C_6H_5I IODOBENZENE, I, 323; II, 351

C_6H_5IO o-IODOPHENOL, I, 326

p-IODOPHENOL, II, 355

IODOSOBENZENE, III, 483; V, 658

$C_6H_5IO_2$ IODOXYBENZENE, III, 485; V, 665

C_6H_5NO NITROSOBENZENE, III, 668

$C_6H_5NO_2$ NICOTINIC ACID, I, 385

PICOLINIC ACID HYDROCHLOR- IDE, III, 740

$C_6H_5NO_3$ 6-HYDROXYNICOTINIC ACID, IV, 532

m-NITROPHENOL, I, 404

$C_6H_5N_3$ 1,2,3-BENZOTRIAZOLE, III, 106

PHENYL AZIDE, III, 710

$C_6H_5N_3O_4$ 2,4-DINITROANILINE, II, 221

2,6-DINITROANILINE, IV, 364

C_6H_6 BENZENE, V, 998

$C_6H_6AsNO_5$ p-NITROPHENYLARSONIC ACID, III, 665

C_6H_6IN p-IODOANILINE, II, 347

$C_6H_6N_2O_2$ NICOTINAMIDE-1-OXIDE, IV, 704

o-NITROANILINE, I, 388

N-NITROSO-N-PHENYLHYDROXYL- AMINE, I, 177

$C_6H_6N_2O_3$ 2-AMINO-4-NITROPHENOL, III, 82

3-METHYL-4-NITROPYRIDINE-1- OXIDE, IV, 654

$C_6H_6N_4O_4$ 2,4-DINITROPHENYLHYDRA- ZINE, II, 228

C_6H_6OS 2-ACETOTHIENONE, III, 14; II, 8

$C_6H_6O_2$ CATECHOL, I, 149

5-METHYLFURFURAL, II, 393

$C_6H_6O_3$ 3-METHYL-2-FUROIC ACID, IV, 628

PHLOROGLUCINOL, I, 455

$C_6H_6O_4$ MUCONIC ACID, III, 623

DIMETHYL ACETYLENEDICAR-BOXYLATE, IV, 329

$C_6H_6O_6$ ACONITIC ACID, II, 12

HEXAHYDROXYBENZEBE, V, 595

C_6H_6S THIOPHENOL, I, 504

2-VINYLTHIOPHENE, IV, 980

$C_6H_6S_2$ 1,2-DIMERCAPTOBENZENE, V, 419

C_6H_6Se SELENOPHENOL, III, 771

$C_6H_7AsO_3$ PHENYLARSONIC ACID, II, 494

$C_6H_7AsO_4$ SODIUM p-HYDROXYPHENYL-ARSONATE, I, 490

C_6H_7NO 1-METHYL-2-PYRIDONE, II, 419

PHENYLHYDROXYLAMINE, I, 445

$C_6H_7NO_3S$ ORTHANILIC ACID, II, 471

$C_6H_7N_3O_2$ 1,2-DIAMINO-4-NITROBEN-ZENE, III, 242

C_6H_8 1,3-CYCLOHEXADIENE, V, 285

1,3,5-HEXATRIENE, V, 608

$C_6H_8AsNO_3$ ARSANILIC ACID, I, 70

$C_6H_8Cl_2O$ 2-OXA-7,7-DICHLORONOR-CARANE, V, 874

$C_6H_8N_2$ PHENYLHYDRAZINE, I, 442

o-PHENYLENEDIAMINE, II, 501

o-PHENYLENEDIAMINE DIHY-DROCHLORIDE, II, 502

$C_6H_8N_2O_2$ CROTYL DIAZOACETATE, V, 258

$C_6H_8N_4O_2$ 2,4,5-TRIAMINONITROBEN-ZENE, V, 1067

C_6H_8O 2-CYCLOHEXENONE, V, 294

$C_6H_8O_2$ 1,4-CYCLOHEXANEDIONE, V, 288

DIHYDRORESORCINOL, III, 278

2-METHYLCYCLOPENTANE-1,3-DIONE, V, 747

SORBIC ACID, III, 783

$C_6H_8O_3$ β-METHYLGLUTARIC ANHY-DRIDE, IV, 630

$C_6H_8O_4$ 1,1-CYCLOBUTANEDICARBOX-

YLIC ACID, III, 213

$C_6H_8O_6$ TRICARBALLYLIC ACID, I, 523

C_6H_9ClO 2-CHLOROCYCLOHEXANONE, III, 188

C_6H_9N 2,4-DIMETHYLPYRROLE, II, 217

2,5-DIMETHYLPYRROLE, II, 219

$C_6H_9N_3$ bis-(β-CYANOETHYL)AMINE, III, 93

2-(DIMETHYLAMINO)PYRIMIDINE, IV, 336

$C_6H_9N_3O_2$ l-HISTIDINE MONOHYDRO-CHLORIDE, II, 330

C_6H_{10} n-BUTYLACETYLENE, IV, 117

BIALLYL, III, 121

CYCLOHEXENE, I, 183; II, 152

2,3-DIMETHYL-1,3-BUTADIENE, III, 312

$C_6H_{10}Br_2$ 1,2-DIBROMOCYCLOHEXANE, II, 171

$C_6H_{10}ClN$ N-CHLOROCYCLOHEXYLI-DENEIMINE, V, 208

$C_6H_{10}Cl_3N$ N,N-DIETHYL-1,2,2-TRI-CHLOROVINYLAMINE, V, 387

$C_6H_{10}N_2$ 3,3-PENTAMETHYLENEDIAZIR-INE, V, 897

$C_6H_{10}N_2O_2$ t-BUTYL DIAZOACETATE, V, 179

1,2-CYCLOHEXANEDION DIOXIME, IV, 229

$C_6H_{10}N_2O_4$ 3,4-DINITRO-3-HEXENE, IV, 372

ETHYL AZODICARBOXYLATE, III, 375; IV, 411

$C_6H_{10}O$ CYCLOHEXENE OXIDE, I, 185

CYCLOPENTANECARBOXALDE-HYDE, V, 320

MESITYL OXIDE, I, 345

$C_6H_{10}O_2$ ACETONYLACETONE, II, 219

3-METHYLPENTANE-2,4-DIONE, V, 785

β-METHYL-δ-VALEROLACTONE, IV, 677

$C_6H_{10}O_3$ ALLYL LACTATE, III, 46

n-BUTYL GLYOXYLATE, IV, 124

2,5-DIHYDRO-2,5-DIMETHOXY-

FURAN, V, 403
ETHYL ACETOACETATE, I, 235
$C_6H_{10}O_4$ ADIPIC ACID, I, 18
ETHYL OXALATE, I, 261
β-METHYLGLUTARIC ACID, III,
591
$C_6H_{10}O_6$ D-GULONIC-γ-LACTONE, IV,
506
$C_6H_{10}S$ CYCLOHEXENE SULFIDE, IV,
232
$C_6H_{11}Br$ CYCLOHEXYL BROMIDE, II,
247
$C_6H_{11}BrO_2$ α-BROMO-n-CAPROIC ACID,
I, 115; II, 95
α-BROMOISOCAPROIC ACID, II,
95; III, 523
α-BROMO-β-METHYLVALERIC
ACID, II, 95
ETHYL γ-BROMOBUTYRATE, V,
545
METHYL 5-BROMOVALERATE, III,
578
$C_6H_{11}ClO$ 2-CHLOROCYCLOHEXANOL,
I, 158
$C_6H_{11}I$ IODOCYCLOHEXANE, IV, 543
$C_6H_{11}N$ DIALLYLAMINE, I, 201
$C_6H_{11}NO$ CYCLOHEXANONE OXIME, II,
76, 314
1,5-DIMETHYL-2-PYRROLIDONE,
III, 328
5,5-DIMETHYL-2-PYRROLIDONE,
IV, 357
2-KETOHEXAMETHYLENEIMINE,
II, 371
$C_6H_{11}NO_2$ 3-ACETAMIDO-2-BUTANONE,
IV, 5
$C_6H_{11}NO_4$ ETHYL α-NITROBUTYRATE,
IV, 454
$C_6H_{11}NS$ 2-METHYLMERCAPTO-N-
METHYL-$Δ^2$-PYRROLINE, V, 780
$C_6H_{12}I_2$ 1,6-DIIODOHEXANE, IV, 323
$C_6H_{12}N_2$ DIETHYLAMINOACETONI-
TRILE, III, 275
$C_6H_{12}N_2O_4$ ETHYL HYDRAZODICAR-
BOXYLATE, III, 375
$C_6H_{12}N_2O_4S_2$ l-CYSTINE, I, 194

$C_6H_{12}O$ n-HEXALDEHYDE, II, 323
METHYL iso-BUTYL KETONE, I,
352
METHYL n-BUTYL KETONE, I, 352
PINACOLONE, I, 462
$C_6H_{12}O_2$ t-BUTYL ACETATE, III, 141
n-CAPROIC ACID, II, 417, 475
trans-1,2-CYCLOHEXANEDIOL, III,
217
DIACETONE ALCOHOL, I, 199
KETENE DIETHYLACETAL, III, 506
3-METHYLPENTANOIC ACID, II,
416
$C_6H_{12}O_3$ ETHYL ETHOXYACETATE, II,
261
dl-ISOPROPYLIDENEGLYCEROL,
III, 502
ISOPROPYL LACTATE, II, 365
$C_6H_{12}O_6$ D-MANNOSE, III, 541
$C_6H_{13}BrO_2$ BROMOACETAL, III, 123
$C_6H_{13}ClO$ HEXAMETHYLENE CHLO-
ROHYDRIN, III, 446
$C_6H_{13}F$ n-HEXYL FLUORIDE, IV, 525
$C_6H_{13}N$ CYCLOHEXYLAMINE, II, 319
2,2-DIMETHYLPYRROLIDINE, IV,
354
$C_6H_{13}NO_2$ α-AMINO-n-CAPROIC ACID, I,
48
ε-AMINOCAPROIC ACID, II, 28; IV,
39
α-AMINODIETHYLACETIC ACID,
III, 66
GLYCINE t-BUTYL ESTER, V, 586
dl-ISOLEUCINE, III, 495
dl-LEUCINE, III, 523
$C_6H_{13}NO_5$ d-GLUCOSAMINE HYDRO-
CHLORIDE, III, 430
$C_6H_{14}BrN$ β-DIETHYLAMINOETHYL
BROMIDE HYDROBROMIDE, II,
92
$C_6H_{14}ClP$ CHLORODIISOPROPYLPHOS-
PHINE, V, 211
$C_6H_{14}N_2O_2$ dl-LYSINE DIHYDRO-
CHLORIDE, II, 375
dl-LYSINE MONOHYDRO-
CHLORIDE, II, 375

$C_6H_{14}N_4O_2$ *d*-ARGININE HYDRO-
CHLORIDE, II, 49
$C_6H_{14}O$ *n*-HEXYL ALCOHOL, I, 306
$C_6H_{14}O_2$ ACETAL, I, 1
HEXAMETHYLENE GLYCOL, II,
325
3-METHYL-1,5-PENTANEDIOL, IV,
660
PINACOL HYDRATE, I, 459
$C_6H_{14}O_2S$ DIETHYL MERCAPTOACE-
TAL, IV, 295
$C_6H_{14}O_4S$ *n*-PROPYL SULFATE, II, 112
$C_6H_{14}S$ *n*-PROPYL SULFIDE, II, 547
$C_6H_{14}Zn$ DI-*n*-PROPYL ZINC, II, 187
$C_6H_{15}BF_4O$ TRIETHYLOXONIUM FLUO-
BORATE, V, 1080
$C_6H_{15}NO$ β-DIETHYLAMINOETHYL AL-
COHOL, II, 183
$C_6H_{15}NO_2$ AMINOACETAL, III, 50
$C_6H_{15}O_3P$ TRIETHYL PHOSPHITE, IV,
955
$C_6H_{18}N_3P$ HEXAMETHYLPHOSPHOROUS
TRIAMIDE, V, 602
C_6N_4 TETRACYANOETHYLENE, IV, 877
C_6N_4O TETRACYANOETHYLENE OX-
IDE, V, 1007
$C_7HCl_5O_2$ PENTACHLOROBENZOIC
ACID, V, 890
$C_7H_3Br_3O_2$ 2,4,6-TRIBROMOBENZOIC
ACID, IV, 947
$C_7H_3HgNO_4$ ANHYDRO-2-HYDROXY-
MERCURI-3-NITROBENZOIC
ACID, I, 56
$C_7H_3N_3O_8$ 2,4,6-TRINITROBENZOIC
ACID, I, 543
$C_7H_4BrNO_4$ 2-BROMO-3-NITROBENZOIC
ACID, I, 125
C_7H_4BrNS *m*-BROMOPHENYL ISOTHIO-
CYANATE, I, 448
p-BROMOPHENYL ISOTHIOCYAN-
ATE, I, 448
$C_7H_4ClNO_3$ *p*-NITROBENZOYL CHLO-
RIDE, I, 394
C_7H_4ClNS *p*-CHLOROPHENYL ISOTHIO-
CYANATE, I, 165; V, 223
$C_7H_4Cl_2O$ *o*-CHLOROBENZOYL CHLOR-

IDE, I, 155
$C_7H_4HgO_2$ ANDYDRO-*o*-HYDROXYMER-
CURIBENZOIC ACID, I, 57
$C_7H_4INO_4$ 2-IODO-3-NITROBENZOIC
ACID, I, 126
C_7H_4INS *p*-IODOPHENYL ISOTHIO-
CYANATE, I, 448
$C_7H_4I_2O_3$ 2-HYDROXY-3,5-DIIODOBEN-
ZOIC ACID, II, 343
4-HYDROXY-3,5-DIIODOBENZOIC
ACID, II, 344
$C_7H_4N_2O_2$ BENZENEDIAZONIUM-2-CAR-
BOXYLATE, V, 54
p-NITROBENZONITRILE, III, 646
$C_7H_4N_2O_3$ *p*-NITROPHENYL ISOCYA-
NATE, II, 453
$C_7H_4N_2O_5$ 2,4-DINITROBENZALDEHYDE,
II, 223
$C_7H_4N_2O_6$ 2,5-DINITROBENZOIC ACID,
III, 334
3,5-DINITROBENZOIC ACID, III,
337
$C_7H_4N_4O_3$ *m*-NITROBENZAZIDE, IV, 715
$C_7H_4O_3$ *o*-PHENYLENE CARBONATE, IV,
788
$C_7H_4O_4S$ *o*-SULFOBENZOIC ANHY-
DRIDE, I, 495
$C_7H_4O_6$ CHELIDONC ACID, II, 126
C_7H_5BrO *m*-BROMOBENZALDEHYDE,
II, 132
p-BROMOBENZALDEHYDE, II, 89,
442
$C_7H_5BrO_4$ 2,4-DIHYDROXY-5-BROMO-
BENZOIC ACID, II, 100
$C_7H_5ClHgO_2$ *p*-CHLOROMERCURIBEN-
ZOIC ACID, I, 159
$C_7H_5ClN_2$ *o*-CHLOROPHENYLCYANA-
MIDE, IV, 172
3-CHLOROINDAZOLE, III, 476
C_7H_5ClO *m*-CHLOROBENZALEHYDE, II,
130
p-CHLOROBENZALDEHYDE, II, 133
$C_7H_5ClO_2$ *o*-CHLOROBENZOIC ACID, II,
135
$C_7H_5Cl_3Hg$ PHENYL(TRICHLORO-
METHYL)MERCURY, V, 969

C$_7$H$_5$Cl$_5$ 1,2,3,4,5-PENTACHLORO-5-ETHYLCYCLOPENTADIENE, V, 893

C$_7$H$_5$FO BENZOYL FLUORIDE, V, 66

C$_7$H$_5$FO$_2$ p-FLUOROBENZOIC ACID, II, 299

C$_7$H$_5$IO$_2$ p-IODOBENZOIC ACID, I, 325
m-IODOBENZOIC ACID, II, 353

C$_7$H$_5$NO 3-(2-FURYL)ACRYLONITRILE, V, 585

C$_7$H$_5$NO$_3$ o-NITROBENZALDEHYDE, III, 641; V, 825
p-NITROBENZALDEHYDE, II, 441

C$_7$H$_5$NO$_4$ m-NITROBENZOIC ACID, I, 391
p-NITROBENZOIC ACID, I, 392

C$_7$H$_5$NS PHENYL ISOTHIOCYANATE, I, 447

C$_7$H$_5$N$_3$O$_2$ 5-NITROINDAZOLE, III, 660

C$_7$H$_6$BrNO$_2$ p-NITROBENZYL BROMIDE, II, 443

C$_7$H$_6$ClNO$_3$ 2-HYDROXY-5-NITROBENZYL CHLORIDE, III, 468

C$_7$H$_6$Cl$_2$O p-CHLOROPHENOXYMETHYL CHLORIDE, V, 221

C$_7$H$_6$F$_2$ α,α-DIFLUOROTOLUENE, V, 396

C$_7$H$_6$INO$_2$ 5-IODOANTHRANILIC ACID, II, 349

C$_7$H$_6$N$_2$ BENZIMIDAZOLE, II, 65
2-CYANO-6-METHYLPYRIDINE, V, 269
INDAZOLE, III, 475; IV, 536; V, 650

C$_7$H$_6$N$_2$O 3-CYANO-6-METHYL-2(1)-PYRIDONE, IV, 210
INDAZOLONE, III, 476

C$_7$H$_6$N$_2$O$_5$ 3,5-DINITROANISOLE, I, 219

C$_7$H$_6$N$_2$S 2-MERCAPTOBENZIMIDAZOLE, IV, 569

C$_7$H$_6$N$_4$O 1,1-CARBONYLDIIMIDAZOLE, V, 201

C$_7$H$_6$OS THIOBENZOIC ACID, IV, 924

C$_7$H$_6$O$_2$ BENZOIC ACID, I, 363
m-HYDROXYBENZALDEHYDE, III, 453, 564

C$_7$H$_6$O$_2$S THIOSALICYLIC ACID, II, 580

C$_7$H$_6$O$_3$ FURYLACRYLIC ACID, III, 425
p-HYDROXYBENZOIC ACID, II, 341

PERBENZOIC ACID, I, 431
PEROXYBENZOIC ACID, V, 904
PROTOCATECHUALDEHYDE, II, 549

C$_7$H$_6$O$_4$ 3,5-DIHYDROXYBENZOIC ACID, III, 288
PROTOCATECHUIC ACID, III, 745
β-RESORCYLIC ACID, II, 557

C$_7$H$_6$O$_5$S ACID AMMONIUM o-SULFOBENZOATE, I, 14

C$_7$H$_7$BF$_4$ TROPYLIUM FLUOBORATE, V, 1138

C$_7$H$_7$Br m-BROMOTOLUENE, I, 133
o-BROMOTOLUENE, I, 135
p-BROMOTOLUENE, I, 136

C$_7$H$_7$BrN$_2$O p-BROMOPHENYLUREA, IV, 49

C$_7$H$_7$BrO 3-BROMO-4-HYDROXYTOLUENE, III, 130

C$_7$H$_7$Cl o-CHLOROTOLUENE, I, 170
p-CHLOROTOLUENE, I, 170

C$_7$H$_7$ClHg p-TOLYMERCURIC CHLORIDE, I, 519

C$_7$H$_7$ClN$_2$S o-CHLOROPHENYLTHIOUREA, IV, 180

C$_7$H$_7$ClOS p-TOLUENESULFINYL CHLORIDE, IV, 937

C$_7$H$_7$ClS p-TOLUENESULFENYL CHLORIDE, IV, 934

C$_7$H$_7$NO o-AMINOBENZALDEHYDE, III, 56
p-AMINOBENZALDEHYDE, IV, 31

C$_7$H$_7$NO$_2$ BENZOHYDROXAMIC ACID, II, 67
m-NITROTOLUENE, I, 415
PHENYLNITROMETHANE, II, 512
POTASSIUM BENZOHYDROXAMATE, II, 67

C$_7$H$_7$NO$_3$ p-NITROBENZYL ALCOHOL, III, 652

C$_7$H$_7$NO$_4$ 3-METHOXY-5-NITROPHENOL, I, 405

C$_7$H$_8$BrN 3-BROMO-4-AMINOTOLUENE, I, 111

C$_7$H$_8$N$_2$ BENZAMIDINE HYDROCHLORIDE, I, 6

C₇H₈N₂O N-NITROSOMETHYLANILINE,
II, 460
PHENYLUREA, I, 453
C₇H₈N₂O₂ 2-AMINO-3-NITROTOLUENE,
IV, 42
o-HYDRAZINOBENZOIC ACID HY-
DROCHLORIDE, III, 475
C₇H₈N₂O₃ 2-NITRO-4-METHOXYANIL-
INE, III, 661
C₇H₈N₂S α-PHENYLTHIOUREA, III, 735
C₇H₈O ANISOLE, I, 58
BENZYL ALCOHOL, II, 591
BICYCLO[2.2.1]HEPTEN-7-ONE, V,
91
p-CRESOL, I, 175
NORTRICYCLANONE, V, 866
C₇H₈OS METHYL PHENYL SULFOXIDE,
V, 791
C₇H₈O₂ 1,4-DIHYDROBENZOIC ACID, V,
400
4,6-DIMETHYLCOUMALIN, IV, 337
C₇H₈O₂S METHYL BENZENESULFIN-
ATE, V, 723
SODIUM p-TOLUENESULFINATE, I,
492
C₇H₈O₃ 2-FURYLMETHYL ACETATE, I,
285
β-FURYLPROPIONIC ACID, I, 313
METHYL 3-METHYL-2-FUROATE,
IV, 649
PYROGALLOL 1-MONOMETHYL
ETHER, III, 759
C₇H₈S m-THIOCRESOL, III, 809
C₇H₉ClO 2-CHLORO-1-FORMYL-1-CY-
CLOHEXENE, V, 215
C₇H₉ClO₂ γ-CHLOROPROPYL ACET-
ATE, III, 203
C₇H₉ClO₃ α-ACETYL-δ-CHLORO-γ-VAL-
EROLACTONE, IV, 10
C₇H₉N 2,6-DIMETHYLPYRIDINE, II, 214
4-ETHYLPYRIDINE, III, 410
C₇H₉NO o-AMINOBENZYL ALCOHOL,
III, 60
C₇H₉NO₃S o-TOLUIDINESULFONIC
ACID, III, 824
C₇H₉N₃O 4-PHENYLSEMICARBAZIDE, I,
450
C₇H₉N₃O₂S 2-MERCAPTO-4-AMINO-5-
CARBETHOXYPYRIMIDINE, IV,
566
C₇H₁₀ NORBORNYLENE, IV, 738
C₇H₁₀N₂ DIALLYLCYANAMIDE, I, 203
2,4-DIAMINOTOLUENE, II, 160
2,4-DIAMINOTOLUENE SULFATE,
II, 160
α-METHYL-α-PHENYLHYDRAZINE,
II, 418·
C₇H₁₀N₂O₂S p-TOLUENESULFONYLHY-
DRAZIDE, V, 1055
C₇H₁₀O DICYCLOPROPYL KETONE, IV,
278
2-METHYL-2-CYCLOHEXENONE,
IV, 162
2-NORBORNANONE, V, 852
NORTRICYCLANOL, V, 863
C₇H₁₀O₂ 2-METHYL-1,3-CYCLOHEXANE-
DIONE, V, 743
C₇H₁₀O₄ ETHYL ACETOPYRUVATE, I,
238
C₇H₁₀O₅ ETHYL OXOMALONATE, I, 266
C₇H₁₀O₆ TRICARBOMETHOXYMETH-
ANE, II, 596
C₇H₁₁BrO₄ ETHYL BROMOMALONATE,
I, 245
C₇H₁₁ClO 2-CHLORO-2-METHYLCY-
CLOHEXANONE, IV, 162
C₇H₁₁N CYCLOHEXYL ISOCYANIDE, V,
300
C₇H₁₁NO 3-ISOQUINUCLIDONE, V, 670
C₇H₁₁NO₂ t-BUTYL CYANOACETATE, V,
171
C₇H₁₂ METHYLENECYCLOHEXANE, IV,
612; V, 751
NORCARANE, V, 855
C₇H₁₂ClO 3-QUINUCLIDONE HYDRO-
CHLORIDE, V, 989
C₇H₁₂O CYCLOHEPTANONE, IV, 221
exo/endo-7-NORCARANOL, V, 859
METHYLENECYCLOHEXANE OX-
IDE, V, 755
5-METHYL-5-HEXEN-2-ONE, V, 767
C₇H₁₂O₂ n-BUTYL ACRYLATE, III, 146

CYCLOHEXANECARBOXYLIC ACID, I, 364

3,4-DIHYDRO-2-METHOXY-4-METHYL-2H-PYRAN, IV, 311

METHYL CYCLOPENTANECARBOXYLATE, IV, 594

γ-n-PROPYLBUTYROLACTONE, III, 742

$C_7H_{12}O_3$ δ-ACETYL-n-VALERIC ACID, IV, 19

β-(TETRAHYDRO-FURYL)PROPIONIC ACID, III, 742

$C_7H_{12}O_4$ β,β-DIMETHYLGLUTARIC ACID, IV, 345

α-ETHYL-α-METHYLSUCCINIC ACID, V, 572

PIMELIC ACID, II, 531

$C_7H_{12}O_5$ HEXAHYDROGALLIC ACID, V, 591

$C_7H_{13}Br$ 4-BROMO-2-HEPTENE, IV, 108

$C_7H_{13}BrO$ α-BROMOHEPTALDEHYDE, III, 127

$C_7H_{13}F_3$ 1,1,1-TRIFLUOROHEPTANE, V, 1082

$C_7H_{13}FO_2$ 2-FLUOROHEPTANOIC ACID, V, 580

$C_7H_{13}NO$ O-METHYLCAPROLACTIM, IV, 588

$C_7H_{13}NO_4$ METHYL γ-METHYL-γ-NITROVALERATE, IV, 652

$C_7H_{14}BrF$ 1-BROMO-2-FLUOROHEPTANE, V, 136

$C_7H_{14}ClNO_4$ DIETHYL AMINOMALONATE HYDROCHLORIDE, V, 376

$C_7H_{14}N_2O$ CYCLOHEXYLUREA, V, 801

$C_7H_{14}O$ CYCLOHEXYLCARBINOL, I, 188

4-HEPTANONE, V, 589

METHYL n-AMYL KETONE, I, 351

4-METHYLHEXANONE-2, I, 352

$C_7H_{14}O_2$ ACROLEIN ACETAL, II, 17; IV, 21

n-HEPTANOIC ACID, II, 315

$C_7H_{14}O_6$ α-METHYL d-GLUCOSIDE, I, 364

α-METHYL MANNOSIDE, I, 371

$C_7H_{15}Br$ n-HEPTYL BROMIDE, II, 247

$C_7H_{15}ClO_2$ β-CHLOROPROPIONALDEHYDE ACETAL, II, 137

$C_7H_{15}N$ 1-AMINO-1-METHYLCYCLOHEXANE, V, 35

N-ETHYLPIPERIDINE, V, 575

$C_7H_{15}NO$ HEPTALDOXIME, II, 313

n-HEPTAMIDE, IV, 513

2-(1-PYRROLIDYL)PROPANOL, IV, 834

$C_7H_{16}O$ 2-HEPTANOL, II, 317

n-HEPTYL ALCOHOL, I, 304

TRIETHYL CARBINOL, II, 602

$C_7H_{16}O_2$ HEPTAMETHYLENE GLYCOL, II, 155

$C_7H_{16}O_3$ ETHYL ORTHOFORMATE, I, 258

$C_7H_{16}O_4$ dl-GLYCERALDEHYDE DIETHYL ACETAL, II, 307

$C_7H_{17}N$ n-HEPTYLAMINE, II, 318

$C_7H_{17}O_3P$ DIISOPROPYL METHYLPHOSPHONATE, IV, 325

$C_8H_3NO_5$ 3-NITROPHTHALIC ANHYDRIDE, I, 410

$C_8H_4Cl_2O_2$ sym. AND unsym. o-PHTHALYL CHLORIDE, II, 528

$C_8H_4F_3NS$ m-(TRIFLUOROMETHYL)PHENYL ISOTHIOCYANATE, I, 448

$C_8H_4N_2O_4$ 4-NITROPHTHALIMIDE, II, 459

$C_8I_4O_3$ TETRAIODOPHTHALIC ANHYDRIDE, III, 796

C_8H_5Br PHENYLBROMOETHYNE, V, 921

$C_8H_5BrO_2$ 3-BROMOPHTHALIDE, V, 145

C_8H_5NO BENZOYL CYANIDE, III, 112

$C_8H_5NO_2$ ISATIN, I, 327

PHTHALIMIDE, I, 457

$C_8H_5NO_3$ ISATOIC ANHYDRIDE, III, 488

$C_8H_5N_3O_4$ 5-NITRO-2,3-DIHYDRO-1,4-PHTHALAZINEDIONE, III, 656

$C_8H_5NO_6$ 3-NITROPHTHALIC ACID, I, 408

4-NITROPHTHALIC ACID, II, 457

$C_8H_5N_3O_4$ 3,5-DINITRO-o-TOLUNITRILE, V, 480

C_8H_6 PHENYLACETYLENE, I, 438; IV, 763

C₈H₆Br₂O *p*-BROMOPHENACYL BROMIDE, I, 127

C₈H₆ClNO₂ ω-CHLOROISONITROSO-ACETOPHENONE, III, 191

C₈H₆Cl₂O DICHLOROACETOPHENONE, III, 538

C₈H₆Cl₃NO α,α,α- TRICHLOROACET-ANILIDE, V, 1074

C₈H₆F₂ β,β-DIFLUOROSTYRENE, V, 390

C₈H₆N₂O₂ BENZOYLENE UREA, II, 79
  p-NITROBENZYL CYANIDE, I, 396
  SODIUM PHENYL-*aci*-NITROACE-TONITRILE, II, 512

C₈H₆N₂O₃ 3-PHENYLSYDNONE, V, 962

C₈H₆N₄O₈ ALLOXANTIN DIHYDRATE, III, 42; IV, 25

C₈H₆O COUMARONE, V, 251

C₈H₆O₂ ISOPHTHALALDEHYDE, V, 668
  PHENYLGLYOXAL, II, 509; V, 937
  o-PHTHALALDEHYDE, IV, 807
  PHTHALIDE, II, 526
  TEREPHTHALALDEHYDE, III, 788

C₈H₆O₃ BENZOYLFORMIC ACID, I, 244; III, 114
  PHTHALALDEHYDIC ACID, II, 513; III, 737

C₈H₆O₄ 2-HYDROXYISOPHTHALIC ACID, V, 617
  PIPERONYLIC ACID, II, 538
  TEREPHTHALIC ACID, III, 791

C₈H₆O₅ MONOPERPHTHALIC ACID, III, 619; V, 805

C₈H₇BrO 3-BROMOACETOPHENONE, V, 117
  p-BROMOACETOPHENONE, I, 109
  2-BROMO-4-METHYLBENZALDE-HYDE, V, 139
  PHENACYL BROMIDE, II, 480

C₈H₇BrO₂ 2-BROMO-3-METHYLBEN-ZOIC ACID, IV, 114

C₈H₇BrO₃ *p*-BROMOMANDELIC ACID, IV, 110

C₈H₇Cl *m*-CHLOROSTYRENE, III, 204

C₈H₇ClO *p*-CHLOROACETOPHENONE, I, 111

C₈H₇ClO₂ CARBOBENZOXY CHLORIDE, III, 167
  α-CHLOROPHENYLACETIC ACID, IV, 169

C₈H₇ClO₂S β-STYRENESULFONYL CHLORIDE, IV, 846

C₈H₇Cl₂OP STYRYLPHOSPHONIC DI-CHLORIDE, V, 1005

C₈H₇N BENZYL CYANIDE, I, 107
  INDOLE, III, 479
  o-TOLUNITRILE, I, 514
  p-TOLUNITRILE, I, 514
  o-TOLYL ISOCYANIDE, V, 1060

C₈H₇NO₂ NITROSTYRENE, I, 413
  m-NITROSTYRENE, IV, 731

C₈H₇NO₃ *o*-NITROACETOPHENONE, IV, 708
  m-NITROACETOPHENONE, II, 434

C₈H₇NO₄ METHYL *m*-NITROBENZOATE, I, 372
  p-NITROPHENYLACETIC ACID, I, 406

C₈H₇N₃O 1-CYANO-3-PHENYLUREA, IV, 213

C₈H₇N₃O₂ 5-AMINO-2,3-DIHYDRO-1,4-PHTHALAZINEDIONE, III, 69

C₈H₈ PHENYLETHYLENE, I, 440

C₈H₈Br₂ *o*-XYLYLENE DIBROMIDE, IV, 984

C₈H₈ClNO₃S *p*-ACETAMINOBENZENE-SULFONYL CHLORIDE, I, 8

C₈H₈N₂S 2-AMINO-6-METHYLBENZO-THIAZOLE, III, 76

C₈H₈N₂ 2-METHYLBENZIMIDAZOLE, II, 66

C₈H₈N₂O₂ ISONITROSOACETANILIDE, I, 327

C₈H₈O ACETOPHENONE, I, 111
  STYRENE OXIDE, I, 494
  o-TOLUALDEHYDE, III, 818; IV, 932
  p-TOLUALDEHYDE, II, 583

C₈H₈O₂ *o*-ANISALDEHYDE, V, 46
  2-FURFURALACETONE, I, 283
  m-METHOXYBENZALDEHYDE, III, 564
  PHENYLACETIC ACID, I, 436

o-TOLUIC ACID, II, 588; III, 820
p-TOLUIC ACID, II, 589; III, 822
C$_8$H$_8$O$_3$ MANDELIC ACID, I, 336; III, 538
    PIPERONYL ALCOHOL, II, 591
    RESACETOPHENONE, III, 761
    cis Δ^4-TETRAHYDROPHTHALIC AN-
      HYDRIDE, IV, 890
C$_8$H$_8$O$_3$S SODIUM β-STYRENESULFON-
    ATE, IV, 846
C$_8$H$_8$O$_4$ DEHYDROACETIC ACID, III, 231
    GALLACETOPHENONE, II, 304
    ISODEHYDROACETIC ACID, IV,
      549
    PHLOROACETOPHENONE, II, 522
    VANILLIC ACID, IV, 972
C$_8$H$_8$O$_7$ DIACETYL-d-TARTARIC ANHY-
    DRIDE, IV, 242
C$_8$H$_9$AsO$_6$ p-ARSONOPHENOXYACETIC
    ACID, I, 75
C$_8$H$_9$Br 4-BROMO-o-XYLENE, III, 138
C$_8$H$_9$BrO β-PHENOXYETHYL BROMIDE,
    I, 436
C$_8$H$_9$ClO m-CHLOROPHENYLMETHYL-
    CARBINOL, III, 200
C$_8$H$_9$IO$_2$ 4-IODOVERATROLE, IV, 547
C$_8$H$_9$N 1,3-DIHYDROISOINDOLE, V, 406
C$_8$H$_9$NO N-METHYLFORMANILIDE, III,
    590
    PHENYLACETAMIDE, IV, 760
    o-TOLUAMIDE, II, 586
C$_8$H$_9$NO$_2$ dl-α-AMINOPHENYLACETIC
    ACID, III, 84
    p-AMINOPHENYLACETIC ACID, I,
      52
    BENZYL CARBAMATE, III, 168
    MANDELAMIDE, III, 536
C$_8$H$_9$NO$_3$ 3-NITRO-4,6-XYLENOL, I, 405
C$_8$H$_9$NO$_3$S p-ACETAMINOBENZENE-
    SULFINIC ACID, I, 7
C$_8$H$_9$N$_3$OS 1-PHENYL-2-THIOBIURET, V,
    966
C$_8$H$_{10}$ cis and trans-1,2-DIVINYLCYCLO-
    BUTANE, V, 528
C$_8$H$_{10}$BrN α-p-BROMOPHENYLETHYL-
    AMINE, II, 505
C$_8$H$_{10}$ClN α-p-CHLOROPHENYLETHYL-

AMINE, II, 505
N-ETHYL-p-CHLOROANILINE, IV,
    420
C$_8$H$_{10}$ClNO PHENACYLAMINE HY-
    DROCHLORIDE, V, 909
C$_8$H$_{10}$N$_2$O p-NITROSODIMETHYLANIL-
    INE HYDROCHLORIDE, II, 223
C$_8$H$_{10}$N$_2$O$_2$ m-NITRODIMETHYLANIL-
    INE, III, 658
C$_8$H$_{10}$N$_2$O$_3$S p-TOLYLSULFONYL-
    METHYLNITROSAMIDE, IV, 943
C$_8$H$_{10}$O 2,7-DIMETHYLOXEPIN, V, 467
    o-METHYLBENZYL ALCOHOL, IV,
      582
    p-TOLYL CARBINOL, II, 590
C$_8$H$_{10}$O$_2$ CREOSOL, IV, 203
C$_8$H$_{10}$O$_2$S METHYL p-TOLYL SULFONE,
    IV, 674
C$_8$H$_{10}$O$_3$S METHYL p-TOLUENESULFON-
    ATE, I, 146
C$_8$H$_{11}$AsN$_2$O$_4$ SODIUM p-ARSONO-N-
    PHENYLGLYCINAMIDE, I, 488
C$_8$H$_{11}$BF$_4$ 2,4,6-TRIMETHYLPYRYLIUM
    TETRAFLUOROBORATE, V, 1112
C$_8$H$_{11}$ClO$_5$ 2,4,6-TRIMETHYLPYRYLIUM
    PERCHLORATE, V, 1106
C$_8$H$_{11}$N 1-CYCLOHEXENYLACETONI-
    TRILE, IV, 234
    6-(DIMETHYLAMINO)FULVENE, V,
      431
    3,4-DIMETHYLANILINE, III, 307
    5-ETHYL-2-METHYLPYRIDINE, IV,
      451
    α-PHENYLETHYLAMINE, III, 717
    β-PHENYLETHYLAMINE, III, 720
    dl-α-PHENYLETHYLAMINE, II, 503
    d-α-PHENYLETHYLAMINE, II, 506
    l-α-PHENYLETHYLAMINE, II, 506
    R(+) and S(−)-α-PHENYLETHYL-
      AMINE, V, 932
C$_8$H$_{11}$NO 1-(α-PYRIDYL)-2-PROPANOL,
    III, 757
C$_8$H$_{11}$NO$_2$ 4-AMINOVERATROLE, II, 44
C$_8$H$_{11}$N$_3$ 1-METHYL-3-p-TOLYLTRI-
    AZENE, V, 797
C$_8$H$_{12}$Cl$_2$ 3,4-DICHLORO-1,2,3,4-TETRA-

METHYLCYCLOBUTENE, V, 370

$C_8H_{12}N_2O_2$ HEXAMETHYLENE DIISO-
CYANATE, IV, 521

$C_8H_{12}O$ 1-ACETYLCYCLOHEXENE, III,
22

3,5-DIMETHYL-2-CYCLOHEXEN-1-
ONE, III, 317

1-ETHYNYLCYCLOHEXANOL, III,
416

$C_8H_{12}O_2$ 5,5-DIMETHYL-1,3-CYCLOHEX-
ANEDIONE, II, 200

DIMETHYLKETENE β-LACTONE
DIMER, V, 456

3-ETHOXY-2-CYCLOHEXENONE, V,
539

$C_8H_{12}O_3$ 2-CARBETHOXYCYCLOPENT-
ANONE, II, 116

$C_8H_{12}O_4$ DIETHYL METHYLENEMAL-
ONATE, IV, 298

ETHYL DIACETYLACETATE, III,
390

$C_8H_{13}ClO$ β-CHLOROVINYL ISOAMYL
KETONE, IV, 186

$C_8H_{13}ClO_2$ 2,2,4-TRIMETHYL-3-OXO-
VALERYL CHLORIDE, V, 1103

$C_8H_{13}N$ KRYPTOPYRROLE, III, 513

2,3,4,5-TETRAMETHYLPYRROLE,
V, 1022

$C_8H_{13}NO$ 7-CYANOHEPTANAL, V, 266

5-METHYLFURFURYLDIMETHYL-
AMINE, IV, 626

C_8H_{14} trans-CYCLOOCTENE, V, 315

$C_8H_{14}O$ CYCLOÖCTANONE, V, 310

CYCLOHEXYL METHYL KETONE,
V, 775

2,4,4-TRIMETHYLCYCLOPENTAN-
ONE, IV, 957

$C_8H_{14}O_2$ 1-ACETYLCYCLOHEXANOL,
IV, 13

1-METHYLCYCLOHEXANECAR-
BOXYLIC ACID, V, 739

2,2,5,5-TRIMETHYLTETRAHYDRO-
3-KETOFURAN, V, 1024

$C_8H_{14}O_3$ t-BUTYL ACETOACETATE, V,
155

$C_8H_{14}O_4$ DIETHYL SUCCINATE, V, 993

ETHYL HYDROGEN ADIPATE, II,
277

β-ETHYL-β-METHYLGLUTARIC
ACID, IV, 441

ETHYL METHYLMALONATE, II,
279

$C_8H_{14}O_4S$ METHYL β-THIODIPROPION-
ATE, IV, 669

$C_8H_{15}N$ 2-ETHYLHEXANONITRILE, IV,
436

$C_8H_{15}NO_5$ DIACETONAMINE HYDRO-
GEN OXALATE, I, 196

$C_8H_{16}Cl_2O$ 4,4'-DICHLORODIBUTYL
ETHER, IV, 266

$C_8H_{16}O$ OCTANOL, V, 872

$C_8H_{16}O_2$ n-BUTYL n-BUTYRATE, I, 138

BUTYROIN, II, 114

2-CYCLOHEXYLOXYETHANOL, V,
303

ISOBUTYROIN, II, 115

3-METHYLHEPTANOIC ACID, V,
762

$C_8H_{16}O_4$ ETHYL DIETHOXYACETATE,
IV, 427

KETENE DI(2-METHOXYETHYL)
ACETAL, V, 684

$C_8H_{17}Br$ n-OCTYL BROMIDE, I, 30

$C_8H_{17}N$ 1-n-BUTYLPYRROLIDINE, III,
159

$C_8H_{17}NO$ 1-DIETHYLAMINO-3-BUTAN-
ONE, IV, 281

$C_8H_{17}NO_2$ 1-NITROÖCTANE, IV, 724

4-NITRO-2,4,4-TRIMETHYLPENT-
ANE, V, 845

$C_8H_{18}BrN$ β-DIPROPYLAMINOETHYL
BROMIDE HYDROBROMIDE, II,
92

$C_8H_{18}ClN_3$ 1-ETHYL-3-(3-DIMETHYL-
AMINO)PROPYLCARBODIIMIDE
HYDROCHLORIDE, V, 555

$C_8H_{18}NO$ DI-t-BUTYL NITROXIDE, V, 355

$C_8H_{18}O$ METHYL n-HEXYLCARBINOL,
I, 366

d- and l-OCTANOL-2, I, 418

$C_8H_{18}O_2$ $\alpha,\alpha,\alpha',\alpha'$-TETRAMETHYL-
TETRAMETHYLENE GLYCOL, V,
1026

$C_8H_{18}O_3S$ n-BUTYL SULFITE, II, 112

$C_8H_{18}O_4S$ n-BUTYL SULFATE, II, 111

$C_8H_{18}Zn$ DI-n-BUTYLZINC, II, 187

$C_8H_{19}N$ DI-n-BUTYLAMINE, I, 202

$C_8H_{20}Sn$ TETRAETHYLTIN, IV, 881

$C_9H_5Cl_2N$ 4,7-DICHLOROQUINOLINE, III, 272

$C_9H_5F_5$ 2-PHENYLPERFLUOROPRO-PENE, V, 949

$C_9H_6Br_2O_2$ COUMARIN DIBROMIDE, III, 209

$C_9H_6N_2$ INDOLE-3-CARBONITRILE, V, 656

C_9H_6O PHENYLPROPARGYL ALDE-HYDE, III, 731

$C_9H_6O_2$ PHENYLPROPIOLIC ACID, II, 515

$C_9H_6O_3$ COUMARILIC ACID, III, 209

HOMOPHTHALIC ANHYDRIDE, III, 450

C_9H_7N 1-CYANOBENZOCYCLOBUTENE, V, 263

QUINOLINE, I, 478

C_9H_7NO 3-HYDROXYQUINOLINE, V, 635

INDOLE-3-ALDEHYDE, IV, 539

C_9H_7NOS 2α-THIOHOMOPHTHALIM-IDE, V, 1051

$C_9H_7NO_2$ o-CARBOXYPHENYLACETONI-TRILE, III, 174

2-PHENYL-5-OXAZOLONE, V, 946

$C_9H_7NO_3$ o-NITROCINNAMALDEHYDE, IV, 722

$C_9H_7NO_4$ m-NITROCINNAMIC ACID, I, 398

C_9H_8O 2-INDANONE, V, 647

α-HYDRINDONE, II, 336

3-METHYLCOUMARONE, IV, 590

$C_9H_8O_2$ ACETYLBENZOYL, III, 20

HOMOPHTHALIC ACID, V, 612

$C_9H_8O_3$ METHYL BENZOYLFORMATE, I, 244

PHENYLPYRUVIC ACID, II, 519

$C_9H_8O_4$ HOMOPHTHALIC ACID, III, 449

p-HYDROXYPHENYLPYRUVIC ACID, V, 627

$C_9H_8O_2S_2$ THIOBENZOYLTHIOGLY-COLIC ACID, V, 1046

$C_9H_9BF_4N_2O_2$ p-CARBETHOXYBENZENE-DIAZONIUM FLUOBORATE, II, 299

C_9H_9Br CINNAMYL BROMIDE, V, 249

C_9H_9Cl α-CHLOROHYDRINDONE, II, 336

$C_9H_9ClN_2$ 3-(o-CHLOROANIL-INO)PROPIONITRILE, IV, 146

C_9H_9ClO BENZYL CHLOROMETHYL KETONE, III, 119

C_9H_9N 1-METHYLINDOLE, V, 769

C_9H_9NO p-METHOXYPHENYLACETONI-TRILE, IV, 576

3-METHYLOXINDOLE, IV, 657

$C_9H_9NO_2$ 2,6-DIMETHOXYBENZONI-TRILE, III, 293

ISONITROSOPROPIOPHENONE, II, 363

VERATRONITRILE, II, 622

$C_9H_9NO_3$ HIPPURIC ACID, II, 328

$C_9H_9NO_4$ ETHYL m-NITROBENZOATE, I, 373

p-NITROBENZYL ACETATE, III, 650

$C_9H_9NO_5$ 6-NITROVERATRALDEHYDE, IV, 735

$C_9H_9N_3O$ 1-PHENYL-3-AMINO-5-PYRA-ZOLONE, III, 708

$C_9H_9N_5$ BENZOGUANAMINE, IV, 78

C_9H_9O 2-METHYLINDOLE, III, 597

C_9H_{10} CYCLOPROPYLBENZENE, V, 328

PHENYLCYCLOPROPANE, V, 929

$C_9H_{10}BrNO$ 3-BROMO-4-ACETAMINO-TOLUENE, I, 111

$C_9H_{10}NF_3$ m-TRIFLUOROMETHYL-N,N-DIMETHYLANILINE, V, 1085

$C_9H_{10}N_2$ N-2-CYANOETHYLANILINE, IV, 205

$C_9H_{10}N_2O$ METHYLGLYOXAL-ω-PHENYL-HYDRAZONE, IV, 633

$C_9H_{10}N_2O_4$ 2-NITRO-4-METHOXYACET-ANILIDE, III, 661

$C_9H_{10}N_2S$ p-DIMETHYLAMINOPHENYL ISOTHIOCYANATE, I, 448

p-THIOCYANODIMETHYLANILINE,

II, 574

C$_9$H$_{10}$O p-METHYLACETOPHENONE, I, 111

METHYL BENZYL KETONE, II, 389

α-PHENYLPROPIONALDEHYDE, III, 733

C$_9$H$_{10}$O$_2$ HYDROCINNAMIC ACID, I, 311

p-METHOXYACETOPHENONE, I, 111

ω-METHOXYACETOPHENONE, III, 562

o- AND p-PROPIOPHENOL, II, 543

C$_9$H$_{10}$O$_3$ ATROLACTIC ACID, IV, 58

QUINACETOPHENONE MONO-METHYL ETHER, IV, 836

VERATRALDEHYDE, II, 619

C$_9$H$_{10}$O$_4$ SYRINGIC ALDEHYDE, IV, 866

C$_9$H$_{10}$O$_5$ FURFURAL DIACETATE, IV, 489

C$_9$H$_{11}$Br BROMOMESITYLENE, II, 95

C$_9$H$_{11}$BrO γ-PHENOXYPROPYL BROM-IDE, I, 435

C$_9$H$_{11}$F$_3$SO$_4$ 2,4,6-TRIMETHYLPYRY-LIUM TRIFLUOROMETHANE-SULFONATE, V, 1114

C$_9$H$_{11}$NO p-ACETOTOLUIDIDE, I, 111

p-DIMETHYLAMINOBENZALDE-HYDE, I, 214; IV, 331

ETHYL N-PHENYLFORMIMIDATE, IV, 464

C$_9$H$_{11}$NO$_2$ dl-α-AMINO-α-PHENYLPRO-PIONIC ACID, III, 88

dl-β-AMINO-β-PHENYLPROPIONIC ACID, III, 91

CYCLOHEXYLIDENECYANO-ACETIC ACID, IV, 234

ETHYL p-AMINOBENZOATE, I, 240; II, 300

ETHYL 2-PYRIDYLACETATE, III, 413

NITROMESITYLENE, II, 449

dl-β-PHENYLALANINE, II, 489

dl-PHENYLALANINE, III, 705

C$_9$H$_{11}$NO$_3$ l-TYROSINE, II, 613

VERATRIC AMIDE, II, 44

C$_9$H$_{11}$NO$_4$ m-NITROBENZALDEHYDE DI-METHYL ACETAL, III, 644

C$_9$H$_{11}$N$_3$O$_{10}$S TRIMETHYLOXONIUM 2,-4,6-TRINITROBENZENESULFON-ATE, V, 1099

C$_9$H$_{12}$ HEMIMELLITENE, IV, 508

MESITYLENE, I, 341

n-PROPYLBENZENE, I, 471

C$_9$H$_{12}$N$_2$O$_2$ p-ETHOXYPHENYLUREA, IV, 52

C$_9$H$_{12}$N$_2$S 2,6-DIMETHYLPHENYLTHIO-UREA, V, 802

C$_9$H$_{12}$O 3-PHENYL-1-PROPANOL, IV, 798

2,6,6-TRIMETHYL-2,4-CYCLOHEXA-DIENONE, V, 1092

C$_9$H$_{12}$O$_3$ α-GLYCERYL PHENYL ETHER, I, 296

C$_9$H$_{12}$O$_4$ DIMETHYL 3-METHYLENECY-CLOBUTANE-1,2-DICARBOXYL-ATE, V, 459

METHYL BUTADIENOATE, V, 734

C$_9$H$_{13}$N n-ETHYL-m-TOLUIDINE, II, 290

α-p-TOLYLETHYLAMINE, II, 505

C$_9$H$_{13}$NO 1,3-DIHYDRO-3,5,7-TRIMETH-YL-2H-AZEPIN-2-ONE, V, 408

C$_9$H$_{13}$NO$_2$ m-AMINOBENZALDEHYDE DIMETHYLACETAL, III, 59

2,4-DIMETHYL-5-CARBETHOXY-PYRROLE, II, 198

C$_9$H$_{14}$ 3-CYCLOHEXYLPROPINE, I, 191

1,2-CYCLONONADIENE, V, 306

C$_9$H$_{14}$N$_2$ AZELANITRILE, IV, 62

C$_9$H$_{14}$O 2-ALLYLCYCLOHEXANONE, III, 44; V, 25

C$_9$H$_{14}$O$_2$ exo-cis-BICYCLO[3.3.0]OCTANE-2-CARBOXYLIC ACID, V, 93

7,7-DIMETHOXYBICYCLO[2.2.1]-HEPTENE, V, 424

ISOPHORONE OXIDE, IV, 552

C$_9$H$_{14}$O$_3$ ETHYL 2-KETOHEXAHYDRO-BENZOATE, II, 532

C$_9$H$_{14}$O$_4$ DIETHYL ETHYLIDENEMALON-ATE, IV, 293

C$_9$H$_{14}$O$_5$ ETHYL ACETONEDICARBOXY-LATE, I, 237

ETHYL ETHOXALYLPROPIONATE, II, 272

C$_9$H$_{15}$Br 3-CYCLOHEXYL-2-BROMOPRO-
PENE, I, 186

C$_9$H$_{15}$NO PSEUDOPELLETIERINE, IV,
816

C$_9$H$_{15}$NO$_2$ ETHYL n-BUTYLCYANO-
ACETATE, III, 385

C$_9$H$_{15}$NO$_5$ DIETHYL ACETAMIDOMAL-
ONATE, V, 373

C$_9$H$_{15}$NO$_6$ ETHYL N-TRICARBOXYLATE,
III, 415

C$_9$H$_{16}$O$_2$ 2,4-NONANEDIONE, V, 848

C$_9$H$_{16}$O$_3$ ETHYL α-ISOPROPYLACETO-
ACETATE, III, 405

C$_9$H$_{16}$O$_4$ AZELAIC ACID, II, 53
ETHYL t-BUTYL MALONATE, IV,
417

C$_9$H$_{16}$O$_6$ DIETHYL BIS(HYDROXY-
METHYL)MALONATE, V, 446

C$_9$H$_{17}$Cl$_3$ 1,1,3-TRICHLORO-n-NONANE,
V, 1076

C$_9$H$_{17}$NO$_2$ ETHYL α-(1-PYRROLIDYL)-
PROPIONATE, IV, 466

C$_9$H$_{18}$N$_2$ DI-n-BUTYLCYANAMIDE, I, 204

C$_9$H$_{18}$O 4,6-DIMETHYL-1-HEPTEN-4-OL,
V, 452

C$_9$H$_{18}$O$_2$ PELARGONIC ACID, II, 474

C$_9$H$_{19}$N N,N-DIMETHYLCYCLOHEXYL-
METHYLAMINE, IV, 339

C$_9$H$_{20}$IN$_3$ 1-ETHYL-3-(3-DIMETHYL-
AMINO)PROPYLCARBODIIMIDE
METHIODIDE, V, 555

C$_9$H$_{20}$O, DI-n-BUTYL CARBINOL, II, 179

C$_9$H$_{20}$O$_2$ NONAMETHYLENE GLYCOL,
II, 155

C$_9$H$_{20}$O$_4$ ETHYL ORTHOCARBONATE,
IV, 457

C$_9$H$_{21}$O$_3$ β-ETHOXYPROPIONALDEHYDE
ACETAL, III, 371

C$_9$H$_{21}$O$_4$P n-PROPYL PHOSPHATE, II,
110

C$_{10}$Cl$_8$ PERCHLORFULVALENE, V, 901

C$_{10}$H$_6$Cl$_2$O$_4$S$_2$ NAPHTHALENE-1,5-DISUL-
FONYL CHLORIDE, IV, 693

C$_{10}$H$_6$N$_2$O$_4$ 1,4-DINITRONAPHTHALENE,
III, 341

C$_{10}$H$_6$O$_2$ 1,2-NAPHTHOQUINONE, II, 430
1,4-NAPHTHQUINONE, I, 383; IV,
698

C$_{10}$H$_6$O$_3$ 2-HYDROXY-1,4-NAPHTHO-
QUINONE, III, 465

C$_{10}$H$_6$O$_5$S 1,2-NAPHTHOQUINONE-4-SUL-
FONIC ACID SALTS, III, 633

C$_{10}$H$_6$O$_8$ PYROMELLITIC ACID, II, 551

C$_{10}$H$_7$Br α-BROMONAPHTHALENE, I,
121
2-BROMONAPHTHALENE, V, 142

C$_{10}$H$_7$BrHg β-NAPHTHYLMERCURIC
BROMIDE, II, 434

C$_{10}$H$_7$BrO 6-BROMO-2-NAPHTHOL, III,
132

C$_{10}$H$_7$ClHg β-NAPHTHYLMERCURIC
CHLORIDE, II, 432

C$_{10}$H$_7$Cl$_3$HgN$_2$ β-NAPHTHALENEDI-
AZONIUM CHLORIDEMERCURIC
CHLORIDE COMPOUND, II, 432

C$_{10}$H$_7$NO$_2$ α-CYANO-β-PHENYLACRYLIC
ACID, I, 181
NITROSO-β-NAPHTHOL, I, 411
N-PHENYLMALEIMIDE, V, 944

C$_{10}$H$_7$NO$_3$ 2-HYDROXYCINCHONINIC
ACID, III, 456
1-NITRO-2-NAPHTHOL, II, 451

C$_{10}$H$_7$O$_3$ β-BENZOYLACRYLIC ACID, III,
109

C$_{10}$H$_8$BrNO$_2$ β-BROMOETHYLPHTHAL-
IMIDE, I, 120; IV 106

C$_{10}$H$_8$ClN 2-CHLOROLEPIDINE, III, 194

C$_{10}$H$_8$N$_2$ 2,2'-BIPYRIDINE, V, 102

C$_{10}$H$_8$N$_2$O$_2$ 4-NITRO-1-NAPHTHYL-
AMINE, III, 664

C$_{10}$H$_8$N$_2$O$_3$ 6-METHOXY-8-NITRO-
QUINOLINE, III, 568

C$_{10}$H$_8$O$_2$ 2,5-DIHYDROXYACETOPHEN-
ONE, III, 280
2,6-DIHYDROXYACETOPHENONE,
III, 281
4-METHYLCOUMARIN, III, 581
NAPHTHORESORCINOL, III, 637

C$_{10}$H$_8$O$_4$ 4-METHYLESCULETIN, I, 360

C$_{10}$H$_8$S$_2$ 1,5-NAPHTHALENEDITHIOL,
IV, 695

$C_{10}H_9BrO$ α-BROMOBENZALACETONE, III, 125

$C_{10}H_9ClO_3$ ACETYLMANDELYL CHLORIDE, I, 13

$C_{10}H_9N$ LEPIDINE, III, 519
1-METHYLISOQUINOLINE, IV, 641

$C_{10}H_9NO$ 1,2-AMINONAPHTHOL HYDROCHLORIDE, II, 33
1,4-AMINONAPHTHOL HYDROCHLORIDE, I, 49; II, 39
4-METHYLCARBOSTYRIL, III, 580
2-METHYL-4-HYDROXYQUINOLINE, III, 593
α-PHENYLACETOACETONITRILE, II, 487

$C_{10}H_9NO_2$ 3-ACETYLOXINDOLE, V, 12
2-HYDROXY-3-METHYLISOCARBOSTYRIL, V, 623
INDOLE-3-ACETIC ACID, V, 654
1-(p-NITROPHENYL)-1,3-BUTADIENE, IV, 727

$C_{10}H_9NO_4S$ 1-AMINO-2-NAPHTHOL-4-SULFONIC ACID, II, 42

$C_{10}H_{10}$ trans-1-PHENYL-1,3-BUTADIENE, IV, 771
4-PHENYLBUTINE-1, I, 192

$C_{10}H_{10}Br_2O$ BENZALACETONE DIBROMIDE, III, 105

$C_{10}H_{10}ClNO_2$ p-CHLOROACETYLACETANILIDE, III, 183

$C_{10}H_{10}ClN_5$ 2-AMINO-4-ANILINO-6-(CHLOROMETHYL)-s-TRIAZINE, IV, 29

$C_{10}H_{10}Fe$ FERROCENE, IV, 473

$C_{10}H_{10}O$ BENZALACETONE, I, 77
α-TETRALONE, II, 569; III, 798; IV, 898
β-TETRALONE, IV, 903

$C_{10}H_{10}O_3$ β-BENZOYLPROPIONIC ACID, II, 81
ETHYL BENZOYLFORMATE, I, 241
METHYL p-ACETYLBENZOATE, IV, 579

$C_{10}H_{10}O_4$ ACETYLMANDELIC ACID, I, 12

β-(o-CARBOXYPHENYL)PROPIONIC ACID, IV, 136
HYDROQUINONE DIACETATE, III, 452
PHENYLSUCCINIC ACID, I, 451; IV, 804

$C_{10}H_{10}O_6$ 2,5-DIHYDROXY-p-BENZENEDIACETIC ACID, III, 286

$C_{10}H_{10}Ru$ RUTHENOCENE, V, 1001

$C_{10}H_{11}Br$ 4-PHENYL-2-BROMOBUTENE-1, I, 187

$C_{10}H_{11}ClO$ MESITOYL CHLORIDE, III, 555

$C_{10}H_{11}Cl_2N$ β,β-DICHLORO-p-DIMETHYLAMINOSTYRENE, V, 361

$C_{10}H_{11}IO_4$ IODOBENZENE DIACETATE, V, 660

$C_{10}H_{11}NO$ N-BENZYLACRYLAMIDE, V, 73

$C_{10}H_{11}NO_2$ ACETOACETANILIDE, III, 10

$C_{10}H_{11}NO_4$ CARBOBENZOXYGLYCINE, III, 168

$C_{10}H_{11}NS$ p-iso-PROPYLPHENYL ISOTHIOCYANATE, I, 448

$C_{10}H_{11}N_3O_3S$ AMINOMALONONITRILE p-TOLUENESULFONATE, V, 32

$C_{10}H_{12}N_2$ α-N,N-DIMETHYLAMINOPHENYLACETONITRILE, V, 437

$C_{10}H_{12}N_2O_4$ DINITRODURENE, II, 254

$C_{10}H_{12}O$ MESITALDEHYDE, III, 549; V, 49
MESITOIC ACID, V, 706
ar-TETRAHYDRO-α-NAPHTHOL, IV, 887
1-p-TOLYLCYCLOPROPANOL, V, 1058

$C_{10}H_{12}O_2$ DUROQUINONE, II, 254
ETHYL PHENYLACETATE, I, 270
o-EUGENOL, III, 418
MESITOIC ACID, III, 553
o-METHOXYPHENYLACETONE, IV, 573
γ-PHENYLBUTYRIC ACID, II, 499
4-PHENYL-m-DIOXANE, IV, 786
TETRALIN HYDROPEROXIDE, IV, 895

THYMOQUINONE, I, 512

C₁₀H₁₂O₃ 2,5-DIMETHYLMANDELIC ACID, III, 326

C₁₀H₁₂O₄ ETHYL ISODEHYDROACETATE, IV, 549

HOMOVERATRIC ACID, II, 333

C₁₀H₁₂O₅ TRIMETHYLGALLIC ACID, I, 537

C₁₀H₁₃Cl NEOPHYL CHLORIDE, IV, 702

C₁₀H₁₃ClO₂ 4-CHLOROBUTYL BENZOATE, III, 187

C₁₀H₁₃N ac-TETRAHYDRO-β-NAPHTHYLAMINE, I, 499

C₁₀H₁₃NO₂ 2-NITRO-p-CYMENE, III, 653

NITROSOTHYMOL, I, 511

C₁₀H₁₃NO₃ o-n-BUTOXYNITROBENZENE, III, 140

C₁₀H₁₃NO₄ N-METHYL-3,4-DIHYDROXYPHENYLALANINE, III, 586

C₁₀H₁₄ n-BUTYLBENZENE, III, 157

m-CYMENE, V, 332

DURENE, II, 248

ISODURENE, II, 360

C₁₀H₁₄N₂O p-NITROSODIETHYLANILINE, II, 224

C₁₀H₁₄O Δ¹⁽⁹⁾-OCTALONE-2, V, 869

PHENYL t-BUTYL ETHER(Method I), V, 924

PHENYL t-BUTYL ETHER(Method II), V, 926

C₁₀H₁₄O₃ DL-KETOPINIC ACID, V, 689

C₁₀H₁₄O₄ ETHYL 2-KETOCYCLOHEXYLGLYOXALATE, II, 531

TETRAACETYLETHANE, IV, 869

C₁₀H₁₄O₅ DIMETHYL CYCLOHEXANONE-2,6-DICARBOXYLATE, V, 439

C₁₀H₁₅ClO₃S DL-10-CAMPHORSULFONYL CHLORIDE, V, 196

C₁₀H₁₅N 2-AMINO-p-CYMENE, III, 63

α,α-DIMETHYL-β-PHENYLETHYLAMINE, V, 471

2-METHYLBENZYLDIMETHYLAMINE, IV, 585

β-PHENYLETHYLDIMETHYLAMINE, III, 723

C₁₀H₁₅NO AMINOTHYMOL, I, 512

C₁₀H₁₅NO₂ ETHYL (1-ETHYLPROPYLIDENE)CYANOACETATE, III, 399

N-METHYL-2,3-DIMETHOXYBENZYLAMINE, IV, 603

C₁₀H₁₆ ADAMANTANE, V, 16

C₁₀H₁₆N₂ DIAMINODURENE CHLOROSTANNATE, II, 255

SEBACONITRILE, III, 768

TETRAMETHYL-p-PHENYLENEDIAMINE, V, 1018

C₁₀H₁₆O₂ ETHYL CYCLOHEXYLIDENEACETATE, V, 547

SEBACIL, IV, 838

C₁₀H₁₆O₃ ETHYL β,β-PENTAMETHYLENEGLYCIDATE, IV, 459

C₁₀H₁₆O₄ DIETHYL 1,1-CYCLOBUTANEDICARBOXYLATE, IV, 288

ETHYL ETHOXYMETHYLENEMALONATE, III, 395

C₁₀H₁₆O₄S DL-10-CAMPHORSULFONIC ACID, V, 194

C₁₀H₁₆O₅ ETHYL ACETOSUCCINATE, II, 262

C₁₀H₁₆O₆ TRICARBETHOXYMETHANE, II, 594

C₁₀H₁₇NO 1-MORPHOLINO-1-CYCLOHEXENE, V, 808

C₁₀H₁₇NO₂ ω-CYANOPELARGONIC ACID, III, 768

C₁₀H₁₈ DECINE-1, I, 192

C₁₀H₁₈N₂O₄ t-BUTYL AZODIFORMATE, V, 160

C₁₀H₁₈O CYCLODECANONE, IV, 218; V, 277

l-MENTHONE, I, 340

C₁₀H₁₈O₂ SEBACOIN, IV, 840

C₁₀H₁₈O₃ ETHYL n-BUTYLACETOACETATE, I, 248

ETHYL ISOBUTYRYLISOBUTYRATE, II, 270

METHYL 4-KETO-7-METHYLOCTANOATE, III, 601

C₁₀H₁₈O₄ ETHYL ADIPATE, II, 264

ETHYL ISOPROPYLMALONATE, II, 94

$C_{10}H_{18}O_8$ HEXAHYDROGALLIC ACID
AND HEXAHYDROGALLIC ACID
TRIACETATE, V, 591

$C_{10}H_{19}Br$ 2-BROMODECENE-1, I, 187

$C_{10}H_{20}$ trans-4-t-BUTYLCYCLOHEXANOL,
V, 175

$C_{10}H_{20}Br_2$ DECAMETHYLENE BROMIDE,
III, 227

$C_{10}H_{20}N_2O_4$ 2,7-DIMETHYL-2,7-DINI-
TROÖCTANE, V, 445

$C_{10}H_{20}O_2$ 1,2-CYCLODECANEDIOL, IV,
216

PIVALOIN, II, 115

$C_{10}H_{20}O_3$ β-KETOISOÖCTALDEHYDE DI-
METHYL ACETAL, IV, 558

$C_{10}H_{20}O_5$ 2,2,6,6-TETRAMETHYLOLCY-
CLOHEXANOL, IV, 907

$C_{10}H_{20}O_6$ 2,3,4,6-TETRAMETHYL-d-GLU-
COSE, III, 800

$C_{10}H_{22}BrN$ β-DI-n-BUTYLAMINOETHYL
BROMIDE HYDROBROMIDE, II,
92

$C_{10}H_{22}O$ TRI-n-PROPYL CARBINOL, II,
603

$C_{10}H_{22}O_2$ DECAMETHYLENE GLYCOL,
II, 154

$C_{10}H_{22}Zn$ DIISOAMYL ZINC, II, 187

$C_{10}H_{24}N_2$ DECAMETHYLENEDIAMINE,
III, 229

β-DI-n-BUTYLAMINOETHYL-
AMINE, III, 254

$C_{11}H_7N$ α-NAPHTHONITRILE, III, 631

$C_{11}H_7NS$ α-NAPHTHYL ISOTHIOCYA-
NATE, IV, 700

β-NAPHTHYL ISOTHIOCYANATE,
I, 448

$C_{11}H_8N_2O_3$ BENZALBARBITURIC ACID,
III, 39

$C_{11}H_8O$ 1-NAPHTHALDEHYDE, IV, 690

β-NAPHTHALDEHYDE, III, 626

$C_{11}H_8OS$ PHENYL THIENYL KETONE,
II, 520

$C_{11}H_8O_2$ 2-HYDROXY-1-NAPHTHALDE-
HYDE, III, 463

α-NAPHTHOIC ACID, II, 425

β-NAPHTHOIC ACID, II, 428

$C_{11}H_9Cl$ 1-CHLOROMETHYLNA-
PHTHALENE, III, 195

$C_{11}H_9N$ 2-PHENYLPYRIDINE, II, 517

$C_{11}H_9NO_2$ 3-AMINO-2-NAPHTHOIC
ACID, III, 78

AZLACTONE OF α-ACETAMINO-
CINNAMIC ACID, II, 1

$C_{11}H_{10}N_2O_2S_2$ N-(p-ACETYL-
PHENYL)RHODANINE, IV, 6

$C_{11}H_{10}O$ METHYL β-NAPHTHYL ETHER,
I, 59

1-PHENYL-1-PENTEN-4-YN-3-OL,
IV, 792

$C_{11}H_{10}O_2$ 2-ETHYLCHROMONE, III, 387

6-METHOXY-2-NAPHTHOL, V, 918

$C_{11}H_{10}O_3$ α-PHENYLGLUTARIC ANHY-
DRIDE, IV, 790

$C_{11}H_{11}BrO_4$ p-BROMOBENZALDIACET-
ATE, II, 442

$C_{11}H_{11}N$ 2,4-DIMETHYLQUINOLINE, III,
329

$C_{11}H_{11}NO_2$ ETHYL INDOLE-2-CARBOXYL-
ATE, V, 567

ETHYL PHENYLCYANOACETATE,
IV, 461

$C_{11}H_{11}NO_3$ α-ACETAMINOCINNAMIC
ACID, II, 1

$C_{11}H_{11}NO_6$ o-NITROBENZALDIACETATE,
IV, 713

p-NITROBENZALDIACETATE, II,
441; IV, 713

$C_{11}H_{12}Br_2O_2$ ETHYL α,β-DIBROMO-β-
PHENYLPROPIONATE, II, 270

$C_{11}H_{12}FeO$ HYDROXYMETHYLFERRO-
CENE, V, 621

$C_{11}H_{12}N_2O_2$ l-TRYPTOPHANE, II, 612

$C_{11}H_{12}O_2$ ANISALACETONE, I, 78

ETHYL CINNAMATE, I, 252

$C_{11}H_{12}O_3$ γ-BENZOYLBUTYRIC ACID, II,
82

ETHYL BENZOYLACETATE, II, 266;
III, 379; IV, 415

$C_{11}H_{12}O_4$ 2,3-DIMETHOXYCINNAMIC
ACID, IV, 327

C$_{11}$H$_{12}$O$_5$ 3,4-DIMETHOXYPHENYLPY-
RUVIC ACID, II, 335

C$_{11}$H$_{13}$NO 1-METHYL-3-ETHYLOXIN-
DOLE, IV, 620

C$_{11}$H$_{13}$NO$_3$ N-ACETYLPHENYLALANINE,
II, 493

C$_{11}$H$_{13}$NS 2-METHYL-4-iso-PROPYL-
PHENYL ISOTHIOCYANATE, I,
448

C$_{11}$H$_{13}$OP 3-METHYL-1-PHENYLPHOS-
PHOLENE 1-OXIDE, V, 787

C$_{11}$H$_{14}$O$_2$ 1,6-DIOXO-8a-METHYL-1,2,3,4,-
6,7,8a-OCTAHYDRONAPHTHAL-
ENE, V, 486

MESITYLACETIC ACID, III, 557

C$_{11}$H$_{14}$O$_3$ ETHYL β-PHENYL-β-HY-
DROXYPROPIONATE, III, 408

C$_{11}$H$_{14}$O$_4$ METHYL HOMOVERATRATE,
II, 333

C$_{11}$H$_{14}$O$_7$ ETHYL ACETONEDICARBOX-
YLATE, II, 126

C$_{11}$H$_{15}$N 1-PHENYLPIPERIDINE, IV, 795

C$_{11}$H$_{15}$NO 2-t-BUTYL-3-PHENYLOXAZI-
RANE, V, 191

β-DIMETHYLAMINOPROPIOPHEN-
ONE HYDROCHLORIDE, III, 305

C$_{11}$H$_{16}$ n-AMYLBENZENE, II, 47

PENTAMETHYLBENZENE, II, 248

C$_{11}$H$_{16}$N$_2$ 1-BENZYLPIPERAZINE, V, 88

C$_{11}$H$_{16}$O 7-t-BUTOXYNORBORNADIENE,
V, 151

D-2-OXO-7,7-DIMETHYL-1-VINYLBI-
CYCLO[2.2.1]HEPTANE, V, 877

C$_{11}$H$_{16}$O$_2$ 1-ADAMANTANECARBOXYLIC
ACID, V, 20

C$_{11}$H$_{16}$O$_3$ 3,5-DIMETHYL-4-CARBETH-
OXY-2-CYCLOHEXEN-1-ONE, III,
317

C$_{11}$H$_{16}$O$_3$S n-BUTYL p-TOLUENESUL-
FONATE, I, 145

C$_{11}$H$_{17}$NO$_2$ ETHYL (1-ETHYLPRO-
PENYL)-METHYLCYANOACE-
TATE, III, 397

C$_{11}$H$_{18}$O$_2$ 10-UNDECYNOIC ACID, IV, 969

C$_{11}$H$_{18}$O$_3$ 2-CARBETHOXYCYCLOOC-
TANONE, V, 198

C$_{11}$H$_{18}$O$_5$ DIETHYL β-KETOPIMELATE,
V, 384

DIETHYL γ-OXOPIMELATE, IV, 302

ETHYL α-ACETOGLUTARATE, II,
263

6-KETOHENDECANEDIOIC ACID,
IV, 555

C$_{11}$H$_{19}$NO$_2$ METHYL ω-CYANOPELAR-
GONATE, III, 584

C$_{11}$H$_{19}$NO$_3$ N-ACETYLHEXAHYDRO-
PHENYLALANINE, II, 493

C$_{11}$H$_{20}$2 -n-BUTYL-2-METHYLCYCLO-
HEXANONE, V, 187

C$_{11}$H$_{20}$O$_2$ DIISOVALERYLMETHANE, III,
291

4-ETHYL-2-METHYL-2-OCTENOIC
ACID, IV, 444

C$_{11}$H$_{20}$O$_4$ DI-t-BUTYL MALONATE, IV,
261

ETHYL n-BUTYLMALONATE, I, 250

ETHYL sec.-BUTYLMALONATE, II,
417

ETHYL PIMELATE, II, 536

HENDECANEDIOIC ACID, IV, 510

C$_{11}$H$_{21}$NO$_3$ METHYL SEBACAMATE, III,
613

C$_{11}$H$_{21}$NO$_4$ DI-β-CARBETHOXYETHYL-
METHYLAMINE, III, 258

C$_{11}$H$_{22}$O 5,5-DIMETHYL-2-n-PENTYLTE-
TRAHYDROFURAN, IV, 350

UNDECYENYL ALCOHOL, II, 374

C$_{11}$H$_{24}$O$_2$ ACETONE DIBUTYL ACETAL,
V, 5

2-METHYL-2,5-DECANEDIOL, IV,
601

UNDECAMETHYLENE GLYCOL, II,
155

C$_{11}$H$_{26}$N$_2$ γ-DI-n-BUTYLAMINOPROPY-
LAMINE, III, 256

C$_{12}$H$_5$N$_3$O$_7$ 2,4,7-TRINITROFLUOREN-
ONE, III, 837

C$_{12}$H$_6$O$_2$ ACENAPHTHENEQUINONE, III,
1

C$_{12}$H$_6$O$_3$ 1,2-NAPHTHALIC ANHYDRIDE,
II, 423

C$_{12}$H$_8$ BIPHENYLENE, V, 54

$C_{12}H_8B_2F_8N_4$ 4,4'-BIPHENYLENE-*bis*-DI-
AZONIUM FLUOBORATE, II, 188
$C_{12}H_8Br_2$ 4,4'-DIBROMOBIPHENYL, IV,
256
$C_{12}H_8Cl_2$ 3,4-DICHLOROBIPHENYL, V,
51
$C_{12}H_8F_2$ 4,4'-DIFLUOROBIPHENYL, II,
188
$C_{12}H_8NO_3$ NICOTINIC ANHYDRIDE, V,
822
$C_{12}H_8N_2O_2$ 2-NITROCARBAZOLE, V, 829
$C_{12}H_8N_2O_4$ 2,2'-DINITROBIPHENYL, III,
339
$C_{12}H_8N_2O_4S$ *p*-NITROPHENYL SULFIDE,
III, 667
$C_{12}H_8N_2O_4S_2$ DI-*o*-NITROPHENYL DISUL-
FIDE, I, 220
m-NITROPHENYL DISULFIDE, V,
843
$C_{12}H_8OS$ PHENOXTHIN, II, 485
$C_{12}H_8O_3$ 3,4-DIHYDRO-1,2-NAPHTHALIC
ANHYDRIDE, II, 194
$C_{12}H_8O_4$ 2,3-NAPHTHALENEDICARBOX-
YLIC ACID, V, 810
2,6-NAPHTHALENEDICARBOXYLIC
ACID, V, 813
$C_{12}H_9Br$ *p*-BROMOBIPHENYL, I, 113
$C_{12}H_9NO$ 3-BENZOYLPYRIDINE, IV, 88
$C_{12}H_9NO_2$ *m*-NITROBIPHENYL, IV, 718
$C_{12}H_9NO_3$ *o*-NITRODIPHENYL ETHER,
II, 446
p-NITRODIPHENYL ETHER, II, 445
$C_{12}H_9O_4$ 3-CARBETHOXYCOUMARIN,
III, 165
$C_{12}H_{10}Cl_2Se$ DIPHENYLSELENIUM DI-
CHLORIDE, II, 240
$C_{12}H_{10}Hg$ DIPHENYLMERCURY, I, 228
$C_{12}H_{10}I_2$ DIPHENYLIODONIUM IODIDE,
III, 355
$C_{12}H_{10}NS$ 1-METHYL-2-IMINO-β-NA-
PHTHOTHIAZOLINE, III, 595
$C_{12}H_{10}N_2$ AZOBENZENE, III, 103
N-METHYL-1-NAPHTHYLCYANA-
MIDE, III, 608
$C_{12}H_{10}N_2O$ AZOXYBENZENE, II, 57
$C_{12}H_{10}N_2O_3$ 1-NITRO-2-ACETYLAMINO-

NAPHTHALENE, II, 438
$C_{12}H_{10}N_2S$ 1-METHYL-2-AMINO-β-NA-
PHTHOTHIAZOLINE, III, 595
$C_{12}H_{10}O$ ACENAPHTHENOL-7, III, 3
$C_{12}H_{10}O_2$ 3,3'-DIHYDROXYBIPHENYL, V,
412
$C_{12}H_{10}S$ DIPHENYL SULFIDE, II, 242
$C_{12}H_{10}Se$ DIPHENYL SELENIDE, II, 238
$C_{12}H_{11}FeN$ FERROCENYLACETONI-
TRILE, V, 578
$C_{12}H_{11}NO_2$ ETHYL α-CYANO-β-PHENYL-
ACRYLATE, I, 451
1-(2-METHOXYCARBONYL-
PHENYL)PYRROLE, V, 716
$C_{12}H_{11}NO_3$ ETHYL PHENYLCYANOPY-
RUVATE, II, 287
$C_{12}H_{11}N_3$ DIAZOAMINOBENZENE, II, 163
$C_{12}H_{12}$ *trans*-STILBENE, III, 786
$C_{12}H_{12}N_2$ 2-BENZYLAMINOPYRIDINE,
IV, 91
$C_{12}H_{12}N_2O_2S$ 4,4'-DIAMINODIPHENYL-
SULFONE, III, 239
$C_{12}H_{12}N_2S$ 1-METHYL-1-(1-NAPHTHYL)-
2-THIOUREA, III, 609
$C_{12}H_{12}N_2S_2$ *p*-AMINOPHENYL DISUL-
FIDE, III, 86
$C_{12}H_{12}N_4$ 4,4'-DIAMINOAZOBENZENE,
V, 341
$C_{12}H_{12}N_6$ TETRAMETHYLAMMONIUM
1,1,2,3,3-PENTACYANOPROPEN-
IDE, V, 1013
$C_{12}H_{12}O_3$ 1,3,5-TRIACETYLBENZENE,
III, 829
$C_{12}H_{12}O_6$ HYDROXYHYDROQUINONE
TRIACETATE, I, 317
$C_{12}H_{13}N$ α-(β-NAPHTHYL)-ETHYL-
AMINE, II, 505
1,2,3,4-TETRAHYDROCARBAZOLE,
IV, 884
$C_{12}H_{13}NO_2$ *t*-BUTYLPHTHALIMIDE, III,
152
$C_{12}H_{14}NO_2$ ETHYL β-ANILINOCROTON-
ATE, III, 374
$C_{12}H_{14}O$ 2-BENZYLCYCLOPENTANONE,
V, 76
$C_{12}H_{14}O_3$ ETHYL α-PHENYLACETOACET-
ATE, II, 284

PHENYLMETHYLGLYCIDIC ESTER, III, 727

$C_{12}H_{14}O_4$ β-METHYL-β-PHENYLGLU-TARIC ACID, IV, 664

$C_{12}H_{15}N$ JULOLIDINE, III, 504

$C_{12}H_{15}NO$ BENZOYL PIPERIDINE, I, 99

$C_{12}H_{16}$ CYCLOHEXYLBENZENE, II, 151

$C_{12}H_{16}O$ ACETO-p-CYMENE, II, 3

$C_{12}H_{16}O_2$ ETHYL γ-PHENYLBUTYRATE, II, 196

$C_{12}H_{16}O_4$ MONOBENZALPENTAERY-THRITOL, IV, 679

$C_{12}H_{17}NO_2$ ETHYL p-DIMETHYLAMINO-PHENYLACETATE, V, 552

$C_{12}H_{17}NO_4$ 2,4-DIMETHYL-3,5-DICAR-BETHOXYPYRROLE, II, 202

$C_{12}H_{17}NO_5$ d-α-PHENYLMETHYLAMINE ACID l-MALATE, II, 506

$C_{12}H_{17}NO_6$ l-α-PHENYLETHYLAMINE ACID d-TARTRATE, II, 507

$C_{12}H_{18}$ HEXAMETHYLBENZENE, II, 248; IV, 520

$C_{12}H_{18}ClNO_2$ BENZOYLCHOLINE CHLOR-IDE, IV, 84

$C_{12}H_{18}INO_2$ BENZOYLCHOLINE IODIDE, IV, 84

$C_{12}H_{18}N_4O$ 3-n-HEPTYL-5-CYANOCYTO-SINE, IV, 515

$C_{12}H_{18}O$ 2,3,4,5,6,6-HEXAMETHYL-2,4-CYCLOHEXADIEN-1-ONE, V, 598

$C_{12}H_{18}O_4$ DIETHYL Δ^2-CYCLOPENTEN-YLMALONATE, IV, 291

DIETHYL cis-Δ^4-TETRAHYDRO-PHTHALATE, IV, 304

$C_{12}H_{20}$ CYCLOHEXYLIDENECYCLO-HEXANE, V, 297

$C_{12}H_{20}O_2$ CYCLOHEXANONE DIALLYL ACETAL, V, 292

$C_{12}H_{20}O_4$ DIETHYL cis-HEXAHYDRO-PHTHALATE, IV, 304

$C_{12}H_{21}ClO_2$ l-MENTHOXYACETYL CHLORIDE, III, 547

$C_{12}H_{21}NO$ BENZYLTRIMETHYLAMMO-NIUM ETHOXIDE, IV, 98

$C_{12}H_{21}N_3O_3$ HEXAHYDRO-1,3,5-TRIPRO-

PIONYL-s-TRIAZINE, IV, 518

$C_{12}H_{22}$ cis-CYCLODODECENE, V, 281

$C_{12}H_{22}O_2$ $trans$-2-DODECENOIC ACID, IV, 398

$C_{12}H_{22}O_3$ n-CAPROIC ANHYDRIDE, III, 164

l-MENTHOXYACETIC ACID, III, 544

$C_{12}H_{22}O_4$ ETHYL HYDROGEN SEBAC-ATE, II, 276

METHYL HYDROGEN HENDEC-ANEDIOATE, IV, 635

$C_{12}H_{22}O_{11}$ CELLOBIOSE, II, 122

$C_{12}H_{23}NO$ UNDECYL ISOCYANATE, III, 846

$C_{12}H_{24}Sn$ DI-n-BUTYLDIVINYLTIN, IV, 258

$C_{12}H_{25}Br$ n-DODECYL BROMIDE, I, 29; II, 246

$C_{12}H_{26}O$ LAURYL ALCOHOL, II, 372

$C_{12}H_{26}S$ n-DODECYL MERCAPTAN, III, 363

$C_{12}H_{27}AlO_3$ ALUMINUM t-BUTOXIDE, III, 48

$C_{12}H_{27}BO_3$ n-BUTYL BORATE, II, 106

$C_{12}H_{27}O_4P$ n-BUTYL PHOSPHATE, II, 109

$sec.$-BUTYL PHOSPHATE, II, 110

$C_{13}H_4N_4O_9$ 2,4,5,7-TETRANITROFLU-ORENONE, V, 1029

$C_{13}H_8ClN$ 9-CHLOROACRIDINE, III, 54

$C_{13}H_8O_2$ XANTHONE, I, 552

$C_{13}H_9ClN_2O$ p-PHENYLAZOBENZOYL CHLORIDE, III, 712

$C_{13}H_9ClO_3$ p-CHLOROPHENYL SALICYL-ATE, IV, 178

$C_{13}H_9NO$ ACRIDONE, II, 15

$C_{13}H_9NO_2$ 2-NITROFLUORENE, II, 447

$C_{13}H_{10}Cl_2$ BENZOPHENONE DICHLOR-IDE, II, 573

$C_{13}H_{10}N_2$ 9-AMINOACRIDINE, III, 53

DIPHENYLCARBODIIMIDE(Method I), V, 501

DIPHENYLCARBODIIMIDE(Method II), V, 504

DIPHENYLDIAZOMETHANE, III, 351

2-PHENYLINDAZOLE, V, 941

$C_{13}H_{10}N_2O$ PYOCYANINE, III, 753

$C_{13}H_{10}N_2O_2$ p-PHENYLAZOBENZOIC ACID, III, 711

$C_{13}H_{10}N_4$ p-TRICYANOVINYL-N,N-DI-METHYLANILINE, IV, 953

$C_{13}H_{10}O$ BENZOPHENONE, I, 95

$C_{13}H_{10}O_2$ XANTHYDROL, I, 554

$C_{13}H_{10}S$ THIOBENZOPHENONE, II, 573; IV, 927

$C_{13}H_{11}Br$ p-BROMODIPHENYLMETH-ANE, V, 130

$C_{13}H_{11}N$ 2-AMINOFLUORENE, II, 448; V, 30

BENZALANILINE, I, 80

DIPHENYL KETIMINE, V, 520

DIPHENYLMETHANE IMINE HY-DROCHLORIDE, II, 234

$C_{13}H_{11}NO$ 2-AMINOBENZOPHENONE, IV, 34

BENZANILIDE, I, 82

BENZOPHENONE OXIME, II, 70

$C_{13}H_{11}NO_2$ N-PHENYLANTHRANILIC ACID, II, 15

$C_{13}H_{12}$ DIPHENYLMETHANE, II, 232

$C_{13}H_{12}BN$ 10-METHYL-10,9-BORAZARO-PHENANTHRENE, V, 727

$C_{13}H_{12}N_2$ N-PHENYLBENZAMIDINE, IV, 769

$C_{13}H_{12}N_2O$ BENZIL HYDRAZONE, II, 497

sym-DIPHENYLUREA, I, 453

$C_{13}H_{12}N_2O_2$ ETHYL α,β-DICYANO-β-PHENYLPROPIONATE, I, 452

$C_{13}H_{12}N_4S$ DITHIZONE, III, 360

$C_{13}H_{12}O$ BENZOHYDROL, I, 90

$C_{13}H_{12}O_2$ 2-ETHOXY-1-NAPHTHALDE-HYDE, III, 98

ETHYL α-NAPHTHOATE, II, 282

2-METHOXYDIPHENYL ETHER, III, 566

$C_{13}H_{13}N$ BENZYLANILINE, I, 102

$C_{13}H_{13}NO_2$ ETHYL (1-PHENYLETHYLI-DENE)-CYANOACETATE, IV, 463

$C_{13}H_{14}$ 4,6,8-TRIMETHYLAZULENE, V, 1088

$C_{13}H_{14}O_2$ 3-BENZOYLOXYCYCLOHEX-ANE, V, 70

$C_{13}H_{14}O_4$ 6,7-DIMETHOXY-3,4-DIHYDRO-2-NAPHTHOIC ACID, III, 300

ETHYL BENZOYLACETOACETATE, II, 266

γ-PHENYLALLYLSUCCINIC ACID, IV, 766

$C_{13}H_{16}BrNO_3$ ϵ-BENZOYLAMINO-α-BROMOCAPROIC ACID, II, 74

$C_{13}H_{16}O$ BENZALPINACOLONE, I, 81

2-PHENYLCYCLOHEPTANONE, IV, 780

$C_{13}H_{16}O_2$ PHENYLPROPARGYLALDE-HYDE DIETHYL ACETAL, IV, 801

$C_{13}H_{16}O_3$ ETHYL BENZOYLDIMETHY-LACETATE, II, 268

$C_{13}H_{16}O_4$ ETHYL PHENYLMALONATE, II, 288

$C_{13}H_{17}NO_3$ ϵ-BENZOYLAMINOCAPROIC ACID, II, 76

$C_{13}H_{17}NO_4$ 3,5-DICARBETHOXY-2,6-DI-METHYLPYRIDINE, II, 215

$C_{13}H_{18}N_2O_3$ dl-ϵ-BENZOYLLYSINE, II, 374

$C_{13}H_{18}N_4O_2$ BENZYLIDENEARGININE, II, 49

$C_{13}H_{18}O_2$ 3-BENZYL-3-METHYLPENTAN-OIC ACID, IV, 93

$C_{13}H_{19}NO_4$ 1,4-DIHYDRO-3,5-DICAR-BETHOXY-2,6-DIMETHYLPYRI-DINE, II, 214

$C_{13}H_{20}O$ o-n-HEPTYLPHENOL, III, 444

PSEUDOIONONE, III, 747

$C_{13}H_{24}O_2$ trans-2-METHYL-2-DODECEN-OIC ACID, IV, 608

2-METHYLENEDODECANOIC ACID, IV, 616

$G_{13}H_{24}O_3$ sec-BUTYL α-n-CAPROYLPRO-PIONATE, IV, 120

$C_{13}H_{28}O$ TRI-n-BUTYL CARBINOL, II, 603

$C_{13}H_{28}O_2$ TRIDECAMETHYLENE GLY-COL, II, 155

$C_{13}H_{29}N$ LAURYLMETHYLAMINE, IV, 564

$C_{14}H_6Cl_4O_4$ dl-4,4',6,6'-TETRACHLORO-DIPHENIC ACID, IV, 872

$C_{14}H_7ClO_2$ α-CHLOROANTHRAQUIN-ONE, II, 128

$C_{14}H_8Br_2$ 9,10-DIBROMOANTHRACENE,
I, 207

$C_{14}H_8Cl_2O_3$ p-CHLOROBENZOIC ANHY-
DRIDE, III, 29

$C_{14}H_8N_2O_8$ p-NITROBENZOYL PEROX-
IDE, III, 649

$C_{14}H_8O_2$ ANTHRAQUINONE, II, 554
FLUORENONE-2-CARBOXYLIC
ACID, III, 420
PHENANTHRENEQUINONE, IV, 757

$C_{14}H_8O_4$ QUINIZARIN, I, 476

$C_{14}H_8O_5S$ POTASSIUM ANTHRAQUIN-
ONE-α-SULFONATE, II, 539

$C_{14}H_9Br$ 9-BROMOPHENANTHRENE, III,
134

$C_{14}H_9Cl$ 9-CHLOROANTHRACENE, V,
206

$C_{14}H_9NO_2$ 9-NITROANTHRACENE, IV,
711

$C_{14}H_9NO_3$ NITROANTHRONE, I, 390

$C_{14}H_{10}$ DIPHENYLACETYLENE, III, 350;
IV, 377

$C_{14}H_{10}Cl_2O$ 2,2'-DICHLOR-α,α'-EPOXY-
BIBENZYL, V, 358

$C_{14}H_{10}Cl_2O_2$ DI-(p-CHLOROPHENYL)-
ACETIC ACID, III, 270

$C_{14}H_{10}N_2O$ PHENYLBENZOYLDIAZO-
METHANE, II, 496

$C_{14}H_{10}O$ ANTHRONE, I, 60
DIPHENYLKETENE, III, 356

$C_{14}H_{10}O_2$ BENZIL, I, 87
DIPHENALDEHYDE, V, 489
9-FLUORENECARBOXYLIC ACID,
IV, 482

$C_{14}H_{10}O_2S_2$ BENZOYL DISULFIDE, III,
116

$C_{14}H_{10}O_3$ BENZOIC ANHYDRIDE, I, 91
DIPHENALDEHYDIC ACID, V, 493

$C_{14}H_{10}O_3S$ BARIUM 2-PHENANTHRENE-
SULFONATE, II, 482
POTASSIUM 3-PHENANTHRENE-
SULFONATE, II, 482

$C_{14}H_{10}O_4$ DIPHENIC ACID, I, 222

$C_{14}H_{11}ClO$ DESYL CHLORIDE, II, 159

$C_{14}H_{11}N$ DIPHENYLACETONITRILE, III,
347

2-PHENYLINDOLE, III, 725

$C_{14}H_{12}$ 9,10-DIHYDROPHENANTHRENE,
IV, 313; V, 398
1,1-DIPHENYLETHYLENE, I, 226
9-METHYLFLUORENE, IV, 623
cis-STILBENE, IV, 857

$C_{14}H_{12}N_2$ BENZALAZINE, II, 395

$C_{14}H_{12}N_2O_2$ sym-DIBENZOYLHYDRA-
ZINE, II, 208

$C_{14}H_{12}N_2O_4$ p,p'-DINITROBIBENZYL, IV,
367

$C_{14}H_{11}N_3O_2$ β-METHYL-β-PHENYL-α,α'-
DICYANOGLUTARIMIDE, IV, 662

$C_{14}H_{12}N_4$ 1,4-DIPHENYL-5-AMINO-1,2,3-
TRIAZOLE, IV, 380
4-PHENYL-5-ANILINO-1,2,3-TRIA-
ZOLE, IV, 380

$C_{14}H_{12}O$ DESOXYBENZOIN, II, 156
DIPHENYLACETALDEHYDE, IV,
375
trans-STILBENE OXIDE, IV, 860

$C_{14}H_{12}O_2$ ACENAPHTHENOL ACETATE,
III, 3
BENZOIN, I, 94
BENZYL BENZOATE, I, 104
DIPHENYLACETIC ACID, I, 224

$C_{14}H_{12}O_3$ BENZILIC ACID, I, 89
2-p-ACETYLPHENYLHYDROQUIN-
ONE, IV, 15

$C_{14}H_{13}NO_2$ SALICYL-o-TOLUIDE, III, 765

$C_{14}H_{14}$ 3,3'-DIMETHYLBIPHENYL, III,
295

$C_{14}H_{14}Hg$ DI-p-TOLYLMERCURY, I, 231

$C_{14}H_{14}N_2O_2$ α-PHENYL-α-CARBETHOXY-
GLUTARONITRILE, IV, 776

$C_{14}H_{14}O_2$ 3,3'-DIMETHOXYBIPHENYL,
III, 295

$C_{14}H_{14}O_5S_2$ p-TOLUENESULFONIC AN-
HYDRIDE, IV, 940

$C_{14}H_{14}Se$ DI-m-TOLYL SELENIDE, II, 239
DI-o-TOLYL SELENIDE, II, 239
DI-p-TOLYL SELENIDE, II, 239

$C_{14}H_{15}N$ m-TOLYLBENZYLAMINE, III,
827
α-p-XENYLETHYLAMINE, II, 505

$C_{14}H_{15}NO_7S$ 1-METHYLTHIOL-3-PHTHALAMIDOPROPANE-3,3-DI-CARBOXYLIC ACID, II, 385

$C_{14}H_{16}O_4$ ETHYL BENZALMALONATE, III, 377

$C_{14}H_{16}O_5$ DIETHYL BENZOYLMALONATE, IV, 285

$C_{14}H_{17}N$ α-CYCLOHEXYLPHENYL-ACETONITRILE, III, 219

$C_{14}H_{19}BrO_9$ ACETOBROMOGLUCOSE, III, 11

$C_{14}H_{20}ClNO_7$ 2-ACETAMIDO-3,4,6-TRI-O-ACETYL-2-DEOXY-α-D-GLYCO-PYRANOSYL CHLORIDE, V, 1

$C_{14}H_{20}FeIN$ N,N-DIMETHYLAMINO-METHYLFERROCENE METHIODIDE, V, 434

$C_{14}H_{20}N_2$ 1,1'-DICYANO-1,1'-BICYCLO-HEXYL, IV, 273

$C_{14}H_{20}N_4$ 1,1'-AZO-bis-1-CYCLOHEX-ANENITRILE, IV, 66

$C_{14}H_{20}O_8$ ETHYL ETHYLENETETRA-CARBOXYLATE, II, 273

$C_{14}H_{20}O_{10}$ β-d-GLUCOSE-1,2,3,4-TETRA-ACETATE, III, 432

β-d-GLUCOSE-2,3,4,6-TETRAACET-ATE, III, 434

$C_{14}H_{22}N_4$ 1,2-DI-1-(1-CYANO)-CYCLO-HEXYL-HYDRAZINE, IV, 274

$C_{14}H_{22}O_2$ 1,1'-ETHYNYLENE-bis-CY-CLOHEXANOL, IV, 471

$C_{14}H_{22}S_2$ 1,2-BIS(n-BUTYLTHIO)-BENZENE, V, 107

$C_{14}H_{26}O_2$ VINYL LAURATE, IV, 977

$C_{14}H_{26}O_3$ HEPTOIC ANHYDRIDE, III, 28

$C_{14}H_{26}O_4$ ETHYL SEBACATE, II, 277

$C_{14}H_{28}O_2$ MYRISTIC ACID, I, 379

$C_{14}H_{29}Br$ TETRADECYL BROMIDE, II, 247

$C_{14}H_{30}O$ MYRISTYL ALCOHOL, II, 374

$C_{14}H_{30}O_2$ TETRADECAMETHYLENE GLYCOL, II, 155

$C_{15}H_9N$ 9-CYANOPHENANTHRENE, III, 212

$C_{15}H_{10}BrNO_2$ 1-METHYLAMINO-4-BROMOANTHRAQUINONE, III, 575

$C_{15}H_{10}Br_2O_2$ DIBENZOYLDIBROMO-METHANE, II, 244

$C_{15}H_{10}O$ 9-ANTHRALDEHYDE, III, 98

DIPHENYLCYCLOPROPENONE, V, 514

PHENANTHRENE-9-ALDEHYDE, III, 701

$C_{15}H_{10}O_2$ BENZALPHTHALIDE, II, 61

FLAVONE, IV, 478

β-METHYLANTHRAQUINONE, I, 353

$C_{15}H_{10}O_3$ DIPHENYL TRIKETONE, II, 244

DIPHENYL TRIKETONE HYDRATE, II, 244

$C_{15}H_{11}N$ α-PHENYLCINNAMONITRILE, III, 715

$C_{15}H_{11}NO_2$ BENZYLPHTHALIMIDE, II, 83

1-METHYLAMINOANTHRAQUIN-ONE, III, 573

$C_{15}H_{11}NO_4$ $trans$-o-NITRO-α-PHENYLCIN-NAMIC ACID, IV, 730

$C_{15}H_{12}Br_2O$ BENZALACETOPHEN-ONE DIBROMIDE, I, 205

$C_{15}H_{12}N_2O$ 4,5-DIPHENYLGLYOXALONE, II, 231

$C_{15}H_{12}O$ 2-ACETYLFLUORENE, III, 23

BENZALCETOPHENONE, I, 78

$C_{15}H_{12}O_2$ DIBENZOYLMETHANE, I, 205; III, 251

α-PHENYLCINNAMIC ACID, IV, 777

PHENYL CINNAMATE, III, 714

$C_{15}H_{12}O_3$ p-TOLUYL-o-BENZOIC ACID, I, 517

$C_{15}H_{13}NO_2$ BENZOYLACETANILIDE, III, 108; IV, 80

$C_{15}H_{14}$ 1,1-DIPHENYLCYCLOPROPANE, V, 509

$C_{15}H_{14}NO_6Na$ ETHYL SODIUM PHTHAL-IMIDOMALONATE, II, 384

$C_{15}H_{14}O$ BENZYLACETOPHENONE, I, 101

α,α-DIPHENYLACETONE, III, 343

$C_{15}H_{14}O_2$ α,β-DIPHENYLPROPIONIC

ACID, V, 526

2-PHENYLETHYL BENZOATE, V, 336

$C_{15}H_{15}ClO$ BENZHYDRYL β-CHLORO-ETHYL ETHER, IV, 72

$C_{15}H_{15}N$ cis-2-BENZYL-3-PHENYLAZIRI-DINE, V, 83

$C_{15}H_{15}NO$ p-DIMETHYLAMINOBENZO-PHENONE, I, 217

$C_{15}H_{15}NO_2S$ 2-(p-TOLYLSUL-FONYL)DIHYDROISOINDOLE, V, 1064

$C_{15}H_{15}NO_6$ ETHYL PHTHALIMIDOMAL-ONATE, I, 271

$C_{15}H_{15}N_3O_2$ METHYL RED, I, 374

$C_{15}H_{17}N$ N-METHYL-1,2-DIPHENYL-ETHYLAMINE, IV, 605

N-METHYL-1,2-DIPHENYLETHY-LAMINE HYDROCHLORIDE, IV, 605

N-β-NAPHTHYLPIPERIDINE, V, 816

$C_{15}H_{18}O_4$ ETHYL 6,7-DIMETHOXY-3-METHYLINDENE-2-CARBOXYL-ATE, V, 550

$C_{15}H_{20}O_5$ ETHYL α-ACETYL-β-(2,3-DI-METHOXYPHENYL)PRO-PIONATE, IV, 408

$C_{15}H_{24}O_8$ ETHYL PROPANE-1,1,2,3-TETRACARBOXYLATE, I, 272

ETHYL PROPANE-1,1,3,3-TETRACARBOXYLATE, I, 290

$C_{15}H_{26}O_5$ ETHYL ENANTHYLSUCCIN-ATE, IV, 430

$C_{15}H_{30}O_2$ ETHYL n-TRIDECYLATE, II, 292

METHYL MYRISTATE, III, 605

$C_{15}H_{33}BO_3$ n-AMYL BORATE, II, 107

$C_{15}H_{33}O_4P$ n-AMYL PHOSPHATE, II, 110

$C_{16}H_9Br$ 3-BROMOPYRENE, V, 147

$C_{16}H_9N_5O_{11}$ (+)- and (−)-α-(2,4,5,7-TE-TRANITRO-9-FLUORENYLIDE-NEAMINOOXY)PROPIONIC ACID, V, 1031

$C_{16}H_{10}$ DIPHENYLDIACETYLENE, V, 517

3-HYDROXYPYRENE, V, 632

$C_{16}H_{10}O_3$ 5-FORMYL-4-PHENANTHROIC

ACID, IV, 484

$C_{16}H_{11}NO_2$ AZLACTONE OF α-BENZOYL-AMINOCINNAMIC ACID, II, 490

$C_{16}H_{11}NO_4$ α-PHTHALIMIDO-o-TOLUIC ACID, IV, 810

$C_{16}H_{11}O$ 9-ACETYLPHENANTHRENE, III, 26

$C_{16}H_{12}$ 1-PHENYLNAPHTHALENE, III, 729

$C_{16}H_{12}ClNO$ α-(4-CHLOROPHENYL)-γ-PHENYLACETOACETONITRILE, IV, 174

$C_{16}H_{12}N_2$ 2,3-DIPHENYLSUCCINONI-TRILE, IV, 392

$C_{16}H_{12}O$ 9-ACETYLANTHRACENE, IV, 8

$C_{16}H_{12}O_2$ trans-DIBENZOYLETHYLENE, III, 248

2,3-DIMETHYLANTHRAQUIN-ONE, III, 310

$C_{16}H_{13}N$ 2,4-DIPHENYLPYRROLE, III, 358

$C_{16}H_{13}NO$ α-PHENYL-β-BENZOYLPRO-PIONITRILE, II, 498

$C_{16}H_{13}NO_3$ α-BENZOYLAMINOCIN-NAMIC ACID, II, 490

$C_{16}H_{14}$ 1,4-DIPHENYLBUTADIENE, II, 229

1,4-DIPHENYL-1,3-BUTADIENE, V, 499

$C_{16}H_{14}O$ DYPNONE, III, 367

$C_{16}H_{14}O_3$ β-(1-ACENAPHTHOYL)PRO-PIONIC ACID, III, 9

β-(3-ACENAPHTHOYL)PROPIONIC ACID, III, 6

BENZOIN ACETATE, II, 69

$C_{16}H_{14}O_4$ DIPHENYL SUCCINATE, IV, 390

$C_{16}H_{16}$ 1-METHYL-3-PHENYLINDANE, IV, 665

[2.2]PARACYCLOPHANE, V, 883

$C_{16}H_{16}N_2O_2$ DIBENZOYLDIMETHYLHY-DRAZINE, II, 209

$C_{16}H_{16}O_2$ trans-4,4'-DIMETHOXYSTIL-BENE, V, 428

$C_{16}H_{16}O_3$ DEOXYANISOIN, V, 339

$C_{16}H_{18}$ 1,1 DI-*p*-TOLYLETHANE, I, 229

$C_{16}H_{20}N_2$ *N,N'*-DIETHYLBENZIDINE, IV, 283

$C_{16}H_{20}N_6O_{10}S$ ARGININE DINITRO-NAPHTHOLSULFONATE(FLAVIANATE), II, 50

$C_{16}H_{20}O_6$ DIETHYL[*O*-BENZOYL]ETHYL-TARTRONATE, V, 379

$C_{16}H_{21}ClO_{11}$ 2,4,4,5,6-PENTA-*O*-ACETYL-D-GLUCONYL CHLORIDE, V, 887

$C_{16}H_{21}NO_{10}$ PENTAACETYL *d*-GLUCONO-NITRILE, III, 690

$C_{16}H_{22}O_4$ *sec*-OCTYL HYDROGEN PHTHALATE, I, 418

$C_{16}H_{22}O_{12}$ 2,3,4,5,6-PENTA-*O*-ACETYL-D-GLUCONIC ACID, V, 887

$C_{16}H_{32}O_5S$ α-SULFOPALMITIC ACID, IV, 862

$C_{16}H_{33}I$ *n*-HEXADECYL IODIDE, II, 322

$C_{16}H_{34}$ *n*-HEXADECANE, II, 320

$C_{16}H_{34}O$ CETYL ALCOHOL, II, 374
TRI-*n*-AMYL CARBINOL, II, 603

$C_{17}H_{10}O$ BENZANTHRONE, II, 62

$C_{17}H_{12}O_2$ α-BENZYLIDENE-γ-PHENYL-$Δ^γ$-BUTENOLIDE, V, 80

$C_{17}H_{13}NO_4$ *N*-PHTHALYL-L-β-PHENYL-ALANINE, V, 973

$C_{17}H_{14}O$ DIBENZALACETONE, II, 167

$C_{17}H_{15}NO$ β-NAPHTHOL PHENYL-AMINOMETHANE, I, 381

$C_{17}H_{18}O_3$ ETHYL β-HYDROXY-β,β-DI-PHENYLPROPIONATE, V, 564

$C_{17}H_{20}$ 1,1-DIPHENYLPENTANE, V, 523

$C_{17}H_{34}O_2$ METHYL PALMITATE, III, 605

$C_{18}H_{12}$ TRIPHENYLENE, V, 1120

$C_{18}H_{13}N$ *N*-PHENYLCARBAZOLE, I, 547

$C_{18}H_{14}BrP$ DIPHENYL-*p*-BROMOPHEN-YLPHOSPHINE, V, 496

$C_{18}H_{14}O_3$ 2-(*p*-METHOXYPHENYL)-6-PHENYL-4-PYRONE, V, 721

$C_{18}H_{15}Al$ TRIPHENYLALUMINUM, V, 1116

$C_{18}H_{15}B_3O_3$ BENZENEBORONIC ANHYDRIDE, IV, 68

$C_{18}H_{15}ClSe$ TRIPHENYLSELENONIUM CHLORIDE, II, 241

$C_{18}H_{15}N$ TRIPHENYLAMINE, I, 544

$C_{18}H_{15}NO_4$ AZLACTONE OF α-BENZOYL-AMINO-β-(3,4-DIMETHOXY-PHENYL)-ACRYLIC ACID, II, 55

$C_{18}H_{15}Sb$ TRIPHENYLSTIBINE, I, 550

$C_{18}H_{16}O_4$ 1-(*p*-METHOXYPHENYL)-5-PHENYL-1,3,5-PENTANETRIONE, V, 718

$C_{18}H_{18}O_2$ 1,4-DIBENZOYLBUTANE, II, 169

$C_{18}H_{20}N_2O$ 4,4'-BIS(DIMETHYLAMINO)-BENZIL, V, 111

$C_{18}H_{20}O_2$ ETHYL 2,4-DIPHENYLBUTAN-OATE, V, 559

$C_{18}H_{21}NO_6S$ ETHYL 1-METHYLTHIOL-3-PHTHALIMIDOPROPANE-3,3-DI-CARBOXYLATE, II, 384

$C_{18}H_{30}O_2$ LINOLENIC ACID, III, 531

$C_{18}H_{32}ClO$ OLEOYL CHLORIDE, IV, 739

$C_{18}H_{32}O_2$ LINOLEIC ACID, III, 526
STEAROLIC ACID, III, 785; IV, 851

$C_{18}H_{36}O$ OLEYL ALCOHOL, II, 468; III, 671

$C_{18}H_{36}O_4$ 9,10-DIHYDROXYSTEARIC ACID, IV, 317

$C_{18}H_{37}Br$ OCTADECYL BROMIDE, II, 247

$C_{18}H_{38}O_2$ OCTADECAMETHYLENE GLYCOL, II, 155

$C_{19}H_{15}Cl$ TRIPHENYLCHLORO-METHANE, III, 841

$C_{19}H_{15}Cl_2P$ DICHLOROMETHYLENETRI-PHENYLPHOSPHORANE, V, 361

$C_{19}H_{15}Na$ TRIPHENYLMETHYLSODIUM, II, 607

$C_{19}H_{16}$ TRIPHENYLMETHANE, I, 548

$C_{19}H_{16}N_2$ *N,N'*-DIPHENYLBENZAMI-DINE, IV, 383

$C_{19}H_{16}O$ TRIPHENYLCARBINOL, III, 839

$C_{19}H_{16}O_2$ 2,6-DIMETHYL-3,5-DIPHENYL-4*H*-PYRAN-4-ONE, V, 450

$C_{19}H_{18}O_4$ β-CARBETHOXY-γ,γ-DIPHEN-YLVINYLACETIC ACID, IV, 132

$C_{19}H_{24}$ DIMESITYLMETHANE, V, 422

$C_{19}H_{32}O_3S$ *n*-DODECYL *p*-TOLUENESUL-FONATE, III, 366

$C_{20}H_{14}$ 9-PHENYLPHENANTHRENE, V, 952

$C_{20}H_{14}$ TRIPTYCENE, IV, 964

$C_{20}H_{14}Hg$ MERCURY DI-β-NAPHTHYL, II, 381

$C_{20}H_{16}$ TRIPHENYLETHYLENE, II, 606

$C_{20}H_{18}BrP$ VINYL TRIPHENYLPHOS-PHONIUM BROMIDE, V, 1145

$C_{20}H_{20}N_2O_2$ 2-PHENYL-3-n-PROPYLISOX-AZOLIDINE-4,5-cis-DICARBOX-YLIC ACID N-PHENYLIMIDE, V, 957

$C_{20}H_{30}O_2$ ABIETIC ACID, IV, 1
LEVOPIMARIC ACID, V, 699

$C_{20}H_{38}O_4$ DIMETHYL OCTADECANE-DIOATE, V, 463

$C_{21}H_{14}O$ 2,3-DIPHENYLINDONE, III, 353

$C_{21}H_{15}N$ α,β-DIPHENYLCINNAMONI-TRILE, IV, 387

$C_{21}H_{17}N$ α,α,β-TRIPHENYLPROPIONI-TRILE, IV, 962

$C_{21}H_{18}O$ β,β-DIPHENYLPROPIOPHEN-ONE, II, 236

$C_{21}H_{18}O_2$ α,β,β-TRIPHENYLPROPIONIC ACID, IV, 960

$C_{21}H_{19}NO$ 2,3,5-TRIPHENYLISOXAZOLI-DINE, V, 1124

$C_{21}H_{19}NO_3S$ p-TOLUIDINE 2-PHENAN-THRENESULFONATE, II, 483
p-TOLUIDINE 3-PHENANTHRENE-SULFONATE, II, 483

$C_{21}H_{21}Sb$ TRI-p-TOLYLSTIBINE, I, 551

$C_{22}H_{14}O_9$ AMMONIUM SALT OF AURIN TRICARBOXYLIC ACID, I, 54

$C_{22}H_{32}O_4$ 3β-ACETOXYETIENIC ACID, V, 8

$C_{22}H_{42}O$ TRI-n-HEPTYL CARBINOL, II, 603

$C_{22}H_{42}O_2$ BUTYL OLEATE, II, 469
ERUCIC ACID, II, 258

$C_{22}H_{42}O_4$ DOCOSANEDIOIC ACID, V, 533
ETHYL 1,16-HEXADECANEDICAR-BOXYLATE, III, 401

$C_{23}H_{17}BF_4O$ 2,4,6-TRIPHENYLPYRYLIUM TETRAFLUOROBORATE, V, 1135

$C_{23}H_{32}O_3$ 18,20-LACTONE of 3β-ACE-TOXY-20β-HYDROXY-5-PREG-

NENE-18-OIC ACID, V, 692

$C_{23}H_{38}O_4$ nor-DESOXYCHOLIC ACID, III, 234

$C_{23}H_{44}O_4$ CETYLMALONIC ESTER, IV, 141

$C_{23}H_{46}O$ LAURONE, IV, 560

$C_{24}H_{17}NO_2$ 2,4,6-TRIPHENYLNITROBEN-ZENE, V, 1128

$C_{24}H_{17}O$ 2,4,6-TRIPHENYLPHENOXYL, V, 1130

$C_{24}H_{20}AsCl$ TETRAPHENYLARSONIUM CHLORIDE HYDROCHLORIDE, IV, 910

$C_{24}H_{40}O$ CHOLANE-24-AL, V, 242

$C_{25}H_{21}N$ p-AMINOTETRAPHENYL-METHANE, IV, 47

$C_{26}H_{20}$ TETRAPHENYLETHYLENE, IV, 914

$C_{26}H_{22}O_2$ BENZOPINACOL, II, 71
β-BENZOPINACOLONE, II, 73

$C_{26}H_{22}S_2$ DIBENZOHYDRYL DISULFIDE, II, 574

$C_{27}H_{42}O_2$ Δ^4-CHOLESTEN-3,6-DIONE, IV, 189

$C_{27}H_{44}O$ Δ^4-CHOLESTEN-3-ONE, IV, 192, 195
Δ^5-CHOLESTEN-3-ONE, IV, 195
CHOLESTENONE, III, 207

$C_{27}H_{46}O$ CHOLESTANONE, II, 139
CHOLESTEROL, IV, 195

$C_{27}H_{48}O$ DIHYDROCHOLESTEROL, II, 191

$C_{28}H_{38}O_{19}$ α-CELLOBIOSE OCTAACET-ATE, II, 124
β-GENTIOBIOSE OCTAACETATE, III, 428

$C_{28}H_{50}O$ CHOLESTANYL METHYL ETHER, V, 245

$C_{29}H_{20}O$ TETRAPHENYLCYCLOPENTA-DIENONE, III, 806

$C_{29}H_{48}O_2$ CHOLESTERYL ACETATE, II, 193

$C_{30}H_{22}$ p-QUINQUEPHENYL, V, 985

$C_{30}H_{28}N_4$ BIS(1,3-DIPHENYLIMIDAZOLI-DINYLIDENE-2), V, 115

$C_{32}H_{20}O_3$ TETRAPHENYLPHTHALIC AN-

HYDRIDE, III, 807

$C_{34}H_{24}$ 1,2,3,4-TETRAPHENYL-NAPHTHALENE, V, 1037

$C_{34}H_{28}N_6O_{14}S_4$ COUPLING PRODUCT FROM o-TOLUIDINE AND CHICAGO ACID, II, 145

$C_{34}H_{34}N_4O_4$ HEMIN, III, 443

$C_{35}H_{70}O$ STEARONE, IV, 854

$C_{37}H_{28}O$ TRIBIPHENYLCARBINOL, III, 831

$C_{40}H_{52}O_4$ 3,12-DIACETOXY-bisnor-CHOLANYLDIPHENYLETHYLENE, III,237

$C_{42}H_{30}$ HEXAPHENYLBENZENE, V, 604

$C_{45}H_{86}O_6$ TRIMYRISTIN, I, 538

H_3NO HYDROXYLAMINE HYDROCHLORIDE, I, 318

H_4N_2 HYDRAZINE SULFATE, I, 309

SOLVENTS AND REAGENTS

Purification and/or Preparation

Organic Syntheses procedures frequently include notes describing the purification of solvents and reagents. These have been placed in a single index for convenience. The preparations of useful reagents and catalysts, as well as some techniques, determinations, and tests are also included.

At the present time (1975) many pure anhydrous solvents are available commercially. An increasing number of reagents are available as A.C.S. Analytical Reagent Grades or as spectroscopically pure compounds. A great saving in time and effort can be made by use of these purified products which give good results when applied according to *Organic Syntheses* procedures. However, care must be taken to use fresh, full containers of these compounds in those preparations specifying high purity.

A note of caution. Certain compounds such as ethyl ether, dioxane, tetrahydrofuran, aldehydes, and unsaturated compounds form explosive peroxides by reaction with oxygen. Partially filled containers of such compounds form peroxides when kept for a considerable amount of time. These peroxides should be destroyed by use of suitable reducing agents *before* using the purification procedures suggested.

Mention may be made of the fact that valuable criteria of purity are now available due to the perfection of chromatographic techniques and NMR spectroscopy. Recent volumes of *Organic Syntheses* contain such data.

Note. Most of the entries in this special index are notes to preparations. There are numerous other reagents (e.g., sodium amide, sodium ethoxide, sodium methoxide) whose preparations under optimum conditions as to solvent and concentration are given in the main procedures. These may be found by consulting either the Type of Reaction Index (p. 169) where the reagents are shown in brackets, or the General Index.

307

Acetic acid, purification of, **III,** 5; **IV,** 712
Acetone, purification of, **I,** 239, 460; **III,**
    17, 208; **IV,** 211, 577, 837
Acetonitrile, purification of, **I,** 6; **V,** 143,
    250
Acetophenone, purification of, **I,** 79
Acetylene, purification of, **IV,** 117, 187,
    794
Acid number, **IV,** 3
Aldehydes, stabilization against autoxida-
    tion, **IV,** 447
Alizarin indicator, **IV,** 416
Allyl alcohol, analysis of, **I,** 44
Alumina, activation of, **IV,** 796
    chromatographic, activation of, **IV,**
        965
    preparation for chromatography, **IV,**
        818
Aluminum *t*-butoxide, preparation of, **III,**
    48
Aluminum isopropoxide, preparation of,
    IV, 193
1-Amino-8-naphthol-2,4-disulfonic acid
    (Chicago acid) technical, purifi-
    cation of, **II,** 147
Aniline, purification of, **IV,** 769
Anthracene, purification of, **IV,** 9, 712
Anti-foaming agent, **IV,** 138, 685, 856

Beilstein test, use of, **IV,** 199, 725
Benzaldehyde, purification of, **I,** 95, 105,
    253, 414; **II,** 168; **IV,** 778
Benzene, purification of, **II,** 82; **III,** 29,
    208; **IV,** 556, 592, 702, 901; **V,**
    199, 202, 737
1,2,3-Benzothiazole 1,1-dioxide, assay of,
    V, 64
Benzoyl chloride, purification of, **III,** 113
Benzyl cyanide, purification of, **I,** 108
Boron trifluoride, handling of, **V,** 334
Bromine, drying of, **I,** 246; **III,** 124, 524;
    IV, 400, 619
N-Bromoacetamide, determination of pu-
    rity, **IV,** 105
N-Bromoglutarimide, preparation of, **IV,**
    498

N-Bromosuccinimide, preparation of, **IV,**
    922
Butadiene, purification of, **II,** 105
n-Butyl alcohol, purification of, **II,** 469,
    475
t-Butyl alcohol, purification of, **III,** 49,
    143, 145, 507; **V,** 173, 938
    anhydrous, **IV,** 134, 460
t-Butyl hypochlorite, assay of, **V,** 185
    preparation of, **V,** 184

Carbon tetrachloride, anhydrous, **IV,** 935
Caro's acid, preparation of, **III,** 334
Catalysts
    β-alanine, **IV,** 94
    alumina, **IV,** 520, 796
    ammonium acetate, **IV,** 234, 441, 451,
        463
    ammonium nitrate, **IV,** 21
    benzoyl peroxide, **IV,** 109, 430, 921
    boron trifluoride etherate, **IV,** 375, 957
    chromium oxide-calcium carbonate,
        IV, 579
    copper chromite, **IV,** 216, 314, 324,
        337, 678, 857
    cupric acetate, **IV,** 146
    cuprous oxide + silver oxide, **IV,** 494
    diethylamine, **IV,** 205
    ferric chloride, **IV,** 122, 474, 480, 573,
        851, 970
    ferric nitrate, **IV,** 128, 387, 404, 586,
        755, 763, 963
    ferrous sulfate, Skraup's quinoline
        synthesis, **I,** 478
    iron, **IV,** 114
    mercuric acetate, **IV,** 977
    mercuric sulfate, **IV,** 13
    palladium
        on carbon, **III,** 385, 520, 686, 687;
            IV, 306, 537, 886
        on barium carbonate, **III,** 688
        on barium sulfate, **III,** 685
        on calcium carbonate plus
            lead(Lindlar), **V,** 880
    phosphoric acid, **IV,** 350
    piperidine, acetate, **IV,** 210, 409

piperidine-pyridine, **IV**, 327
platinum black, **IV**, 612
platinum oxide, **I**, 463; **IV**, 305, 306
potassium hydroxide, **IV**, 776
pyridine, **IV**, 5, 243, 381, 732
nickel
 Raney, preparation of, **III**, 176, 181;
  IV, 222, 229, 314, 357, 432, 603,
  639, 660, 672
 U.O.P., use of **III**, 278
 WJ-7, preparation of, **V**, 102
rhodium-palladium, preparation of, **V**,
  671
silica gel, **IV**, 64
sodium acetate, **IV**, 669
sodium ethoxide, **IV**, 184
sodium hydroxide, **IV**, 929
p-toluenesulfonic acid, **IV**, 304, 534
triethylamine, **IV**, 730, 777
Triton B, **IV**, 652
urea, **IV**, 276
Zinc iodide, **IV**, 801
Chlorine, anhydrous, **III**, 483; **IV**, 935
 laboratory preparation of, **II**, 45
o-Chlorobenzoic acid technical, purification of, **II**, 16
Chloroform, anhydrous, **III**, 482; **IV**, 265;
  V, 536
Cholesterol, test for, **II**, 193
Chromatographic purification, **IV**, 816, 965
Chromic acid–sulfuric acid reagent, **V**,
  310, 854, 866
Chromic acid solution, titration of, **IV**, 20
Chromium (II) sulfate, assay, **V**, 994
 preparation of, **V**, 993
Claisen's alkali (aqueous-methanolic potas-
  sium hydroxide solution) prepa-
  ration of, **III**, 269, 663; **IV**, 191;
  V, 1095
Copper bronze, activation of, **III**, 339
Copper powder, **IV**, 337, 628, 732
Copper sulfite, preparation of, **III**, 341
Cuprous bromide, preparation of, **II**, 132;
  III, 186
Cuprous oxide, preparation of, **V**, 108
Cyclohexanone, anhydrous, **IV**, 223

Cyclohexene, purification of, **V**, 857
Cyclopentadiene, preparation of, **V**, 415,
  1002

Diazomethane, analysis of solutions of, **II**,
  166; **V**, 247
 preparation of, **II**, 165; **III**, 119, 244;
  IV, 250; **V**, 246, 351
 solution in xylene, **V**, 1101
 Warning! Caution! **II**, 165; **III**, 244;
  IV, 250; **V**, 351
Diethyl malonate, purification of, **IV**, 287,
  632
Diglyme, purification of, **V**, 391
Dihydropyran, purification of, **V**, 875
1,2-Dimethoxyethane, purification of, **V**,
  355, 719, 776, 970, 1002
Dimethylamine, anhydrous, **IV**, 336
Dimethylketene, determination of, **IV**, 349
Dimethyl sulfate, handling of, **V**, 271
Dimethyl sulfoxide, potassium salt, prepa-
  ration of, **V**, 938
Dimethyl sulfoxide, purification, **V**, 243,
  756, 958
Dinitrogen tetroxide, handling of, **V**, 338
2,4-Dinitro-1-naphthol-7-sulfonic acid (Flavianic acid)
 preparation of, **II**, 51
Dioxane, purification of, **IV**, 643, 848; **V**,
  870
Distillation, azeotropic, **IV**, 192, 304, 313,
  416, 682, 901

Ethanol, drying of, **I**, 249, 251, 259; **II**,
  155, 263, 280, 488; **V**, 992
Ethanolamine, commercial, purification of,
  II, 92
Ethyl acetate, purification of, **I**, 227, 253;
  II, 488; **III**, 19
Ethyl acetate-ethanol, preparation of, **I**,
  235
Ethyl benzoate, purification of, **V**, 938
Ethyl bromide, drying of, **II**, 603
Ethyl n-butyrate, purification of, **II**, 115
Ethyl carbonate, purification of, **II**, 283
Ethyl ether, anhydrous, **IV**, 564, 602, 938

ethanol free, **IV,** 822
 peroxide free, **IV,** 207, 926
Ethylene glycol, anhydrous, **IV,** 449
Ethylene glycol dimethyl ether, purification of, **IV,** 475
Ethyl formate, purification of, **II,** 6, 180; **V,** 189
Ethyl isobutyrate, purification of, **II,** 269
Ethyl pyruvate, assay of, **IV,** 469

Filtration, inverted, **II,** 610
Fluoroboric acid, concentration of, **V,** 246
Formaldoxime, preparation of, **V,** 139
Formalin solutions, composition of, **I,** 378
Furan, purification of, **V,** 404

Glycerol, drying of, **I,** 17; **III,** 503
Grignard reagent, test for, **I,** 190
 titration of, **V,** 212

Hexaflorophosphoric acid, handling of, **V,** 135
Hexane, purification of, **IV,** 610
Hinsberg separation, procedure for, **III,** 161
Hydrazine hydrate, concentration of, **II,** 86
Hydrobromic acid, preparation of, **I,** 26
Hydrogen bromide, preparation of, **II,** 338; **III,** 13; **V,** 546
Hydrogen chloride anhydrous, preparation and drying, **I,** 6, 293, 534; **IV,** 171, 607, 751, 927; **V,** 378
Hydrogen fluoride, handling of, **V,** 66, 138, 482
Hydrogen peroxide, assay of, **IV,** 319
Hydrogen peroxide, 90%, handling of, **V,** 368, 599
Hydrogen selenide, generation of, **IV,** 360
Hydroperoxide, quantitative estimate of, **IV,** 896
Hydroxylamine-0-sulfonic acid, assay of, **V,** 898
 purification of, **V,** 44

Iodine number (Wijs titration), **IV,** 318, 853, 978
N-Iodosuccinimide, preparation of, **V,** 664
Ion-exchange resin, treatment of, **IV,** 39, 530
Isopropyl ether, purification of, **III,** 13
Isopropyl nitrite, preparation of, **III,** 192

Kerosene, purification of, **III,** 72
Ketene, determination of, **V,** 682
β-Ketoesters, test for, **IV,** 122

Lead tetraacetate, assay of, **V,** 724
Lead thiocyanate, preparation of, **V,** 1052
Lithium, handling of, **IV,** 888; **V,** 1094, 1121
Lithium amide, preparation of, **V,** 564
Lithium iodide dihydrate, preparation of, **V,** 78

Magnesium methoxide, preparation of, **V,** 439
Magnesium methyl carbonate, preparation of, **V,** 440
Malonitrile, purification of, **V,** 34
Mercuric sulfate, regeneration of, **V,** 321
Methanol, purification of, **I,** 220; **V,** 912
Methylamine, generation of, **IV,** 607
Methylcyclohexane, purification of, **V,** 1143
Methyl ethyl ketone, purification of, **IV,** 86
Methyl lithium, assay of, **V,** 777
 preparation of, **V,** 860
Monoperphthalic acid, assay of, **V,** 806
 preparation of, **III,** 619

Nitrogen, drying of, **IV,** 70, 134, 926
 oxygen-free, preparation of, **IV,** 27; **V,** 81, 363, 1002, 1142
Nitromethane, drying of, **IV,** 223

Ozone, preparation and use of, **III,** 673

Peroxides, test for, **IV,** 830; **V,** 644, 925
Peroxyacetic acid, assay of, **V,** 417, 661
 handling of, **V,** 416

Peroxybenzoic acid, assay of, **V, 906**
Peroxytrifluoroacetic acid, preparation of,
    V, 367, 598
Phenanthrene, purification of, **III,** 134; **IV,**
    313
Phenyllithium, assay of, **V,** 454
Phosgene, purification of, **IV,** 524
o-Phthalaldehyde, color test for, **IV,** 808
Piperidine, purification of, **V,** 576
Potassium amide, preparation of, **IV,** 963;
    V, 12, 188
Potassium t-butoxide, preparation of, **IV,**
    132, 459; **V,** 306, 361, 1061
Potassium fluoride, anhydrous, **IV,** 526
Potassium hydroxide, methanolic, prepara-
    tion, **IV,** 367
Potassium hydroxide powdered, prepara-
    tion of, **II,** 18
Potassium hypochlorite, preparation of,
    IV, 486
Potassium metal, safe handling of, **IV,** 134
Potassium methoxide, preparation of, **III,**
    71
Potassium phthalimide, purification of, **IV,**
    812
Pyridine, drying of, **II,** 518; **III,** 29, 425; **V,**
    243
    removal of pyrrole from, **V,** 105

Quinoline, purification of, **IV,** 733; **V,** 773
Quinoline-Sulfur catalyst poison, prepara-
    tion of, **III,** 629

Ruthenium trichloride, preparation of, **V,**
    1002

Salicylaldehyde, purification of, **III,** 166
Silver chloride, preparation of, **IV,** 85
Silver carbonate, preparation of, **III,** 435
Silver nitrite, preparation of, **IV,** 725
Silver oxide, preparation of, **II,** 20; **IV,**
    548, 919, 972; **V,** 885, 897
Sodium acetylide, preparation of, **III,** 416
Sodium amalgam, preparation of, **IV,** 509
Sodium amide, preparations, See Note p.
    307

Sodium ethoxide, preparations, See Note
    p. 307
Sodium hydride, handling of, **V,** 718
    in oil, handling of, **V,** 198, 379, 753,
    755
Sodium hypobromite, preparation of, **IV,**
    45; **V,** 9
Sodium hypochlorite, preparation of, **II,**
    429; **IV,** 74, 346
Sodium metal, powdered, **III,** 17, 387, 829;
    V, 386
    suspension, preparation of, **V,** 1090
Sodium methoxide, preparations, See Note
    p. 307
Sodium trichloroacetate, preparation and
    purification of, **V,** 970
Stannous bromide, preparation of, **II,** 132
Stannous chloride, anhydrous, **III,** 627
Sulfur, purification of, **IV,** 668
Sulfuric acid, 100%, preparation of, **IV,**
    978
Sulfur trioxide, distillation of, **IV,** 848; **V,**
    228
    liquid, handling of, **IV,** 864

Tetrahydrofuran, purification and drying,
    IV, 259, 356, 474, 796, 966; **V,**
    203
    Warning! Caution!, **V,** 976
Tetralin, purification of, **IV,** 523, 896
β-Tetralone blue test, **IV,** 906
Thionyl chloride, purification of, **II,** 570
Thiophene, recovery of, **II,** 9
Toluene, anhydrous, **III,** 168; **IV,** 194
p-Toluenesulfonyl chloride, purification of,
    V, 181
1,2,3-Triazoles, 1,4-disubstituted-5-amino-,
    titration of, **IV,** 381
Trichloramine, assay of, **V,** 37
    preparation of, **V,** 35
Triethylamine, purification of, **IV,** 557,
    561; **V,** 515
Triethyloxonium fluoborate, preparation
    of, **V,** 1080
Triethylphosphite, purification of, **V,** 894,
    902

Trimethylamine, generation of, **I,** 528; **IV,** 86

Trimethylamine oxide, preparation of, **V,** 872

Trimethyl borate, purification of, **V,** 920

Trimethyloxonium fluoborate, preparation of, **V,** 1096

Triphenylphosphine bromine complex, purification of, **V,** 142, 249

Unsaturation, bromate-bromide titration, **IV,** 849

Water, deaerated, **IV,** 27, 789, 817

Xylene, purification of, **III,** 19, 389; **IV,** 842

Zinc, amalgamated preparation of, **III,** 786; **IV,** 204, 696

Zinc-copper couple, preparation of, **II,** 185; **V,** 330, 855, 857

Zinc dust, purification of, **III,** 74

APPARATUS INDEX

A number of the procedures in *Organic Syntheses* describe the use of special or less common pieces of apparatus and equipment. References to many of them are recorded in this index. Illustrations are indicated by page numbers in boldface type.

Apparatus for
acyloin cyclization, **IV,** 840
addition of butadiene to maleic anhydride, **IV, 891**
assay of *n*-butylmagnesium chloride, **V,** 1143
bromination of *m*-aminobenzoic acid, **IV, 947**
bromination or chlorination using aluminum chloride catalyst, **V,** 117, 120
carbonation of magnesium methoxide, **V,** 439
caustic fusion of vanillin, **IV, 975**
distillation of diethyl methylenemalonate, **IV, 300**
evacuating and introducing nitrogen into reaction vessels, **IV, 133,** 459
handling boron trifluoride, **V,** 600
handling hydrogen fluoride, **V,** 66, 136, 332
handling hydrogen fluoride and boron trifluoride, **V,** 480
introduction of chlorine gas, **V,** 371
monitoring carbon dioxide evolution, **V,** 127
reaction of ethanol with chlorotrifluoroethylene, **IV, 184**

reactions in liquid ammonia, **V,** 12
reactions of cyanogen chloride and sulfur trioxide, **V,** 227
solvent stripping, **V, 974**
transfer of Grignard reagents, **V,** 1058

Apparatus for Preparation of
acrolein, **I, 15**
alkyl bromides, **I, 39**
alloxantin dihydrate, **IV 26,**
bromoacetal, **III, 123**
2-bromophthalide, **III, 738**
1-*n*-butylpyrrolidine, **III, 160**
β-chloropropionaldehyde acetal, **II, 138**
chromous sulfate, **V,** 994
cyclopropyl cyanide, **III, 223**
diazomethane, **V,** 351
hexamethylene chlorohydrin, **III, 446**
isonitrosopropiophenone, **II, 363**
methyl iodide, **II, 401**
p-nitrophenyl isocyanate, **II, 453**
potassium salt of dimethyl sulfoxide, **V, 938**
Raney nickel catalyst, **III, 177; V,** 103
sodium amide, **III, 779; IV,** 117
thiophosgene, **I, 508**
p-tolualdehyde, **II, 584**
Tribromoaniline, **II, 592**

trimethyloxonium 2,4,6-trinitrobenze-
nesulfonate using diazomethane, **V, 1100**
zinc-copper couple, **V,** 857
**Autoclaves, Pressure Reactors; See also Hy-
drogenation Equipment**
Hastelloy-C vessels for fluorine compounds, **V,** 1083
for alkaline rearrangements, **V,** 814
high pressure reactors, **IV,** 311, 530,
930
steel pressure vessel, **III, 517**
stirred or rocking autoclaves for cycloaddition reactions, **V,** 237,
393, 459
use of hydrogenation bombs as reactors, **IV,** 294, 452, 623, 738, 930;
V, 811

Bath, cooling, Dry Ice and acetone, IV, 69,
75, 268, 747, 751
Dry Ice-ethanol, **IV,** 20, 385
Dry Ice and ethylene glycol mono-
methyl ether, **IV,** 240
Dry Ice and methylene chloride,
IV, 369
Dry Ice and trichloroethylene, **IV,**
60, 387, 404
cooling, large scale, **V,** 312
**Bath, heating, fused salt (potassium nitrate
+ sodium nitrite), IV,** 498
wax, **IV,** 488, 683; **V,** 619
Wood's metal, **IV,** 488, 592, 768, 854
Beaker, stainless-steel, for caustic fusion,
IV, 974; **V,** 617
Blender, Waring-type, IV, 111
Bottle, pressure, IV, 261, 418
**Buret, plastic, for dispensing methylene
chloride solution of diazome-
thane, V,** 247

Capillary, sealed, IV, 209, 211, 941
Capillary tube, evacuated, IV, 199
Chromatography, apparatus for continuous,
V, 730

Condenser, Dry Ice-acetone, IV, 118, 119,
258, 269, 296, 755, 851, 970
for low-boiling liquids, **I,117,** 529
low temperature, **V,** 947
partial, Hahn type, **III,** 231
Continuous reactors, IV, 81, 740

Distillation
Apparatus for,
dehydration, **I, 422**
esterification, azeotropic, **I, 262**
laurone, **IV, 562**
low-boiling liquids, **V,** 23, **682**
solids, **III, 133, IV,** 830
sulfur trioxide, **IV,** 848
Columns;
low temperature, **IV,** 686
for methyl esters, Clarke-Rahrs, **III,**
611
Widmer, **III, 423**
Flasks, special types, **I, 130, 501; II,
18, 227; III, 65, 520**
Steam, **I, 282, 479, 507; III, 7, 65**
rapid, **V,** 1020
solids, **IV, 397**
Dropping funnels,
calibrated for benzene solution of
phosgene, **V,** 201
Dry-Ice cooled, **V,** 675
Hershberg, **II, 129; IV,** 70, **747,** 844,
964; **V,** 891

Electrolysis apparatus,
for Kolbe's synthesis, **V,** 363, 446,
464, **466**
for reduction, **I, 486; III, 61; IV,** 149
Evaporator, reduced pressure, I, 427
rotary, **IV,** 614
Extractors, continuous,
of solutions, **I, 277; II, 378, 615; III,
539; IV,** 20, 46, 580; **V,** 630,
615, 835
of solids, **I, 375, 539; IV,** 167, 548,
564, 645

Filter aids, use of, IV, 100, 115, 217, 283,
358, 365, 581, 660, 838, 909

Filtration, inverted, II, 610
Flasks
  addition of solids, **III, 550, IV, 649**
  Claisen, modified, **IV,** 109
  copper, **IV,** 95, 446
  creased, **IV,** 523, 529, 840
  Dewar, **IV,** 724
  jacketed, **IV,** 431
  Morton, **IV,** 68
  sausage, **IV,** 766
  stainless-steel, **IV,** 95, 446, 449
Flow meter, IV, 64, 126
Flow-rate-indicating fluid, Arochlors as, IV,
  126
Funnel, Büchner, cooling jacket for, IV,
  373, 928
  electrically heated, **IV,** 811
  plastic, **V,** 55
  steam-jacketed, **IV,** 561, 856

Gas dispersion tube, IV, 184, 523, 669, 895
Gasometer, acetylene, I, 230
Gas generation apparatus for
  bromine vapors, **II, 340**
  ethyl nitrite, **II, 205**
  hydrogen bromide, **II, 339; V, 546**
  hydrogen cyanide, **I, 314**
  hydrogen sulfide, **IV, 928**
  methyl nitrite, **II, 363**
  nitrogen oxides, **I, 266; V,** 651
Gas traps
  absorption, **I, 97; II, 4; IV,** 44, 63,
    157, 162, 178, 558
  bubbler, mercury, **IV,** 966
  carbon dioxide collection, **V,** 949
  flow regulator, **IV,** 580

Hopper for additions, I, 233
Hydrogenation apparatus, I, 63, 66; IV,
    217, 222, 299, 305, 357, 409,
    603, 660, 888, 930; **V,** 591, 743,
    881, 989, 1132
  Parr, modified for temperature obser-
    vation, **V,** 240

Infrared heat lamp for controlled heating V,
  327
Ion-exchange column, IV, 40

**Ozonizer and accessories, III, 674, 675, 676,
  677, 682; IV,** 485; **V,** 489

Photolysis, apparatus for, V, 298, cf5529,
  952
Pyrolysis apparatus for
  Condensation of tetrahydropyran with
    aniline, **IV, 795**
  Decarboxylation of coumalic acid, **V,**
    982
  Decomposition of diazonium hexa-
    fluorophosphates, **V,** 134
  Elimination of acetate group, **IV, 746;
    V,** 236
  Hofmann decomposition, **V,** 316
  Preparation of,
    Acrylic acid, **III, 30**
    Azelanitrile, **IV, 63**
    Benzyl methyl ketone, **II, 389**
    Bicyclo(2.1.0) pentane, **V,** 99
    Butadiene, **II, 103**
    Ketene from acetone, **I, 331**
      from diketene, **V,** 679
    Methyl butadienoate, **V,** 734
    2-Octanol, **I, 368**
    Thermal cycloaddition, **V,** 425

Rubber dam, IV, 155, 332, 662, 864, 954

Separator, water,
  automatic, for liquids, **I, 151, 262, 277,
    282, 368; III, 382, 503**
  Barrett-type, **IV, 94**
  Dean and Stark, **IV,** 235, 573, 580,
    606
Shaking machine, I, 63
  Parr Instrument Co., **I, 66**
Shaking, manual swirling, IV, 405
Sodium, device for holding IV, 117, 763
Still, evaporative, IV, 794
  special for thiophosgene, **I, 507, 508**

Stirrers, mechanical
  centrifugal, **IV**, 649
  Duplex-Dispersator, **IV**, 70
  gas dispersion, **IV**, 158
  gas-tight bushing, **III, 178**
  glass propeller, **I, 34**
  glycerin-sealed, **IV**, 291, **546**
  Hershberg, **II, 117**
  high-speed, **IV**, 70, 840; **V**, 40, 1027
  magnetic, **IV**, 252, 308, 339, 340, 532,
      662, 794, 870
  mercury sealed, **I, 33**
  metal for caustic fusion, **IV, 975**
  sealed, **IV**, 89, 368, 650, 862
  Stir-O-Vac, **IV**, 70
  sweep-type, **IV**, 573, 598
  Trubore, **IV**, 89, 683, 724, 807, 814,
      829, 947

  Vibromischer for making sodium
      sand, **V**, 76, 1090
    wire, **II, 534; V**, 819
Sublimation apparatus, IV, 153, 880

Tower, soda-lime, IV, 99, 387
Tangential continuous reactors, IV, 81, 740
Tape, heating, IV, 465, 741
Tube, sealed (Carius), V, 1076
  Pyrex® bomb, **V**, 1093

Ultraviolet lamp, IV, 807, 984

**Vacuum "dumb-bell" desiccator (for drying
    large amounts of materials), III,
    845**

CUMULATIVE AUTHOR INDEX

Each preparation was carried out and written directions submitted by the chemists whose names are shown just beneath the equations for each synthesis. Their names were also listed in the front of Annual Volumes 1–49 as "Contributors".

This cumulative author index lists all the chemists who contributed procedures for Annual Volumes 1–49, which were subsequently published in Collective Volumes I, II, III, IV, V.

Abbott, T. W. **I**, 440; **II**, 10, 270, 515
Abell, R. D., **I**, 205
Abramovitch, B., **III**, 142
Ackerman, J. H., **V**, 668
Adams, A. C., **IV**, 415
Adams, E. W., **I**, 459
Adams J. T., **III**, 291, 405
Adams, Roger, **I**, 10, 61, 84, 94, 101, 107, 109, 128, 214, 237, 240, 250, 270, 280, 309, 341, 355, 358, 366, 394, 436, 459, 463, 504, 528, 531; **II**, 287; **V**, 107
Adkins, Homer, **I**, 1, 15; **II**, 262, 206; **III**, 17, 176, 217, 367, 671, 785
Advani, B. G., **V**, 504
Aguilar, V. N., **V**, 572
Ainsworth, C., **IV**, 536; **V**, 1070
Albert, Adrien, **III**, 53
Alberti, C. G., **III**, 371
Albright, Charles F., **IV**, 68
Alexander, Elliot, **III**, 385
Alexander, Kliem, **IV**, 266
Allen, C. F. H., **I**, 156, 205, 226; **II**, 3, 15, 62, 81, 128, 156, 219, 221, 228, 336, 387, 498, 517, 539, 580; **III**,

1, 28, 63, 76, 78, 96, 136, 140, 203, 275, 310, 312, 353, 358, 377, 394, 404, 415, 418, 573, 578, 597, 621, 710, 727, 731, 733, 765, 783, 796, 824, 827, 846; **IV**, 45, 80, 433, 739, 804, 866
Allen, C. Freeman, **IV**, 398, 608, 616,
Allen, George A., **III**, 59
Allen Jr., P., **I**, 77
Allen, R. E., **III**, 200
Alles, Gorden A., **III**, 644, 723, 763
Allinger, Norman L., **IV**, 840
Alt, G. H., **V**, 208
Althausen, Darrell, **II**, 270
Alvardo, A. M., **I**, 3
Amend, W. J., **II**, 325
Amin, G. C., **III**, 280
Amin, J. H., **V**, 36, 580
Amstutz, E. D., **II**, 462
Amundsen, Lawrence, H., **III**, 254, 256
Anderson, Arthur, G., Jr., **IV**, 221
Anderson, B. C., **V**, 235
Anderson, Ernest, **I**, 67
Andreades, S., **V**, 679

Andrus, D. W., **III**, 276, 692, 794
Angyal, S. J., **IV**, 690
Anker, H. S., **III**, 172
Ansbacher, S., **III**, 138,307
Anslow, W. K., **I**, 298
Anspon, Harry, D. **III**, 711, 712
Archer, W. L., **IV**, 331
Arens J. F., **V**, 211
Armendt, B. F., **I**, 417
Arndt, F., **II**, 165, 461; **III**, 231
Arnold, H. R., **II**, 142
Arnold, Richard, T., **II**, 10
Arters, A. A., **III**, 833
Ashley, W. C., **III**, 725
Ashworth, P. J., **IV**, 128
Asscher, M., **V**, 1076
Aston, J. G., **III**, 538
Atkinson, Edward R., **IV**, 872
Atkinson, E. R., **I**, 222
Audrieth, L. F., **III**, 516, 536
Augustine, Robert L., **V**, 869
Ault, Addison, **V**, 932
Ayers, Gilbert B., **I**, 23

Babasinian, V. S., **II**, 466
Babock, G, S, **I**, 85
Babcock, S. H., **II**, 328
Babers, Frank H., **II**, 412, 512
Bach, Robert D., **V**, 315, 1145
Bachman, G. Bryant, **II**, 323
Bachmann, W. E., **I**, 113; **II**, 71, 73; **III**, 839, 841
Backer, H. J., **IV**, 225, 250, 943
Bader, Alfred R., **IV**, 898
Badhwar, I, C., **II**, 304
Bailey, C. F., **II**, 563
Bailey, Philip S., **V**, 489, 493
Baizer, Manuel, **III**, 323
Bak, B., **IV**, 207
Baker, Robert, H., **III**, 141
Baker, Wilson, **I**, 181, 451
Balaban, A. T., **V**, 1106, 1112, 1114
Balasubramanian, M., **V**, 439
Balasubrahmanyam, S.N., **V**, 439
Balcom, Don M., **IV**, 603
Ballard, Donald A., **I**, 89

Ballard, Murel M., **II**, 97
Baltzly, R., **III**, 239
Banks, Charles V., **IV**, 229
Bannard, R. A. B., **IV**, 393
Bantjes, A., **IV**, 534
Bare, Thomas M., **V**, 775
Barger, G., **II**, 384
Barker, W. E., **II**, 156
Barnes, R. P., **III**, 551,555
Barrett, J. H., **V**, 467
Barthel, John W., **IV**, 218
Bartlett, Paul D., **V**, 194, 196, 689
Bass, K. C., **V**, 398
Bast, Klaus, **V**, 650
Baude, Frederic J., **V**, 567
Baum, A. A., **V**, 797
Baumgarten, Henry E., **V**, 909
Bavin, P. M. G., **V**, 30
Baxter, J. F., **III**, 154
Beaber, N. J., **I**, 145
Beal, G. D., **I**, 379, 538
Bean, H. J., **II**, 29
Beaujon, Jan H. R., **V**, 376
Beavers, Ellington M., **V**, 269
Becker, Ernest I., **IV**, 174, 176, 623, 657, 771; **V**, 111, 552
Bedell, S. F., **IV**, 810
Bedoukian, Paul, Z., **III**, 127
Behr, Letha Davies, **II**, 19,562
Behr, Lyell C., **IV**, 342
Bell, Alan, **III**, 78, 96, 310, 312, 404, 415, 846
Bell, E. V., **II**, 223
Bell, E. W., **IV**, 125
Bellis, M. P., **IV**, 722,903
Bender, J. A., **I**, 410
Bender, Myron L., **IV**, 932
Bennett, G. M., **II**, 223
Bennett, Frank, **IV**, 359
Benson, Richard, **III**, 105
Benson, Richard E., **IV**, 588, 746
Benson, W. L., **II**, 273
Benson, W. R., **V**, 663
Berchet, G. J., **II**, 397
Bere, C. M., **I**, 7
Berenbaum, M. B., **IV**, 66, 273, 274

Berger, Alfred, III, 786
Bergstrom, F, W., III, 778
Berlinguet, L., IV, 496
Berndt, A., V, 1128, 1130
Berthold, Robert V., V, 703
Bertz, R. T., IV, 489
Betti, M., I, 381
Beyer, E., V, 365
Bigelow, H. E., II, 57, III, 103
Bigelow, Lucius A., I, 80, 133, 135, 136, 476; II, 244
Biggs, B, S, III, 229, 768
Bill, J. C., IV, 807
Billek, Gerhard, V, 627
Billica, Harry, R., III, 176
Billig, Franklin A., IV, 68
Billman, John, H., III, 360
Bircher, L. J., I, 485
Bisgrove, D. E., IV, 372
Bishop, W. S., III, 229, 584, 613, 768
Bissing, D. E., V, 361
Bistline, R. G., Jr., IV, 862
Bithos, Joe., J. V, 591
Black, Alvin P., II, 412,512
Blackwood, Robert K., IV, 454
Blake, J., IV, 327
Blanchard, K. C., I, 453
Blankley, C. John, V, 258
Blardinelli, Albert J., IV, 311
Blatter, H. M., V, 656
Blegen, James, R. III, 116
Bloch, Henery S., III, 637
Block, Paul, Jr., V, 381, 1031
Blomquist, A. T., IV, 216, 838
Blues, E. T., V, 1141
Boatman, S., V, 187, 767
Boden, H., V, 1029
Bodesheim, Ferdinand, V, 300
Boehme, Werner R., IV, 590
Boekelheide, V., IV, 298, 641
Bogert, Marston, T., I, 220
Böhme, Horst, III, 619
Boisselle, A. P., V, 406
Bollinger, K. M., I, 473
Bordwell, Frederick G., III, 141; IV, 846
Bornstein J., IV, 329, 810;V, 406, 1064

Borum, O. H., IV, 5
Bose, Ajay K., V, 973
Bost, R. W., II, 547, 610; III, 98
Bottini, Albert., V, 121, 124, 541
Bottorff, E. M., V, 687
Boulton, A. J., V, 1112, 1114
Bourgel, M., I, 186, 191, 209
Bourgeois, R. C., IV, 918
Bourns, A. N., IV, 795
Bousquet, E. W., I, 526; II, 313
Boutwell, P. E., III, 605
Bowden, Everett, I, 424; II, 414
Bowen, Douglas M., III, 553
Boyd, Thomas, I, 36
Boyer, J. H., IV, 75
Boyer, J. H., IV, 532; V, 1067
Bramann, G. M., I, 81
Brand, E., II, 49; III, 440
Brand, F. C., III, 440
Brasen, W. R., IV, 508, 582, 585
Braude, E. A., IV, 698, 927
Braun, Charles E., III, 430; IV, 8, 711; V, 887
Braun, Géza, I, 431; II, 122, 124, 256, 308; III, 101
Bredereck, H., V, 794
Bremer, Keith, IV, 777
Brenninger, W., V, 533, 808, 1018
Brent, John T., IV, 342
Breslow, David, III, 408
Breslow, R., V, 514
Brethen, M. R., I, 121, 162; II, 278
Brewster, R. Q., I, 323, 447; II, 347, 445, 574; III, 337
Brooks, L. A., II, 171, 531; III, 471, 698
Brotherton, T. K., V, 816
Brown, B. K., I, 309, 528
Brown, E. A., IV, 960
Brown, George, Bosworth, III, 615
Brown, Herbert C., V, 977
Brown, J. B., IV, 851
Brown, J. F., Jr., IV, 372
Brown, M., V, 1043
Brubaker, M. M., I, 485
Bruce, William F., II, 12, 139, 191; IV, 788

Brüning, Ingrid, **V**, 957, 1124
Bryce-Smith, D. **V**, 993, 1141
Bublitz, D. E., **V**, 1001
Buc, Saul R., **III**, 93; **IV**, 101
Buck, Allen, **III**, 195, 270, 449
Buck, J. S., **II**, 44, 55, 130, 333, 549, 619, 622, 290; **III**, 239
Buehler, C. A., **II**, 341; **III**, 468
Buchdahl, M. R., **IV**, 106
Buchner, B., **IV**, 784
Buckles, Robert E., **IV**, 256, 722, 777, 857, 914
Budde, W. M., **IV**, 31
Budewitz, E. P., **IV**, 748
Bulford, R. H., **II**, 494
Bunnett, J. F., **IV**, 114; **V**, 12, 478, 816
Burckhalter, J. H., **IV**, 333
Bürger, W., **V**, 221
Burgstahler, Albert W., **V**, 251, 591
Buriks, R. S., **V**, 1067
Burks, R. E.Jr., **III**, 785
Burness, D. M., **IV**, 628, 649; **V**, 403
Burnett, Robert E., **II**, 246, 338
Burpitt, R. D., **V**, 277
Burt, Pauline, **I**, 494
Burton, Donald J., **V**, 949
Buswell, R. J., **II**, 276
Butler, John Mann, **IV**, 486
Byers, J. R., **II**, 31, 322, 451
Byers, J. R., Jr., **IV**, 739

Caesar, P. D., **IV**, 693, 695
Cadogan, J. I. G., **V**, 941
Cairns, Theodore L., **II**, 604; **IV**, 588
Calkins, A. E., **III**, 544, 547
Callen, Joseph E., **III**, 26, 134, 212
Calvery, H. O., **I**, 228, 533; **II**, 287
Campaigne, E., **IV**, 31, 331, 918, 919, 921
Campbell, Barbara, K., **III**, 148; **IV**, 117, 763
Campbell, I. D., **V**, 517
Campbell, Kenneth N., **III**, 148, 446; **IV**, 117, 763
Campbell, Tod W., **V**, 499, 501, 985
Cannon, George W., **IV**, 597
Caputo, Joseph A., **V**, 869

Carboni, R. A., **IV**, 877
Carhart, Homer W., **IV**, 436
Carlsmith, Allan, **IV**, 828
Carlson, H. D., **V**, 679
Carmack, Marvin, **IV**, 669
Carpenter, Wayne R., **V**, 288
Carpino, Barbara A., **V**, 157, 166
Carpino, Louis A., **V**, 157, 160, 166
Carter, H. E., **III**, 167, 774, 813; **V**, 946
Carter, K. N., **III**, 154
Carter, P. H., **V**, 339
Casanova, Joseph, Jr., **V**, 273, 772
Cason, James, **III**, 3, 169, 601, 627; **IV**, 510, 555, 630, 635
Castle, J. E., **V**, 393
Castro, C. E., **V**, 993
Cate, W. A., **II**, 341
Catlin, W. E., **I**, 186, 471
Caudle, E. C., **I**, 224
Cava, M. P., **V**, 944
Cejka, L., **I**, 283
Chadhary, Sohan S., **V**, 35
Chadwell, H. M., **I**, 78
Chalmers, William, **II**, 598
Chanan, Henry, **III**, 121
Chao, Tai Siang, **IV**, 380
Chard, S. J., **IV**, 520
Charles Robert G., **IV**, 869
Chaux, R., **I**, 166
Chaykovsky, Michael, **V**, 171,755
Cheetham, H. C., **I**, 70
Chiles, H. M., **I**, 10, 237
Chinn, Leland J., **IV**, 459
Christiansen, W. G., **I**, 490
Chudd, C. C., **IV**, 748
Ciganek, Engelbert, **IV**, 339, 612
Clapp, L. B., **IV**, 372
Clark, John R., **II**, 181; **III**, 260
Clark, R. D., **IV**, 674
Clark, R. Donald,**V**, 456
Clarke, H. T., **I**, 14, 85, 87, 91, 115, 121, 150, 153, 233, 261, 272, 304, 374, 415, 421, 455, 478, 495, 514, 523, 541, 543; **II**, 19, 29, 135, 562, 588; **III**, 172, 226, 690
Cleland, Elizabeth, S., **III**, 360

Clemens, David H., **IV**, 463, 662, 664
Cleveland, E. A., **III**, 339, 60, 227, 465
Clibbens, D. A., **II**, 557
Closson, R. D., **V**, 575
Coan, Stephen B., **IV**, 174, 176
Coburn, E. R., **III**, 313, 696
Cockerille, F. O., **II**, 468
Cocolas, George, **V**, 376
Coffman, Donald D., **III**, 320
Cohen, F. L. **I**, 240
Cohn, E. J., **II**, 120
Coleman, G. H., **I**, 3, 158, 183, 442
Coleman, George H., **II**, 89, 179,
443, 583, 592; **III**, 60, 134, 159, 212, 221,
434, 668, 701, 712
Coleman, Gerald H., **I**, 214
Collins, David, **V**, 166
Collins, Peter M., **V**, 598
Colonge, J., **IV**, 350, 601
Conant, J. B., **I**, 49, 199, 211, 292, 294,
345; **II**, 33
Conard, Charles R., **II**, 167
Conn, M. W., **II**, 547
Conner, R. M., **V**, 478
Connor, Ralph, **II**, 181, 441; **III**, 260, 363
Conrow, Kenneth, **V**, 1138
Constable, E. W., **II**, 610
Converse, S., **I**, 226
Cook, Clinton D., **IV**, 711; **V**, 887
Cook, William B., **III**, 71
Cooke, Giles B., **II**, 406
Cooper, F. C., **IV**, 769
Cope, Arthur C., **II**, 181; **III**, 219, 385,
397, 399, 501; **IV**, 62, 218, 234,
304, 339, 377, 612, 816, 890; **V**,
315
Copenhaver, J. E., **I**, 142
Corbin, Thomas F., **V**, 328
Corey, E. J., **V**, 755
Cornell, John H. Jr., **V**, 422
Cornforth, J. W., **IV**, 467
Corson, B. B., **I**, 49, 179, 241, 336; **II**, 33,
69, 151, 229, 231, 273, 378, 434,
596; **IV**, 884
Cortese, Frank, **II**, 9, 564
Cotter, Robert J., **IV**, 62, 377

Cowan, Dorotha, **V**, 890
Cowper, R. M., **II**, 480
Cox, Gerald J., **II**, 612
Cox, Lucile, **III**, 286
Cox, R. F. B., **II**, 7, 272, 279, 453
Cox, W. M., **II**, 468
Coyner, Eugene C., **IV**, 727
Cragoe, Edward J. Jr., **V**, 635
Craig, David, **II**, 179, 583
Craig, W. E., **IV**, 130
Cram, Donald, J., **III**, 125; **V**, 926
Crandall, J., **V**, 863, 866
Creegan, Francis J., **V**, 623
Cressman, Homer, W. J., **III**, 595, 608,
609, 796
Criegee, R., **V**, 370
Cripps, H. N., **V**, 22, 459
Crocker, Richard E., **IV**, 278
Cromwell, Norman H., **III**, 105, 125
Crosby, Donald G., **V**, 654, 703
Crossley, Frank, **II**, 363; **III**, 599, 617
Crovetti, Aldo J., **IV**, 166, 654, 704
Crowley, Paul., **V**, 157, 160
Croxall, W. J., **IV**, 98
Cruickshank, Philip A., **V**, 555
Culhane, Paul J., **I**, 56, 125, 408
Cullinane, N. M., **IV**, 520
Cupas, C., **V**, 16
Curtin, David Y., **V**, 1092
Curtis, Omer E., JR., **IV**, 278
Cymerman-Craig, J., **IV**, 205, 667, 700; **V**,
88, 339

Dabovich, Thomas, C., **III**, 39
D'Addieco, Alfred A., **IV**, 234
Daeniker, H. U., **V**, 989
Daignault, Ronald A., **V**, 303
Dains, F. B., **I**, 323, 447; **II**, 355
Dakin, H. D., **I**, 149; **II**, 555
Damerell, V. R., **II**, 204
Damschroder, R. E., **III**, 107
Darby, William J., **III**, 460
Darling, S. D., **V**, 918
Daub, Guido H., **IV**, 390
Dauben, Hyp J., Jr., **IV**, 221
Davidsdon, David, **II**, 590

Davidson, L. H., **II**, 480
Davis, Anne W., **I**, 261, 421, 478
Davis, R. B., **IV**, 392
Davis, Robert, **V**, 589
Davis, Tenny L., **I**, 140, 302, 399, 453
Dawkins, C. W. C., **IV**, 520
Dayan, J. E., **IV**, 499
Deacon, B. D., **IV**, 569
Dean, F. H., **V**, 136,580
Deana, A. A., **V**, 944
Deatherage, F. E., **IV**, 851
DeBoer, Charles D., **V**, 528
de Boer, Th. J., **IV**, 225, 250, 943
Deebel, George F., **III**, 468; **IV**, 579
Degering, E. F., **I**, 36
Deghenghi, R., **V**, 645
Dehn, William M., **I**, 89
Denekas, M. O., **III**, 317
DePuy, C. H., **V**, 324, 326, 1058
Dessy, R. E., **IV**, 484
de Stevens, G., **V**, 656
DeTar, D. F., **IV**, 34, 730
Dev, Vasu, **V**, 121
de Vries, G., **V**, 223
Dewar, M. J. S., **V**, 727
Dewar, R. B. K., **V**, 727
DeWitt, C. C., **II**, 25
Diamanti, Joseph, **V**, 198
Dickel, Geraldine B., **IV**, 317
Dickey. J. B., **II**, 21, 31, 60, 173, 175, 242, 322, 451, 494; **III**, 573
Dickinson, C. L., **IV**, 276
Diebold, James L., **IV**, 254
Diehl, Harvey, **III**, 370, 372; **IV**, 229
Dimming, D. A., **III**, 165
Dimroth, K., **V**, 1128, 1130, 1135
Dinsmore, R., **II**, 358
Dobson, A. G., **IV**, 854
Dodge, Ruth A., **I**, 241, 336
Doering, W. E., **III**, 50
Dolliver, Morris A., **II**, 167
Donahoe, Hugh B., **IV**, 157
Donaldson, M. M., **V**, 16
Donleavy, John J., **II**, 422
Dorn, H., **V**, 39
Dornfeld, Clinton A., **III**, 26, 134, 212, 701

Dorsky, Julian, **III**, 538
Douglass, Irwin B., **V**, 709, 710
Doumani, Thomas, F., **III**, 653
Dowbenko, R., **V**, 93
Dox, A. W., **I**, 5, 266
Drake, N. L., **I**, 77; **II**, 406
Drake, W. V., **II**, 196
Dreger, E. E., **I**, 14, 87, 238, 258, 304, 306, 357, 495; **II**, 150
Drummond, P. E., **IV**, 810
Dryden, Hugh L., Jr., **IV**, 816
Dubuis, R., **V**, 880
Duncan, Warren G., **V**, 390
Dunn, M. S., **IV**, 55
Dunnavant, W. R., **IV**, 962; **V**, 526, 564
Durham, Lois J., **IV**, 510, 555, 635
Dutton, G. R., **II**, 109
du Vigneaud, V., **I**, 48
Dyson, G. Malcolm, **I**, 165, 506

Easley, William K., **IV**, 892
Eastham, Jerome F., **IV**, 192
Eatough, Harry, **I**, 80
Eberly, Floyd, **II**, 355
Eck, J. C., **II**, 28, 74, 76, 371, 374
Edens, C. O., **III**, 394, 731
Edgar, Graham, **I**, 172
Egan, Richard, **III**, 270, 449
Eglinton, G., **V**, 517
Ehrenfeld, Louis, **I**, 388
Ehrhart, Wendell A., **V**, 582
Ehrlich, P., **II**, 457
Eilers, K. L., **V**, 326
Eisenbraun, E. J., **V**, 8, 310
Eitel, D. E., **V**, 797
Elam, Edward U., **V**, 1103
Eliel, Ernest L., **IV**, 58, 169, 626; **V**, 175, 303
Ellingboe, E. K., **IV**, 928
Ellis, B. A., **I**, 18
Ellis, Ray C., **IV**, 597
Elser, W., **V**, 780
Elsinger, Fritz, **V**, 76
Embleton, H. W., **IV**, 795
Emerson, Oliver, H., **III**, 151
Emerson, William S., **III**, 831; **IV**, 302, 311, 579, 660, 677, 980

Emmick, Thomas L., **V**, 450
Emmons, W. D., **V**, 191, 367, 547
Endler, Abraham S., **IV**, 657
Endres, Leland, **V**, 285
England, D. C., **V**, 239
Englinton, Geoffrey, **IV**, 404, 755
English, J., Jr., **IV**, 499
Englund, Bruce, **IV**, 184, 423
Erickson, Floyd B., **IV**, 430
Erickson, Ronald E., **V**, 489, 493
Estes, Leland L., **IV**, 62
Evanega, George, **V**, 269
Evans, J. C. W., **II**, 517
Evans, Russell F., **V**, 977
Evans, Wm. Lloyd, **II**, 17, 137, 305, 307;
    III, 432
Evers, W. L., **II**, 408

Faber, E. M., **II**, 576
Fahrenholtz, Susan R., **V**, 151
Falkoff, M. M., **III**, 350
Fanta, Paul E., **III**, 267, 661; **IV**, 844
Farah, Basil Said, **V**, 709
Farlow, Mark W., **II**, 312; **IV**, 521
Farmer, H. H., **IV**, 441
Fatiadi, A. J., **V**, 595, 1011
Fawcett, F. S., **V**, 883
Fawcett, J. S., **IV**, 698
Feely, Wayne, **IV**, 298; **V**, 269
Fegley, Marion, F., **III**, 488; **IV**, 98
Fehnel, Edward A., **IV**, 669
Feldkamp, R. F., **IV**, 671
Feldman, A. M., **V**, 355
Ferretti, Aldo, **V**, 107, 419
Ferris, J. P., **V**, 32, 344
Ferry Clayton W., **II**, 290; **III**, 239
Fetscher, Charles A., **IV**, 735
Feuer, Henry, **IV**, 368, 554
Field, Lamar, **IV**, 674; **V**, 723
Field, R. F., **III**, 317
Fieser, Louis F., **I**, 353, 383, 517; **II**, 35,
    39, 42, 145, 194, 423, 430, 482,
    560, 569; **III**, 6, 98, 465, 590,
    633; **IV**, 189, 195; **V**, 604, 1037
Filler, Robert, **V**, 80
Findlay, J. W. A., **V**, 428
Finelli, A. F., **IV**, 461, 776, 790

Finkel, J. M., **V**, 332
Fischer, Hans, **II**, 198, 202, 217; **III**, 442,
    513
Fischer, Nikolaus, **V**, 877
Fish, M. S., **IV**, 327, 408
Fisk, Milton T., **IV**, 169, 626
Fitch, Howard, M, **III**, 658
Flood, D. T., **II**, 295
Flosdorf, E. W., **I**, 462
Floyd, Don E., **IV**, 141
Fones, William S., **IV**, 293
Fonken, Gunther S., **IV**, 261
Ford, Jared, H., **III**, 34
Ford-Moore, A. H., **IV**, 84, 325, 955
Ford, S. G., **II**, 140, 368, 372, 382
Formo, M. W., **III**, 483, 486
Foster, Duncan G., **III**, 771
Foster, G. L., **II**, 330
Foster, H. M., **IV**, 638
Foster, L. S., **III**, 157
Fox, B. A., **V**, 346
Frank, Robert, L., **III**, 116, 167, 328, 410,
    499, 735, 829; **IV**, 451
Frankenfield, E., **V**, 1018
Franzen, Volker, **V**, 872
Fraser, W. A., **I**, 475
Freeborn, Emeline, **II**, 231
Freeman, Jeremiah P., **IV**, 58; **V**, 839
Freeman, R. C., **V**, 387
Freudenberger, H., **II**, 573
Freudenberg, W., **III**, 541
Friedman, Lester, **III**, 510; **V**, 54, 810,
    1055
Frisell, C., **V**, 642, 924
Frye, John R., **III**, 282
Fujiwara, Kunio, **IV**, 72
Fukushima D. K., **III**, 809
Fuqua, Samuel A., **V**, 390
Furrow, C. L., Jr., **IV**, 242
Fuson, Reynold C., **II**, 169, 260; **III**, 209,
    339, 549, 557

Gaibel, Z. L. F., **V**, 727
Galun, A. B., **V**, 130
Gander, Robert, **III**, 200
Gannon, Walter F., **V**, 294, 539
Gardner, J. H., **II**, 523, 526

Garrison, W. E., Jr., **V,** 463
Gassman, P. G., **V,** 91, 96, 424
Gates, J. W. Jr., **III,** 140, 353, 363, 418
Gaudry, Roger, **III,** 436; **IV,** 496
Gauerke, C. G., **I,** 123
Gaylord, Norman G., **IV,** 178
Gelblum, E., **V,** 355
Gellerson, Hilda E., **IV,** 903
Gerhart, H. L., **II,** 111, 112
Gerold, Corinne, **IV,** 104
Gershbein, Leon, **III,** 458
Gilbert, Leonard S., **V,** 904
Gillespie, H. B., **II,** 489
Gillespie, R. H., **III,** 671
Gillis, Richard G., **IV,** 396
Gilman, Henry, **I,** 145, 188, 361, 471; **II,**
    47, 425
Ginsberg, Helen, **III,** 742
Girard, André, **II,** 85
Girard, F. H., **V,** 572
Giza, Chester A., **V,** 157, 166
Glass, D. B., **III,** 504
Glickman, Samuel A., **IV,** 234
Godt, H. C. Jr., **V,** 893
Gofton, B. F., **IV,** 927
Goheen, D. W., **IV,** 594
Goldstein, Albert, **IV,** 216, 838
Gollis, Morton H., **V,** 422
Gomberg, M., **I,** 113
Gompper, R., **V,** 780
Goodman, Marjorie M., **V,** 162
Gortner, R. A., **I,** 194
Goshorn, R. H., **I,** 36; **IV,** 307
Gösl, R., **V,** 43
Gott, P. Glenn, **V,** 1103
Gottfried, Sidney, **III,** 381
Göwecke, Siegfried, **V,** 166
Gradsten, M. A., **IV,** 518
Graf, Roderich, **V,** 226, 673
Granito, Charles, **V,** 589
Grashey, Rudolf, **V,** 957, 1124
Gray, A. R., **II,** 60, 509
Greenwood, F. L., **III,** 673; **IV,** 108
Greull, Gerhard, **V,** 320
Griffin, K. P., **III,** 242
Gringas, Leopold, **III,** 95

Grob, C. A., **V,** 989
Groening, Theodore, **II,** 445
Gronowitz, S., **V,** 149
Gronwall, Susanne, **V,** 155
Gross, H., **V,** 49, 221, 365
Grummitt, Oliver, **III,** 195, 270, 449, 806,
    807, 833; **IV,** 748, 771; **V,** 320
Grunewald, Gary L., **V,** 982
Guha, Maya, **V,** 384
Guha, P. C., **III,** 623
Gulati, K. C., **II,** 522
Gulen, R., **IV,** 679
Gumprecht, W. H., **V,** 147, 632
Gunstone, F. D., **IV,** 160
Gutenstein, Morris, **V,** 822
Guthrie, J. L., **IV,** 513
Gutsche, C. David, **IV,** 780, 887

Haaf, W., **V,** 20, 739
Hach, Clifford C., **IV,** 229
Haddadin, Makhluf J., **V,** 1037
Haeseler, P. R., **I,** 196
Hafner, K., **V,** 431, 1088, 1106
Hager, F. D., **I,** 58, 224, 248, 351, 544
Hahn, Roger C., **V,** 328
Hain, B. E., **III,** 605
Hall, Luther A. R., **IV,** 333
Hamilton, Cliff S., **IV,** 950
Hamilton, Frances H., **I,** 231, 492, 519
Hamilton, Robert W., **IV,** 576
Hammond, George S., **V,** 528
Hampton, K. Gerald, **V,** 838
Hamrick, Phillip J., **V,** 523
Hancock, Evelyn M., **III,** 219, 397, 399,
    501
Handrick, R. G., **III,** 471
Hankin, Hilda, **III,** 742
Hanslick, Roy S., **II,** 244; **IV,** 788
Hanson, E. R., **I,** 161, 326
Hansuld, Mary K., **IV,** 795
Hargrove, W. W., **V,** 117
Harman, R. E., **IV,** 148
Harris, Arlo D., **V,** 736
Harris, Charles H., **III,** 125
Harris, G. C., **IV,** 1
Harris, S. A., **I,** 241, 336

Harris, T. M., **V**, 187, 718, 848
Harrison, George C., **III**, 370, 372
Harrison, J. G., **V**, 946
Harrisson, R. J., **IV**, 493
Hart, Harold, **IV**, 278; **V**, 598
Harting, Walter, H., **III**, 191
Hartman, R. J., **IV**, 93
Hartman, W. W., **I**, 162, 175, 233, 357,
    455, 541, 543; **II**, 21, 150, 163,
    173, 175, 183, 232, 242, 278,
    322, 404, 418, 438, 440, 451,
    460, 464, 617; **III**, 20, 37, 82,
    650, 652
Hartough, Howard, D., **III**, 14
Hartung, Walter H., **I**, 15; **II**, 363, 543; **V**,
    376
Hartwell, J. L., **II**, 145; **III**, 98, 185
Hasek, Robert H., **V**, 456, 1103
Hasek, W. R., **V**, 1082
Hass, Henry, **II**, 17, 137, 305, 307; **IV**, 932
Hatt, H. H., **II**, 208, 211, 395; **IV**, 854
Hauck, Hans, **V**, 957, 1124
Hauser, C. R., **II**, 67, 268, 607; **III**, 142,
    291, 374, 405, 408, 593, 842; **IV**,
    508, 582, 585, 708, 962; **V**, 187,
    434, 437, 523, 564, 578, 621,
    718, 721, 767, 848
Hays, High R., **V**, 562
Hazen R. K., **I**, 241; **II**, 434
Heaney, H., **V**, 1120
Hearst, Peter J., **IV**, 571
Heator, J. S., **I**, 207
Hedrick, Glen W., **V**, 699
Heidelberger, M., **I**, 153, 488
Heidtke, W. J., **IV**, 960
Heilbron, I. M., **I**, 207
Heininger, S. A., **IV**, 146
Heinzelman, R. V., **IV**, 573
Heisig, G. B., **I**, 54
Helferich, B., **I**, 147, 364; **III**, 428
Helmkamp, G. K., **V**, 1099
Henderson, A., **V**, 355
Hendry, J. L., **II**, 120
Hennion, G. F., **IV**, 683
Hepworth, John D., **V**, 27
Herbst, R. M., **II**, 1, 11, 389, 491, 519

Hering, H., **IV**, 203
Herkes, Frank E., **V**, 949
Herrick, Elbert C., **IV**, 304, 890
Hershberg, E. B., **I**, 168; **II**, 102, 194, 423;
    III, 627
Herzog, Hershel L., **IV**, 753
Hessler, John C., **I**, 438
Hetzner, H. P., **III**, 839
Heusler, K., **V**, 692
Hexner, Peter E., **IV**, 351
Hey, D. H., **V**, 51
Hibbert, Harold, **I**, 494
Hiers, G. S., **I**, 58, 117, 327, 550
Higgins, J. G., **V**, 142, 249
Hill, G. A., **I**, 81, 462
Hill, Julian W., **II**, 53, 325
Hill, Robert M., **I**, 300
Hillebert A., **IV**, 207
Hinegardner, W. S., **I**, 172
Hinga, F. M., **V**, 215
Hixon, R. M., **I**, 511; **II**, 566, 571
Hocker, Jürgen, **V**, 179
Hodgson, H. H., **III**, 341
Hoehn, H. H., **III**, 356
Hoffman, Joseph, **V**, 818
Hoffman, W. F., **I**, 194
Hoffmann, A. K., **V**, 355
Hoffmann, R. W., **V**, 61
Höft, E., **V**, 49, 365
Holden, Raymond F., **III**, 800
Holleman, A. F., **I**, 497, 552, 554
Holmes, H. L., **II**, 428; **III**, 300
Holmgren, A. V., **IV**, 23
Honeywell, G. E., **II**, 89, 443
Hontz, Arthur C., **IV**, 383
Horan, J. E., **V**, 647
Horne, W. H., **II**, 453
Horning, E. C., **III**, 165, 317, 549; **IV**, 144,
    408, 461, 620, 776, 790
Horning, M. G., **III**, 165
Horton, Derek, **V**, 1
House, Herbert O., **IV**, 367, 375, 552, 860,
    957; **V**, 258, 294, 324, 539, 775
Howard, John, John C., **IV**, 42
Howard, J. W., **I**, 475; **II**, 87
Howard, W. L., **V**, 5, 25, 292

Howe, Eugene E., **V**, 373
Howell, Charles F., **IV**, 816
Howk, B. W., **IV**, 801
Hrutfiord, B. F., **V**, 12
Hsueh, C. M., **II**, 416
Huang, E. P. Y., **IV**, 93
Huang, Pao-Tung, **IV**, 66, 274
Hubacher, Max H., **II**, 453
Hudack, N. J., **IV**, 738
Hudrlik, O., **III**, 673
Hudson, B. E., **III**, 142
Hudson, Boyd E. Jr., **III**, 842
Hudson, C. S., **I**, 371
Hufferd, R. W., **I**, 128, 341
Huisgen, Rolf, **V**, 650, 957, 1124
Humphlett, W. J., **IV**, 80, 739
Hünig, S., **V**, 533, 808, 1018
Hunt, Richard H., **IV**, 459
Huntress, E. H., **I**, 168; **II**, 457, 459; **IV**, 329
Hurd, Charles D., **I**, 217, 330; **II**, 541; **III**, 458
Huston, Ralph C., **II**, 97
Hwa, Jesse C. H., **V**, 608
Hymans, W. E., **V**, 863, 866
Hyre, John E., **V**, 684

Icke, Roland N., **III**, 59, 644, 723, 763
Ide, W. S., **II**, 44, 55, 130, 333, 622
Ingersoll, A. W., **I**, 311, 417, 485; **II**, 328, 503, 506
Inglis, J. K. H., **I**, 235, 254
Ipatieff, V. N., **II**, 151
Ireland, Robert E., **V**, 171
Irvine, James, **III**, 326
Isbell, Neville, **II**, 262
Issidorides, C. H., **IV**, 679, 681

Jackson, H. L., **IV**, 438
Jacobs, Thomas L., **III**, 456
Jacobs, W. A., **I**, 153, 488
Jakobsen, Preben, **V**, 70
James, Philip N., **IV**, 539
Janssen, Donald E., **IV**, 547
Jaul, E., **IV**, 307
Jefferson, George D., **II**, 284, 487

Jenkins, D. M., **III**, 538
Jenkins, R. L., **I**, 347, 394
Jenner, E. J., **V**, 1026
John, Joseph P., **V**, 747
Johnson, A. W., **V**, 785, 1022
Johnson, B. A., **V**, 215
Johnson, Carl R., **V**, 791
Johnson, F., **V**, 614
Johnson, F. R., **II**, 449
Johnson, G. Dana, **IV**, 900
Johnson, H. B., **IV**, 804
Johnson, Herbert E., **IV**, 780; **V**, 654, 703
Johnson, Herbert L., **III**, 60
Johnson, John R., **I**, 111, 123, 133, 351, 440, 521; **II**, 8, 106; **III**, 641, 794, 806
Johnson, William S., **IV**, 132, 162, 261, 390, 459; **V**, 245
Johnson, W. W., **II**, 343
Johnston, H. W., **III**, 167
Johnston, J. Derland, **IV**, 576
Johnstone, H. F., **I**, 158, 183
Jolad, S. D., **V**, 139
Jones, E. R. H., **IV**, 404, 755, 792
Jones, J. E., **III**, 98, 590
Jones, Reuben G., **IV**, 824
Jones, Willard J., **V**, 845
Jordon, E. F., Jr., **IV**, 977
Jorgenson, M. J., **V**, 509
Josey, A. D., **V**, 716
Judge, Joseph M., **V**, 255
Julian, P. L., **II**, 391, 487

Kaiser, Edwin M., **V**, 559
Kaiser, E. W., **III**, 209
Kaiser, H., **V**, 1088, 1106
Kalir, A., **V**, 825
Kalish, Joseph, **V**, 471
Kalm, Max J., **IV**, 398, 608, 616
Kamath, P. M., **IV**, 178
Kaminski, F. E., **V**, 263
Kamm, Oliver, **I**, 25, 42, 104, 128, 250, 372, 391, 392, 445
Kamm, W. F., **I**, 104
Kan, R. O., **V**, 612, 1051
Karabinos, J. V., **IV**, 506

Kaslow, C. E., III, 194, 519, 580; IV, 718
Kauer, J. C., IV, 411; V, 1043
Kaufman, Daniel, III, 693
Kaufmann, W. E., I, 258
Kawanisi, M., V, 582
Kehm, Barbara B., IV, 515
Keiser, Jeffrey E., V, 791
Kellert, M. D., IV, 108
Kendall, E. C., I, 21, 131, 246, 256
Kennard, Kenneth C., IV, 950
Kennedy, E. R., II, 351; III, 355, 482, 483, 485
Kent, R. E., III, 490, 493, 591
Kenyon, Joseph, I, 263, 418
Kenyon, R. L., III, 747
Kenyon, William G., V, 559
Kern, J. W., I, 101
Kester, E. B., I, 75
Khan, N. A., IV, 851, 969
Kharasch, Norman, V, 474
Kibler, C. J., III, 28, 108, 610
Kidwell, R. L., V, 918
Kiefer, E. J., V, 22
Kimball, R. H., II, 284, 487, 498
King, Harold, I, 286, 298
King, Harold, S., II, 399
King, Harriette, II, 612
King, Mary S., IV, 88
King, W. B., I, 252
Kirby, R. H., I, 361
Kirchner, Fred K., III, 468
Kirner, W. R., I, 374; II, 136
Kise, Mearl., II, 422
Kitahonoki, Keizo, V, 83
Klair, A., V, 130
Kleiderer, E. C., II, 538
Klein, R. A., V, 1058
Kleinberg, J., III, 516
Kleinberg, Jacob, v, 1001
Kleinfelter, Donald C., V, 852
Klinck, Jane, V, 121
Klingenberg, J. J., IV, 110
Klingsberg, Erwin, III, 119
Kluiber, Rudolph W., IV, 261
Kneisley, J. Wayne, III, 209
Knight, H. B., IV, 895

Knowles, W. S., V, 208
Knox, L. H., V, 194, 196, 689
Kobe, Kenneth A., III, 653
Koch, H., V, 20
Koelsch, C. Frederick, III, 132, 791
Kofod, Helmer, IV, 491
Kogon, Irving C., IV, 182, 336
Kohler, E. P., I, 78
Kohn Earl J., IV, 283
Kohn, E. D., II, 381
Kon, G. A. R., I, 192
König, C., V, 431
Koo, John, IV, 327, 408; V, 550
Korack, M., V, 414
Kormendy, Minerva F., V, 162
Kornblum, Nathan, III, 295; IV, 454, 724; V, 845
Kornfeld, E. C., III, 413
Korosec, Philip S., V, 736
Korpics, C. J., IV, 93
Kosak, Alvin I., III, 14
Koten, I. A., V, 145
Kotera, Katsumi, V, 83
Kovacic, Peter, V, 35
Kowsower, Edward, III, 510
Krantz, Karl W., III, 254
Krapcho, A. Paul, V, 198
Krebs, Edwin, III, 649
Kremers, Edward, I, 511
Krimmel, C. P., IV, 960
Krueger, Paul A., II, 349, 353
Krynitsky, John A., III, 10, 164, 508, 818; IV, 436
Kuehne, M. E., V, 400
Kuhn, Stephen J., V, 66, 480
Kuhn, W. E., II, 447
Kundiger, D., III, 123, 506
Kuryla, William C., V, 648
Kurzer, Frederick, IV, 49, 172, 180, 213, 361, 502, 645, 934, 937; V, 966, 1046
Kyrides, L. P., III, 422

Lacey, A. B., IV, 396
Lachman, Arthur, II, 70, 234
Lack, Ruth E., V, 339

LaLonde, John, **IV,** 757
Lambert, B. F., **V,** 400
Lane, Stanley C., **I,** 140
Lange, N. A., **II,** 79
Lange, Richard M., **V,** 598
Langford, Robert B., **V,** 474
Langkammerer, C. M., **V,** 445
Langley, Wilson D., **I,** 127, 355; **III,** 334
Lapworth, Arthur, **I,** 181, 451
Larsen, E. H., **V,** 379
Lauer, W. M., **I,** 54; **III,** 194, 580
Lavety, J., **V,** 545
Lawesson, Sven-Olov, **V,** 70, 155, 379,
        642, 924
Lawler, H. J., **I,** 222
Lawson, Alexander, **IV,** 645; **V,** 1046
Lawson, J. Keith, **IV,** 892
Lazier, W. A., **I,** 99; **II,** 142, 325
Leal, Joseph R., **IV,** 597
Lednicer, Daniel, **V,** 434, 578, 621
Lee, D. D., **V,** 572
Leermakers, Peter A., **V,** 297
Leete, J. F., **III,** 428
Leffler, M. T., **II,** 24; **III,** 544, 547
Leger, Frank, **II,** 196
Leicester, Henry M., **II,** 238, 240
Leicester, James, **IV,** 525
Leipold, Hans A., **V,** 80
Leiserson, J. L., **III,** 183
Lemmerman, Leo V., **III,** 44
Lespieau, R., **I,** 186, 191, 209
Lesslie, T. E., **IV,** 329
Leston, Gerd., **IV,** 368
LeSuer, William M., **IV,** 919
Letsinger, Robert L., **V,** 450
Leubner, Gerhard W., **IV,** 866
Leuck, G. J., **I,** 283
Levene, P. A., **II,** 5, 88, 288, 320, 545
Levens, Ernest, **IV,** 68
Levin, Nathan, **III,** 191
Levine, R., **III,** 291, 405
Levis, W. W., Jr., **IV,** 307
Levy, Edward F., **III,** 456
Lew, Henry Y., **IV,** 545
Lewis, W. Lee, **I,** 70
Liang, Poe, **III,** 803

Lieber, Eugene, **IV,** 380
Lieberman, S. V., **II,** 441
Liedhegener, Annemarie, **V,** 179
Linden, Gustave B., **III,** 456
Lindlar, H., **V,** 880
Lindsay, R. O., **III,** 710
Linn, W. J., **V,** 1007
Lipinski, Martin, **V,** 300
Liska, John, **V,** 320
Litle, Robert L., **V,** 1055
Lloyd, Winston D., **V,** 699
Locke, J. Michael, **V,** 723
Lockhard, Luther B., **III,** 463
Lockhart, L. B., Jr., **IV,** 784
Loder, D. J., **II,** 282
Loder, J. W., **IV,** 667
Loev, Bernard, **V,** 162
Logan, Ted J., **v,** 969
Logullo, Francis M., **V,** 54
Long, F. A., **II,** 87
Longley, R. I., Jr., **IV,** 302, 311, 660, 677
Lorette, N. B., **V,** 5, 25, 292
Lucas, H. J., **II,** 351; **III,** 355, 482, 483,
        485
Lucier, John J., **V,** 736
Lücke, E., **V,** 533, 808
Lufkin, James E., **IV,** 872
Luijten, J. G. A., **IV,** 881
Lukaszewski, H., **V,** 656
Lund Hakon, **II,** 594
Lutz, Eugene, **V,** 562
Lutz, R. E., **I,** 49; **III,** 248
Lycan, W. H., **II,** 318

McBee, E. T., **V,** 663
McCaleb, Kirtland E., **IV,** 281
McCarthy, W. C., **IV,** 918
Macey, William A. T., **IV,** 525
McCloskey, Allen L., **IV,** 261
McCloskey, Chester M., **III,** 221, 434,
        712, 668
McColm, E. M., **I,** 536
McCormack, W. B., **V,** 787
McCracken, J. H., **III,** 258
McCutcheon, J. W., **III,** 526, 531
McDermott, F. A., **II,** 365

McDonald, Richard N., **V**, 499, 985
McElvain, S. M., **I**, 170, 385, 473; **II**, 114, 214, 272, 279, 419; **III**, 123, 251, 379, 490, 493, 506, 591; **IV**, 463, 662, 664
McEwen, W. L., **I**, 521; **II**, 53, 133; **III**, 227
McEwen, William E., **V**, 1001
McFarland, J. W., **IV**, 940
McGee, John J., **V**, 801
McINtosh, A. V., **III**, 234, 237
Mackay, D. D., **II**, 580
McKee, G. H. W., **II**, 15
McKee, R. A., **II**, 100
McKenzie, B. F., **I**, 21, 131, 246, 256, 335
McKeon, Thomas F., Jr., **IV**, 683
Mackie, R. K., **V**, 941
McKillip, W., **V**, 758
Mckusick, B. C., **IV**, 438, 746, 953
McLachlin, D. M., **III**, 28
McLaughlin, Keith C., **IV**, 824
McLeod, Donald J., **IV**, 510, 555, 635
McLeod, Gerald L., **IV**, 345
Macleod, L. C., **II**, 62
McMahon, Robert E., **IV**, 457
McOmie, J. F. W., **V**, 412
McPhee, Warren D., **III**, 119
McRae, J. A., **IV**, 393
McShane, H, F., **III**, 197
McWherter, P. W., **I**, 245
McWhirter, J., **III**, 714
Magerlein, B. J., **V**, 946
Magnani, Arthur, **III**, 251
Mahadevan, A. P., **III**, 341
Mahood, S. A., **II**, 160
Mallory, F. B., **IV**, 74
Mallory, Frank B., **V**, 952
Mancuso, R. W., **V**, 32
Mansfield, G. H., **IV**, 128
Mansfield, K. T., **V**, 96
Manske, R. H. F., **I**, 404; **II**, 83, 154, 389
Marey, R., **IV**, 350, 601
Mariella, Raymond P., **IV**, 210, 288
Mark, V., **V**, 358, 602, 893, 901
Markham, E., **V**, 785
Marshall, J. L., **V**, 91, 424

Martell, A. E., **V**, 617
Martens, Ted F., **III**, 159
Martin, D. G., **IV**, 162
Martin, E. L., **II**, 499, 501, 560, 569; **III**, 465, 633
Martin, R. J. L., **V**, 175
Marvel, C. S., **I**, 25, 42, 48, 84, 94, 95, 99, 117, 128, 170, 177, 224, 238, 248, 252, 289, 327, 347, 358, 366, 377, 411, 435, 504, 531, 533, 536; **II**, 74, 76, 93, 310, 318, 371, 372, 374, 416, 558, 563, 602; **III**, 366, 495, 523, 705, 822, 848
Mascitti, A., **IV**, 788
Mashburn, T. Arthur, Jr., **V**, 621
Matar, A., **IV**, 681
Matlack, George M., **IV**, 914
Matthews, A. O., **I**, 392
Mauthner, F., **I**, 537
Maxwell, Charles E., **II**, 485; **III**, 305
May, G. E., **II**, 8
Mayberry, Gerald L., **V**, 456
Mayo, Joe, **V**, 166
Meek, John S., **V**, 126
Meerwein, H., **V**, 1080, 1096
Meinert, R. N., **II**, 541
Meinwald, J., **IV**, 738; **V**, 863, 866
Mekler, A. B., **V**, 743
Melby, L. R., **IV**, 276, 953; **V**, 239
Mendenhall, G. David, **V**, 829
Merritt, Charles, Jr., **IV**, 8, 711
Meuwsen, A., **V**, 43
Meyer, G. M., **II**, 288
Meyer, J. D., **II**, 468
Meyer, Karl, **III**, 637
Meyer, Kurt H., **I**, 60, 390
Meyers, Cal Y., **IV**, 39
Meyr, Rudolf, **V**, 300, 1060
Meystre, Ch., **V**, 692
Michaels, R. J., Jr., **IV**, 915
Michelotti, Francis W., **IV**, 29
Mickey, Sally, **III**, 343
Mićović, V. M., **II**, 264
Middleton, W. J., **IV**, 243; **V**, 1013
Mierzwa, Sigmund A. Jr., **V**, 874

Mikol, Gerard J., **V**, 937
Mikulec, Richard A., **IV**, 13
Milas, Nicholas A., **II**, 302
Miles, Marion L., **V**, 718, 721
Millar, I. T., **V**, 1120
Miller, Charles S., **III**, 646
Miller, Ellis, **II**, 543
Miller, G. E., **II**, 576
Miller, Leonard E., **IV**, 39
Miller, Sidney E., **IV**, 141
Miller, Sidney I., **V**, 921
Millet, W. H., **II**, 65
Miner, Laboraties, **I**, 285
Minin, Ronald, **IV**, 182, 336
Minnis, Wesley, **II**, 357, 520
Mintz, M. J., **V**, 184
Mitchell, M. J., **V**, 944
Moffatt, John G., **V**, 242
Moffett, E. W., **II**, 121
Moffett, Robert Bruce, **III**, 234, 237, 562;
    IV, 238, 291, 354, 357, 427, 466,
    605, 652, 834
Möllering, H., **I**, 499
Monagle, J. J., **V**, 501
Moore, A. T., **V**, 586
Moore, James A., **V**, 351
Moore, L. L., **V**, 687
Moore, Maurice, L., **III**, 599, 617
Moreau, C., **I**, 166
Moriconi, Emil J., **V**, 623
Mortensen, Harley E., **V**, 562
Morton, A. A., **III**, 831
Mosher, Harry S., **III**, 568; **IV**, 828
Mowry, David T., **IV**, 486
Moyer, W. W., **II**, 602
Moyle, M., **IV**, 205, 493, 700; **V**, 339
Mozingo, Ralph, **III**, 181, 258, 387, 576,
    685
Müller, Adolf, **II**, 535
Munch-Petersen, Jon, **IV**, 715; **V**, 762
Murphy, Donald M., **IV**, 872
Murphy, M., **V**, 918
Murphy, William S., **V**, 523
Murray, Joseph I., **IV**, 744
Murray, Robert W., **V**, 166
Murray, T. F., **I**, 85, 272, 523

Muth, K., **V**, 944
Myers, R. R., **II**, 462
Myles, J. R., **III**, 831

Nagy, S. M., **III**, 690
Naruse, Norio, **V**, 266
Nasipuri, D., **V**, 384
Natelson, Samuel, **III**, 381
Naylor, C. A., Jr., **II**, 523, 526
Neely, T. A., **V**, 1116
Neeman, M., **V**, 245
Neher, H. T., **I**, 56
Nelson, G. B., **III**, 392
Nenitzescu, Costin D., **II**, 496; **V**, 1106
Nesmajanow, A. N., **II**, 381, 432
Neuman, Fred W., **III**, 73, 519
Neville, Roy G., **V**, 801
Newkirk, D. M., **III**, 538
Newman, Melvin S. **II**, 428; **III**, 502, 631;
    IV, 484; **V**, 486, 743, 1024,
    1029, 1031
Newton, L. W., **III**, 313
Nicholas, R. D., **V**, 16
Nichols, Gust, **III**, 159, 712
Nicolet, B. H., **I**, 410
Nielsen, Arnold T., **V**, 288
Nielsen, D. R., **V**, 414
Niemann, Carl, **III**, 11
Nierenstein, M., **II**, 557
Nieuwland, J. A., **I**, 229
Nightingale, Dorothy, **III**, 42
Nishimura, Tamio, **IV**, 713
Nissen, B. H., **I**, 1
Noble, Paul, Jr., **IV**, 924
Noland, Wayland E., **V**, 567, 833
Noller, C. R., **I**, 109; **II**, 109, 184, 258,
    358, 478, 586; **III**, 835; **IV**, 545,
    571, 603
Nonhebel, D. C., **V**, 206
Normington, J. B., **I**, 205
Norris, James F., **I**, 144, 430, 548
Norton, A. J., **I**, 490
Norton, D. G., **IV**, 348
Norton, Richard V., **V**, 709
Nowak, Robert M., **IV**, 281
Noyes, W. A., **I**, 457; **II**, 108

Oakwood, T. S., **III**, 112, 114
Oehler, Rene, **II**, 276; **III**, 401
Ogliaruso, Michael, **V**, 111
Ohme, Roland, **V**, 897
Ohno, Masaji, **V**, 266, 281
Okamoto, Masaru, **V**, 281
Okuda, Takuo, **IV**, 566
Olah, George A., **V**, 480
Olander, C. P., **I**, 447
Oliver, John J., **II**, 391, 487
Oliveto, Eugene P., **IV**, 104
Olmsted, Alanson W., **I**, 144
Olsen, Robert E., **V**, 541
Olson, Cecil E., **IV**, 898
Opie, J. W., **III**,, 56
Opitz, G., **V**, 877
Oppenauer, R. V., **III**, 207
Orchin, Milton, **III**, 837
Orten, James M., **I**, 300
Orwoll, E. F., **III**, 130, 566
Osterberg, A. E., **I**, 185, 271
Osuga, David T., **V**, 126
Ott, Erwin, **II**, 528
Otterbacher, T. J., **I**, 290; **II**, 317
Overberger, C. G., **III**, 200, 204; **IV**, 29, 66, 182, 273, 274, 336

Pacaud, R. A., **II**, 336
Pagano, A. S., **V**, 191, 367
Page, G. A., **IV**, 136
Palmer, Albert, **II**, 57
Palmer, C. S., **I**, 73, 75, 245
Panella, J. P., **V**, 614
Papa, Domenick, **III**, 742
Pappalardo, Joseph A., **IV**, 186, 558
Paquette, L. A., **V**, 215, 231, 408, 467
Parham, William E., **III**, 395; **IV**, 295; **V**, 874
Paris, G., **IV**, 496
Parris, Chester L., **V**, 73
Parshall, G. W., **V**, 1016
Partridge, M. W., **IV**, 769
Patrick, Tracy M., Jr., **IV**, 430, 980
Patsch, M. R., **V**, 859
Patterson, John M., **V**, 585
Patterson, L. A., **III**, 576

Pattison, F. L. M., **V**, 136, 580
Paufler, Robert M., **V**, 982
Paust, J., **V**, 859
Payne, George B., **V**, 805
Pearl, Irwin A., **III**, 745; **IV**, 972, 974
Pearlman, W. M., **V**, 670
Pearson, David L., **IV**, 221
Pearson, D. E., **III**, 154; **V**, 117, 332, 890
Pedersen, Knud, **V**, 70
Perkins, Edward G., **IV**, 444
Perkins, M. J., **V**, 51
Perry, B. J., **IV**, 325, 955
Perst, H., **V**, 1130
Peter, Hugo H., **IV**, 887
Petersen, James M., **V**, 909
Petersen, R. J., **V**, 929
Peterson, C. J., **V**, 758
Peterson, W. D., **III**, 106, 242, 660
Pettit, D. J., **V**, 1099
Phillips, Donald D., **IV**, 313
Phillipi, E., **II**, 551
Phillips, Ross, **II**, 232, 464, 578; **III**, 337
Piasek, Edmund J., **V**, 80
Pickard, P. L., **V**, 520
Pier, Stanley M., **IV**, 554
Pike, Arthur B., **II**, 284, 487
Pilgrim, Frederick, J., **IV**, 451
Pinkney, P. S., **II**, 116; **III**, 401
Ploss, G., **V**, 431
Pluygers, C. W., **V**, 223
Pope, H. W., **V**, 117
Porter, H. D., **III**, 660, 708
Porter, P. K., **I**, 377, 411, 457
Posner, J., **V**, 514
Potts, K. T., **V**, 769
Powell, J. Roy, **IV**, 213, 934
Powell Garfield, **II**, 449
Powell, S. G., **I**, 168
Price, Charles C., **III**, 86, 174, 272, 649, 664, 667; **IV**, 186, 558, 566, 683; **V**, 255
Price, John A., **IV**, 285
Price, R., **V**, 785, 1022
Prichard, W. W., **III**, 302, 452
Prill, E. A., **II**, 419
Proctor, G. R., **V**, 545

Prosser, Thomas, **IV,** 169
Prout, Franklin S., **III,** 601
Prout, F. S., **IV,** 93; **V,** 572
Puntambeker, S. V., **I,** 524; **II,** 318
Purchase, Earl R., **III,** 430
Puterbaugh, Milton, **I,** 388
Putman, R. E., **V,** 235

Quast, H., **V,** 1018
Quayle, O. R., **I,** 211, 292, 294

Raaen, Vernon, **IV,** 130
Raasch, M. S., **V,** 393
Rabjohn, Norman, **III,** 375, 557; **IV,** 441, 513
Raecke, Bernhard, **V,** 813
Raha, Chittaranjan, **IV,** 263
Rahrs, E. J., **I,** 91; **III,** 650, 652
Rainey, James L., **III,** 17
Rajagopalan, S., **III,** 425; **V,** 139, 504
Ralls, J. O., **II,** 191
Ramachrandran, S., **V,** 486, 743
Raman, P. V. A., **III,** 425
Rand, L., **V,** 663
Rao, C. N. Ramachandra, **IV,** 380
Rapala, Richard J., **III,** 820
Rasmusson, Gary H., **V,** 324
Rassweiler, C. F., **I,** 10
Ratchford, W. P., **III,** 30
Ratts, K. W., **V,** 361
Raube, Richard, **IV,** 288
Rauhut, M. M., **IV,** 114
Ray, F. E., **III,** 23, 420
Raznikiewicz, T., **V,** 149
Read, R. R., **I,** 321, 514; **III,** 157, 444
Redemann, Carl T., **III,** 69, 656
Redemann, C. Ernst, **III,** 11, 59, 69, 244, 644, 656, 763
Redmond, W., **V,** 133
Reed, Donald E., **V,** 351
Reed, L. J., **III,** 395
Reeve, Wilkins, **III,** 693
Regitz, Manfred, **V,** 179
Rehberg, Chessie E., **III,** 33, 46, 146
Reichardt, C., **V,** 1112, 1123, 1130
Reichert, J. S., **I,** 229

Reichle, Walter R., **V,** 1024, 1055
Reid, E. Emmet, **II,** 246, 338, 468, 474; **IV,** 47
Reif, Donald J., **IV,** 375, 860
Reifschneider, Walter, **V,** 107
Reisner, D. B., **IV,** 144
Renfrew, W. B. Jr., **II,** 67, 268, 607
Renoll, Mary, **III,** 502
Reverdin, Frederic, **I,** 219
Reynolds, Delbert, **III,** 432
Reynolds, George A., **III,** 374, 593; **IV,** 15, 633, 708
Reynolds, H. H., **I,** 476
Rhinesmith, Herbert S., **II,** 177
Rice, F. O., **III,** 244
Rice, Rip C., **IV,** 283
Richter, Henry J., **IV,** 482
Rideout, W. H., **V,** 414
Riebsomer, J. L., **III,** 326
Rieche, A., **V,** 49, 365
Riegel, Byron, **III,** 234, 237
Riegel, E. Raymond, **II,** 126
Riener, Edward, E., **IV,** 451
Rietz, Edward. **III,** 851
Rieveschl, George Jr., **III,** 23, 420
Riley, H. A., **II,** 509
Rinderknecht, Heinrich, **V,** 822
Rinehart, Kenneth L., Jr., **IV,** 120, 444
Ringold, Howard J., **IV,** 221
Rinkes, I, J., **II,** 393
Ritchie, Bruce, **III,** 53
Ritter, E. J., **IV,** 307
Ritter, John J., **V,** 471
Robb, Charles M., **V,** 635
Robb, C. M., **III,** 347
Robbins, Frederick, **V,** 341
Roberts, John D., **IV,** 457
Roberts, K. C., **I,** 235
Roberts, R., **III,** 244
Roberts, Royston M., **III,** 272; **IV,** 420, 464
Robertson, G. R., **I,** 396
Robertson, G. Ross, **I,** 52, 138, 406
Robinson, D. B., **III,** 103
Robinson, J., **II,** 47
Robinson, John C. Jr., **III,** 717, 720

Robison, Bonnie L., **IV**, 947
Robison, Michael M., **IV**, 947
Robson, Jean H, **III**, 456
Roebuck, Alan, **III**, 217
Rogers, Crosby U., **IV**, 884
Rogers, Richard J., **III**, 174
Roll, L. J., **II**, 140, 266, 368, 382, 418, 460; **III**, 20
Romanelli, Michael G., **V**, 552
Rondestvedt, Christian S., Jr., **IV**, 766, 846
Ronzio, Anthony R., **III**, 71, 438
Roos, A. T., **I**, 145
Ropp, Gus A., **IV**, 130, 727
Rorig, Kurt, **IV**, 576
Rosen, Milton J., **IV**, 665
Rosendahl, Friedrich, **V**, 300
Rothrock, H. S., **II**, 408
Rousseau, Joseph E., **IV**, 711
Rowland, S. P., **III**, 549
Ruby, Phillip R., **IV**, 786, 798
Ruddy, Wayne A., **III**, 665
Ruhoff, John R., **II**, 102, 246, 292, 315, 338, 468, 474
Russell, Alfred, **III**, 157, 282, 293, 463, 747
Russell, Glen A., **V**, 937
Rutenberg, M. W., **IV**, 620
Rutherford, K. G., **V**, 133
Rydon, H. N., **V**, 586
Ryerson, George D., **IV**, 957
Ryskiewicz, Edward E., **IV**, 831

Sager, W. F., **V**, 595, 1011
Sahyun, Melville R. V., **V**, 926
Saliani, N. A., **II**, 69
Saltman, H., **V**, 658, 660, 665
Salzberg, Paul L., **I**, 46, 119
Sanchez, R. A., **V**, 32, 344
Sandberg, M., **II**, 49
Sandberg, Rune, **V**, 155
Sandborn, L. T., **I**, 111, 133, 340, 526
Sanderson, James J., **III**, 254, 256
Sanderson, T. F., **IV**, 1
Sandin, R. B., **II**, 100, 196, 604
Sandri, Joseph M., **IV**, 278
Sands, Lela, **I**, 67

Sankaran, D. K., **III**, 623
Santilli, Arthur A., **V**, 166
Santurri, Pasco, **V**, 341
Sasse, W. H. F., **V**, 102
Sauer, J. C., **III**, 605; **IV**, 268, 560, 801, 813
Sauer, Robert F., **V**, 145
Saunders, J. H., **III**, 22, 200, 204, 416
Sauter, Frederick J., **V**, 258
Sawyer, R. L., **III**, 276
Saxton, J. E., **V**, 769
Saxton, M. R., **IV**, 78
Sayre, J. L., **II**, 596
Scanlan, John T., **IV**, 317
Scarrow, J. A., **II**, 387
Schaefer, G. F., **IV**, 31
Schaefer, J. P., **V**, 142, 249, 285
Schaefer, W., **I**, 147
Schaeffer, J. R., **IV**, 19
Schäfer, W., **I**, 364
Schaffner, P. V. L., **II**, 160
Scheifele, H. J., Jr., **IV**, 34
Schenk, Erwin, **III**, 742
Schiemann, G, **II**, 188, 299,
Schiemenz, G. P., **V**, 496
Schiessler, R. W., **V**, 647
Schildneck, P. R., **II**, 236, 411
Schipp, Hubert, **V**, 813
Schlatter, M. J., **III**, 223
Schleyer, Paul von R., **V**, 16, 852
Schloss, E. L. Jr., **II**, 457
Schmidt, A. G., **II**, 266
Schmitz, Ernst, **V**, 897
Schmitz, William R., **III**, 328
Schmutzler, R., **V**, 218, 1005
Schneider, H. J., **IV**, 98
Schneider, William P., **IV**, 132
Schoellkopf, U., **V**, 751, 859
Schoen, Kurt L, **IV**, 623
Schroeder, E. F., **II**, 17, 137, 305, 307
Schroeder, Wesley, **II**, 574
Schultz, Everett M., **III**, 343
Schultz, Harry P., **IV**, 364; **V**, 589
Schulze, F, **II**, 425
Schurink, H. B. J., **I**, 425; **II**, 476
Schuster, R. E., **V**, 772

Schwartz, George, **III**, 332
Schwartzman, Louis H., **IV**, 471
Schwarz, R., **IV**, 203
Schwarz, William W., **V**, 1116
Schweizer, Edward, E., **V**, 874, 1145
Scott, James E., **V**, 772
Scott, R. W., **I**, 179; **II**, 379
Scott, W. J., **II**, 128, 539
Searle, N. E., **IV**, 424
Searles, Scott, **V**, 562
Sedlak, J., **IV**, 108
Segur, J. B., **I**, 372, 391
Seidl, Helmit, **V**, 957, 1124
Seikel, Margaret K., **III**, 262
Seitz, Arnold H., **V**, 54
Sekera, V. C., **III**, 366
Sellas, J. T., **IV**, 702
Semon, Waldo L., **I**, 318
Semon, W. L., **II**, 204
Seth, S. R., **II**, 522
Seven, Raymond P., **III**, 499
Seyferth, Dietmar, **IV**, 258; **V**, 452
Seymour, Dexter, **III**, 456
Shah, N. M., **III**, 280; **IV**, 836
Shapiro, S. H., **II**, 531
Sharefkin, J. G., **V**, 658, 660, 665
Sharkey, W. H., **V**, 235, 445, 734
Shaw, B. L., **IV**, 404
Shechter, Harold, **IV**, 321, 323, 543; **V**, 328
Sheehan, John C., **V**, 555
Sheehan, J. T., **III**, 541
Sheibley, F. E., **II**, 79
Shemin, D., **II**, 1, 11, 330, 491, 519
Shenoy, P. K., **V**, 142, 249
Shepard, Inella G., **V**, 839
Sheppard, O. E., **II**, 440, 617; **III**, 37
Sheppard, William A., **V**, 396, 843, 959, 1085
Sherman, Wm. R., **IV**, 247
Shibe, W. J., **III**, 260
Shields, J. E., **V**, 406, 1064
Shildneck, P. R., **II**, 345
Shoffner, J. P., **V**, 572
Short, William A., **IV**, 706
Shriner, R. L., **I**, 101, 463; **II**, 140, 200,

266, 366, 368, 382, 453, 459, 538, 560; **III**, 73, 519, 562, 725, 737, 786; **IV**, 242, 786, 798, 910
Shrivers, J. C., **III**, 142
Siegel, Elaine, **V**, 904
Silloway, H. L., **III**, 82
Silverstein, Robert M., **IV**, 831; **V**, 390
Simmons, H. E., **V**, 855
Simons, J. K., **IV**, 78
Simril, V. L., **III**, 157
Sims, Homer, **V**, 608
Singer, Alvin, **II**, 214
Sisti, A. J., **V**, 46
Sjostedt, George, **III**, 95
Skattebøl, Lars, **IV**, 792; **V**, 306
Skell, P. S., **V**, 929
Skelly, W. G., **IV**, 154
Skorcz, J. A., **V**, 263
Smart, B. W., **IV**, 55
Smiles, S., **I**, 7, 8
Smith, C. W., **IV**, 348
Smith, Douglas S., **IV**, 377
Smith, L. A., **II**, 242, 438
Smith, Lee Irvin, **II**, 95, 248, 254, 360; **III**, 56, 151, 302, 350, 673
Smith, L. H., **III**, 793
Smith, Lowell R., **V**, 204
Smith, Newton R., **IV**, 201, 337, 549, 731
Smith, P. A. S., **IV**, 75, 819
Smith, Paul V., **III**, 410, 735
Smith, Peter A. S., **V**, 612, 829, 1051
Smith, R. D., **V**, 855
Smith, Ronald Dean, **IV**, 218
Smith, Walter T., Jr., **IV**, 345, 702
Smolin, Edwin M., **III**, 715; **IV**, 387
Snell, J. M.; **III**, 788
Snell, John M., **II**, 114
Snoddy, A. O., **IV**, 19, 529
Snyder, H. R., **II**, 171, 333, 489, 531; **III**, 471, 698, 717, 720, 798; **IV**, 539, 638
Soffer, M. D., **IV**, 903
Soine, T. O., **IV**, 106
Sokol, Phillip E., **V**, 706
Sollazzo, R., **III**, 371
Solomon, S., **V**, 306

Solomonika, Eugen, **II,** 496
Somerville, L. F., **II,** 81
Sommer, Nolan, B., **III,** 519
Sommers, Armiger H., **III,** 148, 446
Sonneborn, H., **I,** 90
Spangler, F. W., **III,** 203, 377, 621; **IV,** 433
Spartberg, M. S., **II,** 558
Spassow, A., **III,** 144, 390
Speer, John H., **III,** 39
Sperry, W. M., **I,** 95
Speziale, A. John, **IV,** 401; **V,** 204, 361, 387
Spillane, Leo, Jr., **III,** 302
Sprinzak, Yair, **IV,** 91
Sroog, C. E., **IV,** 271
Staab, Heinz A., **V,** 201
Stacy, Gardner W., **III,** 86, 667; **IV,** 13
Stampfli, J. G., **II,** 175
Starkey, E. B., **II,** 225
Starkey, Edgar B., **III,** 665
Starr, Donald, **II,** 566, 571
Staudinger, H, **II,** 573
Staunton, J., **V,** 8
Stearns, James A., **III,** 833
Steiger, Robert E., **III,** 66, 84, 88, 91
Stein, Allan R., **V,** 1092
Stenberg, J. F., **IV,** 564
Stephens, Verlin C., **IV,** 333
Stephenson, Emily F. M., **III,** 475; **IV,** 984
Stevenson, H. B., **V,** 459, 734
Stewart, Jesse, **I,** 8
Stewart, Roberta A., **IV,** 903
Stirton, A. J., **IV,** 862
St. John, Nina B., **II,** 425
Stone, Herman, **IV,** 321, 323, 543
Stormont, R. T., **II,** 7
Story, Paul R., **V,** 151
Straley, J. M., **IV,** 415
Strube, R. E., **IV,** 6, 417, 967
Struck, H. C., **II,** 560
Stuart, Frank A., **III,** 668
Stubbings, Robert, **V,** 341
Stull, Arthur, **I,** 220
Sugasawa, Shigehiko, **IV,** 72
Sukornick, Bernard, **V,** 1074

Summers, R. M., **IV,** 718
Supniewski, J. V., **I,** 46, 119
Surrey, Alexander R., **III,** 535, 753, 759
Surrey, A. R., **V,** 668
Suter, C. M., **II,** 111, 112, 485; **III,** 288
Sveda, M., **III,** 536
Swaminathan, S., **V,** 743, 747
Swann, Sherlock, Jr., **II,** 276; **III,** 401, **V,** 463
Swern, Daniel, **IV,** 317, 895, 977, **V,** 904
Synerholm, Martin E., **III,** 187

Talaty, C. N., **V,** 504
Talbot, R. H., **II,** 258
Talbot, William F., **II,** 592
Tannenbaum, A. L., **I,** 435
Tarbell, D. S., **III,** 267, 661, 809; **IV,** 136, 285, 807, 924
Taylor, Edward C., **IV,** 166, 247, 654, 704; **V,** 582
Taylor, E. R., **I,** 115, 150, 153, 415; **II,** 135, 588
Taylor, G. A., **IV,** 688
Taylor, Harold M., **V,** 437
Teague, Peyton C., **IV,** 706
Tebbins, W. G., **III,** 293
Teeter, H. M., **IV,** 125
Teeters, W. O., **IV,** 518
Telinski, Thomas J., **IV,** 576
Teranishi, Roy, **IV,** 192
Terasawa, Isao, **V,** 266
Terry, Paul H., **V,** 166
Tertiuk, W., **V,** 966
Tetaz, J. R., **IV,** 690
Thacher, A. W., **V,** 509
Thal, A. F., **I,** 107, 270, 436
Thatte, S. D., **V,** 166
Thayer, F. K., **I,** 12, 117, 398
Thelen, R., **II,** 551
Thielen, L. E., **IV,** 960
Thirtle, John R., **III,** 136
Thoman, Charles J., **V,** 962
Thompson, Ralph B., **II,** 10
Thompson, R. B., **III,** 278
Thorpe, J. F., **I,** 192
Threlfall, T. L., **V,** 346

Thurman, N., **I**, 231, 519
Thweatt, J. G., **V**, 277
Tibbetts, Fred, **V**, 166
Tichelaar, G. R., **IV**, 93
Tipson, R. Stuart, **IV**, 25
Tobie, Walter C., **I**, 23
Todd, David, **V**, 617
Todd, H. R., **II**, 200
Tolbert, T. L., **V**, 520
Tompkins, S. W., **II**, 106
Totter, John R., **III**, 460
Towles, H. V., **IV**, 266
Trevoy, L. W., **III**, 300
Tsang, S. M., **III**, 641
Tucker, S. Horwood, **IV**, 160
Tuley, W. F., **I**, 289; **III**, 822
Tullar, B. F., **IV**, 671, 921
Turk, Amos, **III**, 121
Turner, A. B., **V**, 428
Turner, Leslie, **IV**, 828
Turro, Nicholas J., **V**, 297, 528
Tuttle, Neal, **I**, 199
Tuttle, Ned, **I**, 345
Tuzun, Celel, **V**, 111
Tyson, F. T., **III**, 479

Ugi, Ivar, **V**, 300, 1060
Ulbricht, T. L. V., **IV**, 566
Underwood, H. W., Jr., **II**, 553
Ungnade, H. E., **III**, 130, 566; **IV**, 724
Urquhart, G. G., **III**, 363

VanAllan, J. A., **III**, 1, 63, 76, 275, 353,
    394, 597, 726, 733, 765, 783,
    824, 827; **IV**, 15, 21, 245, 569,
    633
Vandenberg, G. E., **V**, 946
van der Kerk, G. J. M., **V**, 223
Van Der Kerk, G. L. M., **IV**, 881
van der Plas, H. C., **V**, 977
Vanderwerf, Calvin A., **III**, 44; **IV**, 157
van Tamelen, Eugene E., **IV**, 10, 232
Van Zyl, G., **IV**, 10
Varland, Robert H., **III**, 829
Vassel, B., **IV**, 154
Vaughan, Herbert W, Jr., **V**, 1116

Vaughan, W. R., **III**, 329; **IV**, 594
Venkataraman, K., **II**, 304, 522
Venkataramani, P. S., **V**, 747
Vesley, George F., **V**, 297
Vielt, E. B., **II**, 416
duVigneaud, V., **II**, 93
Villani, Frank J., **IV**, 88
Vliet, E. B., **I**, 201, 203, 317, 360, 482
Voaden, Denys J., **V**, 962
Vofsi, D., **V**, 1076
Vogel, Arthur I., **IV**, 525
Vogel, K., **V**, 1135
Vogt, Paul J., **IV**, 420, 464
Voit, Axel, **II**, 594
von Braun, J., **I**, 428
Voong, Sing-Tuh, **III**, 664
Voorhees, V., **I**, 61, 280, 463
Vöpel, K. H., **V**, 431
Vose, C. E., **I**, 179; **II**, 379
Voskuil, W., **V**, 211
Vyas, G. N., **IV**, 836

Waddey, Walter E., **III**, 560
Wade, Robert H., **IV**, 221
Wadsworth, W. S. Jr., **V**, 547
Wagner, E. C., **II**, 65; **III**, 323, 488; **IV**,
    383
Wagner, William S., **IV**, 892
Wakefield, B. J., **V**, 998
Wakeman, Nellie, **I**, 511
Walker, G. N., **IV**, 327, 408
Walker, Joseph T., **II**, 169
Walling, C., **V**, 184
Wallingford, V. H., **II**, 349, 353
Walsh, W. L., **II**, 553
Walter, L. A., **III**, 757
Walti, A., **II**, 5, 545
Wann, R. E., **V**, 893
Wanzlick, H. W., **V**, 115
Ward, A. M., **II**, 159
Ward, E. R., **III**, 341
Ward, H. P., **III**, 244
Ward, J. A., Jr., **IV**, 392
Ward, M. L., **III**, 549
Warnhoff, E. W., **IV**, 162
Waser, E. B. H., **I**, 499

Washburn, Robert M., **IV**, 68
Wasson, Richard L., **IV**, 552, 957
Waugh, T. D., **III**, 438
Wawzonek, Stanley, **III**, 715; **IV**, 387, 681; **V**, 758
Wayne, Winston, **III**, 367
Webb, Carl N., **I**, 82, 217
Weber, K. H., **III**, 379
Webster, E. R., **IV**, 433
Weichselbaum, T. E., **II**, 384
Weijlard, John, **IV**, 124
Weil; J. K., **IV**, 862
Weiner, Michael A., **V**, 452
Weiner, Nathan, **II**, 279, 376
Weinstock, J., **IV**, 641
Weisgerber, C. A., **III**, 112, 114
Weiss, Marvin, **II**, 590
Weiss, Richard, **II**, 61; **III**, 729
Weissberger, A., **III**, 108, 183, 504, 610, 708, 788
Welch, E., **III**, 725
Wells, F. B., **II**, 221
Wendel, Kurt, **V**, 201
Wendland, Ray, **IV**, 757
Wenner, Wilhelm, **IV**, 23, 760
Werber, Frank X., **III**, 798
Werner, Newton W., **V**, 273
Wertheim, E., **II**, 471
West, D. E., **V**, 412
West, Edward S., **III**, 800
West, Harold D., **III**, 774, 813
Weston, Arthur W., **III**, 288; **IV**, 915
Whaley, A. M., **I**, 142
Wheeler, A. S., **I**, 450
Wheeler, Norris G., **IV**, 256, 857
Wheeler, T. S., **I**, 102, 296; **IV**, 478
White, Emil., **V**, 336, 797
White, R. A., **IV**, 700
Whitehead, Calvert W., **IV**, 515
Whiting, M. C., **IV**, 128, 404, 755, 792
Whitmore, Frank C., **I**, 56, 159, 161, 231, 325, 326, 401, 492, 519; **II**, 282, 317, 408; **III**, 568
Whitmore, Marion G., **I**, 401
Whyte, Donald Edward, **IV**, 234
Wiberg, Kenneth B., **III**, 197, 811

Wieland, P., **V**, 692
Wieleseck, R., **V**, 921
Wilds, Alfred L., **IV**, 281
Wiley, D. W., **V**, 1013
Wiley, Richard H., **III**, 560, 853; **IV**, 5, 201, 337, 351, 549, 731
Wilkinson, G., **IV**, 473, 476
Willard, Constance, **IV**, 831
Williams, Bill, **III**, 337
Williams, J. K., **V**, 459
Williams, Jonathan W., **III**, 10, 164, 508, 626, 818
Williamson, S. M., **V**, 12, 816
Willson, F. G., **I**, 102, 296
Wilson, C. V., **III**, 28, 358, 573, 575, 578; **IV**, 547, 564
Wilson, J. G., **IV**, 690
Wilson, J. W., **III**, 267
Wilson, W. C., **I**, 274, 276
Wilt, James W., **IV**, 254
Winberg, H. E., **V**, 883
Windus, Wallace, **II**, 136, 345, 411
Winkelmüller, W., **II**, 188, 299
Wisansky, W. A., **III**, 138, 307
Wisegarver, Burnett B., **III**, 59, 644, 723
Wiselogle, F. Y., **I**, 90
Wittcoff, Harold, **IV**, 907
Witten, Benjamin, **IV**, 47
Witten, Charles H., **III**, 818
Wittenbrook, Lawrence S., **V**, 231
Wittig, George, **IV**, 964; **V**, 60, 751
Witzemann, E. J., **II**, 17, 137, 305, 307
Wojcik, Bruno H., **II**, 260, 262
Wolf, Calvin N., **IV**, 45, 910
Wolf, Frank J., **III**, 737; **IV**, 124
Wollensak, J., **V**, 575
Womack, Ennis B., **III**, 392, 714
Wood, Clelia S., **V**, 952
Wood, Ernest H., **III**, 641
Wood, J. H., **III**, 98, 286
Wood, John Jr., **III**, 444
Woodburn, H. M., **IV**, 271
Woodruff, Eugene H., **III**, 581
Woods, G. Forrest, **III**, 470; **IV**, 471
Woodward, Gladys, E., **I**, 159, 325, 408
Woodward, R. B., **III**, 50, 413, 453

Woollett, G. H., **II,** 343
Woolfolk, E. O., **III,** 837
Worden, Leonard R., **V,** 251
Worrall, David E., **I,** 413
Wynberg, Hans, **IV,** 295, 534
Wysong, Robert D., **V,** 332

Yanko, William, **III,** 568
Yeaw, J. S., **I,** 336
Young, D. M., **II,** 219
Young, R. J., **V,** 88

Zambito, Arthur J., **V,** 373
Zartman, Walter, **II,** 606

Zaugg, Harold E., **III,** 820
Zaweski, Edward F., **V,** 324
Zeidman, Blossom, **III,** 328
Ziegler, Gene R., **V,** 921
Ziegler, K., **I,** 314
Zimmerman, F. J., **II,** 549
Zimmerman, Howard E., **V,** 892
Zingaro, Ralph, **IV,** 359
Zoellner, E. A., **I,** 524
Zubek, A., **V,** 39
Zuidema, G. D., **IV,** 10
Zurquiyah, A., **V,** 993
Zwilgmeyer, F., **II,** 126

CUMULATIVE GENERAL INDEX

A large number of the syntheses in *Collective Volumes* I through V involve two or more steps and procedures for the preparation of the intermediates are given. In earlier volumes, each intermediate was written up as a separate preparation. Later, the intermediates were described in paragraphs with an italicized name. These intermediates and the methods for preparing them have proven to be extremely useful and hence this general index contains not only the Title Name of each preparation but also the names of the intermediates. Whenever complete preparative directions are given, the names are shown in CAPITAL LETTERS with the *Collective Volume* and page numbers in bold faced type. The name of a compound in ordinary light faced type together with a page number in bold faced type indicates directions, usually adequate but not in full detail, for preparing the substance named. A name in light faced type indicates a compound or an item mentioned in connection with a preparation. In certain syntheses, intermediates which are immediately used up in the next step are not listed in this general index but are shown in Index 3; Type of Reaction (diazonium salts, Grignard reagents, organo-metallic compounds, etc.)

The systematic Chemical Abstracts names, shown just beneath the title name of each preparation were those preferred in the year of publication. They were not included in the General Indices of *Collective Volumes* I, II, III, or IV. However, the General Index of *Collective Volume* V does have both kinds of names for its preparations. It is for this reason, the special Contents Indices, Nos. 1 and 2 were compiled. These list common Title Names, the latest Chemical Abstracts Index names, and Chemical Abstracts Registry Numbers for all five Collective Volumes. This involves some duplication but insures locating preparations regardless of name.

This cumulative general index does not repeat information concerning: Types of Reactions, Types of Compounds, Formulas, Solvents and Special Reagents, Apparatus, and Authors. These special indices,

Nos. 3, 4, 5, 6, 7, or 8 are valuable in planning synthetic work and should be consulted whenever a specific compound is desired which has not yet been described in *Organic Syntheses*.

ABIETIC ACID, IV, 1
  diamylamine salt of, IV, 2
Acenaphthene, III, 1, 3, 6
ACENAPHTHENEQUINONE, III, 1
ACENAPHTHENOL, III, 7, 3
ACENAPHTHENOL ACETATE, III, 3
β-(1-ACENAPHTHOYL)PROPIONIC
  ACID, III, 9
β-(3-ACENAPHTHOYL)PROPIONIC
  ACID, III, 6
ACETAL, I, 1
Acetaldehyde, I, 1, 21, 425; II, 406; III,
  200, 317, 591, 757; IV, 294; V,
  364
ACETAMIDE, I, 3; II, 462; III,332; IV,
  104
ACETAMIDINE HYDROCHLORIDE,
  I, 5
ACETAMIDOACETONE, V, 27
p-Acetamidobenzaldehyde, IV, 32
3-ACETAMIDO-2-BUTANONE, IV, 5;
  V, 28
2-Acetamido-2-deoxy-D-glucose, V, 1
2-ACETAMIDO-3,4,6-TRI-O-ACETYL-
  2-DEOXY-α-D-GLUCOPYR-
  ANOSYL CHLORIDE, V, 1
p-ACETAMINOBENZENESULFINIC
  ACID, I, 7; III, 239
p-ACETAMINOBENZENESUL-
  FONYL CHLORIDE, I, 7, 8
α-ACETAMINOCINNAMIC ACID, II,
  1, 491, 519
ACETANILIDE, I, 8, 332; III, 183
ACETANILIDE,α,α,α-TRICHLORO-,
  V, 1074
ACETIC ACID, DIAZO-, t-BUTYL
  ESTER, V, 179
ACETIC ACID, DIAZO-, trans-2-BU-
  TENYL ESTER, V, 258
ACETIC ACID, p-DIMETHYLAMIN-

OPHENYL-, ETHYL
  ESTER, V, 552
ACETOACETANILIDE, III, 10, 580
ACETOACETIC ACID, t-BUTYL
  ESTER, V, 155
o-Acetoacetochloranilide, V, 12
ACETOBROMOGLUCOSE, III, 11, 434
ACETO-p-CYMENE, II, 3
Acetol, II, 5, 546
1-Acetonaphthone oxime, V, 86
ACETONE-ANIL, III, 329
ACETONE CARBOXYMETHOXIME,
  III, 172
ACETONE CYANOHYDRIN, II, 7; III,
  323, 324, 560; V, 839
ACETONE CYANOHYDRIN NI-
  TRATE, V, 839
ACETONE DIBUTYL ACETAL, V, 5
ACETONEDICARBOXYLIC ACID, I,
  10, 237; IV, 816
Acetone dimethyl acetal, V, 5, 293
Acetonedioxalic ester, II, 127
Acetone oxime, V, 1031
Acetonitrile, I, 5; II, 522; III, 71
ACETONYLACETONE, II, 219
ACETOPHENONE, I, 79, 111; II, 434,
  480, 503, 510; III, 88, 251, 305,
  367, 538, 717, 727; IV, 59, 463;
  V, 120, 1135
ACETOPHENONE, 2-AMINO-, HY-
  DROCHLORIDE, V, 909
ACETOPHENONE, 3-BROMO, V, 117
ACETOPHENONE, 4'-METHOXY-2-
  (β-METHOXYPHENYL)-, V,
  339
Acetophenone oxime, III, 92; V, 86
Acetophenone phenylhydrazone, III, 725
2-ACETOTHIENONE, II, 8; III, 14
  p-Acetotoluide, I, 111
o-Acetotoluidide, V, 650

ACETOXIME, **I, 318; III,** 172; **IV,** 932
3β-ACETOXY-5-ANDROSTENE-17β-
  CARBOXYLIC ACID, **V, 8**
3β-ACETOXYETIENIC ACID, **V, 8**
3β-ACETOXY-18-HYDROXY-18,20β-
  OXIDE-5-PREGNENE, **V,
  693**
3β-ACETOXY-20β-HYDROXY-5-
  PREGNENE, **V, 692**
3β-ACETOXY-18-IODO-18,20β-OXIDO-
  5-PREGNENE, **V, 693**
3β-ACETOXY-18-IODO-5-PREGNENE,
  V, 692
Acetoxymethyl methyl sulfoxide, **V, 792**
3β-Acetoxy-5-pregnen-20-one, **V, 8, 692**
ACETYLACETONE, **III, 16, 513; IV,**
  351, 869; **V,** 720
β-Acetylaminonaphthalene, **II, 438**
2-Acetylamino-3-nitrotoluene, **IV, 44**
2-Acetylamino-5-nitrotoluene, **IV, 44**
N-(p-ACETYLAMINO-
  PHENYL)RHODANINE, **IV,
  6**
9-ACETYLANTHRACENE, **IV, 8**
ACETYLBENZOYL, **III, 20**
Acetylbenzoyl dioxime, **III, 21**
α-Acetyl-γ-butyrolactone, **IV, 597**
Acetyl chloride, **I, 12; II, 4, 9, 560; III,**
  142, 144, 237, 390, 841; **IV, 8,**
  580
α-ACETYL-δ-CHLORO-γ-VALERO-
  LACTONE, **IV, 10**
1-ACETYLCYCLOHEXANOL, **IV, 13**
1-ACETYLCYCLOHEXENE, **III, 22**
Acetylene, **I, 229; III, 320, 416, 854; IV,**
  117, 186, 187, 793
ACETYLENEDICARBOXYLIC ACID,
  II, 10
Acetylenes, 1-halo-, **V, 883, 923**
2-ACETYLFLUORENE, **III, 23, 420**
Acetyl fluoride, **V, 69**
N-Acetylglucosamine, **V, 1**
ACETYLGLYCINE, **II, 1, 11**
N-ACETYLHEXAHYDROPHENYL-
  ALANINE, **II, 493**
N-ACETYLISATIN, **III, 456**

ACETYLMANDELIC ACID, **I, 12**
ACETYLMANDELYL CHLORIDE, **I,
  12**
ACETYL METHYLUREA, **II, 462**
3-ACETYLOXINDOLE, **V, 12**
9-ACETYLPHENANTHRENE, **III, 26**
N-Acetylphenylalanine, **II, 493**
2-p-ACETYLPHENYLHYDROQUIN-
  ONE, **IV, 15**
2-p-ACETYLPHENYLHYDROQUIN-
  ONE DIACETATE, **IV, 17**
2-p-ACETYLPHENYLQUINONE, **IV,
  16**
Acetyl-o-toluidine, **III, 597**
δ-ACETYL-n-VALERIC ACID, **IV, 19**
ACID AMMONIUM o-SULFOBEN-
  ZOATE, **I, 14, 495**
ACID ANHYDRIDES, **III, 28**
Acid fluorides, **V, 66**
Acid potassium sulfate, **I, 16**
Acids, esterification, See Index No. 3,
  Reaction Types
ACONITIC ACID, **II, 12**
ACRIDONE, **II, 15; III, 53**
ACROLEIN, **I, 15, 44, 166; II, 138; III,**
  371; **IV, 21, 312, 794, 817**
ACROLEIN ACETAL, **II, 17, 307; IV,
  21**
ACRYLAMIDE, N-BENZYL, **V, 73**
ACRYLIC ACID, **III, 30, 33**
Acrylic acid esters, **III, 147**
ACRYLONITRILE, **III, 93, 458; IV, 146,**
  147, 205, 776; **V, 39, 73**
Acylations, See Index No. 3, Reaction
  Types
ADAMANTANE, **V, 16**
1-ADAMANTANECARBOXYLIC
  ACID, **V, 20, 742**
Addition Reactions, See Index No. 3,
  Reaction Types
ADIPIC ACID, **I, 18, 192; II, 169, 264;
  III, 623**
Adipyl azide, **IV, 820**
ADIPYL HYDRAZIDE, **IV, 819**
Ajinomoto, **I, 287**
Alanine, **IV, 5**

β-ALANINE, **II, 19; III, 34; IV, 94; V,**
    573
dl-ALANINE, **I, 21**
Aldehyde-collidine, **IV, 452**
Aldehydes, aromatic, **V,** 49, 217, 244, 873
Alkoxyanilines, **III,** 141
(β-Alkoxyethyl)triphenylphosphonium
    bromides, **V,** 1147
Alkoxynitrobenzenes, **III,** 141
t-Alkylamines from *t*-alkanes, **V,** 38
1-Alkyl-5-aminopyrazoles, **V,** 42
1-Alkyl-2-arylsulfonyl-5-amino-4-pyrazoles,
    V, 42
Alkylation Reactions, See Index No. 3,
    Reaction Types
ALKYL BROMIDES, **I, 25; V,** 126
Alkyl chlorides, **V,** 390
ALKYLENE BROMIDES, **I, 25**
Alkyl halides, **V,** 151
Alkylhydrazines, **V,** 899
2-Alkyl-1-indanones, **V,** 625
Alkyllepidines, **III,** 521
Alkyl mercaptans, **III, 364**
5-Alkyl-1,2,3,4,5-pentachlorocyclopenta-
    dienes, **V,** 895
Alkylphenols, **III, 445**
N-Alkylpropargylamines, **V,** 544
N-Alkylpyrrolidines, **III,** 163
Alkylresorcinols, **III, 445**
1-Alkyl-3-*p*-tolyltriazenes, **V,** 800
ALLANTOIN, **II, 21**
ALLENE, **V, 22**
Allenes, **V,** 309
1-(1-Allenimino)-2-hydroxy-3-butene, **V,**
    544
ALLOXAN MONOHYDRATE, **III, 37,**
    39; IV, 23, 25, 26, 28
ALLOXANTIN DIHYDRATE, **III,** 37,
    42; IV, 25
ALLYL ALCOHOL, **I, 27, 42; III, 46**
ALLYLAMINE, **II, 24**
·ALLYLAMINE, 2-BROMO-, **V, 121**
ALLYLAMINE, 2-BROMO-*N*-ETHYL-,
    V, 124
Allylbenzene, **IV,** 766
ALLYL BROMIDE, **I, 27,** 46, 203, 521;

    III, 44, 45, 418; **IV,** 749, 767
α-ALLYL-β-BROMOETHYL ETHYL
    ETHER, **IV, 750**
Allyl chloride, **III,** 121; **IV,** 751
ALLYL CYANIDE, **I, 46; III,** 851, 852
2-ALLYLCYCLOHEXANONE, **III, 44;**
    V, 25
Allyl isothiocyanate, **II, 24**
ALLYL LACTATE, **III, 46**
Allyllithium, **V,** 453
Allylmagnesium bromide, **IV, 749; V,** 608
Allylmagnesium chloride, **IV,** 751
5-Allyl-1,2,3,4,5-pentachlorocyclopenta-
    diene, **V,** 896
Allyltriphenyltin, **V, 453**
Alumina, **IV,** 520, 816
Alumina, activated, **I,** 184; **III,** 276, 313;
    IV, 244, 796, 818, 965
Aluminum, **II,** 599; **III,** 48
    amalgamated, **II,** 232, **233; V,** 33
Aluminum bromide, **IV,** 960
ALUMINUM *t*-BUTOXIDE, **III, 48,**
    207, 367
Aluminum chloride, as catalyst, See Index
    No. 6, Catalysts
Aluminum chloride-phosphorus oxychlor-
    ide complex, **IV,** 784
ALUMINUM ETHOXIDE, **II, 598, 599;**
    III, 49
Aluminum isopropoxide, **III,** 49; **IV,** 193
Aluminum metal, wool, **V,** 1116
Aluminum selenide, **IV,** 360
ALUMINUM, TRIPHENYL-, **V, 1116**
Amberlite IR-4B resin, **IV,** 39
Amberlite IR-120 resin, **IV,** 530
Amidation Reactions, See Index No. 3,
    Reaction Types and No. 4,
    Types of Compounds
N-Amination, by hydroxylamine-*O*-sul-
    fonic acid, **V,** 43 by trialkyl
    phosphates, **V,** 1087
Amination with trichloramine, **V,** 35
AMINOACETAL, **III, 50**
p-Aminoacetanilide, **IV,** 7; **V,** 341
AMINOACETONE HYDRO-
    CHLORIDE, **V, 27**

AMINOACETONE SEMICARBA-
ZONE HYDROCHLORIDE,
V, 27
AMINOACETONITRILE HYDRO-
GEN SULFATE, I, 298
p-Aminoacetophenone, IV, 16
9-AMINOACRIDINE, III, 53
2-AMINO-4-ANILINO-6-(CHLORO-
METHYL)-s-TRIAZINE, IV,
29
o-AMINOBENZALDEHYDE, III, 56
m-AMINOBENZALDEHYDE DI-
METHYL ACETAL, III, 59,
564
p-AMINOBENZALDEHYDE, IV, 31
azine, IV, 33
oxime, IV, 33
phenylhydrazone, IV, 33
m-Aminobenzoic acid, IV, 948
p-Aminobenzoic acid, III, 711; V, 670
2-AMINOBENZOPHENONE, IV, 34
o-AMINOBENZYL ALCOHOL, III, 60
o-AMINOBIPHENYL, V, 829
nitration of, V, 830
reaction with boron trichloride, V, 727
2-AMINO-5-BROMO-3-NITROPYRI-
DINE, V, 347
2-AMINO-5-BROMOPYRIDINE, V, 346
γ-AMINOBUTYRIC ACID, II, 25
α-AMINO-n-CAPROIC ACID, I, 48
ε-AMINOCAPROIC ACID, II, 28; IV,
39
ε-Aminocaproic acid hydrochloride, IV, 39
4-Amino-3-chlorophenol, IV, 150
4-AMINOCYCLOHEXANE CARBOX-
YLIC ACID, V, 670
2-Aminocyclohexanone, hydrochloride of,
V, 915
2-Aminocyclopentanone, hydrochloride of,
V, 915
2-AMINO-p-CYMENE, III, 63
2-Amino-2-deoxy-D-glucose, V, 2
2-Amino-3,5-dibromopyridine, V, 346
α-AMINODIETHYLACETIC ACID,
III, 66
5-AMINO-2,3-DIHYDRO-1,4-PHTHAL-

AZINEDIONE, III, 69
1-Amino-2,4-dimethylpyridinium iodide, V,
45
1-Amino-2,6-dimethylpyridinium ioide, V,
45
4-AMINO-2,6-DIMETHYLPYRIMI-
DINE, III, 71
2-Amino-4,4-dimethyl-1-tetralone, hydrochloride, V, 915
N-β-Aminoethylethylenimine, IV, 434
β-Aminoethylsulfuric acid, IV, 433
2-AMINOFLUORENE, II, 448; V, 30
AMINOGUANIDINE BICARBON-
ATE, III, 73, 95
3-Amino-2-heptanone, hydrochloride, V,
915
α-AMINOISOBUTYRIC ACID, II, 29
α-Aminoketones, V, 914
Aminomalononitrile, V, 33
AMINOMALONONITRILE p-TOLU-
ENESULFONATE, V, 32, 344
2-AMINO-6-METHYLBENZOTHIA-
ZOLE, III, 76
1-AMINO-1-METHYLCYCLOHEX-
ANE, V, 35
1-(Aminomethyl)cyclohexanol, acetic acid
salt, IV, 224
2-Amino-2-methyl-1-propanol, III, 148; V,
699
1-Amino-2-methylpyridinium iodide, V, 45
2-AMINO-4-METHYLTHIAZOLE, II,
31
3-AMINO-2-NAPHTHOIC ACID, III,
78
1-Amino-8-naphthol-2,4-disulfonic acid, II,
146
1-Amino-8-naphthol-3,6-disulfonic acid, II,
149
1,2-AMINONAPHTHOL HYDRO-
CHLORIDE, II, 33, 431; IV, 53
1,4-AMINONAPHTHOL HYDRO-
CHLORIDE, I, 49, 383; II, 39
1-AMINO-2-NAPHTHOL-4-SULFONIC
ACID, II, 42; III, 633
1-Amino-8-naphthol-4-sulfonic acid, II, 149
2-Amino-5-naphthol-7-sulfonic acid, II, 149

2-Amino-8-naphthol-6-sulfonic acid, **II**, 149
α-Aminonitriles, **V**, 438
2-Amino-5-nitroanisole, **IV**, 82
o-AMINO-*p*-NITROBIPHENYL, **V, 830**
2-AMINO-4-NITROPHENOL, **III, 82**
2-Amino-5-nitropyrimidine, **IV**, 845
2-AMINO-3-NITROTOLUENE, **IV, 42,
44**
2-Amino-5-nitrotoluene, **III, 660; IV, 44**
o-Aminophenol, **IV**, 570
p-Aminophenol, **II**, 355
dl-α-AMINOPHENYLACETIC ACID,
III, 84
p-AMINOPHENYLACETIC ACID, **I,
52**
2-Amino-2-phenylacetophenone, **V**, 915
p-AMINOPHENYL DISULFIDE, **III,
86**
dl-α-AMINO-α-PHENYLPROPIONIC
ACID, **III, 88**
dl-β-AMINO-β-PHENYLPROPIONIC
ACID, **III, 91**
1-Aminopiperidinium hydrogen oxalate, **V**,
45
AMINO-2-PROPANONE, SEMICAR-
BAZONE HYDROCHLOR-
IDE, **V, 27**
β-AMINOPROPIONITRILE, **III**, 35, **93**
3(5)-AMINOPYRAZOLE, **V, 39**
3-AMINO-3-PYRAZOLINE SULFATE,
V, 39
2-Aminopyridine, **III**, 136; **IV**, 91; **V**, 346
3-AMINOPYRIDINE, **IV, 45**
1-AMINOPYRIDINIUM IODIDE, **V,
43**
2-Aminopyrimidine, **IV**, 182
1-Aminoquinolinium iodide, **V**, 45
2-Amino-1-tetralone, hydrochloride, **V**, 915
p-AMINOTETRAPHENYLMETHANE,
IV, 47
p-Aminotetraphenylmethane hydrochloride, **IV**, 47
AMINOTHYMOL, **I, 512**
3-AMINO-1*H*-1,2,4-TRIAZOLE, **III, 95**
4-AMINO-4*H*-1,2,4-TRIAZOLE, **III, 96**
4-Amino-1,2,4-triazole hydrochloride, **III,
97**

3-AMINO-2,4,6-TRIBROMOBENZOIC
ACID, **IV, 947**
4-Amino-2,2,4-trimethylpentane, **V**, 845
1-Amino-2,4,6-trimethylpyridinium iodide,
V, 45
6-Aminouracil, **IV**, 248
α-Aminovalerophenone, hydrochloride, **V**,
915
4-AMINOVERATROLE, **II, 44**
Ammonia, **I**, 177, 381; **II**, 30, 215, 221
Ammonium 1-adamantanecarboxylate, **V**,
20
Ammonium carbonate, **I**, 3, 458; **II**, 219;
III, 54, 323
Ammonium chloride, **II**, 29, 266
Ammonium chloroplatinate, **I**, 466
Ammonium dichromate, **II**, 142, 555
AMMONIUM DITHIOCARBAMATE,
III, 763
Ammonium formate, **II**, 503
AMMONIUM IODIDE,
[3[[(ETHYLIMINO)METHYL-
ENE]AMINO]-PROPYL]-
TRIMETHYL-, **V, 555**
Ammonium 1,2-naphthoquinone-4-sulfo-
nate, **III**, 465
Ammonium nitrate, **I**, 302; **IV**, 21
Ammonium nitrate, explosion hazard of,
V, 589
Ammonium pentachlorobenzoate, **V**, 891
AMMONIUM SALT OF AURIN TRI-
CARBOXYLIC ACID, **I, 54**
AMMONIUM SALT OF N-NITROSO-
N-PHENYLHYDROXYL-
AMINE, **I, 177**
Ammonium sulfide, **I**, 53
Ammonium thiocyanate, **II**, 555, 574; **III**,
735; **IV**, 180
Ammonium vanadate, **I**, 18
Ammonolysis, **V**, 582
n-Amylacetylene, **IV**, 119
n-Amyl alcohol, **II**, 107
t-Amyl alcohol, **II**, 408
t-Amylamine, **III, 155**
N-Amylaniline, **IV, 284**
n-AMYLBENZENE, **II**, 47; **IV, 522**
n-AMYL BORATE, **II, 107**

n-Amyl bromide, **II, 323, 324**
n-Amyl fluoride, **IV, 527**
n-Amyl iodide, **II, 402**
3-Amyl-4-methyl-1-phenyl-5-pyrazolone, **IV, 122**
n-Amyl phosphate, **II, 110**
s-Amyl sulfide, **II, 548**
t-Amylurea, **III, 155**
Anethole, **IV, 787**
Anhydrides, **V, 824**
ANHYDRO-*o*-HYDROXYMERCURI-BENZOIC ACID, **I, 57**
ANHYDRO-2-HYDROXYMERCURI-3-NITROBENZOIC ACID, I, **56,** 125
Anilides, **III, 766**
Aniline, **I,** 49, 70, 80, 82, 103, 287, 328, 332, 442, 447, 478; **II,** 15, 163, 239, 347, 351, 494, 592; **III,** 10, 108, 329, 330, 374, 375; **IV,** 77, 81, 349, 383, 384, 464, 634, 764, 769, 796; **V,** 941, 944, 1074
ANILINE, 4,4'-AZODI-, **V, 341**
Aniline benzenesulfonate, **IV, 206**
Aniline hydrochloride, **I,** 453; **II,** 295; **IV,** 47, 48, 205
ANISALACETONE, **I, 78**
Anisaldehyde, **I,** 78; **IV,** 605
o-ANISALDEHYDE, **V, 46**
o-Anisidine, **III, 186**
p-Anisidine, **III, 662**
p-Anisidine hydrochloride, **IV, 53**
Anisoin, **V, 339**
ANISOLE, **I, 58**
ANISOLE, *p*-α-DICHLORO-, **V, 221**
Anisyl alcohol, **IV, 576**
p-Anisyl *t*-butyl ether, **V, 925**
Anisyl chloride, **IV, 577**
Anthracene, **I,** 207; **II,** 554; **III, 98; IV,** 8, 9, 711, 965; **V,** 206
ANTHRACENE, 9-CHLORO, **V, 206**
ANTHRACENE, 9,10-DIHYDRO-, **V, 398**
9-ANTHRALDEHYDE, **III, 98**
Anthranilic acid, **I,** 222, 374; **II,** 79, 349, 581; **III,** 60, 475, 488; **IV,** 34, 872; **V,** 55, 57

Anthranils, **V, 943**
Anthraquinone, **I,** 60; **II,** 62, 540, **554; IV,** 712, 758
ANTHRONE, **I, 60,** 390
Anti-foam agent, **I,** 426, 475
Antimony trichloride, **I, 550**
Antioxidant, **I,** 441; **II,** 36, 40
1-APOCAMPHANECARBOXYLIC ACID, 2-OXO-, **V, 689**
Aqua regia, **II, 566**
D-ARABINOSE, **III, 101**
l-ARABINOSE, **I, 67**
Arenes, polyalkyl, **V, 601**
Arenesulfonyl chlorides, **V, 42**
ARGININE DINITRONAPHTHOL-SULFONATE, **II, 50**
d-ARGININE HYDROCHLORIDE, **II, 49**
ARSANILIC ACID, **I, 70, 488**
Arsenic acid, **I,** 70, 490
Arsenic oxide, **III, 568**
Arsenic trichloride, **IV, 910**
Arsenious oxide, **I,** 73, 267; **II,** 57, 494
ARSENOACETIC ACID, **I, 74**
ARSONOACETIC ACID, **I, 73**
p-ARSONOPHENOXYACETIC ACID, **I, 75**
Arylarsonic acids, **III, 666**
Arylbenzenes, **V, 51**
2-Arylbenzotriazoles, **V, 943**
Aryl *t*-butyl ethers, **V, 644**
Arylcyclopropanes, **V, 931**
Aryl halides, **V, 142**
2-Arylindazoles, **V, 943**
Arylmethylcarbinols, **III, 201**
Arylsulfonylhydrazones, **V, 262**
ARYLUREAS, **IV, 49**
Arynes, **V, 263**
Asbestos, **IV, 940**
Asbestos-sodium silicate stoppers, **I, 20**
DL-ASPARTIC ACID, **IV, 55**
ATROLACTIC ACID, **IV, 58**
3-AZABICYCLO(2.2.2)OCTAN-2-ONE, **V, 670**
7-Azaindole-3-carbonitrile, **V, 657**
AZELAIC ACID, **II, 53; IV,** 64, 65
Azelanitrile, **IV, 62**

2*H*-AZEPIN-2-ONE, 1,3-DIHYDRO-3,-
  5,7-TRIMETHYL-, **V, 408**
2-AZETIDINONE, 4,4-DIMETHYL-1-
  SULFONYL CHLORIDE
  AND 2-AZETIDINONE, 4,4-
  DIMETHYL, **V, 673**
2-Azetidinones, **V,** 678
o-Azidobiphenyls, **V,** 832
o-AZIDO-*p*-NITROBIPHENYL, **V, 830**
AZIRIDINE, 2-BENZYL-3-PHENYL-,
  cis-, **V, 83**
AZIRIDINE, 1-ETHYL-2-METHYL-
  ENE, **V, 541**
Aziridines, **V,** 86
AZLACTONE OF α-ACETAMINO-
  CINNAMIC ACID, **II, 1**
AZLACTONE OF α-BENZOYLAMIN-
  OCINNAMIC ACID, **II, 490**
AZLACTONE OF α-BENZOYL-
  AMINO-β-(3,4-DIMETHOXY-
  PHENYL)-ACRYLIC ACID,
  II, 55, 333
AZOBENZENE, **III, 103**
Azo compounds, **V,** 343
1,1'-AZO-*bis*-1-CYCLOHEXANENI-
  TRILE, **IV, 66,** 273, 275
2,2'-Azo-*bis*-isobutyronitrile, **IV,** 67, 273,
  275
AZOXYBENZENE, **II, 57**
Azulenes, **V,** 1088, 1091
AZULENE, 4,6,8-TRIMETHYL-, **V,**
  1088

BARBITURIC ACID, **II, 60,** 440; **III,** 39;
  IV, 23
Barium carbonate, **III,** 289, 541
Barium chloride, **II,** 482; **III,** 214; **IV,** 507,
  529, 676
Barium D-gulonate, **IV,** 506
Barium hydroxide, **I,** 192, 199, 299; **II,** 26;
  III, 35, 542, 588, 615, 638, 685;
  IV, 177, 225, 506, 636
Barium nitrate, **II,** 142
BARIUM 2-PHENANTHRENESULF-
  ONATE, **II, 482**
Bayberry wax, **III,** 605

Beckmann Fission, **V,** 267
Behenic acid, **IV,** 742
BENZALACETONE, **I, 77; III,** 105; **IV,**
  312
BENZALACETONE DIBROMIDE, **III,**
  105, 125
BENZALACETOPHENONE, **I, 78,** 101,
  205; **II,** 237, 498; **IV,** 312; **V,**
  1135
BENZALACETOPHENONE DIBROM-
  IDE, **I, 205**
BENZALANILINE, **I, 80**
BENZALAZINE, **II, 395**
BENZALBARBITURIC ACID, **III, 39**
Benzaldehyde, **I,** 77, 79, 80, 81, 94, 105,
  181, 252, 336, 381, 413; **II,** 1,
  49, 168, 395, 490; **III,** 39, 84,
  377, 408, 644, 715, 827; **IV,** 392,
  605, 606, 679, 777, 933; **V,** 80,
  130, 191, 390, 396, 437, 500,
  736, 758, 1124
BENZALDEHYDE, *o*-NITRO-, **V, 825**
Benzaldehydes, halogen- and methyl-, **V,**
  141
BENZALDEHYDE, 2,4,6-TRIMETH-
  YL-, **V, 49**
Benzaldoxime, **V,** 505
4-Benzal-2-phenyl-5-oxazolone, **IV,** 83
BENZALPHTHALIDE, **II, 61; III,** 353
BENZALPINACOLONE, **I, 81**
Benzal-*bis*-thiobenzoate, **IV,** 925
BENZAMIDINE HYDROCHLORIDE,
  I, 6
BENZANILIDE, **I, 82,** 217; **IV,** 383, **384**
BENZANTHRONE, **II, 62**
peri-Benzazulene, **V,** 1091
BENZENE, **V, 998**
Benzeneazo-α-naphthol, **I,** 50
BENZENE, *o*-BIS(BUTYLTHIO)-, **V,**
  107
BENZENEBORONIC ACID, **IV, 69, 71**
BENZENEBORONIC ANHYDRIDE,
  IV, 68
BENZENE, BROMOETHYNYL-, **V, 921**
BENZENE, 1-BROMO-2-FLUORO-, **V,**
  133

BENZENE, 3-BROMOPROPENYL-, V, 249

BENZENE, CYCLOPROPYL-, V, 328, 929

BENZENEDIAZONIUM, o-CAR-BOXY-, HYDROXIDE, IN-NER SALT, V, 54

BENZENEDIAZONIUM-2-CARBOXYL-ATE, V, 54

decomposition of, V, 55

derivatives, V, 59

dry, detonation of, V, 54, 57

Benzenediazonium chloride, I, 49; IV, 634

Benzenediazonium fluoborate, II, 295

BENZENE, 1,3-DICHLORO-2-NITRO, V, 367

o-BENZENEDITHIOL, V, 419

BENZENE, HEXAPHENYL-,V, 604

BENZENEHEXOL, V, 595

BENZENE, 1-IODO-2,4-DINITRO-, V, 478

BENZENE, IODOSO-, V, 658

BENZENE, IODOSO-, DIACETATE, V, 660

BENZENE, 2-NITRO-1,3,5-TRI-PHENYL, V, 1128

BENZENESULFENYL CHLORIDE, 2,4-DINITRO-, V, 474

BENZENESULFENYL TRIFLUOR-IDE, V, 859

BENZENESULFINIC ACID, METHYL ESTER, V, 723

BENZENESULFINYL FLUORIDE, V, 396

BENZENESULFONYL CHLORIDE, I, 84, 504; III, 161; IV, 676, 753

1,2,4-BENZENETRIAMINE, 5-NITRO-, V, 1067

Benzhydrol, IV, 73

Benzhydryl chloride, IV, 963; V, 527

BENZHYDRYL β-CHLOROETHYL ETHER, IV, 72

Benzidine, II, 149, 188; IV, 283, 284

Benzidine dihydrochloride, IV, 284

BENZIL, I, 87; II, 497; III, 357, 806; IV, 377; V, 605

Benzil, 4,4'-bis(diethylamino)-, V, 113

BENZIL, 4,4'-BIS(DIMETHYL-AMINO),- V, 111

Benzil, 4,4'-bis(di-n-propylamino),- V, 113

Benzil dihydrazone, IV, 377

BENZIL HYDRAZONE, II, 496, 497

BENZILIC ACID, I, 89, 224; IV, 482

Benzil monohydrazone, III, 357

BENZIMIDAZOLE, II, 65

Benzocyclobutenes, 1-substituted, V, 265

1,2-Benzo-3,4-dihydrocarbazole, IV, 885

BENZOFURAN, V, 251

BENZOFURAZAN OXIDE, IV, 74

BENZOGUANAMINE, IV, 78

BENZOHYDROL, I, 90

BENZOHYDROXAMIC ACID, II, 67

BENZOHYDROXAMOYL CHLORIDE, V, 505

BENZOIC ACID, I, 82, 91, 363; III, 288, 337; IV, 286

Benzoic acid, Birch reduction, V, 400

BENZOIC ACID, DITHIO-, CARBOXY-METHYL ESTER, V, 1046

BENZOIC ACID, PENTACHLORO-, V, 890

BENZOIC ACID, 2-PHENYLETHYL ESTER, V, 336

BENZOIC ACID, 2,4,6-TRIMETHYL-, V, 706

BENZOIC ANHYDRIDE, I, 91

BENZOIC-CARBONIC ANHYDRIDE, ETHYL ESTER, IV, 286

BENZOIN, I, 87, 89, 94; II, 69, 159, 231; III, 787; V, 1076

BENZOIN ACETATE, II, 69

Benzonitrile, I, 6; IV, 79, 123, 437, 769; V, 520

BENZONITRILE, 2-METHYL-3,5-DINITRO-, V, 480

Benzonitrile oxide, V, 505

Benzo[c]phenanthrene, V, 956

BENZOPHENONE, I, 90, 95; II, 70, 71, 573,606; III, 352; IV, 133, 388, 914, 927; V, 364, 509, 564, 604

BENZOPHENONE DICHLORIDE, II, 573

Benzophenone hydrazone, **III**, 352
BENZOPHENONE OXIME, **II**, 70, 234
BENZOPINACOL, **II**, 71, 73
β-BENZOPINACOLONE, **II**, 73
BENZOPYRAZOLE, **V**, 650
p-Benzoquinone, **III**, 286; **IV**, 16, 152
1,2,3-BENZOTHIADIAZOLE 1,1-
  DIOXIDE, **V**, 60, 64, 65
N-(2-Benzothiazolyl)urea, **V**, 804
1,2,3-BENZOTRIAZOLE, **III**, 106
Benzotrichloride, **V**, 1046
BENZOYLACETANILIDE, **III**, 108;
  IV, 80
Benzoylacetone, **V**, 718
β-BENZOYLACRYLIC ACID, **III**, 109
ε-BENZOYLAMINO-α-BROMO-
  CAPROIC ACID, **II**, 74, 374
ε-BENZOYLAMINOCAPROIC ACID,
  II, 74, 76
α-BENZOYLAMINOCINNAMIC
  ACID, **II**, 490
γ-Benzoylbutyric acid, **II**, 82
Benzoyl chloride, **I**, 100, 101; **III**, 112, 116,
  187, 735; **IV**, 84, 384, 415, 478,
  642, 924
BENZOYLCHOLINE CHLORIDE, **IV**,
  84, 87
BENZOYLCHOLINE IODIDE, **IV**, 84,
  85
Benzoylcholine picrate, **IV**, 87
BENZOYL CYANIDE, **III**, 112, 114
2-BENZOYL-1-CYANO-1-METHYL-
  1,2-DIHYDROISOQUIN-
  OLINE, **IV**, 642
BENZOYL DISULFIDE, **III**, 116
BENZOYLENE UREA, **II**, 79
BENZOYL FLUORIDE, **V**, 66, 336
BENZOYLFORMIC ACID, **I**, 244; **III**,
  114
β-Benzoyllactic acid, **III**, 111
dl-ε-BENZOYLLYSINE, **II**, 374
Benzoyl-2-methoxy-4-nitroacetanilide, **IV**,
  82
o-BENZOYLOXYACETOPHENONE,
  IV, 478, 479
BENZOYLOXYCYCLOHEXENE, **V**,

70
Benzoyloxy group, **V**, 380
Benzoyloxylation, **V**, 70
Benzoyl peroxide, **I**, 431; **IV**, 109, 430,
  921; **V**, 93, 379
  analysis of, **I**, 433
  warning, **I**, 432
BENZOYL PIPERIDINE, **I**, 99, 428
β-BENZOYLPROPIONIC ACID, **II**, 81,
  499; **V**, 80
3-BENZOYLPYRIDINE, **IV**, 88
4-Benzoylpyridine, **IV**, 89
N-Benzylacetamide, **V**, 75
BENZYLACETOPHENONE, **I**, 101
N-BENZYLACRYLAMIDE, **V**, 73
Benzyl alcohol, **I**, 104; **II**, 591; **III**, 167;
  IV, 91; **V**, 73
Benzylamine, **IV**, 780
2-BENZYLAMINOPYRIDINE, **IV**, 91
N-BENZYLANILINE, **I**, 102; **IV**, 92,
  284
BENZYL BENZOATE, **I**, 104
Benzyl t-butyl ether, **V**, 925
BENZYL CARBAMATE, **III**, 168
2-BENZYL-2-CARBOMETHOXY-
  CYCLOPENTANONE, **V**, 76
Benzyl chloride, **I**, 103, 107, 471; **II**, 47,
  83, 232, 606; **III**, 705; **IV**, 94,
  99, 606, 963
Benzyl chloroformate, **III**, 167
BENZYL CHLOROMETYL KETONE,
  III, 119
BENZYL CYANIDE, **I**, 107, 270, 396,
  436; **II**, 287, 487, 512; **IV**, 175,
  380, 387, 392, 461, 760
2-BENZYLCYCLOPENTANONE, **V**,
  76
9-Benzyl-1-decalone, **V**, 190
Benzyldimethylcarbinyl hypochlorite, **V**,
  186
BENZYLIDENE ARGININE, **II**, 49
N-Benzylidenebutylamine, **V**, 736
N-BENZYLIDENE ETHYLAMINE, **V**,
  758
N-BENZYLIDENEMETHYLAMINE,
  IV, 605

α-BENZYLIDENE-γ-PHENYL Δ^{β,γ}-BU-
 TENOLIDE, **V, 80**
Benzyl isocyanide, **IV, 388; V, 302**
1-Benzylisoquinoline, **IV, 644**
Benzyllithium, **V, 455**
BENZYLMAGNESIUM CHLORIDE,
 I, 187, 471; IV, 94, 606
Benzyl mercaptan, **V, 474**
2-Benzyl-2-methylcyclohexanone, **V, 190**
3-BENZYL-3-METHYLPENTANENI-
 TRILE, **IV, 95**
3-BENZYL-3-METHYLPENTANOIC
 ACID, **IV, 93**
2-Benzylperfluoropropene, **V, 951**
cis-2-Benzyl-3-PHENYLAZIRIDINE, **V,
 83**
BENZYL PHTHALIMIDE, **II, 83**
1-BENZYLPIPERAZINE, **V, 88**
1-Benzylsulfinyl-2-propanone, **V, 792**
Benzyl sulfoxide, **V, 792**
Benzyltrimethylammonium chloride, **IV, 99**
BENZYLTRIMETHYLAMMONIUM
 ETHOXIDE, **IV, 98**
Benzyltrimethylammonium hydroxide, **IV,
 652; V, 605**
BENZYLTRIMETHYLAMMONIUM
 IODIDE, **IV, 585, 586**
Benzylurea, **V, 804**
Benzyne, **IV, 964; V, 65, 928, 1037, 1042**
BETAINE HYDRAZIDE HYDRO-
 CHLORIDE, **II, 85**
Betaine hydrochloride, **IV, 154**
Biacenaphthylidenedione, **III, 2**
Biacetyl, **IV, 343**
BIACETYL MONOXIME, **II, 205**
BIALLYL, **III, 121**
Bibenzyl, **IV, 96, 97, 257**
BIBENZYL, α,α'-EPOXY-2,2'-DI-
 CHLORO-, **V, 358**
Bicyclo(2.2.1)hepta-2,5-diene, **V, 863**
BICYCLO(4.1.0)HEPTANE, **V, 855**
BICYCLO(2.2.1)HEPTEN-7-ONE, **V, 91**
Bicyclohexyl, **V, 696**
BICYCLOHEXYLIDENE, **V, 297**
exo-cis-BICYCLO(3.3.0)OCTANE-2-
 CARBOXYLIC ACID, **V, 93**

BICYCŁO(4.2.0)OCTA-1,3,5-TRIENE-7-
 CARBONITRILE, **V, 263**
BICYCLOPENTADIENYLIDENE,
 OCTACHLORO-, **V, 901**
BICYCLO(2.1.0)PENTANE, **V, 96**
Biguanide, **IV, 30**
m,m'-BIPHENOL, **V, 412**
Biphenyl, **IV, 256**
2-BIPHENYLCARBOXYLIC ACID, 2'-
 FORMYL-, **V, 493**
4,4'-Biphenyldicarboxylic acid, **V, 815**
BIPHENYL, 3,4-DICHLORO-, **V, 51**
BIPHENYL, 2,2'-DIFORMYL-, **V, 489**
BIPHENYLENE, **V, 54, 59**
4,4'-BIPHENYLENE-bis-DIAZONIUM
 FLUOBORATE, **II, 188**
p-Biphenylyl isothiocyanate, **IV, 701**
1-(2-Biphenylyl)urea, **IV, 51**
1-(4-Biphenylyl)urea, **IV, 51**
2,2'-BIPYRIDINE, **V, 102**
2,2'-Bipyridines, **V, 106**
2,2'-Biquinolines, **V, 106**
Birch reduction, **V, 400, 402, 467**
BIS(10,9-BORAZAROPHENAN-
 THRYL)-ETHER, **V, 727, 729**
1,2-BIS(N-BUTYLTHIO)BENZENE, **V,
 107**
1,4-Bis(n-butylthio)benzene, **V, 109**
Bis(chloromethyl)ether, hazard note, **V,
 218**
Bis-3,4-dichlorobenzoyl peroxide, **V, 51,
 52**
4,4'-Bis(diethylamino)benzil, **V, 113**
Bis(2-diethylaminoethyl)sulfoxide, **V, 792**
14,4'-BIS(DIMETHYLAMINO)-
 BENZIL, **V, 111**
4,4'-Bis(dimethylamino)benzophenone, **V,
 529**
Bis(dimethylamino)methane, **V, 434**
Bis-(2,4-Dinitrophenyl)disulfide, **V, 476**
Δ^{2,2}-BIS(1,3-DIPHENYLIMIDAZOLI-
 DINE) **V, 115**
BIS(1,3-DIPHENYLIMIDAZOLIDI-
 NYLIDENE-2-), **V, 115**
4,4'-Bis(di-n-propylamino)benzil, **V, 113**
Bis(N-methyl-N-nitroso)terephthalamide,

V, 351
1,4-Bis(3-methyl-4-phenylbutadienyl)-ben-
    zene, **V**, 987
1,4-Bis-[4-(3-nitrophenyl)butadienyl]-ben-
    zene, **V**, 987
1,4-BIS-(4-PHENYLBUTADIENYL)-
    BENZENE, **V**, **986**
1,2-Bis(phenylthio)benzene, **V**, 109
1,4-Bis(phenylthio)benzene, **V**, 109
1,4-Bis[4-(*p*-tolyl)butadienyl]benzene, **V**,
    987
p-Bis(trifluoromethyl)benzene, **V**, 1084
2,4-Bis(trifluoromethyl)chlorobenzene, **V**,
    1084
BIURET, 1-PHENYL-2-THIO-, **V**, **966**
Blood, **III**, 442
Blood corpuscle paste, **II**, 330
Boric acid, **I**, 476; **II**, 106, 188, 226, 295,
    299; **IV**, 866
Boron tribromide, **V**, 412
Boron trichloride, **V**, 727
Boron trifluoride, **III**, 16, 406; **V**, 332, 480,
    598, 601
Boron trifluoride 2etherate, **IV**, 375, 957;
    V, 1080
BROMAL, **II**, **87**
Bromination, See Index No. 2, Reaction
    Types; Halogenation
BROMOACETAL, **III**, **123**, 506
N-BROMOACETAMIDE, **IV**, **104**; **V**,
    136
3-BROMO-4-ACETAMINOTOLUENE,
    I, **112**
BROMOACETIC ACID, **III**, 172, 381
BROMOACETONE, **II**, 5, **88**
p-BROMOACETOPHENONE, **I**, **109**,
    127; **IV**, 110
3-BROMOACETOPHENONE, **V**, 117
Bromoacetyl fluoride, **V**, 69
2-BROMOALLYLAMINE, **V**, **121**
N-(2-BROMOALLYL)ETHYLAMINE,
    V, **124**
2-BROMOALLYLHEXAMINIUM
    BROMIDE, **V**, 121
3-BROMO-4-AMINOTOLUENE, **I**, **111**,
    133; **III**, 130

o-Bromoaniline, **V**, 133, 1120
m-Bromoaniline, **I**, 449; **III**, 131, 186
p-Bromoaniline, **I**, 113, 448; **IV**, 49
o-Bromoanisole, **III**, **186**
9-Bromoanthracene, **V**, 207
α-BROMOBENZALACETONE, **III**, **125**
α-Bromobenzalacetophenone, **III**, **126**
3-Bromobenzaldehyde, **V**, 120
m-BROMOBENZALDEHYDE, **II**, **132**
p-BROMOBENZALDEHYDE, **II**, **89**,
    442; **IV**, **933**
    bisulfite addition product, **II**, 90
p-Bromobenzaldiacetate, **II**, **442**
Bromobenzene, **I**, 109, 226, 228, 363, 550;
    II, 518; **III**, 157, 238, 353, 413,
    562, 566, 729, 757, 771, 807,
    839; **IV**, 912; **V**, 131, 520, 924,
    926
o-BROMOBENZENEDIAZONIUM
    HEXAFLUOROPHOS-
    PHATE, **V**, **133**
p-Bromobenzoic acid, **II**, 90
p-BROMOBIPHENYL, **I**, **113**
1-Bromobutane, **V**, 848
3-Bromo-4-*t*-butylacetophenone, **V**, 120
α-Bromo-*n*-butyric acid, **IV**, 621
α-BROMO-γ-BUTYROLACTONE, **V**,
    255
α-Bromo-*n*-butyryl chloride, **IV**, 621
α-BROMO-*n*-CAPROIC ACID, **I**, 48,
    115; **II**, **95**
m-Bromochlorobenzene, **III**, 200
1-Bromo-3-chlorocyclobutane, **V**, 129
m-Bromocinnamic acid, **IV**, **733**
α-BROMOCINNAMIC ALDEHYDE,
    III, **731**
α-BROMOCINNAMIC ALDEHYDE
    ACETAL, **III**, **732**
Bromocyclohexane, **III**, 220
BROMOCYCLOPROPANE, **V**, **126**
2-BROMODECENE, **I**, **187**, 192
Bromodiisopropylphosphine, **V**, 212
2-Bromododecanoic acid, **IV**, 398
2-BROMOETHANOL, **I**, **117**
β-BROMOETHYLAMINE HYDRO-
    BROMIDE, **II**, **91**, 564

(2-Bromoethyl)benzene, V, 559
β-BROMOETHYLPHTHALIMIDE, I,
    119; IV, 106
vic-Bromoflourides from olefins, V, 139
Bromofluorination of olefins, V, 582
1-BROMO-2-FLUOROBENZENE, V,
    133
o-Bromofluorobenzene, V, 1121
1-BROMO-2-FLUOROHEPTANE, V,
    136
Bromoform, I, 357
N-BROMOGLUTARIMIDE, IV, 498
α-BROMOHEPTALDEHYDE, III, 127
α-BROMOHEPTALDEHYDE DI-
    METHYL ACETAL, III, 128
4-BROMO-2-HEPTENE, IV, 108
1-Bromo-2,5-hexadiene, V, 610
3-Bromo-1,5-hexadiene, V, 610
Bromohexadiene(mixture), V, 609
3-BROMO-4-HYDROXYTOLUENE,
    III, 130
5-Bromoindole-3-carbonitrile, V, 657
o-Bromoiodobenzene, V, 1120
α-BROMOISOBUTYRYL BROMIDE,
    IV, 348
α-BROMOISOCAPROIC ACID, II, 95;
    III, 523
α-BROMOISOVALERIC ACID, II, 93;
    III, 848
p-BROMOMANDELIC ACID, IV, 110
BROMOMESITYLENE, II, 95, 360; III,
    554, 556
α-BROMO-β-METHOXY-n-BUTYRIC
    ACID, III, 813
2-Bromo-4-methoxyhydrocinnamonitrile,
    V, 265
6-Bromo-2-methoxynaphthalene, V, 920
3-Bromo-4-methylacetophenone, V, 120
2-Bromo-4-methylaniline, V, 140
2-BROMO-4-METHYLBENZALDE-
    HYDE, V, 139
2-Bromo-5-methylbenzaldehyde, V, 141
2-Bromo-4-methylbenzenediazonium chlor-
    ide, V, 140
2-BROMO-3-METHYLBENZOIC
    ACID, IV, 114

2-Bromo-3-methylthiophene, IV, 923
α-BROMO-β-METHYLVALERIC
    ACID, II, 95; III, 496
α-BROMONAPHTHALENE, I, 121; II,
    282, 425; III, 631; V, 145
2-Bromonaphthalene, V, 142
6-BROMO-2-NAPHTHOL, III, 132
m-BROMONITROBENZENE, I, 123
2-BROMO-3-NITROBENZOIC ACID,
    I, 125
2-BROMO-4-NITROTOLUENE, IV, 114
1-Bromoöctane, IV, 724
2-Bromopentane, II, 478
p-Bromophenacylamine, hydrochloride, V,
    915
p-BROMOPHENACYL BROMIDE, I,
    127
9-BROMOPHENANTHRENE, III, 134,
    212, 702
β-Bromophenetole, V, 1145
m-Bromophenol, III, 131
o-BROMOPHENOL, II, 97
p-BROMOPHENOL, I, 128
α-BROMO-α-PHENYLACETONE, III,
    343
α-BROMO-α-PHENYLACETONI-
    TRILE, III, 347
4-(m-Bromophenyl)-2-butanone, V, 769
4-(o-Bromophenyl)-2-butanone, V, 769
α-p-Bromophenylethylamine, II, 505
p-Bromophenylglyoxal, V, 940
m-BROMOPHENYL ISOTHIOCYAN-
    ATE, I, 449
p-BROMOPHENYL ISOTHIOCYAN-
    ATE, I, 448; IV, 701; V, 226
p-BROMOPHENYLMETHANE, V, 130
1-(p-Bromophenyl)-3-phenyl-2-propanone,
    IV, 177
1-(p-BROMOPHENYL)-1-PHENYL-2,-
    2,2-TRICHLOROETHANE,
    V, 130
1-Bromo-3-phenylpropane, V, 328
α-BROMO-β-PHENYLPROPIONIC
    ACID, III, 705
m-Bromophenylurea, IV, 51
o-Bromophenylurea, IV, 51

p-BROMOPHENYLUREA, **IV, 49**

2-BROMOPHTHALIDE, **III, 737**

3-BROMOPHTHALIDE, **V, 145**

α-Bromopropionic acid, **I,** 23

β-BROMOPROPIONIC ACID, **I, 131,** 246

3-Bromopropiophenone, **V,** 120

γ-Bromopropylphthalimide, **II, 84; III,** 256. 257

3-BROMOPYRENE, **V, 147,** 632

2-BROMOPYRIDINE, **III, 136; V,** 145

3-Bromopyridine, **V,** 145

8-Bromoquinoline, **V,** 145

4-BROMORESORCINOL, **II, 100**

β-Bromostyrene, **I,** 438

m-Bromostyrene, **IV, 733**

N-Bromosuccinimide, **IV,** 108, 254, 921; **V,** 145, 160, 825

2-Bromothiophene, **V,** 642

3-BROMOTHIOPHENE, **V, 149**

3-Bromo-4-tolualdehyde, **V,** 120

m-BROMOTOLUENE, **I, 133**

o-BROMOTOLUENE, **I, 135; V,** 145

p-BROMOTOLUENE, **I, 136,** 551; **II,** 89, 442; **V,** 635, 1058

3-Bromo-2,2,3-trifluoropropionic acid, **V,** 242

5-Bromo-2-vinylthiophene, **IV, 983**

4-BROMO-*o*-XYLENE, **III, 138,** 307

Bronze powder, **IV,** 914

Brucine, **I,** 419

Butadiene, **IV,** 728, 890; **V,** 528

1,3-BUTADIENE, **II, 102**

BUTADIENE, 1-CHLORO-1,4,4-TRI-FLUORO-, **V, 235**

1,3-BUTADIENE, 1,4-DIPHENYL-, **V, 499**

Butadiene monoxide, **IV,** 12

BUTADIENOIC ACID, METHYL ES-TER, **V, 734**

BUTADIYNE, DIPHENYL-, **V, 517**

2-Butanone, **V,** 572, 747

Δ^{α,β}-BUTENOLIDE, **V, 255**

Butenolides, **V,** 82

o-(*n*-BUTOXY)NITROBENZENE, **III, 140**

7*t*-BUTOXYNORBORNADIENE, **V, 151**

1-*n*-Butoxy-3-oxabicyclo(4.4.0)-3-decene, **IV, 312**

2-*t*-BUTOXYTHIOPHENE, **V, 642**

t-BUTYL ACETATE, **III, 141; IV, 263; V,** 566

α-*t*-Butylacetoacetate, **III, 407**

t-BUTYL ACETOACETATE, **V, 155**

n-BUTYLACETYLENE, **IV, 117**

n-BUTYL ACRYLATE, **III, 146**

s-Butyl acrylate, **V,** 764

n-Butyl alcohol, **I,** 28, 29, 37, 138, 140, 142, 145; **II,** 106, 108, 109; **III,** 146; **IV,** 750, 798; anhydrous, **II,** 112, 468, **469,** 474, **475**

s-Butyl alcohol, **I,** 38; **IV,** 122; **V,** 762, 764

t-Butyl alcohol, **I,** 38, 144; **III,** 48, 141, 142, 144, 152, 479, 506; **IV,** 126, 135, 776; **V,** 20, 162, 172, 926, 938, 1026, 1106, 1112, 1114; anhydrous, **IV,** 132, 264, 399, 459, 617

N-Butylallenimine, **V, 544**

n-Butylamine, **II, 319; IV,** 284, 573; **V,** 736

s-Butylamine, **II,** 319

t-BUTYLAMINE, **III, 148, 153; V,** 191

t-BUTYLAMINE HYDROCHLORIDE, **III, 153**

BUTYLAMINE, *N*-METHYL-, **V, 736**

N-*n*-Butylaniline, **IV,** 147, 284

t-BUTYL AZIDOACETATE, **V, 586,** 587

t-BUTYL AZIDOFORMATE, **V, 157**

t-BUTYL AZODIFORMATE, **V, 160**

N-*t*-BUTYLBENZALDIMINE, **V, 191**

n-BUTYLBENZENE, **III, 157**

t-Butyl *o*-benzoylbenzoate, **IV, 263**

n-BUTYL BORATE, **II, 106**

n-BUTYL BROMIDE, **I,** 28, 37, 248, 250; **II,** 179; **III,** 140, 157; **IV,** 118, 726; **V,** 763

s-BUTYL BROMIDE, **I,** 38; **II,** 359; **III,** 495

t-BUTYL BROMIDE, **I,** 38

t-Butyl bromoacetate, **IV, 263**

s-Butyl α-bromopropionate, IV, 121, 122
n-BUTYL n-BUTYRATE, I, 138
s-BUTYL α-n-CAPROYLPROPION-
ATE, IV, 120
n-BUTYL CARBAMATE, I, 140
t-BUTYL CARBAMATE, V, 162
t-BUTYL CARBAZATE, V, 157, 160,
166
p-t-Butylcatechol, III, 204
n-Butyl Cellosolve, IV, 684
n-BUTYL CHLORIDE, I, 142, 363
s-BUTYL CHLORIDE, I, 143, 361
t-BUTYL CHLORIDE, I, 144, 524
t-Butyl chloroacetate, IV, 263; V, 586
t-Butyl α-chloropropionate, IV, 263
s-Butyl cinnamate, V, 764
s-BUTYL CROTONATE, V, 762
t-BUTYL CYANOACETATE, V, 171
t-Butyl cyanoformate, V, 160
trans-4-t-BUTYLCYCLOHEXANOL, V,
175
4-t-Butylcyclohexanone, V, 175
n-Butyl cyclohexenyl ether, IV, 312
9-n-Butyl-1-decalone, V, 190
n-Butyl diazoacetate, IV, 426
t-BUTYL DIAZOACETATE, V, 179
t-BUTYL α-DIAZOACETOACETATE,
V, 180
s-Butyl β,β-dimethylacrylate, V, 764
t-Butyl β,β-dimethylglutarate, IV, 263
t-Butyl 2,4-diphenylbutanoate, V, 561
t-Butyl 2,3-diphenylpropionate, V, 561
t-Butyl esters, III, 143
n-Butyl ether, II, 276, 478
9-n-Butylfluorene, IV, 625
t-Butyl glutarate, IV, 263
n-BUTYL GLYOXYLATE, IV, 124
Butylhydrazinium hydrogen sulfate, V, 45
t-BUTYLHYDRAZODIFORMATE, V,
160
t-Butyl β-hydroxy-β,β-diphenylpropionate,
V, 566
t-BUTYL HYPOCHLORITE, IV, 125;
V, 183, 184
n-Butyl iodide, II, 402; IV, 322, 726
t-Butyl iodide, IV, 324

n-Butyl isocyanide, V, 302
s-Butyl isocyanide, V, 774
t-Butyl isocyanide, V, 302
1-Butylisoquinoline, IV, 644
Butylketene dimer, IV, 563
n-BUTYLMAGNESIUM BROMIDE, I,
307; V, 763
n-BUTYL MAGNESIUM CHLORIDE,
UNSOLVATED, V, 1141
s-BUTYLMAGNESIUM CHLORIDE,
I, 361
t-BUTYLMAGNESIUM CHLORIDE,
I, 524
s-Butylmethacrylate, V, 764
2-n-BUTYL-3-METHYLCYCLOHEX-
ANONE, V, 187
t-Butyl 2-methylenedodecanoate, IV, 617
s-Butyl 3-methylheptanoate, V, 764
t-BUTYL S-METHYLTHIOCARBON-
ATE, V, 166
n-BUTYL NITRITE, I, 177; II, 108; III,
191
n-BUTYL OLEATE, II, 468, 469
5-sec-Butyl-1,2,3,4,5-pentachlorocyclopen-
tadiene, V, 896
5-Butyl-1,2,3,4,5-pentachlorocyclopenta-
diene, V, 896
t-Butyl perbenzoate, V, 70, 152, 642, 924
p-t-Butylperoxybenzoic acid, V, 908
t-BUTYL PHENYL CARBONATE, V,
168
t-Butyl 2-phenylhexanoate, V, 561
m-t-Butylphenylmethylcarbinol, III, 201
2-t-BUTYL-3-PHENYLOXAZIRANE,
V, 191
t-Butyl 2-phenylpropionate, V, 561
p-t-Butylphenyl salicylate, IV, 179
n-BUTYL PHOSPHATE, II, 109
s-Butyl phosphate, II, 110
t-BUTYLPHTHALIMIDE, III, 152
n-Butylpropiolaldehyde diethyl acetal, IV,
802
1-n-BUTYLPYRROLIDINE, III, 159
t-Butyl succinate, IV, 263
n-BUTYL SULFATE, II, 111
n-Butyl sulfide, II, 548

n-BUTYL SULFITE, **II,** 111, **112**
t-Butylthiourea, **V, 804**
s-Butyl tiglate, **V,** 764
n-BUTYL *p*-TOLUENESULFONATE,
   I, 145; **II,** 47
N-Butyl-*m*-toluidine, **II,** 291
t-Butyl *N*-trifluoroacetylcarbamate, **V,** 164
n-Butyltrivinyltin, **IV,** 260
t-BUTYLUREA, **III, 151; V, 804**
n-Butyl vinyl ether, **IV,** 312
2-Butyne, **V,** 370
2-BUTYNOIC ACID, **V, 1043**
2-BUTYN-1-OL, **IV, 128**
n-Butyraldehyde, **III,** 385; **V,** 957
BUTYRCHLORAL, **IV, 130**
n-Butyric acid, **I,** 148; **V,** 589
BUTYRIC ACID, γ-BROMO-, ETHYL
   ESTER, **V, 545**
BUTYRIC ACID, 2,4-DIPHENYL-,
   ETHYL ESTER, **V, 559**
BUTYROIN, **II, 114**
γ-Butyrolactone, **IV,** 278, 496, 898; **V,** 255,
   545
n-BUTYRYL CHLORIDE, **I, 147**
n-Butyryl fluoride, **V,** 69

Cadmium chloride, **III,** 601; **V,** 814
Calcium carbide, **IV,** 471
Calcium carbonate, **II,** 182; **IV,** 157, 428,
   579, 922; **V,** 880
Calcium chloride, **I,** 1
Calcium cyanamide, **I,** 203; **IV,** 645
Calcium formate, **IV,** 908
Calcium hydride, **IV,** 460, 657; **V,** 173, 719
Calcium hydroxide, **I,** 425; **II,** 256
Calcium hypochlorite, **IV,** 486
Calcium malonate, **II, 377**
Calcium oxide, **I,** 425; **II,** 216; **IV,** 659, 907
Calcium picrate, **IV,** 87
Calcium sulfate, **III,** 356
DL-Camphor, **V,** 194
D-10-Camphorsulfonic acid, **V,** 878
DL-10-CAMPHORSULFONIC ACID,
   V, 194
D-10-Camphorsulfonyl chloride, **V,** 877,
   878

D-L-10-CAMPHORSULFONYL CHLO-
   RIDE, **V, 196,** 689
Cane sugar, **I,** 335
γ-Caprilactone, **IV, 432**
n-Caproic acid, **I,** 115; **II,** 417, 475; **III,**
   164
n-CAPROIC ANHYDRIDE, **III, 164**
ε-Caprolactam, **IV,** 39, 588
n-Capronitrile, **IV,** 121, 123
Caproyl chloride, **IV,** 562
Caproyl fluoride, **V,** 69
CAPRYL ALCOHOL, **I, 366**
Carbamates, **V,** 165
CARBAMIC ACID, *t*-BUTYL ESTER,
   V, 162
CARBAZIC ACID, *t*-BUTYL ESTER,
   V, 166
Carbazole, **I,** 547
CARBAZOLE, 2-NITRO-, **V, 829**
Carbazoles, **V,** 832, 943
p-CARBETHOXYBENZENEDIAZON-
   IUM FLUOBORATE, **II, 299**
γ-Carbethoxybutyryl chloride, **V,** 384
3-CARBETHOXYCOUMARIN, **III, 165**
2-Carbethoxycyclodecanone, **V,** 200
2-Carbethoxycyclododecanone, **V,** 200
2-Carbethoxycyclononanone, **V,** 200
2-CARBETHOXYCYCLOOCTA-
   NONE, **V, 198**
2-CARBETHOXYCYCLOPENTA-
   NONE, **II, 116**
β-CARBETHOXY-γ,γ-DIPHENYLVI-
   NYLACETIC ACID, **IV, 132**
t-Carbinamines, See *t*-Alkylamines, **V,** 38
Carbitol, **III,** 240; **IV,** 251
Carbobenzoxyalanine, **III,** 169
CARBOBENZOXY CHLORIDE, **III,**
   167
CARBOBENZOXYGLYCINE, **III, 168**
CARBODIIMIDE, [3-(DIMETHYLA-
   MINO)-PROPYL]ETHYL-,
   HYDROCHLORIDE, **V, 555**
CARBODIIMIDE, DIPHENYL-, **V,**
   501, 504
Carbodiimides, from isocyanates, **V,** 503
   from ureas, **V,** 557

p-Carbomethoxybenzaldehyde, **IV, 933**
2-Carbomethoxycyclopentanone, **V,** 76
ω-CARBOMETHOXYPELARGONYL
    CHLORIDE, **III, 613**
β-CARBOMETHOXYPROPIONYL
    CHLORIDE, **III, 169**, 602
δ-Carbomethoxyvaleryl chloride, **IV,** 555,
    556
Carbonation, **V,** 890
Carbon bisulfide, **II,** 543
Carbon dioxide, **I,** 362, 525; **II,** 426, 557;
    III, 35, 413, 554, 556, 737; **IV,**
    1, 130, 131, 296, 318, 410, 671,
    927
Carbon disulfide, **I,** 447, 448, 506; **III,** 23,
    183, 360, 394, 599, 763; **IV,** 25,
    28, 244, 570, 967; **V,** 223
Carbon filter mat, **I,** 491
Carbon monoxide, **II,** 584, **585; IV,** 142
Carbon tetrachloride, **I,** 96, 548; **III,** 105,
    128, 134, 578, 800, 842; **IV,** 109,
    186, 285, 309, 708, 862, 922,
    934, 950, 961
Carbonylation, **V,** 742
1,1'-CARBONYLDIIMIDAZOLE, **V,**
    201
o-CARBOXYCINNAMIC ACID, **IV,**
    136
2-Carboxy-3,5-dichlorophenyldiazonium
    chloride, **IV,** 873
3-Carboxy-4-hydroxyquinoline, **IV,** 83
Carboxylation, **V,** 22, 443, 739, 742
Carboxylic acids, **V,** 538, 742, 1084
CARBOXYMETHOXYLAMINE HE-
    MIHYDROCHLORIDE, **III,**
    172
o-CARBOXYPHENYLACETONI-
    TRILE, **III, 174**
4-(*o*-Carboxyphenyl)-5,6-benzocoumarin,
    IV, 138
β-(*o*-CARBOXYPHENYL)PROPIONIC
    ACID, **IV,** 136
CASEIN, **II, 120,** 612
Castor oil, **I,** 366; **II,** 53
Catalyst, copper chromite, **III,** 473, 693
CATALYST, PALLADIUM-BARIUM

    SULFATE, **III,** 552, 628, **685**
    PALLADIUM-CARBON, **III, 385,**
    519, 686
Catalyst, platinum oxide, **III,** 501
    Raney nickel, **III,** 59, 63, 149, 229,
    278, 328, 358, 717, 720, 794, 827
CATALYST, RANEY NICKEL, W-2,
    III, 181
Catalyst, Raney nickel, W-5, **III, 180**
CATALYST, RANEY NICKEL, W-6,
    III, 176
Catalyst, Raney nickel, W-7, **III,** 179
    U.O.P. nickel, **III,** 278
CATECHOL, **I, 149; IV,** 778, 788
CELLOBIOSE, **II, 122**
α-CELLOBIOSE OCTAACETATE, **II,**
    122, **124**
Cellosolve, **IV,** 115
Cerebrosterol, **IV,** 200
Ceric acetate, **III,** 1, 2
Ceric carbonate, **III,** 2
Cerous chloride, **III,** 2
Cetyl alcohol, **II,** 322, **374**
CETYL IODIDE, **II,** 320, **322**
Cetylmalonic acid, **IV,** 143
CETYLMALONIC ESTER, **IV, 141**
Charcoal, pine, **II,** 551
    spruce, **II,** 551
CHELIDONIC ACID, **II, 126**
Chicago acid, **II,** 146
    purification of, **II,** 147
Chitin, **III,** 430
Chloral, **II,** 598
Chloral hydrate, **I,** 327; **II,** 181; **III,** 260
Chloramine, **V,** 411
p-Chloranisole, **V,** 221
Chlorination, See Index No. 3, under
    Halogenation
Chlorine, **V,** 37, 370, 709, 710
Chloroacetal, **III,** 50
CHLOROACETAMIDE, **I, 153,** 488; **IV,**
    144; **V,** 204
Chloroacetic acid, **I,** 73, 76, 181, 254, 300,
    401; **II,** 260, 263, 328, 376, 397;
    III, 545, 763, 854; **IV,** 967; **V,**
    251

o-Chloro-acetoacetanilide, **V,** 12
Chloroacetone, **II,** 32; **III,** 332
CHLOROACETONITRILE, **IV,** 144
o-Chloroacetophenone, **IV,** 709
3-Chloroacetophenone, **V,** 120
p-CHLOROACETOPHENONE, **I,** 111;
    IV, 112
p-CHLOROACETYLACETANILIDE,
    III, 183
Chloroacetyl chloride, **III,** 183, 192
Chloroacetyl fluoride, **V,** 69
α-CHLOROACETYL ISOCYANATE,
    V, 204
9-CHLOROACRIDINE, **III, 54**
10-Chloro-9-alkylanthracenes, **V,** 207
2-Chloroallylamines, **V,** 544
m-Chloroaniline, **III,** 186, 272
o-Chloroaniline, **III,** 185; **IV,** 146, 180
p-Chloroaniline, **I,** 165; **IV,** 420; **V,** 223
o-Chloroaniline hydrochloride, **IV,** 146
o-Chloroaniline thiocyanate, **IV,** 180
3-(*o*-CHLOROANILINO)-
    PROPIONITRILE, **IV,** 146
9-CHLOROANTHRACENE, **V, 206**
α-CHLOROANTHRAQUINONE, **II,**
    128
10-Chloro-9-arylanthracenes, **V,** 207
m-CHLOROBENZALDEHYDE, **II, 130;**
    V, 120
o-Chlorobenzaldehyde, **I,** 155; **V,** 358
p-CHLOROBENZALDEHYDE, **II, 133**
Chlorobenzene, **I,** 111; **III,** 76; **IV,** 364,
    700, 878, 910; **V,** 998
o-CHLOROBENZOIC ACID, **II,** 15, **135**
p-Chlorobenzoic acid, **II,** 134; **III,** 29
p-CHLOROBENZOIC ANHYDRIDE,
    III, 29
p-Chlorobenzonitrile, **V,** 657
CHLORO-*p*-BENZOQUINONE, **IV,** 148
o-CHLOROBENZOYL CHLORIDE, **I,**
    155
p-Chlorobenzoyl chloride, **III, 29**
2-(*p*-Chlorobenzyl)cyanoacetic acid, **V,** 264
N-CHLOROBETAINYL CHLORIDE,
    IV, 154
p-Chlorobiphenyl, **III,** 831

m-Chlorobromobenzene, **III, 186**
o-CHLOROBROMOBENZENE, **III, 185**
p-Chlorobromobenzene, **V,** 145
1-Chlorobutane, **V,** 1141
3-CHLORO-2-BUTENE-1-OL, **IV,** 128
4-CHLOROBUTYL BENZOATE, **III,**
    187
bis-4-Chlorobutyl ether, **IV,** 529
γ-CHLOROBUTYRONITRILE, **I, 156;**
    II, 26; **III,** 221, 224
m-Chlorocinnamic acid, **IV,** 733
o-Chlorocinnamic acid, **IV,** 733
p-Chlorocinnamic acid, **IV, 733**
α-CHLOROCROTONALDEHYDE, **IV,**
    131
2-CHLOROCYCLOHEXANOL, **I, 158,**
    185; **IV,** 233
2-CHLOROCYCLOHEXANONE, **III,**
    188; **IV,** 594
2-CHLORO-1-CYCLOHEXENEALDE-
    HYDE, **V, 215**
N-CHLOROCYCLOHEXYLIDENE-
    IMINE, **V, 208**
2-CHLOROCYCLOOCTANONE OX-
    IME HYDROCHLORIDE, **V,**
    266
trans-2-CHLOROCYCLOPENTANOL,
    IV, 157
3-CHLOROCYCLOPENTENE, **IV, 238,**
    291
Chlorodialkylphosphines, **V,** 213
Chlorodi-*s*-butylphosphine, **V,** 214
Chlorodi-*t*-butylphosphine, **V,** 214
Chlorodicyclohexylphosphine, **V,** 214
Chlorodifluoroacetic acid, **V,** 950
Chlorodiisobutylphosphine, **V,** 214
CHLORODIISOPROPYLPHOSPHINE,
    V, 211
1-CHLORO-2,6-DINITROBENZENE,
    IV, 160
exo/endo-7-(2-CHLOROETHOXY)-
    BICYCLO(4.1.0)-HEPTANE,
    V, 859
2-CHLOROETHYL BENZOATE, **IV, 84**
α-CHLOROETHYL ETHYL ETHER,
    IV, 748

β-CHLOROETHYL METHYL SULF-
IDE, **II, 136,** 385
Chloroform, **I,** 258; **II,** 312; **III,** 101, 209,
234, 463, 482, 811; **IV,** 265, 548,
864, 918; **V,** 93, 130, 1076, 1078
2-CHLORO-1-FORMYL-1-CYCLO-
HEXENE, **V, 215**
10-Chloro-9-halogenoanthracenes, **V,** 207
α-CHLOROHYDRINDENE, **II, 336**
o-Chlorohydrocinnamonitrile, **V,** 263, 264
4-Chloro-3-hydroxybutyronitrile, **V,** 614
7-CHLORO-4-HYDROXY-3-QUINOLI-
NECARBOXYLIC ACID,
III, 272
3-CHLOROINDAZOLE, **III, 476**
α²-CHLOROISODURENE, **III, 557**
ω-CHLOROISONITROSOACETO-
PHENONE, **III, 191**
2-CHLOROLEPIDINE, **III, 194,** 519
p-Chloromandelic acid, **IV, 112**
p-CHLOROMERCURIBENZOIC
ACID,**I,** 159, 325
o-CHLOROMERCURIPHENOL, **I, 161,**
326
2-Chloromercurithiophene, **II,** 9
CHLOROMETHANESULFONYL
CHLORIDE, **V, 232**
2-Chloro-4-methylbenzaldehyde, **V,** 141
2-CHLORO-2-METHYLCYCLOHEX-
ANONE, **IV, 162**
2-Chloromethyl-4,6-diamino-s-triazine, **IV,**
30
bis-CHLOROMETHYL ETHER, **IV, 101**
1-CHLOROMETHYLNAPHTHAL-
ENE, **III, 195; IV,** 690
2-Chloro-6-methylphenyl isocyanide, **V,**
1063
Chloromethylphosphonic dichloride, **V,**
218
CHLOROMETHYLPHOSPHONO-
THIOIC DICHLORIDE, **V,**
218
2-CHLOROMETHYLTHIOPHENE, **III,**
197, 811
2-CHLORONICOTINONITRILE, **IV,**
166

m-CHLORONITROBENZENE, **I, 162;**
IV, 151
o-Chloronitrobenzene, **I,** 221; **III,** 339; **IV,**
149
p-Chloronitrobenzene, **III,** 86, 240, 667
9-Chloro-10-nitro-9,10-dihydroanthracene,
IV, 711
1-Chloro-1-nitroethane, **IV,** 374
1-Chloro-1-nitropropane, **I,** 372
2-Chloro-2-nitropropane, **IV,** 374
5-CHLORO-2-PENTANONE, **IV, 597**
m-Chloroperbenzoic acid, **V,** 467
p-Chlorophenacylamine, hydrochloride, **V,**
915
p-Chlorophenol, **I,** 476; **IV,** 178
p-CHLOROPHENOXYMETHYL
CHLORIDE, **V, 221**
α-CHLOROPHENYLACETIC ACID,
IV, 169
p-Chlorophenylacetonitrile, **IV,** 174, 578
4-(m-Chlorophenyl)-2-butanone, **V,** 769
4-(o-Chlorophenyl)-2-butanone, **V,** 769
4-(p-Chlorophenyl)-2-butanone, **V,** 769
o-CHLOROPHENYLCYANAMIDE,
IV, 172
α-p-Chlorophenylethylamine, **II, 505**
o-Chlorophenylhydroxylamine, **IV,** 151
o-Chlorophenyl isocyanide, **V,** 1063
o-Chlorophenyl isothiocyanate, **IV,** 701
p-CHLOROPHENYL ISOTHIOCYAN-
ATE, **I, 165; V, 223**
m-CHLOROPHENYLMETHYLCAR-
BINOL, **III, 200,** 204
6-(p-Chlorophenyl)-6-methyl-4-pyrone, **V,**
722
α-(4-CHLOROPHENYL)-γ-PHENYL-
ACETOACETONITRILE, **IV,**
174, 176
1-(p-Chlorophenyl)-5-phenyl-1,3,5-pentane-
trione, **V,** 720
1-(p-CHLOROPHENYL)-3-PHENYL-2-
PROPANONE, **IV, 176**
2-(p-Chlorophenyl)-6-phenyl-4-pyrone, **V,**
722
o-Chlorophenyl salicylate, **IV, 179**
p-CHLOROPHENYL SALICYLATE,

IV, 178
o-CHLOROPHENYLTHIOUREA, IV,
172, **180**
o-Chlorophenylurea, IV, 51
Chloropicrin, IV, 457
Chloroplatinic acid, I, 463
β-CHLOROPROPIONALDEHYDE
ACETAL, II, 17, **137**
β-CHLOROPROPIONIC ACID, I, **166**
γ-CHLOROPROPYL ACETATE, III,
203, 836
1-Chloropyrene, V, 207
2-CHLOROPYRIMIDINE, IV, **182**, 336
Chloropyrosulfonyl isocyanate, V, 727
CHLOROPYRUVIC ACID, V, **636**
m-CHLOROSTYRENE, III, **204**; IV, **733**
o-Chlorostyrene, IV, **733**
p-Chlorostyrene, IV, **733**
N-Chlorosuccinimide, IV, 255
Chlorosulfonic acid, I, 8, 85; IV, 102
m-Chlorosulfonylbenzoic acid, IV, 697
CHLOROSULFONYL ISOCYANATE,
V, **226**, 230, 673
2-Chloro-1,1,4,4-tetrafluorobutadiene, V,
239
2-CHLOROTHIIRANE 1,1-DIOXIDE,
V, **231**
4-Chloro-2-thiohomophthalimide, V, 1053
o-CHLOROTOLUENE, I, 170; II, 135
p-CHLOROTOLUENE, I, 170; II, 133
3-Chlorotoluquinone, IV, 152
1-CHLORO-1,4,4-TRIFLUOROBUTA-
DIENE, V, **235**
1-CHLORO-2,3,3-TRIFLUOROCY-
CLOBUTENE, V, **394**
2-CHLORO-2,3,3-TRIFLUOROCY-
CLOBUTYL ACETATE, V,
235
Chlorotrifluoroethylene, IV, 185; V, 239
2-CHLORO-1,1,2-TRIFLUOROETHYL
ETHYL ETHER, IV, **184**, 423
1-Chloro-1,4,4-trifluoro-2-phenylbutadiene,
V, 239
3-Chloro-2,2,3-trifluoropropionamide, V,
241
3-CHLORO-2,2,3-TRIFLUOROPRO-

PIONIC ACID, V, **239**
Chlorourea, see Monochlorourea, IV, 157,
158
β-CHLOROVINYL ISOAMYL
KETONE, IV, **186**, 558
β-Chlorovinyl isobutyl ketone, IV, **188**
β-Chlorovinyl isohexyl ketone, IV, **188**
β-Chlorovinyl methyl ketone, IV, **188**
5-Chloro-2-vinylthiophene, IV, **983**
CHOLANE-24-AL, V, **242**
Cholane-24-al 2,4-dinitrophenylhydrazone,
V, 243
Cholane-24-ol, V, 242
CHOLESTANE, 3β-Methoxy-, V, **245**
Cholestanol, IV, 200
CHOLESTANONE, II, **139**
CHOLESTANYL METHYL ETHER,
V, **245**
Δ⁴-CHOLESTEN-3,6-DIONE, IV, **189**
Δ⁷-CHOLESTENOL, IV, **200**
CHOLESTENONE, III, **207**
Δ⁴-CHOLESTEN-3-ONE, IV, **192**, 195,
198
Δ⁵-CHOLESTEN-3-ONE, IV, 195, **198**
CHOLESTEROL, II, 191; III, 207; IV,
189, 192, **195**, 196
CHOLESTEROL DIBROMIDE, IV, **195**
CHOLESTERYL ACETATE, II, 191,
193; IV, **199**
Cholesteryl chromate, IV, 189
Choline, IV, 148
Chromic anhydride- See Chromium triox-
ide (CrO₃)
Chromium(III)oxide, IV, 579
CHROMIUM(II) SULFATE, V, **993**
Chromium trioxide (CrO₃), II, 336, 441;
III, 39, 134, 235, 641; IV, 19,
23, 698, 713, 757, 813; V, 310,
324, 694, 866
Chrysene, V, 956
Cinnamaldehyde, II, 230; III, 731; IV, 312,
722, 771, 793; V, 364, 513, 929,
985
Cinnamic acid, I, 311, 440; III, 91, 714;
IV, **961**
Cinnamic acid dibromide, IV, 960

CINNAMIC ACID, α-(β-HYDROXY-
STYRYL)-γ-LACTONE, V,
80
Cinnamic acids and derivatives, V, 512
Cinnamyl alcohol, V, 249
CINNAMYL BROMIDE, V, **249**
CITRACONIC ACID, II, **140**
CITRACONIC ANHYDRIDE, II, **140,**
382
Citral, III, 747
Citric acid, I, 10
Citric acid monohydrate, II, 12, 368
Claisen's alkali, III, 268, 662; IV, 190, **191**
Claisen flask, modified, II, 18, 227
Clayton yellow paper, III, 81
2,4,6-Collidine, IV, 163; V, 77
Condensation Reactions, See Index No. 3,
Type of Reaction
Condenser, partial, I, 184
Copper, I, 135, 136, 545; III, 330, 566
Copper acetate, III, 16, 292, 390
Copper bronze, I, 133; II, 356; III, 339
Copper carbonate, III, 460
Copper chromite, IV, 216, 314, 324, 337,
678, 857
COPPER CHROMITE CATALYST, II,
142, 325
Copper oxide, II, 15; III, 473
Copper powder, II, 62, 226, 381, 433, 445,
446; IV, 337, 628, 732, 878, 914;
sheet, IV, 149
Copper sulfate, I, 46, 87, 136, 222, 514; II,
131, 353, 446, 494, 526; III, 166,
186, 341; IV, 873; V, 261
Copper wire, III, 307
Corn cobs, I, 280
Cotton, II, 124; absorbent, III, 18, 19
COUMALIC ACID, IV, **201,** 532; V,
982, 983
COUMARILIC ACID, III, **209;** V, 253
Coumarin, III, 209
COUMARIN DIBROMIDE, III, **209**
COUMARONE, V, **251**
COUPLING OF o-TOLIDINE AND
CHICAGO ACID, II, **145**
Coupling Reactions, See Index No. 3,
Type of Reaction
Crab shells, III, 430
Creatine hydrate, I, 172
CREATININE, I, **172;** III, 587
CREATININE PICRATE, I, **173**
CREATININE ZINC CHLORIDE, I,
173
CREOSOL, IV, **203**
p-CRESOL, I, **175;** V, 635
Crotonaldehyde, III, 697, 783; IV, 130,
311, 312, 794
Crotonaldehyde diethyl acetal, IV, **22**
Crotonic acid, III, 813; V, 762
γ-CROTONOLACTONE, V, **255**
CROTYL DIAZOACETATE, V, **258**
Crystal violet, V, 112, 113
Cumyl hypochlorite, V, 186
1-(p-Cumyl)-1-propene, IV, 787
CUPFERRON, I, **177**
Cupric acetate, V, 517
Cupric acetate monohydrate, IV, 146, 147,
838
Cupric bromide, IV, 121
Cupric chloride, IV, 728; V, 98, 106
Cupric nitrate trihydrate, II, 142
CUPROUS BROMIDE, I, **136;** II, **132;**
III, 186; V, 70, 151
CUPROUS n-BUTYLMERCAPTIDE,
V, **107**
CUPROUS CHLORIDE, I, **170;** II, **131,**
583; III, 307, 665; IV, 160; V,
765
CUPROUS CYANIDE, I, **46, 514;** III,
112, 212, 631
Cuprous oxide, IV, 493; V, 107
Cyanamide, IV, 213, **645**
CYANOACETAMIDE, I, **179;** II, **379;**
III, 535, 591; IV, 210, 662
Cyanoacetic acid, IV, 234, 254; V, 171,
585
CYANOACETIC ACID, t-BUTYL
ESTER, V, **171**
Cyanoacetyl chloride, V, 172
Cyanoacetylurea, IV, 248
p-Cyanobenzaldehyde, II, 443; IV, **933**
p-Cyanobenzaldiacetate, IV, 714

1-CYANOBENZOCYCLOBUTENE, V, 263

1-CYANO-2-BENZOYL-1,2-DIHY-DROISOQUINOLINE, IV, 641

m-Cyanocinnamic acid, IV, 733

2-Cyano-4,6-dimethylpyridine, V, 272

3-Cyano-5,6-dimethyl-2(1)-pyridone, IV, 212

bis-(β-CYANOETHYL)AMINE, III, 93

tris-(β-Cyanoethyl)amine, III, 94

N-2-CYANOETHYLANILINE, IV, 205

N-2-Cyanoethyl-p-anisidine, IV, 206

N-2-Cyanoethyl-m-chloroaniline, IV, 206

β-CYANOETHYLHYDRAZINE, V, 39

2-Cyano-3-ethyl-3-methylhexanoic acid, IV, 97

N,N'-bis-2-Cyanoethyl-o-phenylenedi-amine, IV, 206

N,N'-bis-2-Cyanoethyl-p-phenylenedi-amine, IV, 206

Cyanogen, IV, 209

CYANOGEN BROMIDE, II, 150; III, 608

Cyanogen chloride, V, 226

CYANOGEN IODIDE, IV, 207

N-Cyanoguanidine, IV, 502

7-CYANOHEPTANAL, V, 266

3-Cyano-6-isobutyl-2(1)-pyridone, IV, 212

1-Cyanoisoquinoline, V, 272

1-Cyano-5-methoxybenzocyclobutene, V, 265

2-CYANO-6-METHYLPYRIDINE, V, 269

3-CYANO-6-METHYL-2(1)-PYRI-DONE, IV, 210

1-Cyano-3-α-naphthylurea, IV, 215

ω-CYANOPELARGONIC ACID, III, 768

9-CYANOPHENANTHRENE, III, 27, 212

α-CYANO-β-PHENYLACRYLIC ACID, I, 181, 451

1-CYANO-3-PHENYLUREA, IV, 213

2-Cyanopyridine, V, 272

4-Cyanopyridine, V, 272

2-Cyanoquinoline, V, 272

m-Cyanostyrene, IV, 733

Cyclization Reactions, See Index No. 3, Type of Reaction

CYCLOBUTANECARBOXYLIC ACID, III, 213; IV, 289; V, 273

1,1-CYCLOBUTANEDICARBOXYLIC ACID, III, 213

CYCLOBUTENE,1,2,3,4-TETRA-METHYL-3,4-DICHLORO-, V, 370, 372

CYCLOBUTYLAMINE, V, 273

Cyclobutyl isocyanide, V, 774

Cyclodecane, IV, 219

1,2-CYCLODECANEDIOL (cis and trans), IV, 216

1,2-Cyclodecanedione, IV, 838

CYCLODECANONE, IV, 218, 220; V, 277

Cyclodecanone semicarbazone, IV, 220

Cyclododecanone, V, 279

cis,trans,trans-1,5,9-Cyclododecatriene, V, 281

cis-CYCLODODECENE, V, 281
    silver nitrate complex, V, 283

CYCLOHEPTANONE, IV, 221; V, 279
    bisulfite adduct, IV, 226

Cycloheptatriene, V, 1138

CYCLOHEPTATRIENECARBON-IUMFLUOBORATE, V, 1138

1,3-CYCLOHEXADIENE, V, 285

2,5-CYCLOHEXADIENE-1-CARBOX-YLIC ACID, V, 400

2,4-CYCLOHEXADIEN-1-ONE,2,3,4,-5,6,6-HEXAMETHYL-, V, 598

Cyclohexadienones, V, 1095

2,4-Cyclohexadienones, V, 601

2,4-CYCLOHEXADIEN-1-ONE,2,6,6-TRIMETHYL-, V, 1092

CYCLOHEXANIMINE,N-CHLORO-, V, 208

Cyclohexane, IV, 314; V, 1000

Δ'-α-CYCLOHEXANEACETIC ACID, ETHYL ESTER, V, 547

CYCLOHEXANECARBOXYLIC

ACID, **I, 364; IV,** 339; **V,** 742
Cyclohexanecarboxylic acid chloride, **IV,** 340
CYCLOHEXANECARBOXYLIC ACID,1-METHYL-, **V, 739**
3,4,5-TRIACETOXY-, **V, 591**
3,4,5-TRIOL, **V, 591**
trans-1,2,-CYCLOHEXANEDIOL, **III, 217**
1,2-CYCLOHEXANEDIONE, **IV, 229**
1,4-CYCLOHEXANEDIONE, **V, 288**
1,2-CYCLOHEXANEDIONE DIOXIME, **IV, 229**
1,3-CYCLOHEXANEDIONE,2-METHYL-, **V, 743**
CYCLOHEXANE,METHYLENE-, **V, 751**
Cyclohexanol, **I,** 18, 19, **183; II,** 152; **III,** 246; **IV,** 324
Cyclohexanol acetate, **V,** 696
CYCLOHEXANOL,4-*t*-BUTYL,*trans*-, **V, 175**
Cyclohexanone, **II,** 76, 532; **III,** 44, 189, 417; **IV,** 192, 221, 225, 229, 234, 274, 459, 471, 536, 537, 781, 884, 907; **V,** 215, 303, 439, 547, 751, 755, 808, 897
CYCLOHEXANONE,2-ALLYL-, **V, 25**
2-BUTYL-3-METHYL-, **V, 187**
CYCLOHEXANONE DIALLYL ACETAL, **V, 292**
CYCLOHEXANONE-2,6-DICARBOXYLIC ACID, DIMETHYL ESTER, **V, 439**
Cyclohexanone 2,4-dinitrophenylhydrazone, **IV,** 14
CYCLOHEXANONE OXIME, **II, 76, 314,** 371
CYCLOHEXENE, **I,** 158, **183; II,** 102, 151, **152,** 172; **III,** 217; **IV,** 543; **V,** 70, 286, 320, 855, 857
CYCLOHEXENE OXIDE, **I, 185; IV,** 232
CYCLOHEXENE SULFIDE, **IV, 232**
2-Cyclohexenol, **V,** 314
2-CYCLOHEXEN-1-OL, BENZOATE,

V, 70
2-CYCLOHEXENONE, **V, 294,** 314
1-CYCLOHEXENYLACETONITRILE, **IV, 234**
Cyclohexylamine, **II,** 319; **V,** 208, 301, 801
CYCLOHEXYLAMINE, 1-METHYL-, **V, 35**
CYCLOHEXYLBENZENE, **II,** 151
Cyclohexyl bromide, **I,** 186; **II, 247**
3-CYCLOHEXYL-2-BROMOPROPENE, **I, 186,** 191
CYCLOHEXYLCARBINOL, **I,** 188
Cyclohexylchloride, **I,** 188, 364
N-Cyclohexylformamide, **V,** 300, 301
Cyclohexylglyoxal, **V,** 940
CYCLOHEXYLIDENECYANOACETIC ACID, **IV, 234**
CYCLOHEXYLIDENECYCLOHEXANE, **V, 297**
Cyclohexyl iodide, **IV,** 324, **543**
CYCLOHEXYL ISOCYANIDE, **V, 300**
CYCLOHEXYLMAGNESIUM BROMIDE, **I,** 186
CYCLOHEXYLMAGNESIUM CHLORIDE, **I,** 188
Cyclohexylmethylacetylene, **IV,** 802
CYCLOHEXYL METHYL KETONE, **V, 775**
Cyclohexylmethylpropiolaldehyde diethyl acetal, **IV,** 802
2-CYCLOHEXYLOXYETHANOL, **V, 303**
2-Cyclohexylperfluoropropene, **V,** 951
α-CYCLOHEXYLPHENYLACETONITRILE, **III,** 219
3-CYCLOHEXYLPROPINE, **I,** 191
CYCLOHEXYLUREA, **V, 801**
1,2-CYCLONONADIENE, **V, 306**
1,2-Cyclononanedione, **IV,** 839
Cyclononanone, **V,** 279
cis-Cyclononene, **V,** 308
cis,cis-1,5-Cyclooctadiene, **V,** 93
Cyclooctanol, **V,** 310
CYCLOOCTANONE, **IV,** 227; **V,** 198, 277, **310**
Cyclooctene, **V,** 266, 306

trans-CYCLOOCTENE, **V, 315**
N-(1-Cyclooctene-1-yl)pyrrolidine, **V,** 277
CYCLOPENTADIENE, **IV,** 238, 239,
    474, 476; **V,** 64, 96, 414, **415,**
    1001, 1002, 1088
CYCLOPENTADIENE,1,2,3,4,5-PEN-
    TACHLORO-5-ETHYL-, **V,**
    893
CYCLOPENTADIENYLSODIUM, **V,**
    432, **1089**
CYCLOPENTANECARBOXALDE-
    HYDE, **V, 320**
Cyclopentanecarboxaldehyde trimer, **V,**
    322
Cyclopentanecarboxylic acid, **V,** 742
1,3-CYCLOPENTANEDIONE,2-
    METHYL-, **V, 747**
Cyclopentanol, **II,** 153
CYCLOPENTANONE, **I,** 192, 290
CYCLOPENTENE, **II,** 153; **IV,** 158
CYCLOPENTENEDIOL, **V, 414**
2-Cyclopentene-1,4-diol, **V,** 416
3-Cyclopentene-1,2-diol, **V,** 415
2-CYCLOPENTENE-1,4-DIONE, **V, 324**
2-CYCLOPENTENE-1-ONE, **V, 326**
Cyclopenteno(*b*)-6-(*p*-methoxyphenyl)-4-
    pyrone, **V,** 723
2-CYCLOPENTENONE, **V, 326**
Cyclopropane,aryl, **V,** 931
CYCLOPROPANE, BROMO-, **V, 126**
CYCLOPROPANECARBOXYLIC
    ACID, **III, 221; V,** 126
CYCLOPROPANE, 1,1-DIPHENYL-,
    V, 509
Cyclopropanols, **V,** 861, 1060
CYCLOPROPYLBENZENE, **V,** 328,
    929
CYCLOPROPYL CYANIDE, **III, 223**
Cyclotetradecanone, **V,** 279
m-CYMENE, **V, 332**
p-Cymene, **II,** 4; **III,** 653, 822
CYSTEIC ACID MONOHYDRATE,
    III, 226
l-CYSTINE, **I, 194; III,** 226

DDQ. See 2,3-Dichloro-5,6-dicyano-1,4-
    benzoquinone, **V,** 428

Deamination, of amines, **V,** 337
Decachlorobi-2,4-cyclopentadienyl, **V,** 901
Decalin, **IV,** 537
cis-9-Decalincarboxylic acid, **V,** 742
trans-9-Decalincarboxylic acid, **V,** 742
DECAMETHYLENE BROMIDE, **III,**
    227
DECAMETHYLENEDIAMINE, **III,**
    229
DECAMETHYLENE GLYCOL, **II, 154;**
    III, 227
Decanoyl fluoride, **V,** 69
Decarboxylation, See Index No. 3, Type
    of Reaction
Dechlorination, See Index No. 3, Type of
    Reaction; Dehalogenation
DECINE, **I, 192**
n-Decyl bromide, **IV,** 618
Decyl diazoacetate, **IV,** 426
n-Decyl fluoride, **IV, 527**
Decylketene dimer, **IV,** 560, **561**
Dehydration, See Index No. 3, Type of
    Reaction
DEHYDROACETIC ACID, **III, 231**
Dehydrobenzene, See Benzyne, **V,** 65,
    928, 1037, 1042
7-Dehydrocholesterol, **IV,** 200
Dehydrogenation, See Index No. 3, Type
    of Reaction
Dehydrohalogenation, See Index No. 3,
    Type of Reaction
Delepine reaction, **V,** 121
DEOXYANISOIN, **V, 339**
Deoxybenzoin, **V,** 340
Deoxypiperoin, **V,** 340
DESOXYBENZOIN, **II, 156**
Desoxycholic acid, **III,** 237
nor-DESOXYCHOLIC ACID, **III, 234**
Desylamine, **V,** 915
DESYL CHLORIDE, **II, 159**
Dextrose, **II,** 58
Diacetalylamine, **III,** 51
DIACETONAMINE HYDROGEN OX-
    ALATE, **I, 196**
DIACETONE ALCOHOL, **I, 199,** 345
3,12-DIACETOXY-*bis*-*nor*-CHOLANYL-
    DIPHENYLETHYLENE, **III,**
    234, **237**

N,N'-Diacetyl-4,6-dimethyl-m-xylene-α,α'-diamine, **V**, 75

Diacetyl monoxime, **V**, 1022

DIACETYL-d-TARTARIC ANHY-DRIDE, **IV**, 242

N,N'-Diacrylyl-p-xylene-α,α'-diamine, **V**, 75

1,2-Dialkenylcyclobutanes, **V**, 532

N,N-Dialkyl-1,2,2-trichlorovinylamines, **V**, 389

DIALLYLAMINE, **I**, 201

DIALLYLCYANAMIDE, **I**, 201, 203

Diallylcyclohexanone, **III**, 45

Diallyldiphenyltin, **V**, 454

DIALURIC ACID MONOHYDRATE, **IV**, 28

4,4'-DIAMINOAZOBENZENE, **V**, 341

DIAMINOBIURET, **III**, 404

2,3-DIAMINO-5-BROMOPYRIDINE, **V**, 347

2,5-DIAMINO-3,4-DICYANOTHIO-PHENE, **IV**, 243

1,5-DIAMINO-2,4-DINITROBEN-ZENE, **V**, 1068

4,4'-DIAMINODIPHENYLSULFONE, **III**, 239

DIAMINODURENE CHLOROSTAN-NATE, **II**, 255

2,4-DIAMINO-6-HYDROXYPYRIMI-DINE, **IV**, 245

DIAMINOMALEONITRILE, **V**, 344

1,2-DIAMINO-4-NITROBENZENE, **III**, 242

2,4-Diamino-6-phenyl-s-triazine, **IV**, 78

2,3-DIAMINOPYRIDINE, **V**, 346

2,4-DIAMINOTOLUENE, **II**, 160

2,4-DIAMINOTOLUENE SULFATE, **II**, 160

Diaminouracil bisulfite, **IV**, 248

DIAMINOURACIL HYDROCHLOR-IDE, **IV**, 247

α,α'-Diamino-m-xylene, **V**, 668

Diammonium hydrogen phosphate, **V**, 656

Diamylamine, **IV**, 2

Di-n-amyl ketone, **IV**, 562

2,4-Di-t-amylphenoxyacetyl chloride, **IV**, 742

1,2-Dianilinoethane, See N,N'-Diphenyle-thylenediamine, **V**, 115

Dianisalacetone, **I**, 78

Dianisidine, **II**, 149

o-Dianisidine, **III**, 295

Diarylmethanes, **V**, 133

2,3-DIAZABICYCLO(2.2.1)HEPT-2-ENE, **V**, 97

Diazirines, **V**, 899

DIAZOAMINOBENZENE, **II**, 163

Diazo compounds, **V**, 182

Diazoethane, **V**, 879

DIAZOMETHANE, **II**, 165; **III**, 119, 244; **IV**, 225, 250 **V**, 232, 245, **246**, 247, **351**, 352, 877, 1089, 1101

1-Diazo-2-methylpropane, **V**, 879

Diazonium xanthates, detonation of, **V**, 1050

Diazotization, See Index No. 3. Type of Reaction

"Diazo transfer" reaction, **V**, 182

DIBENZALACETONE, **II**, 167

Dibenzalpentaerythritol, **IV**, 680

DIBENZ[c,e][1,2]AZABORINE,5,6-DIHYDRO-6-METHYL-, **V**, 727

Dibenzeneborinic acid, **IV**, 69, 70

Dibenzhydryl ether, **IV**, 73

Dibenzohydryl disulfide, **II**, 574

1,4-DIBENZOYLBUTANE, **II**, 169

DIBENZOYLDIBROMOMETHANE, **II**, 244

DIBENZOYLDIMETHYLHYDRA-ZINE, **II**, 209

$trans$-DIBENZOYLETHYLENE, **III**, 248

DIBENZOYLHYDRAZINE, **II**, 208

DIBENZOYLMETHANE, **I**, 205; **III**, 251

Dibenzoylmethane, **II**, 244

Dibenzylcadmium, **IV**, 96

Dibenzyl ketone, **II**, 806; **V**, 84, 450, 514, 605

Dibenzyl ketoxime, **V**, 83, **84**

N,N'-Dibenzyl-p-phenylenediamine, **IV**, 92

Dibenzylthiourea, **V,** 804
N,*N*-Dibromoacetamide, **IV,** 105
DIBROMOACETONITRILE, **IV,** 254
3,4-Dibromoacetophenone, **V,** 120
2,6-DIBROMO-4-AMINOPHENOL
    CHLOROSTANNATE, **II,**
    175
2,6-DIBROMOANILINE, **III,** 262
9,10-DIBROMOANTHRACENE, **I,** 207
m-Dibromobenzene, **III,** 186
o-Dibromobenzene, **V,** 108
p-Dibromobenzene, **V,** 496
α,α′-Dibromobibenzyl, **IV,** 257
4,4′-Dibromobibenzyl, **IV,** 257
9,9-DIBROMOBICYCLO[6,1,0]-NON-
    ANE, **V,** 306
4,4′-DIBROMOBIPHENYL, **IV, 256**
α,γ-Dibromobutyryl bromide, **V,** 256
5α,6β-DIBROMOCHOLESTAN-3-ONE,
    IV, 197
1,2-DIBROMOCYCLOHEXANE, **II,**
    171; **V, 285, 286**
α,α′-DIBROMODIBENZYL KETONE,
    V, 514
4,5-DIBROMO-1,2-DIMETHYL-1,2-
    EPOXYCYCLOHEXANE, **V,**
    469
2,2′-Dibromo-α,α′-epoxybibenzyl, **V,** 360
3,3-Dibromo-α,α′-epoxybibenzyl, **V,** 360
1,2-Dibromoethane, See Ethylene bro-
    mide, **V,** 890
α,β-DIBROMOETHYL ETHYL
    ETHER, **IV, 749**
10,11-Dibromohendecanoic acid, **IV,** 970
Dibromohydrocinnamic acid, **IV,** 960, **961**
DIBROMOMALONONITRILE-
    POTASSIUM BROMIDE
    COMPLEX, **IV, 877**
3,5-Dibromo-4-methylacetophenone, **V,**
    120
2,6-DIBROMO-4-NITROPHENOL, **II,**
    173, 175
Dibromopentaerythritol, **IV,** 683
1,3-DIBROMO-1-PHENYLPROPANE,
    V, 328
2,3-DIBROMOPROPENE, **I, 186, 187,**

209
2,3-Dibromopropene, **V,** 121, 124
2,6-DIBROMOQUINONE-4-CHLOR-
    OIMIDE, **II,** 175
Dibromostearic acid, **IV,** 852
α,β-DIBROMOSUCCINIC ACID, **II,** 10,
    177
DI-*n*-BUTYLAMINE, **I, 202; III,** 159, 256
β-DI-*n*-BUTYLAMINOETHYLAMINE,
    III, 254
β-DIBUTYLAMINOETHYL BROM-
    IDE HYDROBROMIDE, **II, 92**
Di-*n*-butylaminoethyl bromide hydrobrom-
    ide, **III,** 254
γ-DI-*n*-BUTYLAMINOPROPYL-
    AMINE, **III, 256**
N,*N*′-Dibutylbenzidine, **IV,** 284
DI-*n*-BUTYL CARBINOL, **II,** 179
DI-*n*-BUTYLCYANAMIDE, **I,** 202, 204
Di-*t*-butyl β,β-dimethylglutarate, **IV, 263**
DI-*n*-BUTYLDIVINYLTIN, **IV, 258**
Di-*n*-butyl ether, **IV,** 322
Di-*t*-butyl glutarate, **IV, 263**
Di-*s*-butyl-α-(2-hexyl)-β-methylglutarate,
    V, 765
1,1-Dibutylhydrazinium hydrogen oxalate,
    V, 45
N,*N*-Di-*t*-butylhydroxyamine, **V,** 357
DI-*t*-BUTYL MALONATE, **IV, 261**
DI-*t*-BUTYL NITROXIDE, **V, 355**
Di-*t*-butyl succinate, **IV, 263**
Di-*n*-butyl *d*-tartrate, **IV,** 124
Di-*n*-butyltin dichloride, **IV,** 258
N,*N*-Di-*n*-butyl-*p*-toluenesulfonamide, **V,**
    1066
Di-*t*-butylurea, **III,** 155
DI-*n*-BUTYL ZINC, **II, 187**
Dibutyrolactone, **IV,** 278, 279
2,5-DICARBETHOXY-1,4-CYCLO-
    HEXANEDIONE, **V, 288,** 289
3,5-DICARBETHOXY-2,6-DIMETH-
    YLPYRIDINE, **II, 215**
DI-β-CARBETHOXYETHYLMETHYL-
    AMINE, **III, 258,** 259
N,*N*′-Dicarbethoxyputrescine, **IV, 822**
2,2′-Dicarboxydiazoaminobenzene, **V,** 56

Dichloramine-B, **V,** 911

α,γ-DICHLOROACETONE, **I, 211**

α,α-DICHLOROACETAMIDE, **III, 260**

DICHLOROACETIC ACID, **II, 181**

Dichloroacetic acid, **IV,** 272, 427

Dichloroacetonitrile, **IV, 255**

DICHLOROACETOPHENONE, **III, 538**

Dichloroacetyl chloride, **IV,** 271

Dichloroacetyl fluoride, **V,** 69

3,5-DICHLORO-2-AMINOBENZOIC ACID, **IV, 872**

2,6-DICHLOROANILINE, **III, 262; V,** 367

2,3-Dichlorobenzaldehyde, **V,** 141

2,4-Dichlorobenzaldehyde, **V,** 141

2,6-Dichlorobenzaldehyde, **V,** 364

m-Dichlorobenzene, **V,** 1067

o-Dichlorobenzene, **III,** 1, 98; **IV,** 523, 766

N,N-Dichlorobenzenesulfonamide, **V,** 911

3,4-DICHLOROBIPHENYL, **V, 51**

1,4-Dichlorobutane, **IV,** 893

1,3-Dichloro-2-butene, **IV,** 128, 684

N,N-Dichloro-*t*-butylamine, **V,** 186

p-Di(chloro-*t*-butyl)benzene, **IV,** 703

Dichloro-*t*-butylphosphine, **V,** 214

Dichlorocarbene, **V,** 362, 874, 971

2,4-Dichlorocinnamic acid, **IV, 733**

3,4-Dichlorocinnamic acid, **IV, 733**

N,N-DICHLOROCYCLOHEXYL-AMINE, **V, 208**

1,2-Dichlorocyclopentene, **IV,** 159

Dichlorocyclopropanes, **V,** 971

4,4'-DICHLORODIBUTYL ETHER, **IV, 266**

2,3-Dichloro-5,6-dicyano-1,4-benzoquinone, **V,** 428

1,1-DICHLORO-2,2-DIFLUOROETHY-LENE, **IV, 268**

3,3-Dichloro-2,2-difluoropropionic acid, **V,** 242

β,β-DICHLORO-*p*-DIMETHYLAMI-NOSTYRENE, **V, 361**

3,4-Dichloro-α,β-dimethylcinnamic acid, **V,** 513

1,5-DICHLORO-2,4-DINITROBEN-ZENE, **V, 1067**

2,2'-DICHLORO-α,α'-EPOXYBIBEN-ZYL, **V, 358**

2,2-DICHLOROETHANOL, **IV, 271**

1,1-Dichloroethylene, **V,** 364

1,7-Dichloro-4-heptanone, **IV,** 279

3,5-DICHLORO-4-HYDROXYBEN-ZOIC ACID, **III, 268**

Dichloromethyl 2-chloroethyl ether, **V,** 859, 860

2-(Dichloromethylene)bicyclo[3.3.0]-octane, **V,** 95

DICHLOROMETHYLENETRIPHEN-YLPHOSPHORANE, **V, 361**

DICHLOROMETHYL METHYL ETHER, **V,** 49, **365**

2,6-DICHLORONITROBENZENE, **V, 367**

1,1-Dichloro-1-nitroethane, **IV,** 374

7,7-DICHLORO-2-OXABICYCLO[4.1]-HEPTANE, **V, 874**

2,6-Dichloro-1,4,3,5-oxathiadiazine-4,4-dioxide, **V,** 227

2,6-DICHLOROPHENOL, **III, 267**

2,4-Dichlorophenoxymethyl chloride, **V,** 222

DI-(*p*-CHLOROPHENYL)ACETIC ACID, **III, 270; V,** 370

1,1-Di-(*p*-chlorophenyl)-2,2-dichloroethy-lene, **III, 271**

1-(3,4-Dichlorophenyl)-1,2-dimethylcyclo-propane, **V,** 513

N,N-Dichloro-α-phenylethylamine, **V, 909**

Dichlorophenylphosphine, **V,** 787

1,1-Di-(*p*-chlorophenyl)-2,2,2-trichloro-ethane, **III, 270**

1,3-Dichloro-2-propanone, **V,** 1058

2,3-Dichloropropene, **V,** 22

4,7-DICHLOROQUINOLINE, **III,** 272

2,5-Dichloroquinone, **IV, 152**

3,5-DICHLOROSULFANILAMIDE, **III, 262**

3,4-DICHLORO-1,2,3,4-TETRAMETH-YLCYCLOBUTENE, **V, 370**

o,α-Dichlorotoluene, **V,** 264

Dicyandiamide, **IV,** 79, 502

1,1'-DICYANO-1,1'-BICYCLOHEXYL, IV, 273

1,2-DI-1-(1-CYANO)CYCLO-HEXYLHYDRAZINE, IV, 66, 274, 275

Dicyanodiamide, I, 302

4,4'-Dicyano-α,α'-epoxybibenzyl, V, 360

α,α'-DICYANO-β-ETHYL-β-METHYL-GLUTARIMIDE, IV, 441

DICYANOKETENE ETHYLENE ACETAL, IV, 276

α,α-DICYANO-β-METHYLGLUTAR-AMIDE, III, 591

1,2-Di-2-(2-cyano)propylhydrazine, IV, 275

Dicyclohexylamine, III, 199

1,4-Dicyclohexylbenzene, II, 153

Dicyclohexylcarbodiimide, V, 242

Dicyclopentadiene, IV, 238, 475, 738; V, 16

Dicyclopentadienylnickel, IV, 477

DICYCLOPROPYL KETONE, IV, 278

DIELS ACID, IV, 191

Diels-Alder Reaction. See Index No. 3., Type of Reaction; Addition and Cycloaddition.

Di-(p-ethoxyphenyl)urea, IV, 53

DIETHYL ACETAMIDOMALONATE, V, 373

Diethyl acetylenedicarboxylate, IV, 330; V, 986

DIETHYL α-ACETYL-β-KETOPIME-LATE, V, 384

Diethyl adipate, IV, 324, 819

Diethylamine, I, 291; II, 183, 214; III, 275; IV, 147, 205, 476; V, 387

Diethylamine hydrochloride, IV, 281

DIETHYLAMINOACETONITRILE, III, 275

1-DIETHYLAMINO-3-BUTANONE, IV, 281

β-DIETHYLAMINOETHYL ALCO-HOL, II, 183

β-DIETHYLAMINOETHYL BRO-MIDE HYDROBROMIDE, II, 92

[β-(Diethylamino)ethyl]triphenyl-phosphonium bromide, V, 1147

DIETHYL AMINOMALONATE, V, 376

DIETHYL AMINOMALONATE HY-DROCHLORIDE, V, 377

1,1-bis(Diethylaminomethyl)acetone, IV, 281

p-Diethylaminophenyl isocyanide, V, 1063

3-Diethylaminopropionitrile, IV, 206

γ-Diethylaminopropylamine, III, 257

Diethylammonium chloride, V, 1076

N,N-Diethylaniline, IV, 955; V, 113

N,N-Diethylaniline hydrochloride, IV, 955

DIETHYL AZODICARBOXYLATE, IV, 411; V, 96

Diethyl benzalmalonate, IV, 804

Diethylbenzene, V, 1039

N,N'-DIETHYLBENZIDINE, IV, 283

DIETHYL(O-BENZOYL)ETHYL-TARTRONATE, V, 379

DIETHYL BENZYLMALONATE, III, 705; IV, 285

DIETHYL BIS(HYDROXYMETHYL)-MALONATE, V, 381

Diethyl bromoacetal, IV, 295, 297

DIETHYL s-BUTYLMALONATE, III, 495

Diethyl carbonate, IV, 461; V, 198

DIETHYL CETYLMALONATE, IV, 141

Diethyl chloroacetal, IV, 404

Diethylcyanamide, IV, 360

DIETHYL-1,1-CYCLOBUTANEDI-CARBOXYLATE, IV, 288

DIETHYL Δ²-CYCLOPENTENYL-MALONATE, IV, 291

DIETHYL 2,3-DIAZABICYCLO[2.2.1]-HEPTANE-2,3-DICARBOX-YLATE, V, 97

DIETHYL 2,3-DIAZABICYCLO[2.2.1]-HEPT-5-ENE-2,3-DICAR-BOXYLATE, V, 96

DIETHYL α,δ-DIBROMOADIPATE, III, 623

Diethylene glycol, III, 270; IV, 313, 510, 753

Diethylene glycol monoethyl ether, **IV**, 252
Diethyl ethoxymethylenemalonate, **IV**, 298, 299
DIETHYL ETHYLIDENEMALONATE, **IV, 293**
Diethyl ethylmalonate, **V**, 379
Diethyl ethylphosphonate, **IV, 326**
Diethyl fumarate, **I**, 272; **IV**, 486; **V**, 993
DIETHYL HEPTANOYLSUCCINATE, **IV, 430**
DIETHYL cis-HEXAHYDROPHTHALATE, **IV, 304**
Diethyl hydrogen phosphite, **IV**, 956
DIETHYL ISONITROSOMALONATE, **V, 373**, 377
Diethyl ketone, **III**, 66, 399
DIETHYL-β-KETOPIMELATE, **V, 384**
Diethyl maleate, **IV**, 430
Diethyl malonate, **I**, 245, 250, 267, 272, 290; **II**, 60, 201, 279, 474, 594; **III**, 165, 214, 377, 395, 495, 637, 705; **IV**, 285, 288, 291, 293, 294, 417, 631, 708; **V**, 373, 381
DIETHYL MERCAPTOACETAL, **IV, 295**
DIETHYL METHYLENEMALONATE, **IV, 298**
Diethyl methylmalonate, **IV**, 300, 618
Diethyl methylphosphonate, **IV**, 326
Diethyl o-nitrobenzoylmalonate, **IV**, 709
Diethyl oxalate, **IV**, 141, 744; **V**, 567, 687, 747
DIETHYL γ-OXOPIMELATE, **IV, 302**
Diethyl phthalate, **IV**, 11, 449
N,N-Diethylselenourea, **IV**, 360
Diethyl sodium phthalimidomalonate, **IV**, 55
DIETHYL SUCCINATE, **IV**, 133; **V**, 288, 637, 687, **993**
Diethyl sulfate, **I**, 471
Diethyl sulfoxide, **V**, 792
DIETHYL-cis-Δ⁴-TETRAHYDROPHTHALATE, **IV, 304**, 305
DIETHYLTHIOCARBAMYL CHLORIDE, **IV, 307**

N,N-Diethyl-p-toluenesulfonamide, **V**, 1066
N,N-DIETHYL-2,2,2-TRICHLOROACETAMIDE, **V, 387**
N,N-DIETHYL-1,2,2-TRICHLOROVINYLAMINE, **V, 387**, 390
DIETHYL ZINC, **II, 184**
4,4'-DIFLUOROBIPHENYL, **II, 188**
2,2'-Difluoro-α,α'-epoxybenzil, **V**, 360
1,1-Difluoro-2(2-furyl)ethylene, **V**, 392
β,β-Difluoro-α-furylethylene, **V**, 392
β,β-Difluoro-p-methoxystyrene, **V**, 392
1,1-Difluoro-1-octene, **V**, 392
1,1-DIFLUORO-2-PHENYLETHYLENE, **V, 390**
β,β-DIFLUOROSTYRENE, **V, 390**
2,2-DIFLUOROSUCCINIC ACID, **V, 383**
α,α-DIFLUOROTOLUENE, **V, 396**
4,4-Diformyl-α,α'-epoxybibenzyl, **V**, 360
Diglyme, **V**, 390, 391
9,10-DIHYDROANTHRACENE, **V, 398**
1,4-Dihydrobenzamide, **V**, 403
1,4-DIHYDROBENZOIC ACID, **V, 400**
3,4-Dihydro-2-n-butoxy-2H-pyran, **IV, 312**
DIHYDROCHOLESTEROL, **II**, 139, 191; **V**, 245
1,4-DIHYDRO-3,5-DICARBETHOXY-2,6-DIMETHYLPYRIDINE, **II, 215**
1,4-Dihydro-3,5-dimethoxybenzamide, **V**, 403
1,4-Dihydro-3,5-dimethoxybenzoic acid, **V**, 403
2,5-DIHYDRO-3,5-DIMETHOXYFURAN, **V, 403**
3,4-Dihydro-4,6-diphenyl-2-ethoxy-2H-pyran, **IV, 312**
3,4-Dihydro-2-ethoxy-5-ethyl-4-n-propyl-2H-pyran, **IV, 312**
3,4-Dihydro-2-ethoxy-4-furyl-2H-pyran, **IV, 312**
3,4-Dihydro-2-ethoxy-6-methyl-4-phenyl-2H-pyran, **IV, 312**
3,4-Dihydro-2-ethoxy-2-methyl-2H-pyran, **IV, 312**

3,4-Dihydro-2-ethoxy-4-methyl-2*H*-pyran,
IV, 312
3,4-Dihydro-2-ethoxy-5-methyl-2*H*-pyran,
IV, 312
3,4-Dihydro-2-ethoxy-6-methyl-2*H*-pyran,
IV, 312
3,4-Dihydro-2-ethoxy-4-phenyl-2*H*-pyran,
IV, 312
3,4-Dihydro-2-ethoxy-2*H*-pyran, IV, 312
1,3-DIHYDROISOINDOLE, V, 406
1,4-Dihydro-3-methoxybenzamide, V, 403
1,4-Dihydro-2-methoxybenzoic acid, V,
402
3,4-DIHYDRO-2-METHOXY-4-
METHYL-2*H*-PYRAN, IV,
311, 660
3,4-Dihydro-2-methoxy-2*H*-pyran, IV, 312
3,4-DIHYDRO-1,2-NAPHTHALIC AN-
HYDRIDE, II, 194, 423
3,4-Dihydro-2-naphthoic acid, III, 302
5,8-Dihydro-1-naphthol, IV, 888
9,10-DIHYDROPHENANTHRENE, IV,
313
Dihydropyran, IV, 499; V, 874, 875
2,3-DIHYDROPYRAN, III, 276, 470,
795
DIHYDRORESORCINOL, III, 278; V,
539, 743, 745
DIHYDRORESORCINOL, MONO-
ETHYL ETHER, V, 539
1,4-Dihydro-*o*-toluic acid, V, 402
1,3-DIHYDRO-3,5,7-TRIMETHYL-2*H*-
AZEPIN-2-ONE, V, 408
2,5-DIHYDROXYACETOPHENONE,
III, 280
2,6-DIHYDROXYACETOPHENONE,
III, 281
5,5-DIHYDROXYBARBITURIC
ACID, IV, 23
2,5-DIHYDROXY-*p*-BENZENEDI-
ACETIC ACID, III, 286
2,4-DIHYDROXYBENZOIC ACID, II,
100, 557
3,5-DIHYDROXYBENZOIC ACID, III,
288
3,3'-DIHYDROXYBIPHENOL, V, 412

2,4-DIHYDROXY-5-BROMOBENZOIC
ACID, II, 100
2,3-Dihydroxy-4-butyrolactone, V, 257
3,12-Dihydroxy-*nor*-cholanyldiphenylcarbi-
nol, III, 238
DIHYDROXYCYCLOPENTENE, V,
414
6,7-DIHYDROXY-4-METHYL-1,2-
BENZOPYRONE, I, 360
p-(2,5-DIHYDROXYPHENYL)-
ACETOPHENONE, IV, 15
Dihydroxystearic acid, IV, 742
9,10-DIHYDROXYSTEARIC ACID
(LOW-MELTING ISOMER),
IV, 317
9,10-Dihydroxystearic acid, high-melting
isomer, IV, 320
Diimide, V, 281
1,4-DIIODOBUTANE, IV, 321, 369
1,6-DIIODOHEXANE, IV, 323, 370
2,6-DIIODO-*p*-NITROANILINE, II,
196, 604
1,5-Diiodopentane, IV, 370
1,3-Diiodopropane, IV, 370
2,5-Diiodothiophene, II, 358
DI-ISOAMYL ZINC, II, 187
Diisobutylene, V, 818
Diisobutylthiocarbamyl chloride, IV, 310
Diisopropyl ether, IV, 322
Diisopropyl ethylphosphonate, IV, 326
DIISOPROPYL METHYLPHOS-
PHONATE, IV, 325
Diisopropylthiocarbamyl chloride, IV, 310
DIISOVALERYLMETHANE, III, 291
Diketene, V, 155, 679
7,16-Diketodocosanedioic acid, V, 536
2,4-Diketones, V, 851
DIMEDONE, II, 200
1,2-DIMERCAPTOBENZENE, V, 419
1,4-Dimercaptobenzene, V, 420
4,4'-Dimercaptobiphenyl, V, 420
2,5-Dimercaptotoluene, V, 420
DIMESITYLMETHANE, V, 422
2,3-Dimethoxybenzaldehyde, IV, 327, 408,
603
2,6-DIMETHOXYBENZONITRILE,

III, 293
4,4'-Dimethoxybibenzyl, **V, 428**
7,7-DIMETHOXYBICYCLO(2.2.1)-
HEPTENE, **V**, 91, **424**
3,3'-DIMETHOXYBIPHENYL, **III, 295;**
V, 412
4,4-Dimethoxy-2-butanone, **IV**, 629, 649;
V, 794
2,3-DIMETHOXYCINNAMIC ACID,
IV, 327
3,4-Dimethoxycinnamic acid, **IV, 733**
3,8-DIMETHOXY-4,5,6,7-DIBENZO-1,2-
DIOXACYCLOOCTANE,
V, 493, 494, 495
2,2'-Dimethoxydiethyl ether(diglyme), **V,**
391
6,7-DIMETHOXY-3,4-DIHYDRO-2-
NAPHTHOIC ACID, **III, 300**
1,2-Dimethoxyethane, **IV, 475**
1,2-Dimethoxyethane, (monoglyme), **V,**
719, 970, 1001
5,8-Dimethoxylepidine, **III, 521**
6,7-Dimethoxy-2-naphthoic acid, **III, 302**
2,3-Dimethoxyphenylacetamide, **IV, 762**
3,4-Dimethoxyphenylacetamide, **IV, 762**
4-(3,4-Dimethoxyphenyl)-*m*-dioxane, **IV,**
787
1,5-Di(*p*-methoxyphenyl)-1,3,5-pentane-
trione, **V, 721**
1-(3',4'-Dimethoxyphenyl)-1-propene, **IV,**
787
3,4-DIMETHOXYPHENYLPYRUVIC
ACID, **II, 335**
trans-4,4'-DIMETHOXYSTILBENE, **V,**
428
7,7-DIMETHOXY-1,2,3,4-TETRA-
CHLOROBICYCLO-
(2.2.1)HEPT-2-ENE, **V, 425**
5,5-DIMETHOXY-1,2,3,4-TETRA-
CHLOROCYCLOPENTA-
DIENE, **V, 425**
2,5-Dimethoxytetrahydrofuran, **V, 716**
2,4-DIMETHYL-3-ACETYL-5-CAR-
BETHOXYPYRROLE, **III,**
514
DIMETHYL ACETYLENEDICAR-

BOXYLATE, **IV, 329**
β,β-DIMETHYLACRYLIC ACID, **III,**
302
1,3-Dimethyladamantane, **V,** 19
Dimethylamine, **IV,** 336, 340, 893; **V,** 602
Dimethylamine, aqueous, **IV,** 336, 361,
626, 719
Dimethylamine hydrochloride, **II,** 211; **III**
305
p-DIMETHYLAMINOBENZALDE-
HYDE, **I, 214; IV, 331; V,** 46,
362
m-Dimethylaminobenzoic acid, **V,** 554
p-Dimethylaminobenzoic acid, **V,** 554
p-N,N-Dimethylaminobenzonitrile, **V,** 657
p-DIMETHYLAMINOBENZOPHEN-
ONE, **I, 217**
β-Dimethylaminoethanol, **IV,** 333
β-DIMETHYLAMINOETHYL BROM-
IDE HYDROBROMIDE, **II, 92**
β-DIMETHYLAMINOETHYL CHLOR-
IDE HYDROCHLORIDE, **IV,**
333
6-(DIMETHYLAMINO)FULVENE, **V,**
431
4-DIMETHYLAMINO-2'-METHOXY-
BENZHYDROL, **V, 46**
N,N-Dimethylaminomethoxymethylium
methyl sulfate. See Dimethyl-
formamide-dimethyl sulfate
complex, **V,** 431
N,N-DIMETHYLAMINOMETHYL-
FERROCENE METHIOD-
IDE, **V, 434** 578, 621
p-Dimethylaminophenylacetic acid, **V,** 554
α-*N,N*-DIMETHYLAMINOPHENYL-
ACETONITRILE, **V, 437**
p-DIMETHYLAMINOPHENYL ISO-
THIOCYANATE, **I, 449**
N-(*p*-DIMETHYLAMINOPHENYL)-α-
(*o*-NITRO-
PHENYL)NITRONE, **V, 826**
β-DIMETHYLAMINOPROPIOPHEN-
ONE HYDROCHLORIDE,
III, 305
N-[3-(Dimethylamino)propyl]-*N'*-ethyl-

urea, **V, 555**
2-(DIMETHYLAMINO)PYRIMIDINE,
IV, 336
o-Dimethylaminotoluene, **V, 554**
p-Dimethylaminotoluene, **V, 554**
Dimethylaniline, **I,** 214, 217, 374; **II,** 223, 574
N,N-DIMETHYLANILINE, **III,** 142, 268, 658; **IV,** 264, 331, 953; **V,** 111, 172
2,6-Dimethylaniline, **V, 802**
3,4-DIMETHYLANILINE, **III, 307**
2,3-DIMETHYLANTHRAQUINONE, **III, 310**
2,5-Dimethylbenzoic acid, **V, 708**
N-(2,4-Dimethylbenzyl)-acetamide, **V, 75**
2,3-Dimethylbenzyl acetate, **IV, 584**
2,3-Dimethylbenzyl alcohol, **IV, 584**
N,N,-Dimethylbenzylamine, **IV, 585; V,** 609
2,3-Dimethylbenzyldimethylamine, **IV,** 508, **587**
2,3-Dimethylbenzylethyldimethyl-ammonium bromide, **IV, 583**
2,3-Dimethylbenzyltrimethylammonium iodide, **IV,** 508, 587
3,3'-DIMETHYLBIPHENYL, **III, 295**
2,3-DIMETHYL-1,3-BUTADIENE, **III,** 310, **312**
2,2-Dimethylbutanoic acid, **V, 22**
2,3-Dimethyl-2-butene, **IV, 544**
2,2-Dimethylbutyric acid, **V, 742**
3,5-DIMETHYL-4-CARBETHOXY-2-CYCLOHEXEN-1-ONE, **III. 317**
2,4-DIMETHYL-5-CARBETHOXY-PYRROLE, **II, 198**
Dimethyl chloroacetal, **IV,** 297, 406
4,6-DIMETHYLCOUMALIN, **IV, 337**
Dimethylcyanamide, **IV, 359**
1,2-DIMETHYL-1,4-CYCLOHEXA-DIENE, **V, 467**
N,N-DIMETHYLCYCLOHEXANE-CARBOXAMIDE, **IV, 339,** 340
5,5-DIMETHYL-1,3-CYCLOHEXANE-DIONE, **II, 200**

2,2-Dimethylcyclohexanone, **V, 190**
DIMETHYL CYCLOHEXANONE-2,6-DICARBOXYLATE, **V, 439**
3,5-DIMETHYL-2-CYCLOHEXEN-1-ONE, **III, 317**
N,N-DIMETHYLCYCLOHEXYL-METHYLAMINE, **IV,** 339, 612
N,N-Dimethylcyclohexylmethylamine-*N*-oxide, **IV, 612**
2,2-Dimethylcyclopropanol, **V, 861**
2,3-Dimethylcyclopropanol, **V, 861**
Di-(2-methylcyclopropyl), ketone, **IV, 280**
2,4-DIMETHYL-3,5-DICARBETHOXY-PYRROLE, **II, 202,** 217
2,2'-Dimethyl-5,5'-Dimethoxybiphenyl, **III, 297**
2,7-DIMETHYL-2,7-DINITRO-ÖCTANE, **V, 445**
2,6-DIMETHYL-3,5-DIPHENYL-4*H*-PYRN-4-ONE, **V, 450**
Dimethyldivinyltin, **IV, 260**
1,2-DIMETHYL-1,2-EPOXYCY-CLOHEX-4-ENE, **V, 468**
Dimethyl ether, **V,** 1096, 1099
2,2-DIMETHYLETHYLENIMINE, **III, 148**
DIMETHYLETHYNYLCARBINOL, **III, 320**
Dimethylformamide, **IV,** 147, 163, 331, 382, 454, 484, 540, 811, 831, 893, 953; **V,** 215, 431
DIMETHYLFORMAMIDE-DI-METHYL SULFATE COM-PLEX, **V, 431**
Dimethyl fumarate, **IV, 487**
2,5-Dimethylfuran, **II, 220**
DIMETHYLFURAZAN, **IV, 342**
β,β-DIMETHYLGLUTARIC ACID, **IV, 345**
DIMETHYLGLYOXIME, **II, 204; IV,** 342
Dimethyl hendecanedioate, **IV, 635**
4,6-DIMETHYL-1-HEPTEN-4-OL, **V, 452**
Dimethyl *cis*-hexahydrophthalate, **IV, 306**
2,5-Dimethyl-3-hexyne-2,5-diol, **V, 1024**
5,5-DIMETHYLHYDANTOIN, **III, 323**

unsym.-DIMETHYLHYDRAZINE, **II, 213**

sym.-DIMETHYLHYDRAZINE DIHY-DROCHLORIDE, **II, 208**

unsym.-DIMETHYLHYDRAZINE HY-DROCHLORIDE, **II, 211**

N,N-DIMETHYLHYDROXYLAMINE HYDROCHLORIDE, **IV, 612**

2,3-Dimethyl-2-iodobutane, **IV, 544**

DIMETHYLKETENE, **IV, 348**

DIMETHYL KETENE β-LACTONE DI-MER, **V, 456, 1103**

2,5-DIMETHYLMANDELIC ACID, **III, 326**

Dimethyl mercaptoacetal, **IV, 298**

DIMETHYL 3-METHYLENECYCLO-BUTANE-1,2-DICARBOXY-LATE, **V, 459, 734**

N,N-Dimethyl-1-(methylthio)-1-butenyl-amine, **V, 784**

N,N-Dimethyl-1-(methyl-thio)propenylamine, **V, 784**

N,N-Dimethyl-1-(methylthio)vinylamine, **V, 784**

2,3-Dimethylnaphthalene, **V, 810**

Dimethyl-α-naphthylamine, **III, 608**

DIMETHYLNITROSOAMINE, **II, 211**

DIMETHYL OCTADECANEDIOATE, **V, 463**

N,N-Dimethyloctylamine, **V, 315, 317**

2,7-DIMETHYLOXEPIN, **V, 467**

DIMETHYLOXOSULFONIUM METHYLIDE, **V, 755, 757**

2,4-Dimethyl-2-pentacosenoic acid, **IV, 611**

3,3-Dimethylpentane-2,4-dione, **V, 785**

5,5-DIMETHYL-2-*n*-PENTYLTETRA-HYDROFURAN, **IV, 350**

α,α-Dimethyl-β-phenethyl alcohol, **V, 471**

α,α-DIMETHYL-β-PHENETHYL-AMINE, **V, 471**

1,2-Dimethyl-1-phenylcyclopropane, **V, 513**

2,6-Dimethylphenyl isocyanide, **V, 1063**

2,6-Dimethyl-3-phenyl-4*H*-pyran-4-one, **V, 452**

2,6-DIMETHYLPHENYLTHIOUREA, **V, 802**

N,N-Dimethyl-1,3-propanediamine, **V, 555**

2,2-DIMETHYL-1-PROPANOL, **V, 818**

3,5-DIMETHYLPYRAZOLE, **IV, 351**

2,6-DIMETHYLPYRIDINE, **II, 214**

4,6-Dimethyl-1,2-pyrone, **IV, 551**

2,4-DIMETHYLPYRROLE, **II, 198, 217**

2,5-DIMETHYLPYRROLE, **II, 219**

2,2-DIMETHYLPYRROLIDINE, **IV, 354**

2,2-Dimethylpyrrolidine hydrochloride, **IV, 355**

1,5-DIMETHYL-2-PYRROLIDONE, **III, 328**

5,5-DIMETHYL-2-PYRROLIDONE, **IV, 355, 357**

2,4-DIMETHYLQUINOLINE, **III, 329**

2′,3′′′-Dimethylquinquephenyl, **V, 987**

4,4′′′′′-Dimethylquinquephenyl, **V, 987**

Dimethyl sebacate, **IV, 840**

N,N-DIMETHYLSELENOUREA, **IV, 359**

Dimethyl sulfate, **I, 58, 537; IV, 588, 674, 837; V, 271, 431, 736, 1018, 1021**

Dimethyl sulfoxide, **IV, 147, 455; V, 752, 926, 938**

Dimethyl *cis*-Δ⁴-tetrahydrophthalate, **IV, 306**

2,4-DIMETHYLTHIAZOLE, **III, 332**

Dimethylthiocarbamyl chloride, **IV, 310**

4,6-Dimethyl-1α-thiophthalimide, **V, 1053**

N,N-Dimethyl-1,2,2-trichlorovinylamines, **V, 389**

1,2-Dimethyl-1-[*m*-(trifluoromethyl)-phen-yl]cyclopropane, **V, 513**

asym-DIMETHYLUREA, **IV, 361**

Di-α-naphthylthiourea, **IV, 701**

2,4-DINITROANILINE, **II, 221; III, 242**

2,6-DINITROANILINE, **IV, 160, 364**

3,5-DINITROANISOLE, **I, 219**

2,4-DINITROBENZALDEHYDE, **II, 223**

m-Dinitrobenzene, **III, 293**

o-DINITROBENZENE, **II, 226**

p-DINITROBENZENE, **II, 225**

2,4-Dinitrobenzenesulfenyl bromide, **V, 476**

2,4-DINITROBENZENESULFENYL
   CHLORIDE, **V, 474**
2,5-DINITROBENZOIC ACID, **III, 334**
3,5-DINITROBENZOIC ACID, **III, 337**
2,4-Dinitrobenzylidene-*p*-aminodimethyl-
   aniline, **II,** 224
p,p'-DINITROBIBENZYL, **IV, 367**
2,2'-DINITROBIPHENYL, **III, 339**
1,4-DINITROBUTANE, **IV, 368**
2,3-Dinitro-2-butene, **IV, 374**
2,4-Dinitrochlorobenzene, **II,** 221, 228; **V,**
   474, 478
DINITRODURENE, **II, 254**
3,3'-Dinitro-α,α'-epoxybibenzyl, **V,** 360
Dinitrogen tetroxide, **V,** 337, 651, 652
1,6-Dinitrohexane, **IV, 370**
3,4-DINITRO-3-HEXENE, **IV, 372**
2,4-DINITROIODOBENZENE, **V, 478**
1,8-Dinitro-*p*-menthane, **V,** 847
1,4-DINITRONAPHTHALENE, **III, 341**
2,4-Dinitro-1-naphthol-7-sulfonic acid, **II,**
   50
1,5-DINITROPENTANE, **IV, 370**
2,4-Dinitrophenol, **III,** 82
2,4-DINITROPHENYL BENZYL SULF-
   IDE, **V, 474**
DI-*o*-NITROPHENYL DISULFIDE, **I,**
   220; II, 455, 471
2,4-DINITROPHENYLHYDRAZINE,
   II, 228; IV, 14
2,6-Dinitrophenylhydrazine, **II, 229**
2,4-Dinitrophenylhydrazine reagent, **V,** 253
1,1-Dinitropropane, **IV, 372**
1,3-Dinitropropane, **IV, 370**
2,4-Dinitrophenylsulfur chloride, **II, 456**
2,4-Dinitrotoluene, **II,** 160, 223
3,5-DINITRO-*o*-TOLUNITRILE, **V, 480**
Dioxane, **III,** 149, 182, 303, 438, 638; **IV,**
   229, 578, 642, 652, 846
Dioxane sulfotrioxide, **IV,** 846
1,4-DIOXASPIRO[4,5]DECANE, **V, 303**
1,6-DIOXO-8*A*-METHYL-1,2,3,4,6,7-8,8*A*-
   OCTOHYDRONAPH-
   THALENE, **V, 486**
4-(3,4-Dioxymethylenephenyl)-*m*-dioxane,
   IV, 787

Dipentaerythritol, **IV,** 754
DIPHENALDEHYDE, **V, 489**
DIPHENALDEHYDIC ACID, **V, 493**
   methyl ester, **V,** 495
DIPHENIC ACID, **I, 222; IV,** 757; **V,**
   495
DIPHENYLACETALDEHYDE, **IV,**
   375
Diphenylacetaldehyde 2,4-dinitrophenyl-
   hydrazone, **IV,** 376
Diphenylacetamidine, **IV,** 385
DIPHENYLACETIC ACID, **I, 224**
α,α-DIPHENYLACETONE, **III, 343**
DIPHENYLACETONITRILE, **III, 347;**
   IV, 963
Diphenylacetyl chloride, **III,** 120
DIPHENYLACETYLENE, **III, 350; IV,**
   377; V, 604, 606
Diphenylamine, **I,** 545
1,4-DIPHENYL-5-AMINO-1,2,3-TRI-
   AZOLE, **IV, 380**
N,N'-DIPHENYLBENZAMIDINE, **IV,**
   383
Diphenylbenzamidine hydrochloride, **IV,**
   383
Diphenyl-*m*-bromophenylphosphine, **V,**
   498
DIPHENYL-*p*-BROMOPHENYL
   PHOSPHINE, **V, 496**
1,4-DIPHENYLBUTADIENE, **II, 229**
1,4-DIPHENYL-1,3-BUTADIENE, **V,**
   499
2,4-Diphenylbutane-1,4-sultone, **IV,** 849
trans-1,4-Diphenylbut-1-en-3-yne, **V,** 519
α,β-Diphenylbutyric acid, **V,** 527
DIPHENYLCARBODIIMIDE, **V, 501,**
   504
Diphenyl-*p*-chlorophenylphosphine, **V,** 498
α,β-DIPHENYLCINNAMONITRILE,
   IV, 387
1,1-DIPHENYLCYCLOPROPANE, **V,**
   509
DIPHENYLCYCLOPROPENONE, **V,**
   514
DIPHENYLDIACETYLENE, **V, 517**
DIPHENYLDIAZOMETHANE, **III, 351**

Diphenyldichloromethane, **IV, 914**
Diphenyl disulfide, **V,** 723
Diphenyldivinyltin, **IV,** 260
Diphenyl ether, **III,** 31
1,1-DIPHENYLETHYLENE, **I, 226; V,**
    1126
N,N-Diphenylethylenediamine, **V,** 115
1,1-Diphenyl-3-ethylpentane, **V,** 525
N,N'-Diphenylformamidine, **IV,** 465
4,5-DIPHENYLGLYOXALONE, **II, 231**
1,7-Diphenyl-1,3,5,7-heptanetetraone, **V,**
    720
Diphenylhydrazine, **III,** 102
2,3-DIPHENYLINDONE, **III, 353**
DIPHENYLIODONIUM-2-CARBOXY-
    LATE, **V, 1037**
DIPHENYLIODONIUM IODIDE, **III,**
    355
DIPHENYLKETENE, **III, 356**
DIPHENYL KETIMINE, **V, 520**
DIPHENYLMERCURY, **I, 228; V,** 1116
DIPHENYLMETHANE, **II, 232**
Diphenylmethane, **V,** 523
DIPHENYLMETHANE IMINE HY-
    DROCHLORIDE, **II, 234**
1,1-Diphenyl-2-methylcyclopropane, **V,**
    513
DIPHENYLMETHYLENIMINE, **V,**
    520
N,α-DIPHENYLNITRONE, **V, 1124**
1,1-Diphenylnonane, **V,** 525
1,1-DIPHENYLPENTANE, **V, 523**
1,5-Diphenyl-1,3,5-pentanetrione, **V,** 720,
    721
p-Diphenylphosphinobenzoic acid, **V,** 498
1,3-Diphenyl-2-propanone, **IV, 177**
α,β-DIPHENYLPROPIONIC ACID, **V,**
    526
β,β-DIPHENYLPROPIOPHENONE,
    II, 236
2,6-Diphenyl-4-pyrone, **V,** 722
2,4-DIPHENYLPYRROLE, **III, 358**
DIPHENYL SELENIDE, **II, 238,** 240
DIPHENYLSELENIUM DICHLOR-
    IDE, **II, 240**

DIPHENYL SUCCINATE, **IV, 390**
2,3-DIPHENYLSUCCINONITRILE,
    IV, 392
meso-α,α'-Diphenylsuccinonitrile, **IV, 394**
DIPHENYL SULFIDE, **II, 242**
Diphenyl sulfoxide, **IV, 89**
DIPHENYL TRIKETONE, **II, 244**
DIPHENYL TRIKETONE HYDRATE,
    II, 244
sym.-DIPHENYLUREA, **I, 453**
DIPHOSPHINE, TETRAMETHYL-,
    DISULFIDE, **V, 1016**
α,γ-Diphthalimidopropane, **II, 84**
Dipotassium 1,8-naphthalenedicarboxylate,
    V, 813
β-DIPROPYLAMINOETHYL BRO-
    MIDE HYDROBROMIDE,
    II, 92
γ-Di-*n*-propylaminopropylamine, **III, 257**
N,N-Di-*n*-propylaniline, **V,** 113
DI-*n*-PROPYL ZINC, **II, 187**
DISODIUM 7,16-DIKETODOCOSA-
    NEDIOATE, **V, 534**
Disodium monohydrogen phosphate dode-
    cahydrate, **IV,** 808, 816
DISPIRO(5.1.5.1)TETRADECANE-7,14-
    DIONE, **V, 297**
N,N-Disubstituted formamidines, **V,** 584
N,N-Disubstituted thioureas, **V,** 804
N,N-Disubstituted ureas, **V,** 804
DISULFIDE, BIS-(*m*-NITRO-
    PHENYL), **V, 843**
p-DITHIANE, **IV, 396**
2,4-DITHIOBIURET, **IV, 503, 504**
Dithiosalicylic acid, **II, 580**
DITHIZONE, **III, 360**
Di-*p*-tolylacetylene, **IV, 378**
1,1-DI-*p*-TOLYLETHANE, **I, 229**
DI-*p*-TOLYLMERCURY, **I, 231**
1,4-Di-*p*-tolyl-3-methyl-1,4-pentazadiene,
    V, 799
Di-*m*-tolyl selenide, **II, 240**
Di-*o*-tolyl selenide, **II, 240**
Di-*p*-tolyl selenide, **II, 240**
1,3-Di-*p*-tolyltriazene, **V,** 799
Di(trimethylsilylmethyl)divinyltin, **IV,** 260

cis- And *trans*-1,2-DIVINYLCYCLOBUT-
    ANE, **V, 528**
DOCOSANEDIOIC ACID, **V, 533**
Dodecanoic acid, **IV, 398**
Dodecanol, **III,** 366
trans-2-DODECENOIC ACID, **IV,** 398
3-Dodecenoic acid, **IV,** 400
n-DODECYL ALCOHOL, **II,** 246, **372**
n-DODECYL BROMIDE, **I, 29; II, 246,**
    292; **III, 363**
n-Dodecyl fluoride, **IV,** 527
n-DODECYL MERCAPTAN, **III,** 363
5-Dodecyl-1,2,3,4,5-pentachlorocyclopen-
    tadiene, **V,** 896
n-DODECYL *p*-TOLUENESULFO-
    NATE, **III, 366**
Dowtherm, **III,** 273, 593; **IV,** 672
DULCIN, **IV, 52**
DURENE, **II, 248,** 254; **V,** 604
Durenecarboxylic acid, **V,** 708
DUROQUINONE, **II, 254**
DYPNONE, **III, 367**

Eicosanedioic acid, **V,** 538
Elaidic acid, **IV,** 320
Electrolysis, Kolbe synthesis, **V,** 445, 463
Elimination Reactions, See Index No. 3.
    Type of Reaction
Emulsin, **III,** 428
Enamines, **V,** 277, 809
Enanthaldehyde, **IV,** 430
(−)-Ephedrine, **V,** 1033
EPIBROMOHYDRIN, **II, 257**
EPICHLOROHYDRIN, **I, 233; II, 256;**
    IV, 11, 12; **V,** 614, 1080
1-(Epithioethyl)-7,7-dimethyl-2-norborna-
    none *S,S*-dioxide, **V,** 877
Epoxides, **V,** 360
α,α′-Epoxy-di(1-naphthyl)ethane, **V,** 360
α,α′ Epoxydipyridylethane, **V,** 360
α,α′-Epoxydithienyl, **V,** 360
ERUCIC ACID, **II, 258**
Esterification, See Index No. 3. Type of
    Reaction
1,2-Ethanediol, **V,** 303
1,2-ETHANEDITHIOL, **IV,** 396; **401**

ETHANETETRACARBONITRILE,
    1,2-EPOXY, **V, 1007**
Ethanolamine, **II,** 91; **III,** 501; **IV,** 107; **V,**
    1018, 1021
ETHANOL, 2-(CYCLOHEXYLOXY)-,
    V, 303
ETHANOL, 2-NITRO-, **V, 833,** 838
ETHER, *t*-BUTYL PHENYL, **V, 924,**
    926
ETHOXYACETIC ACID, **II, 260**
ETHOXYACETYLENE, **IV, 404**
3-ETHOXY-2-CYCLOHEXENONE, **V,**
    294, **539**
2-Ethoxy-3,4-dihydro-2*H*-pyran, **IV,** 816
2-(2-Ethoxyethoxy)ethanol, **V,** 351
β-Ethoxyethyl alcohol, **III,** 370
β-ETHOXYETHYL BROMIDE, **III,**
    370, 372
ETHOXYMAGNESIUMMALONIC
    ESTER, **IV, 285, 708**
2-ETHOXY-1-NAPHTHALDEHYDE,
    III, 98
p-Ethoxyphenyl isothiocyanate, **V,** 226
m-Ethoxyphenylurea, **IV, 51**
o-Ethoxyphenylurea, **IV, 51**
p-ETHOXYPHENYLUREA, **IV,** 51–52
β-Ethoxypropionaldehyde, **III, 372**
β-ETHOXYPROPIONALDEHYDE
    ACETAL, **III, 371**
β-ETHOXYPROPIONITRILE, **III, 372**
Ethyl acetate, **I,** 226, 235, 252; **II,** 487; **III,**
    18, 19; **IV,** 348, 355, 409, 835,
    888; **V,** 564
ETHYL ACETOACETATE, **I, 235,** 248,
    360; **II,** 202, 214, 262, 266, 422;
    III, 232, 282, 317, 374, 379, 390,
    405, 513, 581; **IV,** 11, 408, 415,
    549, 592, 634, 635, 638
Ethyl α-acetoglutarate, **II, 263**
ETHYL ACETONEDICARBOXY-
    LATE, **I, 237**
Ethyl acetonedioxalate, **II,** 127
p-Ethylacetophenone, **IV,** 580
ETHYL ACETOPYRUVATE, **I, 238**
ETHYL ACETOSUCCINATE, **II, 262**
Ethyl β-acetotricarballylate, **II,** 263

ETHYL α-ACETYL-β-(2,3-DIMETH-
OXYPHENYL)-
ACRYLATE, **IV, 408**
ETHYL α-ACETYL-β-(2,3-DIMETH-
OXYPHENYL)-
PROPIONATE, **IV, 408; V,** 550
Ethyl α-acetyl-β-(3,4-dimethoxy-
phenyl)propionate, **IV, 410**
Ethylacetylene, **IV,** 118
Ethyl acrylate, **III,** 31
ETHYL ADIPATE, **II,** 116, **264,** 325
N-ETHYLALLENIMINE, **V, 541**
Ethylamine, **V,** 124, 754
ETHYLAMINE, *N*-METHYL-, **V, 758**
ETHYL *p*-AMINOBENZOATE, **I, 240;**
II, 299, **300**
N-Ethylaniline, **IV,** 147, 284, **422**
ETHYL β-ANILINOCROTONATE, **III,**
374, 593
ETHYL AZODICARBOXYLATE, **III,**
375; **IV, 411;** warning, **V,** 544
ETHYL BENZALMALONATE, **III,**
377; **IV,** 83
Ethylbenzene, **I,** 231; **IV,** 580
Ethyl benzoate, **II,** 67; **III,** 251, 380, 839;
V, 937, 938
p-Ethylbenzoic acid, **IV, 580**
ETHYL BENZOYLACETATE, **II, 266;**
III, 108, **379; IV,** 80, 82, **415,**
635
ETHYL BENZOYLACETOACETATE,
II, 266
ETHYL BENZOYLDIMETHYLACET-
ATE, **II, 268**
ETHYL BENZOYLFORMATE, **I, 241**
ETHYL *N*-BENZYLCARBAMATE, **IV,**
780
ETHYL-3-BENZYL-2-CYANO-3-
METHYLPENTANOATE,
IV, 94, 95
Ethyl benzyl ketone, **II,** 391
ETHYL BROMIDE, **I, 29, 36; II,** 184,
198, 290, 602; **III,** 440, 554; **IV,**
582, 792, 881
ETHYL BROMOACETATE, **III, 381,**
408; **IV,** 456

Ethyl α-bromobutyrate, **IV,** 454
ETHYL γ-BROMOBUTYRATE, **V, 545**
ETHYL BROMOMALONATE, **I, 245,**
271; **II, 273**
Ethyl α-bromopropionate, **IV,** 122, 445,
466; **V,** 1031
ETHYL β-BROMOPROPIONATE, **I,**
246; **II,** 263
ETHYL *n*-BUTYLACETOACETATE,
I, 248, 351
N-Ethylbutylamine, **V,** 738
ETHYL *n*-BUTYLCYANOACETATE,
III, 385
Ethyl *s*-butylcyanoacetate, **IV,** 96
ETHYL *s*-BUTYLIDENECYANOACET-
ATE, **IV, 94**
ETHYL *n*-BUTYLMALONATE, **I, 250**
Ethyl *s*-butylmalonate, **II,** 416, **417**
ETHYL *t*-BUTYL MALONATE, **IV,**
417
Ethyl *n*-butyrate, **II,** 114
O-Ethylcaprolactim, **IV, 589**
Ethyl carbamate, **V,** 645
ETHYL α-CARBETHOXY-β-*m*-CHLOR-
OANILINOACRYLATE, **III,**
272
Ethyl carbonate, **II,** 282, 602; **III,** 831
purification of, **II,** 283
Ethyl chloroacetate, **I,** 153; **II,** 85, 262,
263; **III,** 727, 751; **IV,** 29, 30,
55, 459
Ethyl α-chloroacetoacetate, **IV,** 591, **592**
N-ETHYL-*p*-CHLOROANILINE, **IV,**
420
Ethyl-2-(*o*-chlorobenzyl)cyanoacetate, **V,**
264
Ethyl chlorocarbonate, see Ethyl chloro-
formate
ETHYL CHLOROFLUOROACETATE,
IV, 185, **423**
N-ETHYL-*p*-CHLOROFORMANIL-
IDE, **IV, 420,** 421
Ethyl chloroformate, **II,** 198, 278, 595; **III,**
375; **IV,** 286, 411, 780
ETHYL α-CHLOROPHENYLACET-
ATE, **IV, 169**

Ethyl *N-p*-chlorophenylformimidate, **IV**, 422

2-ETHYLCHROMONE, **III, 387**

ETHYL CINNAMATE, **I, 252; II,** 270

Ethyl crotonate, **III,** 615; **V,** 762

ETHYL CYANOACETATE, **I,** 179, **254;** **III,** 286, 385, 399, 708; **IV,** 94, 246, 247, 441, 463, 567; **V,** 264, 572

Ethyl β-cyanoacrylate, **IV,** 488

Ethyl 2-cyano-3-ethyl-3-methylhexanoate, **IV, 97**

Ethyl 2-cyano-3-methyl-3-phenylpentanoate, **IV,** 97

ETHYL α-CYANO-β-PHENYLACRYLATE, **I, 451**

Ethyl α-cyclohexylacetoacetate, **III, 407**

ETHYL CYCLOHEXYLIDENEACETATE, **V, 547**

ETHYL DIACETYLACETATE, **III, 390**

ETHYL DIAZOACETATE, **III, 392; IV, 424**

ETHYL α,β-DIBROMO-β-PHENYLPROPIONATE, **II, 270,** 515

Ethyl di-*n*-butylamine, **V,** 577

Ethyl dichloroacetate, **IV,** 272

ETHYL 3,5-DICHLORO-4-HYDROXYBENZOATE, **III, 267**

Ethyl 2,3-dicyano-3-methylpentanoate, **V,** 573

ETHYL α,β-DICYANO-β-PHENYLPROPIONATE, **I, 452**

ETHYL DIETHOXYACETATE, **IV, 427**

ETHYL-1,3-DIHYDROXY-2-NAPHTHOATE, **III, 638**

ETHYL 6,7-DIMETHOXY-3-METHYLINDENE-2-CARBOXYLATE, **V, 550**

ETHYL *p*-DIMETHYLAMINOPHENYLACETATE, **V, 552**

1-ETHYL-3-(3-DIMETHYLAMINO)-PROPYLCARBODIIMIDE, **V, 555,** 556

1-ETHYL-3-(3-DIMETHYLAMINO)-PROPYLCARBODI-

IMIDE HYDROCHLORIDE, **V, 555**

Ethyl α,β-dimethylcinnamate, **V,** 513

ETHYL 2,5-DIMETHYLPHENYLHYDROXYMALONATE, **III, 326**

Ethyl α,α-dimethyl-β-phenyl-β-hydroxypropionate, **III, 409**

Ethyl α,β-dimethyl-*m*-(trifluoromethyl)cinnamate, **V,** 513

ETHYL 2,4-DIPHENYLBUTANOATE, **V, 559**

Ethyl 2,3-diphenylpropionate, **V,** 561

ETHYL ENANTHYLSUCCINATE, **IV, 430**

Ethylene, **IV,** 738; **V,** 425, 575

Ethylene bromide, **V,** 890

ETHYLENE BROMOHYDRIN, **I, 117**

Ethylene carbonate, **V,** 562

Ethylene chlorohydrin, **I,** 256; **II,** 183, 346, 576; **IV,** 73, 84

ETHYLENE CYANOHYDRIN, **I,** 131, **256,** 321

Ethylenediamine, **III,** 394

Ethylene dibromide, **I,** 120; **II,** 558; **IV,** 396, 402

Ethylene dichloride, **III,** 535; **IV,** 831, 846, 940

Ethylene diisothiouronium bromide, **IV,** 402

5,5'-Ethylene-1,2-*bis*-(2-furyl)-ethane, **V,** 886

Ethylene glycol, **III,** 154, 240; **IV,** 95, 276, 449, 525, 684

Ethylene glycol dimethyl ether, **IV,** 475

Ethylene glycol monomethyl ether, **IV,** 240, 506

Ethylene oxide, **I,** 117, 307

ETHYLENE SULFIDE, **V, 562**

5,5'-Ethylene-1,2-*bis*-(2-thienyl)ethane, **V,** 886

ETHYLENE THIOUREA, **III, 394**

ETHYLENIMINE, **IV, 433**

Ethyl α-ethoxalyl-γ-phenylbutyrate, **II,** 195

ETHYL ETHOXALYLPROPIONATE, **II, 272,** 279

Ethyl α-ethoxalylstearate, sodio derivative, **IV,** 142

ETHYL ETHOXYACETATE, **II, 261**
Ethyl ethoxymethylenecyanoacetate, **IV,**
566, 567
ETHYL ETHOXYMETHYLENEMAL-
ONATE, **III,** 272, **395**
ETHYL ETHYLENETETRACARBOX-
YLATE, **II, 273**
Ethyl 4-ethyl-2-methyl-3-hydroxyoctan-
oate, **IV, 447**
Ethyl 4-ethyl-2-methyl-2-octenoate, **IV,**
446, **448**
Ethyl 4-ethyl-2-methyl-3-octenoate, **IV,**
446, **448**
ETHYL (1-ETHYLPRO-
PENYL)METHYLCYANO-
ACETATE, **III, 397**
ETHYL (1-ETHYLPROPYLI-
DENE)CYANOACETATE,
III, 398, **399**
9-Ethylfluorene, **IV, 625**
Ethyl formate, **II,** 5; **III,** 96, 300, 829; **IV,**
210, 536, 537; **V,** 187, 189, 301
Ethyl fumarate, **IV,** 431
Ethyl glycidyl ether, **IV,** 12
Ethyl glycinate hydrochloride, **IV,** 424
Ethyl glyoxylate, **IV,** 125
N-Ethylheptylamine, **V,** 738
ETHYL 1,16-HEXADECANEDICAR-
BOXYLATE, **III, 401**
2-Ethylhexanal, **IV,** 445
2-Ethylhexanamide, **IV, 436**
2-Ethylhexanoic acid, **IV,** 436
2-ETHYLHEXANONITRILE, **IV, 436**
Ethyl n-hexylamine, **V,** 577
2-Ethylhexyl diazoacetate, **IV,** 426
5-(2-Ethylhexyl)-1,2,3,4,5-pentachloro-
cyclopentadiene, **V,** 896
ETHYL HYDRAZINECARBOXY-
LATE, **III, 404**
Ethylhydrazinium hydrogen oxalate, **V,** 45
ETHYL HYDRAZODICARBOXY-
LATE, **III,** 375; **IV, 411,** 412
Ethyl hydrogen adipate, **II,** 277; **III,** 171
Ethyl hydrogen malonate, **IV,** 418
ETHYL HYDROGEN SEBACATE, **II,**
276; III, 171, 401

Ethyl 4-hydroxybenzoate, **III,** 268
ETHYL β-HYDROXY-β,β-DIPHEN-
YLPROPIONATE, **V, 564**
4-ETHYL-4-HYDROXY-2-METHYL-
OCTANOIC ACID, γ-LAC-
TONE, **IV, 447,** 449
ETHYL INDOLE-2-CARBOXYLATE,
V, 567
Ethyl iodide, **II,** 184, **402; IV,** 326, 438
Ethyl isobutyrate, **II,** 268
Ethyl isobutyrylisobutyrate, **II, 270**
Ethyl isocyanate, **V,** 301, 555, 774
ETHYL ISOCYANIDE, **IV, 438**
ETHYL ISODEHYDROACETATE, **IV,**
549
Ethyl isonicotinate, **V,** 989
Ethyl isopropenyl ether, **IV,** 312
ETHYL α-ISOPROPYLACETOACET-
ATE, **III, 405**
ETHYL α-(ISOPROPYLIDENE-
AMINOOXY)PROPIONATE,
V, 1031
Ethyl isopropylmalonate, **II,** 93, 94
Ethyl isovalerate, **III,** 291
ETHYL 2-KETOCYCLOHEXYL-
GLYOXALATE, **II, 532**
ETHYL 2-KETOHEXAHYDROBEN-
ZOATE, **II, 532**
Ethyl lactate, **III,** 516; **IV,** 467
Ethyl laurate, **II,** 373
ETHYL LINOLEATE, **III, 527**
ETHYL LINOLENATE, **III, 532**
Ethylmagnesium bromide, **IV, 792; V,** 1058
Ethyl maleate, **IV,** 430
Ethyl malonate (see Diethyl malonate)
ETHYL MANDELATE, **IV, 169**
N-Ethyl-N-methylaniline, **V,** 577
ETHYL N-METHYLCARBAMATE, **II,**
278, 464
ETHYL 3-METHYLCOUMARILATE,
IV, 590, 591
β-ETHYL-β-METHYLGLUTARIC
ACID, **IV, 441**
3-Ethyl-3-methylhexanenitrile, **IV,** 97
3-Ethyl-3-methylhexanoic acid, **IV, 97**

ETHYL METHYLMALONATE, **II, 279**
4-Ethyl-2-methyl-1,4-octanolide, **IV, 447**
4-ETHYL-2-METHYL-2-OCTENOIC
    ACID, **IV, 444,** 449
Ethyl α-methyl-β-phenylcinnamate, **V,** 513
5-ETHYL-2-METHYLPYRIDINE, **IV,**
    451
α-ETHYL-α-METHYLSUCCINIC
    ACID, **V, 572**
ETHYL 1-METHYLTHIOL-3-
        PHTHALIMIDOPROPANE-
        3,3-DICARBOXYLATE, **II,**
        384
ETHYL α-NAPHTHOATE, **II, 282**
Ethyl nitrite, **II, 204**
ETHYL m-NITROBENZOATE, **I, 373**
Ethyl p-nitrobenzoate, **I,** 240
ETHYL α-NITROBUTYRATE, **IV, 454**
Ethyl α-nitrocaproate, **IV, 456**
Ethyl α-nitroisobutyrate, **IV, 456**
Ethyl α-nitroisovalerate, **IV, 456**
Ethyl α-nitrophenylacetate, **IV, 456**
Ethyl p-nitrophenylacetate, **V,** 552
ETHYL o-NITROPHENYLPYRU-
        VATE, POTASSIUM SALT,
        V, 567
Ethyl α-nitropropionate, **IV, 456**
ETHYL N-NITROSO-N-BENZYL-
        CARBAMATE, **IV, 780**
Ethyl oleate, **III,** 671
ETHYL ORTHOCARBONATE, **IV, 457**
ETHYL ORTHOFORMATE, **I, 258; II,**
        323; **III,** 395, 702, 732; **IV,** 21,
        420, 465, 516, 567, 801, 802
Ethyl orthosilicate, **IV,** 22
ETHYL OXALATE, **I,** 238, **261; II,** 127,
        194, 272, 287, 288, 532; **III,** 510
ETHYL OXALYLSUCCINATE, **III,**
        510
ETHYL OXOMALONATE, **I, 266; III,**
        326
Ethyl pentadecylate, **II, 294**
ETHYL β,β-PENTAMETHYLENE-
        GLYCIDATE, **IV, 459**
Ethyl pentane-1,1,5,5-tetracarboxylate, **III,**
        215

N-Ethylpentylamine, **V,** 738
Ethyl α-phenoxyacetoacetate, **IV,** 591
ETHYL PHENYLACETATE, **I, 270; II,**
        288; **IV,** 174; **V,** 559
ETHYL α-PHENYLACETOACETATE,
        II, 284
ETHYL PHENYLACETYLMALON-
        ATE, **III, 637**
ETHYL γ-PHENYLBUTYRATE, **II,**
        194, **196**
ETHYL PHENYLCYANOACETATE,
        IV, 461, 776
ETHYL β-PHENYL-β-CYANO-
        PROPIONATE, **IV, 804**
ETHYL PHENYLCYANOPYRU-
        VATE, **II, 287**
ETHYL (1-PHENYLETHYLI-
        DENE)CYANOACETATE,
        IV, 463, 662
Ethyl β-phenylethyl ketone, **II, 391**
ETHYL N-PHENYLFORMIMIDATE,
        IV, 464
Ethyl 2-phenylhexanoate, **V,** 561
ETHYL β-PHENYL-β-HYDROXY-
        PROPIONATE, **III, 408**
ETHYL PHENYLMALONATE, **II, 288**
Ethyl 2-phenylpropionate, **V,** 561
Ethyl phenyl sulfoxide, **V,** 792
Ethyl phthalate, **II,** 155
ETHYL PHTHALIMIDOMALONATE,
        I, 271; II, 384
ETHYL PIMELATE, **II, 536**
N-ETHYLPIPERIDINE, **V, 575**
ETHYL PROPANE-1,1,2,3-TETRA-
        CARBOXYLATE, **I, 272,** 523
ETHYL PROPANE-1,1,3,3-TETRA-
        CARBOXYLATE, **I, 290**
N-Ethylpropargylamine, **V, 544**
Ethyl propionate, **II,** 272; **III,** 387
α-Ethyl-β-n-propylacrolein, **IV,** 312
4-ETHYLPYRIDINE, **III, 410**
ETHYL 2-PYRIDYLACETATE, **III,**
        413
ETHYL α-(1-PYRROLI-
        DYL)PROPIONATE, **IV, 466,**
        834

ETHYL PYRUVATE, **IV, 467**
  sodium bisulfite adduct, **IV, 469**
Ethyl sebacate, **II**, 154, 276, **277**
ETHYL SODIUM PHTHALIMIDO-
  MALONATE, **II, 384**
Ethyl stearate, **IV**, 141, **142**
Ethyl succinate, **III**, 510
Ethyl tartrate, **IV**, 125
S-ETHYLTHIOUREA HYDROBROM-
  IDE, **III, 440**
N-ETHYL-*m*-TOLUIDINE, **II**, 290; **IV,**
  422
Ethyl-*o*-toluidine, **V**, 577
ETHYL *N*-TRICARBOXYLATE, **III,**
  415
Ethyl trichloroacetate, **V**, 874
ETHYL *n*-TRIDECYLATE, **II, 292**
Ethyl β-uraminocrotonate, **II, 422**
Ethyl γ-veratrylbutyrate, **III**, 300
Ethyl vinyl ether, **IV**, 312, 817
1-ETHYNYLCYCLOHEXANOL, **III,**
  22, **416; IV**, 13
1,1'-ETHYNYLENE-*bis*-CYCLOHEX-
  ANOL, **IV, 471**
ETHYNYLMAGNESIUM BROMIDE,
  IV, 792
o-EUGENOL, **III, 418**

Ferric alum, **IV**, 648
Ferric chloride, **II**, 255, 430; **IV**, 122, 474,
  480, 573, 851, 970; **V**, 1058,
  1076
Ferric nitrate, **III**, 44, 219, 291; **IV**, 128,
  387, 404, 586, 755, 763, 963
FERROCENE, **IV, 473; V**, 434
FERROCENYLACETONITRILE, **V,**
  578
Ferrous chloride, **IV**, 474, 476
Ferrous sulfate, **I**, 478; **III**; 57; **V**, 1026
Filter mat, asbestos-Norite, **II**, 14
Filter mat, carbon, **I**, 491
Filter pad, flannel, **II**, 13
Filtration, inverted, **II**, 610
Filtros plate, **I**, 11
Flavianic acid, **II**, 50
FLAVONE, **IV, 478; V**, 723

Fluoboric acid, **II**, 188, 225, **226, 297**, 299
Fluorene, **II**, 447; **III**, 24, 25; **IV**, 313, 624
9-FLUORENECARBOXYLIC ACID,
  IV, 482
Fluorenone, **III**, 837
FLUORENONE-2-CARBOXYLIC
  ACID, **III, 420**
9-FLUORENONE, 2,4,5,7-TETRA-
  NITRO-, **V, 1029**
2-FLUORENYLAMINE, **V**, 30
2-Fluoroacetic acid, **V**, 582
Fluoroacetyl fluoride, **V**, 69
2-Fluoroalkanoic acids, **V**, 582
p-Fluorobenzaldehyde, **V**, 142
FLUOROBENZENE, **II, 295**
p-FLUOROBENZOIC ACID, **II, 299**
Fluoroboric acid, **V**, 245, 1135, 1138
o-Fluorobromobenzene, **IV**, 965
Fluorobutylsulfur trifluoride, **V**, 961
2-Fluorobutyric acid, **V**, 582
2-Fluorodecanoic acid, **V**, 582
2-FLUOROHEPTANOIC ACID, **V,**
  580, 581
2-FLUOROHEPTYL ACETATE, **V, 580**
4-Fluoro-2-nitrobenzaldehyde, **V**, 828
5-Fluoro-2-nitrobenzaldehyde, **V**, 828
6-Fluoro-2-nitrobenzaldehyde, **V**, 828
2-Fluorooctanoic acid, **V**, 582
4-(*m*-Fluorophenyl)-2-butanone, **V**, 769
2-(*p*-Fluorophenyl)perfluoropropene, **V**,
  951
α-(4-Fluorophenyl)-γ-phenylacetoacetoni-
  trile, **IV**, 175
1-(*p*-Fluorophenyl)-3-phenyl-2-propanone,
  IV, 177
2-Fluoropropionic acid, **V**, 582
2-Fluoroundecanedioate, **V**, 582
Foam prevention, **II**, 175, 182
Formaldehyde, **I**, 54, 214, 290, 347, 355,
  378; **II**, 214, 590, 610; **III**, 198,
  275, 436, 460, 558, 685, 686,
  723; **IV**, 469, 626, 786, 908; **V**,
  381, 424, 552
Formalin, **II**, 214, 590, 610
Formalin, methanol content, **I**, 378 (see
  also Formaldehyde)

FORMALDOXIME, **V, 139**, 140
Formamide, **V**, 794
FORMAMIDINE ACETATE, **V, 582**
Formic acid, **I**, 42; **II**, 65; **III**, 33, 95, 217, 590, 723; **IV**, 317, 907; **V**, 20, 647, 739, 1062, 1070
FORMIC ACID, AZIDO-, BUTYL ESTER, **V, 157**
FORMIC ACID, AZODI-, DI-*t*-BUTYL ESTER, **V, 160**
o-Formotoluide, **III**, 479
6-Formyl-6-*n*-butyl-6-methylcyclohexanone, **V**, 190
2-Formyl-1-decalone, dianion, **V**, 190
N-FORMYL-α,α-DIMETHYL-β-PHENETHYLAMINE, **V, 471**, 472
α-Formylethyl methyl ketone, sodium salt, **IV**, 212
2-Formyl-6-methylcyclohexanone, dianion, **V**, 187
Formylmethyl isobutyl ketone, sodium salt, **IV**, 212
5-FORMYL-4-PHENANTHROIC ACID, **IV, 484**
o-FORMYLPHENOXYACETIC ACID, **V, 251**
p-Formylstyrene, **IV**, 733
1-Formyl-3-thiosemicarbazide, **V, 1070**
2-Formyl-2,4,4-trimethylcyclopentanone, **IV, 958**
Fructose, **III**, 460
FUMARAMIDE, **IV, 486**
FUMARIC ACID, **II**, 177, **302**
FUMARONITRILE, **IV, 486**
FUMARYL CHLORIDE, **III**, 248, **422**
FURAN, **I**, 274; **II**, 566; **V**, 403
2-FURANACRYLONITRILE, **V, 585**
2-FURANCARBOXYLIC ACID, **I**, 274, **276**
FURAN, 2,5-DIHYDRO-2,5-DIMETHOXY-, **V, 403**
3(2)-FURANONE, 4,5-DIHYDRO-2,2,-5,5-TETRAMETHYL-, **V, 1024**
FURFURAL, **I**, 276, 280, 283; **II**, 302; **III**, 425, 426; **IV**, 489, 493, 688;

V, 585
2-FURFURALACETONE, **I, 283**
FURFURAL DIACETATE, **IV, 489**
FURFURYL ACETATE, **I, 285**
FURFURYL ALCOHOL, **I**, 276; **IV**, 491
S-2-Furfurylisothiourea, **IV**, 491
2-FURFURYL MERCAPTAN, **IV, 491**
FUROIC ACID, **I**, 276; **III**, 621; **IV, 493**
β-Furylacrolein, **IV**, 312
FURYLACRYLIC ACID, **III, 425**, 742; **IV**, 302
β-Furylacrylic acid, **I**, 313
3-(2-FURYL)ACRYLONITRILE, **V, 585**
2-FURYLCARBINOL, **I**, 276, 285
  stabilization by urea, 1, 279
2-FURYLMETHYL ACETATE, **I, 285**
β-FURYLPROPIONIC ACID, **I, 313**

GALLACETOPHENONE, **II, 304**
Gallic acid, **I**, 537
Gamma acid, **II**, 149
Gas absorption trap, **II**, 4
Gelatine, **II**, 49
β-GENTIOBIOSE OCTAACETATE, **III, 428**
Gilman's catalyst, **II**, 360, **361**
GIRARD'S REAGENT, **II, 85**
Glaser oxidative coupling, **V**, 519
GLUCONIC ACID, PENTAACETYL-, D-, **V, 887**
GLUCONYL CHLORIDE, PENTAACETYL, D-, **V, 887**
GLUCOPYRANOSYL CHLORIDE, 2-ACETAMIDO-2-DEOXY-, TRIACETATE α-D-, **V, 1**
D-Glucosamine, See 2-Amino-2-deoxy-D-glucose, **V**, 2
d-GLUCOSAMINE HYDROCHLORIDE, **III, 430**
d-Glucose, **I**, 364; **III**, 11, 428, 432, 690, 800
β-*d*-GLUCOSE-1,2,3,4-TETRAACETATE, **III, 432**
β-*d*-GLUCOSE-2,3,4,6-TETRAACETATE, **III, 434**
d-GLUTAMIC ACID, **I, 287**

GLUTAMIC ACID HYDROCHLO-
RIDE, **I, 286**
Glutaraldehyde, **IV**, 816
GLUTARIC ACID, **I, 289; IV, 496;**
monoamide, **IV**, 497
GLUTARIC ACID, 2-OXO-, **V, 687**
Glutaric anhydride, **II**, 82
GLUTARIMIDE, **IV, 496, 498**
GLUTARONITRILE, 3-HYDROXY-,
V, 614
Gluten flour, **I**, 286
dl-GLYCERALDEHYDE, **II, 305**
dl-GLYCERALDEHYDE ACETAL, **II,**
305, **307**
Glycerol, **I**, 16, 42, 292, 294, 473, 478; **II,**
63, 308; **III**, 502, 526, 568; **IV,**
478, 546, 866
GLYCEROL α,γ-DIBROMOHYDRIN,
II, 257, 308
GLYCEROL α,γ-DICHLORO-
HYDRIN, **I,** 211, 233, **292; II,**
256; **IV,** 12
GLYCEROL α-MONOCHLORO-
HYDRIN, **I, 294,** 297
α-GLYCERYL PHENYL ETHER, **I, 296**
GLYCINE, **I, 298; II,** 11; **III,** 441; **IV,** 5,
57
GLYCINE *t*-BUTYL ESTER, **V,** 27, **586**
Glycine ethyl ester, **III,** 392
GLYCINE ETHYL ESTER
HYDROCHLORIDE, **II, 310**
GLYCINONITRILE, *N,N*-DI-
METHYL-2-PHENYL-, **V,**
437
Glycolic acid, **V,** 654
GLYCOLONITRILE, **III, 436**
Glyoxal, **V,** 1011
GLYOXAL BISULFITE, **III, 438**
Glyoxalic acid, **V,** 258
GLYOXAL, PHENYL-, **V, 937**
Glyoxals, **V,** 940
Glyoxal-sodium bisulfite, **IV,** 824
GLYOXYLIC ACID CHLORIDE *p*-
TOLUENESULFONYL-
HYDRAZONE, **V,** 259, 260
GLYOXYLIC ACID *p*-TOLUENESUL-

FONYLHYDRAZONE, **V, 258**
Grignard reactions, See Index No. 3, Type
of Reaction
Guaiacol, **I,** 150; **III,** 566
GUAIACOL ALLYL ETHER, **III, 418**
Guanidine, **IV,** 845
Guanidine hydrochloride, **IV,** 246
GUANIDINE NITRATE, **I, 302,** 399
warning note, **V,** 589
Guanidine thiocyanate, **IV,** 504
GUANIDOACETIC ACID, **III, 440**
GUANYLTHIOUREA, **IV, 502**
D-GULONIC-γ-LACTONE, **IV, 506**

H acid, **II,** 149
Hair, **I,** 194
N-(2-Haloallyl)alkylamines, **V,** 126
Halogenation, See Index No. 3, Type of
Reaction
HEMIMELLITENE, **IV, 508**
HEMIN, **III, 442**
HENDECANEDIOIC ACID, **IV, 510,**
636
unsym.-HEPTACHLOROPROPANE, **II,**
312
Heptadecanoic acid, **V,** 538
Heptaldehyde, **I,** 304; **II,** 313, 315
Heptaldehyde diacetate, **III,** 129
HEPTALDEHYDE ENOL ACETATE,
III, 127
HEPTALDOXIME, **II, 313,** 318
Heptalene, **V,** 1091
HEPTAMETHYLENE GLYCOL, **II,**
155
n-HEPTAMIDE, **IV, 513**
HEPTANE, 1-BROMO-2-FLUORO-, **V,**
136
HEPTANE, 1,1,1-TRIFLUORO-, **V, 1082**
n-HEPTANOIC ACID, **II,** 315; **IV,** 513;
V, 1082
HEPTANOIC ACID, 2-FLUORO-, **V,**
580
HEPTANOIC ACID, 3-METHYL-, **V,**
762
2-HEPTANOL, **II, 317**
HEPTANOL, 7-CYANO, **V, 266**

2-HEPTANONE, I, 351; V, 769
4-HEPTANONE, V, 589
Heptanoyl fluoride, V, 69
o-Heptanoylphenol, III, 444
1-Heptene, V, 136
2-Heptene, IV, 108
1-HEPTEN-4-OL, 4,6-DIMETHYL-, V, 452
Heptoic acid, III, 28
HEPTOIC ANHYDRIDE, III, 28
Heptoyl chloride, III, 28
n-HEPTYL ALCOHOL, I, 304
n-HEPTYLAMINE, II, 318; IV, 515
n-Heptyl bromide, II, 247, 474
3-n-HEPTYL-5-CYANOCYTOSINE, IV, 515
2-Heptyl-5,5-dimethyltetrahydrofuran, IV, 351
9-n-Heptylfluorene, IV, 625
n-Heptyl fluoride, IV, 527
Heptylmagnesium bromide, I, 187
o-n-HEPTYLPHENOL, III, 444
N-n-HEPTYLUREA, IV, 515
3-n-HEPTYLUREIDOMETHYLENE-MALONONITRILE, IV, 515
Hershberg stirrer, II, 117
HEXABROMOSTEARIC ACID, III, 531
Hexachloroacetone, V, 1074, 1075
Hexachlorobenzene, V, 890
Hexachlorocyclopentadiene, V, 425, 893
n-HEXADECANE, II, 320
Hexadecanedioic acid, V, 538
n-Hexadecyl fluoride, IV, 527
n-HEXADECYL IODIDE, II, 320, 322
1,3-HEXADIEN-3-OL, V, 608
Hexaethylbenzene, V, 600
Hexaethyl-2,4-cyclohexadienone, V, 600
Hexaethylphosphorous triamide, V, 360
1,1,1,10,10,10-Hexafluorodecane, V, 1084
1,1,1,6,6,6-Hexafluoro-3-hexene, V, 1084
Hexafluorophosphoric acid, V, 133
HEXAHYDROGALLIC ACID, V, 591
HEXAHYDROGALLIC ACID TRI-ACETATE, V, 591
HEXAHYDRO-1,3,5-TRIPROPIONYL-

s-TRIAZINE, IV, 518
HEXAHYDROXYBENZENE, V, 595
n-HEXALDEHYDE, II, 323
Hexaldehyde acetal, II, 324
HEXAMETHYLBENZENE, II, 248; IV, 520; V, 598
2,3,4,5,6,6-HEXAMETHYL-2,4-CY-CLOHEXADIEN-1-ONE, V, 598
Hexamethylene bromide, III, 228
HEXAMETHYLENE CHLOROHY-DRIN, III, 446
Hexamethylenediamine, IV, 522
HEXAMETHYLENEDIAMMONIUM CHLORIDE, IV, 522
HEXAMETHYLENE DIISOCY-ANATE, IV, 521
HEXAMETHYLENE GLYCOL, II, 325; III, 446
Hexamethylenetetramine, III, 811; IV, 690, 866, 918; V, 121, 668
HEXAMETHYLPHOSPHOROUS TRIAMIDE, V, 358, 360, 602
Hexane, IV, 610
1,6-Hexanediol, IV, 323
2,5-HEXANEDIOL, 2,5-DIMETHYL-, V, 1026
2,4-Hexanedione, V, 851
HEXAPHENYLBENZENE, V, 604
1,3,5-HEXATRIENE, V, 608, 611
1-Hexene, IV, 544
4-Hexen-1-yn-3-ol, IV, 794
n-Hexylacetylene, IV, 119
n-HEXYL ALCOHOL, I, 306; IV, 527
N-Hexylallenimine, V, 544
N-Hexylaniline, IV, 284
n-Hexyl bromide, IV, 525, 726
n-Hexyl diazoacetate, IV, 426
9-n-Hexylfluorene, IV, 625
n-HEXYL FLUORIDE, IV, 525
1-Hexyne, IV, 802
Hinsberg separation, III, 161
HIPPURIC ACID, II, 55, 328, 490; V, 946
l-HISTIDINE MONOHYDROCHLO-RIDE, II, 330

Hofmann decomposition, **V**, 886
Hofmann elimination, **V**, 316
HOMOPHTHALIC ACID, **III**, **449**; **V**, **612**
HOMOPHTHALIC ANHYDRIDE, **III**, **450**
Homophthalimide, **III**, 175
HOMOVERATRIC ACID, **II**, **333**
"HTH" (Calcium hypochlorite), **II**, 429
Hydantoin, **V**, 627
HYDRACRYLIC ACID, **I**, **321**
HYDRACRYLIC ACID, 3,3-DI-PHENYL, ETHYL ESTER, **V**, **564**
Hydrazine, **III**, 97, 153, 352, 357, 375, 404, 514; **V**, 39, 281, 899, 929
Hydrazine hydrate, **I**, 450; **II**, 85; **IV**, 377, 411, 413, 510, 537, 819; **V**, 30, 533, 1055, 1057
asym-Hydrazines, **V**, 44
HYDRAZINE SULFATE, **I**, **309**; **II**, 208, 228, 395, 497; **III**, 75, 656; **IV**, 274, 351
o-HYDRAZINOBENZOIC ACID HYDROCHLORIDE, **III**, **475**
2,2'-Hydrazo-bis-isobutyronitrile, **IV**, **275**
Hydrazoic acid, **IV**, 77; **V**, 273
α-HYDRINDONE, **II**, **336**
Hydriodic acid, **II**, 490; **III**, 476, 588; **V**, 843
HYDROBROMIC ACID, **I**, **26**, 117, 131, 135, 150; **II**, 30, 91, 132; **III**, 136, 185, 263, 312, 692, 754, 815; **IV**, 681
HYDROCHLORIDE OF 3-BROMO-4-AMINOTOLUENE, **I**, **112**
HYDROCINNAMIC ACID, **I**, **311**
Hydrocinnamoyl chloride, **III**, 120
Hydrofluoric acid, **II**, 188, 226, 295, 299
α-HYDROFORMAMINE CYANIDE, **I**, **355**
Hydrogenation, See Index No. 3, Type of Reaction; Reduction
HYDROGEN BROMIDE, **II**, 246, 338; **III**, 12, 13, 227, 433, 576
HYDROGEN CHLORIDE, **I**, **293**

Hydrogen cyanide, **IV**, 274, 393
HYDROGEN CYANIDE (ANHYDROUS), **I**, **314**
HYDROGEN CYANIDE TETRAMER, **V**, **344**
Hydrogen fluoride, anhydrous, **V**, 66, 136, 138, 332, 480
Hydrogen iodide, **IV**, 324
Hydrogen peroxide, **I**, 149; **II**, 44, 333, 586; **III**, 87, 217, 262, 263, 619, 759; **IV**, 317, 552, 612, 655, 704, 759, 911; **V**, 112, 367, 598, 647, 806, 818, 914, 1007, 1026, 1028. See also Index No. 3, Type of Reaction; Oxidation
Hydrogen selenide, **IV**, 359
Hydrogen sulfide, **I**, 53, 512; **II**, 28, 331, 548, 573, 610; **III**, 42, 67, 116, 242, 438, 460, 609; **IV**, 25, 27, 28, 244, 502, 669, 924, 927, 929, 967; **V**, 1046
Hydrolysis Reactions, See Index No. 3, Type of Reaction
Hydroquinone, **I**, 482; **II**, 553; **III**, 31, 33, 146, 173, 312, 452, 499, 576, 854; **IV**, 187, 311, 447, 670, 778; **V**, 50
HYDROQUINONE DIACETATE, **III**, 280, **452**
o-Hydroxyacetophenone, **III**, 387; **IV**, 478
2-Hydroxy-3-alkylisocarbostyrils, **V**, 625
m-HYDROXYBENZALDEHYDE, **III**, **453**, **564**
p-Hydroxybenzaldehyde, **V**, 627
5-(*p*-HYDROXYBENZAL)HYDANTOIN, **V**, **627**
p-HYDROXYBENZOIC ACID, **II**, **341**
6-Hydroxybicyclo[3.1.0]hexane, **V**, 861
9-Hydroxybicyclo[6.1.0]nonane, **V**, 861
8-Hydroxybicyclo[5.1.0]octane, **V**, 861
10-Hydroxy-10,9-borazarophenanthrene, **V**, 727
4-HYDROXY-1-BUTANESULFONIC ACID SULTONE, **IV**, **529**
2-Hydroxy-3-butylisocarbostyril, **V**, 626
2-Hydroxy-3-*t*-butylisocarbostyril, **V**, 626

β-Hydroxy-β-(o-carboxyphenyl)propionic acid lactone, **IV**, 138
24-Hydroxycholesterol, **IV**, 200
25-Hydroxycholesterol, **IV**, 200
2-HYDROXYCINCHONINIC ACID, **III**, 456
3-HYDROXYCINCHONNINIC ACID, **V**, 636
4-Hydroxycrotononitrile, **V**, 614
2-Hydroxycyclodecanone, **IV**, 840
2-Hydroxycyclononanone, **IV**, 839
o-HYDROXYDIBENZOYL-METHANE, **IV**, 479
2-HYDROXY-3,5-DIIODOBENZOIC ACID, **II**, 343
4-Hydroxy-3,5-diiodobenzoic acid, **II**, 344
2-Hydroxy-3-ethylisocarbostyril, **V**, 626
β-(2-HYDROXYETHYLMER-CAPTO)PROPIONITRILE, **III**, 458
β-HYDROXYETHYL METHYL SULF-IDE, **II**, 136, 345
β-Hydroxyethylphthalimide, **IV**, 107
3β-HYDROXYETIENIC ACID, **V**, 8
Hydroxyformoxystearic acids, **IV**, 318
3-HYDROXYGLUTARONITRILE, **V**, 614
HYDROXYHYDROQUINONE TRI-ACETATE, **I**, 317, 360
2-HYDROXYISOPHTHALIC ACID, **V**, 617
2-Hydroxy-3-isopropylisocarbostyril, **V**, 626
HYDROXYLAMINE HYDROCHLOR-IDE, **I**, 318, 328
Hydroxylamine hydrochloride, **II**, 67, 70, 313, 622; **III**, 91, 664, 690; **V**, 645
Hydroxylamine-O-sulfonic acid, **V**, 43, 897
Hydroxylammonium chloride, **IV**, 230, 873
Hydroxylammonium sulfate, **I**, 222
2-Hydroxy-3-methoxybenzaldehyde, **III**, 759
2-Hydroxy-3-methylbenzoic acid, **V**, 617
2-HYDROXYMETHYLENECYCLO-HEXANONE, **IV**, 536, 537

HYDROXYMETHYLFERROCENE, **V**, 621
4(5)-HYDROXYMETHYLIMIDAZOLE HYDROCHLORIDE, **III**, 460
2-HYDROXY-3-METHYLISOCAR-BOSTYRIL, **V**, 623
3-Hydroxy-2-methylpropiophenone, **V**, 624
2-HYDROXY-1-NAPHTHALDE-HYDE, **III**, 463
3-Hydroxy-2-naphthoic acid, **III**, 79
2-HYDROXY-1,4-NAPHTHOQUIN-ONE, **III**, 465
1-(2-Hydroxy-1-naphthyl)urea, **IV**, 53
6-HYDROXYNICOTINIC ACID, **IV**, 532
2-HYDROXY-5-NITROBENZYL CHLORIDE, **III**, 468
18-HYDROXY-18,20β-OXIDO-5-PREG-NENE, **V**, 693
5-HYDROXYPENTANAL, **III**, 470
α-HYDROXYPHENAZINE, **III**, 754
p-Hydroxyphenylarsonic acid, **I**, 75
p-HYDROXYPHENYLPYRUVIC ACID, **V**, 627
β-HYDROXYPROPIONIC ACID, **I**, 321
2-Hydroxy-3-propylisocarbostyril, **V**, 626
3-HYDROXYPYRENE, **V**, 632
3-HYDROXYQUINOLINE, **V**, 635
3-HYDROXYTETRAHYDROFURAN, **IV**, 534
2-HYDROXYTHIOPHENE, **V**, 642
3-Hydroxy-2,2,4-trimethyl-3-pentenoic acid β-lactone, **V**, 456
HYDROXYUREA, **V**, 645
δ-Hydroxyvaleraldehyde, **IV**, 500
Hypophosphorous acid, **III**, 296; **IV**, 948

IMIDAZOLE, **III**, 471
IMIDAZOLE, 1,1'-CARBONYLDI-, **V**, 201
Imidazolium chloride, **V**, 203
3-Imino-1-arylsulfonylpyrazolidines, **V**, 42
3-IMINO-1-(p-TOLYLSULFONYL)-PYRAZOLIDINE, **V**, 40
1,2-Indanediol monoformate, **V**, 647

2-INDANONE, V, **647**
INDAZOLE, III, **475**; IV, **536**; V, **650**
2*H*-INDAZOLE, 2-PHENYL-, V, **941**
INDAZOLONE, III, **476**
Indene, II, 336; III, 449; IV, 475
INDENE, 6,6-DIMETHOXY-3-
    METHYL-2-CARBETHOXY-
    , V, **550**
INDOLE, III, **479**; IV, 540
INDOLE-3-ACETIC ACID, V, **654**
INDOLE-3-ALDEHYDE, IV, **539**
INDOLE-3-CARBONITRILE, V, **656**
Indole-3-carboxaldehyde, V, 656
INDOLE-2-CARBOXYLIC ACID,
    ETHYL ESTER, V, **567**
Indole, conversion to indole-3-acetic acid,
    V, 654, 769
INDOLE, 1-METHYL-, V, **769**
Iodination Reactions, See Index No. 3,
    Type of Reaction; Halogena-
    tion
Iodine, I, 126, 224, 323, 325, 326, 345; II,
    73, 322, 344, 347, 357, 399; III,
    116, 138, 329, 796; IV, 207, 209,
    469, 545, 548, 601, 749, 870, 934
    recovery from sodium iodide, II, 348
Iodine chloride, IV, 978
IODINE MONOCHLORIDE, II, 196,
    197, 343, **344**, 349
p-Iodoacetophenone, IV, 112
p-IODOANILINE, II, **347**
5-IODOANTHRANILIC ACID, II, **349**,
    353
p-Iodobenzaldehyde, V, 142
IODOBENZENE, I, **323**, 545; II, **351**;
    III, **482**, 485; V, 660, 665
IODOBENZENE DICHLORIDE, III,
    482, 483, 486
m-IODOBENZOIC ACID, II, **353**
p-IODOBENZOIC ACID, I, **325**
IODOCYCLOHEXANE, IV, 324, **543**
2-IODOETHYL BENZOATE, IV, **84**, 85
Iodoform, I, 358
2-Iodohexane, IV, 544
p-Iodomandelic acid, IV, **112**
6-Iodo-3-methylbenzaldehyde, V, 142

2-Iodo-4-methylbenzaldehyde, V, 142
2-IODO-3-NITROBENZOIC ACID, I,
    126
2-Iodo-5-nitrothiophene, IV, 547
o-IODOPHENOL, I, **326**
p-IODOPHENOL, II, **355**
p-IODOPHENYL ISOTHIOCYAN-
    ATE, I, **449**
Iodosoarenes, V, 659
IODOSOBENZENE, III, 355, **483**, 484,
    485; V, **658**
IODOSOBENZENE DIACETATE, V,
    660
4-Iodosobiphenyl, V, 659
4-Iodosobiphenyl diacetate, V, 662
o-Iodosophenetole, V, 659
o-Iodosophenetole diacetate, V, 662
m-Iodosotoluene, V, 659
o-Iodosotoluene, V, 659
p-Iodosotoluene, V, 659
m-Iodosotoluene diacetate, V, 662
o-Iodosotoluene diacetate, V, 662
p-Iodosotoluene diacetate, V, 662
2-Iodoso-*m*-xylene, V, 659
4-Iodoso-*m*-xylene, V, 659
2-Iodoso-*p*-xylene, V, 659
2-Iodoso-*m*-xylene diacetate, V, 662
4-Iodoso-*m*-xylene diacetate, V, 662
2-Iodoso-*p*-xylene diacetate, V, 662
N-IODOSUCCINIMIDE, V, **663**
2-IODOTHIOPHENE, II, **357**; IV, 545,
    667
4-IODOVERATROLE, IV, **547**
Iodoxyarenes, V, 667
IODOXYBENZENE, III, 355, 484, 485;
    V, **665**
o-Iodoxybenzoic acid, V, 667
4-Iodoxybiphenyl, V, 667
o-Iodoxyphenetole, V, 667
m-Iodoxytoluene, V, 667
o-Iodoxytoluene, V, 667
p-Iodoxytoluene, V, 667
2-Iodoxy-*m*-xylene, V, 667
4-Iodoxy-*m*-xylene, V, 667
2-Iodoxy-*p*-xylene, V, 667
Ion-exchange resins, IV, 39, 529

Iron, **II**, 160, 472
IRON[(CYANOMETHYL)CYCLO-
    PENTADIENYL]CYCLO-
    PENTADIENYL-, **V, 578**
IRON, CYCLOPENTADIENYL[(DI-
    METHYLAMINOETHYL)-
    CYCLOPENTADIENYL]-,
    METHIODIDE, **V, 434**
IRON, CYCLOPENTADIENYL-
    [(HYDROXYMETHYL)-
    CYCLOPENTADIENYL]-,
    V, 621
Iron filings, **I**, 304; **III**, 138; **IV**, 225, 240
Iron powder, **I**, 123; **IV**, 114, 474, 573; **V**,
    347, 590
Iron selenide, **IV**, 360
Iron sulfide, **IV**, 928
ISATIN, **I**, 327; **III**, 456; **V**, 636
ISATOIC ANHYDRIDE, **III, 488**
Isoamylacetylene, **IV**, 119
Isoamyl alcohol, **I**, 28, 500; **II**, 535
N-Isoamylaniline, **IV**, 284, **422**
ISOAMYL BROMIDE, **I**, 27; **III**, 601
Isoamyl iodide, **II, 402**
Isoamyl nitrite, **II, 109**
1(3H)-ISOBENZOFURANONE, 3-
    BROMO-, **V, 145**
Isobutene, **V**, 673
Isobutenylbenzene, **V**, 472
Isobutyl alcohol, **II**, 358; **IV**, 324
N-Isobutylaniline, **IV**, 284
ISOBUTYL BROMIDE, **II, 358**
Isobutylene, **IV**, 261, 418
Isobutyl iodide, **II, 402**; **IV**, 324, 726
5-Isobutyl-1,2,3,4,5-pentachlorocyclopen-
    tadiene, **V**, 896
ISOBUTYRAMIDE, **III, 490**, 493
Isobutyranilide, **IV**, 349
Isobutyric acid, **III**, 490; **IV**, 348
Isobutyroin, **II**, 115
ISOBUTYRONITRILE, **III, 493**
ISOBUTYRYL CHLORIDE, **III, 490**
Isobutyryl fluoride, **V**, 69
Isocaproic acid, **III**, 523
Isocaproyl chloride, **IV**, 186

ISOCARBOSTYRIL, 2-HYDROXY-3-
    METHYL-, **V, 623**
Isocyanates, acyl, **V**, 205
ISOCYANIC ACID, ANHYDRIDE
    WITH CHLOROACETIC
    ACID, **V, 204**
ISOCYANIC ACID, ANHYDRIDE
    WITH CHLOROSULFONIC
    ACID, **V, 226**
Isocyanides, **V**, 300, 774, 1060
ISODEHYDROACETIC ACID, **IV**,
    337, **549**
ISODURENE, **II, 360**; **V**, 601
Isodurenol, **V**, 601
ISOINDOLINE, **V, 406**
2-ISOINDOLINEACETIC ACID, α-,
    BENZYL-1,3-DIOXO-, L-, **V,
    973**
ISOINDOLINE, 2-p-TOLYLSUL-
    FONYL-, **V, 1064**
dl-ISOLEUCINE, **III, 495**
Isonicotinic acid, **IV**, 89
ISONITROSOACETANILIDE, **I, 327**
ISONITROSOACETO-p-TOLUIDINE,
    I, 330
ISONITROSOPROPIOPHENONE, **II,
    363**; **III**, 20
Isophorone, **IV, 552**
ISOPHORONE OXIDE, **IV**, 552, 957
ISOPHTHALALDEHYDE, **V, 668**
ISOPHTHALIC ACID, 2-HYDROXY-,
    V, 617
Isoprene, **III**, 499; **V**, 787
ISOPRENE CYCLIC SULFONE, **III,
    499**
Isopropyl acetate, **V**, 566
Isopropyl alcohol, **I**, 37; **II**, 71; **III**, 173,
    245, 405; **IV**, 956
    anhydrous, **II**, 365
N-Isopropylallenimine, **V, 544**
2-ISOPROPYLAMINOETHANOL, **III,
    501**
N-Isopropylaniline, **IV**, 147
ISOPROPYL BROMIDE, **I**, 37; **II, 359**,
    366, 406

Isopropyl ether, III, 12, 13
Isopropyl β-hydroxy-β,β-diphenylpropion-
    ate, V, 566
Isopropyl hypochlorite, V, 186
(+)- and (−)-α-(ISOPROPYLIDENE-
    AMINOOXY)PROPIONIC
    ACID, V, 1033
(−)-EPHEDRIN SALTS, V, 1033
DL-α-(ISOPROPYLIDENEAMI-
    NOOXY)-PROPIONIC
    ACID, V, 1032
dl-ISOPROPYLIDENEGLYCEROL,
    III, 502
Isopropyl iodide, IV, 322, 325
Isopropyl isocyanide, V, 302
ISOPROPYL LACTATE, II, 365
Isopropylmagnesium chloride, V, 211, 212
Isopropyl nitrite, III, 192
5-Isopropyl-1,2,3,4,5-pentachlorocyclopen-
    tadiene, V, 896
p-Isopropylphenylacetamide, IV, 762
p-ISOPROPYLPHENYL ISOTHIO-
    CYANATE, I, 449
4-(4-p-Isopropylphenyl)-5-methyl-m-diox-
    ane, IV, 787
ISOPROPYL THIOCYANATE, II, 366
m-ISOPROPYLTOLUENE, V, 332
N-Isopropyl-m-toluidine, II, 291
Isoquinoline, IV, 642
[3(2H)-ISOQUINOLONE, 1,4-DIHY-
    DRO-1-THIOXO-], V, 1051
3-ISOQUINUCLIDONE, V, 670
Isosafrole, IV, 787
ISOTHIOCYANIC ACID, p-CHLORO-
    PHENYL ESTER, V, 223
Isovaleric acid, III, 848
β-ISOVALEROLACTAM, V, 673
β-ISOVALEROLACTAM-N-SUL-
    FONYL CHLORIDE, V, 673
Isovaleryl fluoride, V, 69
ISOXAZOLIDINE, 2,3,5-TRIPHENYL-
    V, 1124
Isoxazolidines, V, 1126
ITACONIC ACID, II, 368; IV, 554, 672
ITACONIC ANHYDRIDE, II, 140, 368

ITACONYL CHLORIDE, IV, 554
Ivory waste, I, 371

J acid, II, 149
JULOLIDINE, III, 504

Kerosene, III, 71
Ketals, V, 305
KETENE, I, 330; III, 164, 508; V, 679,
    682, 683
Ketene S,N-acetals, V, 783
KETENE BIS(2-METHOXY-
    ETHYL)ACETAL, V, 684
KETENE DIETHYLACETAL, III, 506
KETENE DIMER, III, 10, 508
KETENE DI(2-METHOXY-
    ETHYL)ACETAL, V, 684
Ketimines, V, 520
β-Ketobutyraldehyde dimethyl acetal, IV,
    559, 651
γ-Ketocapric acid, IV, 432
α-KETOGLUTARIC ACID, III, 510; V,
    687
6-KETOHENDECANEDIOIC ACID,
    IV, 510, 555
2-KETOHEXAMETHYLENIMINE, II,
    28, 371; IV, 39
β-KETOISOÖCTALDEHYDE DI-
    METHYL ACETAL, IV, 558
KETONE, CYCLOHEXYL METHYL,
    V, 775
Ketones, V, 217, 244, 590, 768, 868, 936
Ketones, α-amino, V, 915
DL-KETOPINIC ACID, V, 689
Kieselguhr, IV, 769, 940
Kolbe electrolysis, V, 445, 463
Kröhnke reaction, V, 828
KRYPTOPYRROLE, III, 513

LACTAMIDE, III, 516
β-Lactams, V, 230, 678
Lactic acid, II, 365; III, 46, 47; V, 656
18,20-LACTONE OF 3β-ACETOXY-
    20β-HYDROXY-5-PREG-
    NENE-18-OIC ACID, V, 692

Lathosterol, **IV**, 200
Lauraldehyde, **V**, 364
Lauric acid, **IV**, 561, 977
LAURONE, **IV**, 560
Lauroyl chloride, **III**, 846; **IV**, 560, **561**
LAURYL ALCOHOL, **I**, 29; **II**, 246, **372**
LAURYL BROMIDE, **I**, 29; **II**, 246, 292; **III**, 364
LAURYL MERCAPTAN, **III**, **363**
LAURYLMETHYLAMINE, **IV**, **564**
LAURYL p-TOLUENESULFONATE, **III**, **366**
Lead acetate, **III**, 68, 438; **V**, 880
Lead acetate trihydrate, **IV**, 173
Lead dioxide, **III**, 753; **V**, 617
Lead electrodes, **III**, 60
Lead hydroxide, **II**, 28; **III**, **68**
Lead nitrate, **I**, 447, 448
Lead oxide, **I**, 22
Lead oxide (red lead), **III**, 4, 5
Lead sulfide, **IV**, 173
Lead tetraacetate, **III**, 5; **IV**, 124; **V**, 692, 723
Lead thiocyanate, **V**, 1052
LEPIDINE, **III**, **519**
dl-LEUCINE, **III**, **523**
LEVOPIMARIC ACID, **V**, **699**
LEVULINIC ACID, **I**, 335; **III**, 328
Ligroin, **III**, 159
Lime-nitrogen, **I**, 203
LINOLEIC ACID, **III**, **526**
LINOLENIC ACID, **III**, **531**
Linseed oil, **III**, 531
Litharge, **II**, 28, 230
Lithium, **II**, 518; **III**, 413, 757; **IV**, 887; **V**, 564, 1092, 1120
Lithium aluminum tri-t-butoxyhydride, **V**, 692
Lithium aluminum hydride, **IV**, 271, 340, 355, 474, 564, 794, 834; **V**, 83, 86, 175, 294, 303, 304, 510, 512, 970, 976, 1022
Lithium amide, **V**, 564
Lithium chloride, **IV**, 163
LITHIUM 2,6-DIMETHYLPHEN-OXIDE, **V**, **1092**

Lithium ethoxide, **V**, 500
Lithium hydride, **V**, 775
Lithium iodide, **V**, 77
dl-LYSINE HYDROCHLORIDES, **II**, **374**

Magnesium, **V**, 912, 998, 1141. See also Index No. 3. Type of Reaction; Grignard Reactions
Magnesium ethoxide, **IV**, 11
Magnesium oxide, **IV**, 419, 854
Magnesium sulfate, **IV**, 467, 469
MALEANILIC ACID, **V**, **944**
Maleic acid, **I**, 64
Maleic anhydride, **III**, 109, 422, 807; **IV**, 313, 766, 890, 965; **V**, 459, 944
Maleic anhydride-anthracene adduct, **IV**, 965
MALEIMIDE, N-PHENYL-, **V**, **944**
MALEONITRILE, DIAMINO-, **V**, **344**
Malic acid, **IV**, 201
l-Malic acid, **II**, 506
  recovery of, **II**, 508
MALONIC ACID, **II**, 376; **III**, 425, 783; **IV**, 261, 263, 327, 732
MALONIC ACID, ACETAMIDO-, DIETHYL ESTER, **V**, **373**
  AMINO-, DIETHYL ESTER, HY-DROCHLORIDE, **V**, **376**
  BIS(HYDROXYMETHYL)-, DI-ETHYL ESTER, **V**, **381**
  ETHYLHYDROXY-, DIETHYL ESTER, BENZOATE, **V**, **379**
Malonic ester (see Diethyl malonate)
  ethoxymagnesium derivative, solution of, **IV**, **294**
MALONONITRILE, **II**, 379; **III**, 535; **IV**, 516, 877; **V**, 32, 34, 1013, 1014
MALONONITRILE, AMINO-, p-TOL-UENESULFONATE, **V**, **32**
MALONYL DICHLORIDE, **IV**, **263**, **264**
MANDELAMIDE, **III**, **536**
MANDELIC ACID, **I**, 12, 241, **336**; **III**, 536, **538**; **IV**, 169

Manganese dioxide, **IV**, 825
D-MANNOSE, **III, 541**
Melamine, **IV**, 79
Menthol, **I**, 340; **III**, 544
l-MENTHONE, **I, 340**
l-MENTHOXYACETIC ACID, **III, 544,** 547
l-MENTHOXYACETYL CHLORIDE, **III, 547**
2-MERCAPTO-4-AMINO-5-CARBETH-OXYPYRIMIDINE, **IV, 566**
2-MERCAPTOBENZIMIDAZOLE, **IV, 569**
2-Mercaptobenzoxazole, **IV**, 570
2-Mercaptoethanol, **III**, 458
2-MERCAPTO-4-HYDROXY-5-CY-ANOPYRIMIDINE, **IV, 566**
2-MERCAPTOPYRIMIDINE, **V, 703**
Mercuric acetate, **I**, 56, 161; **III**, 774, 812; **IV**, 896, 977
Mercuric chloride, **I**, 158, 459, 519; **II**, 330, 433, 499, 599; **III**, 48, 786; **IV**, 204; **V**, 362
Mercuric oxide, **II**, 357, 496, 540; **III**, 351, 356, 854; **IV**, 13, 378, 548; **V**, 126, 1024
Mercuric sulfate, **I**, 229; **V**, 320, 321
Mercury, **I**, 554; **II**, 551; **IV**, 509, 896
MERCURY DI-β-NAPHTHYL, **II, 381**
Mercury valve, **III**, 189
MESACONIC ACID, **II, 382**
MESITALDEHYDE, **III, 549**; **V, 49**
MESITOIC ACID, **III, 553**; **V, 706**
Mesitol, **V**, 601
MESITOYL CHLORIDE, **III, 552, 555**
MESITYLACETIC ACID, **III, 557**
Mesitylacetonitrile, **III**, 558
MESITYLENE, **I**, 231, **341**; **II**, 95, 449; **III**, 549, 557; **V**, 49, 601, 706
Mesityl isocyanide, **V**, 1063
MESITYL OXIDE, **I**, 196, **345**; **II**, 201; **III**, 244, 303; **V**, 1108
Mesquite gum, **I**, 67
Methacrolein, **IV**, 312
METHACRYLAMIDE, **III, 560**
Methacrylic acid, **III**, 32

Methallylbenzene, **V**, 472
Methallyl chloride, **IV**, 702; **V**, 767
5-Methallyl-1,2,3,4,5-pentachlorocyclopen-tadiene, **V**, 896
METHANE, (*p*-BROMOPHENYL)-PHENYL-, **V, 130**
METHANE, DIMESITYL-, **V, 422**
METHANESULFINYL CHLORIDE, **V, 709, 710**
Methanesulfonic acid, **IV**, 571; **V**, 904
METHANESULFONYL CHLORIDE, **IV, 571**
dl-METHIONINE, **II, 384**
METHONE, **II**, 200; **IV**, 346
METHOXYACETONITRILE, **II, 387**; **III, 563**
p-METHOXYACETOPHENONE, **I, 111**
ω-METHOXYACETOPHENONE, **III, 562**
Methoxyacetylene, **IV, 406**
m-METHOXYBENZALDEHYDE, **III, 564**
o-Methoxybenzaldehyde, **IV**, 573
N-(4-Methoxybenzyl)-acetamide, **V**, 75
p-Methoxybromobenzene, **V**, 145
4-Methoxy-3-buten-2-one, **IV**, 651
1-(2-METHOXYCARBONYL-PHENYL)-PYRROLE, **V, 716**
p-Methoxycinnamic acid, **IV**, 733
2-METHOXYCYCLOOCTANONE OX-IME, **V, 267**
p-Methoxy-β,β-difluorostyrene, **V**, 392
7-Methoxy-3,4-dihydro-2-naphthoic acid, **III**, 302
2-METHOXYDIPHENYL ETHER, **III, 566**
5-(3-METHOXY-4-HYDROXYBEN-ZAL)-CREATININE, **III, 587**
6-Methoxylepidine, **III**, 521
1-METHOXY-2-METHYLPYRIDI-NIUM METHYL SULFATE, **V, 270**
6-METHOXY-2-NAPHTHOL, **V, 918**
3-METHOXY-5-NITROPHENOL, **I, 405**
6-METHOXY-8-NITROQUINOLINE, **III, 568**

p-Methoxyphenacylamine, hydrochloride, **V, 915**

α-METHOXYPHENAZINE, **III, 753**

p-Methoxyphenylacetamide, **IV, 762**

o-METHOXYPHENYLACETONE, **IV, 573**

p-METHOXYPHENYLACETONI-
TRILE, **IV, 576**

2-(*m*-Methoxyphenyl)cycloheptanone, **IV, 783**

2-(*o*-Methoxyphenyl)cycloheptanone, **IV, 783**

2-(*p*-Methoxyphenyl)cycloheptanone, **IV, 783**

4-(*p*-Methoxyphenyl)-*m*-dioxane, **IV, 786**

p-Methoxyphenylglyoxal, **V, 940**

p-Methoxyphenyl isocyanide, **V, 1063**

o-Methoxyphenylmagnesium bromide, **V, 46**

1-(*o*-METHOXYPHENYL)-2-NITRO-1-
PROPENE, **IV, 573**

1-(*p*-METHOXYPHENYL)-5-PHENYL-
1,3,5-PENTANETRIONE, **V, 718, 721**

1-(*p*-Methoxyphenyl)-3-phenyl-2-propa-
none, **IV, 177**

2-(*p*-METHOXYPHENYL)-6-PHENYL-
4-PYRONE, **V, 721**

p-Methoxyphenylurea, **IV, 51, 53**

Methoxyquinone, **IV, 153**

o-Methoxystyrene, **IV, 733**

p-Methoxystyrene, **IV, 733**

4'-Methoxy-5,6,7,8-tetrahydroflavone, **V, 723**

5-Methoxy-2*a*-thiohomophthalimide, **V, 1053**

p-METHYLACETOPHENONE, **I, 111; III, 791**

4-METHYL-7-ACETOXYCOUMARIN, **III, 283**

METHYL *p*-ACETYLBENZOATE, **IV, 579**

Methylacetylene, **V, 1043**

Methyl acrylate, **III, 33, 146, 576, 774; IV, 669**

1-Methyladamantane, **V, 19**

Methyl-1-adamantanecarboxylate, **V, 21**

Methylal, **III, 468**

N-Methylallenimine, **V, 544, 760**

N-Methylallylamine, **V, 760**

Methylamine, **II, 278, 397, 461; III, 244, 258, 328, 573, 599; IV, 603, 606, 943**

METHYLAMINE HYDROCHLOR-
IDE, **I, 347; IV, 607, 816**

1-METHYLAMINOANTHRAQUIN-
ONE, **III, 573, 575**

4'-Methyl-2-aminobenzophenone, **IV, 38**

1-METHYLAMINO-4-BROMOAN-
THRAQUINONE, **III, 575**

1-METHYL-2-AMINO-β-NAPHTHO-
THIAZOLINE, **III, 595**

N-Methylaminopyrimidine, **IV, 336**

METHYL *n*-AMYL KETONE, **I, 351; II, 317**

Methylaniline, **II, 460; III, 590; IV, 147, 422, 621, 916, 917**

Methylaniline hydrochloride, **IV, 621**

Methyl anisate, **V, 718**

Methyl anthranilate, **V, 716**

β-METHYLANTHRAQUINONE, **I, 353**

METHYL AZODICARBOXYLATE, **IV, 414**

Methyl benzalmalonate, **III, 378**

METHYL BENZENESULFINATE, **V, 723**

2-Methylbenzimidazole, **II, 66**

Methyl benzoate, **I, 372**

METHYL BENZOYLFORMATE, **I, 244**

o-METHYLBENZYL ACETATE, **IV, 582, 583**

o-METHYLBENZYL ALCOHOL, **IV, 582**

N-Methylbenzylamine, **IV, 605**

2-(α-Methylbenzyl)aziridine, **V, 86**

2-METHYLBENZYLDIMETHYL-
AMINE, **IV, 582, 585**

Methyl benzyl ether, **IV, 782**

2-Methylbenzylethyldimethylammonium
bromide, **IV, 582**

METHYL BENZYL KETONE, **II, 389**

p-METHYLBENZYLTRIMETHYLAM-

MONIUM BROMIDE, **V, 884**
2-Methylbenzyltrimethylammonium iodide, **IV,** 587
Methyl borate, **IV,** 68, 70
Methyl borate-methanol azeotrope, **IV,** 70
10-METHYL-10,9-BORAZAROPHEN-ANTHRENE, **V, 727**
Methyl bromide, **II,** 279, **280; IV,** 601
2-Methyl-4-bromobenzaldehyde, **V,** 141
2-Methyl-5-bromobenzaldehyde, **V,** 142
3-Methyl-4-bromobenzaldehyde, **V,** 142
N-Methyl-α-bromo-*n*-butyranilide, **IV,** 621
METHYL α-BROMO-β-METHOXY-PROPIONATE, **III, 774**
METHYL β-BROMOPROPIONATE, **III, 576**
METHYL 5-BROMOVALERATE, **III, 578**
METHYL BUTADIENOATE, **V, 734**
3-Methyl-3-butenamide-*N*-sulfonyl chloride, **V,** 673, 676
N-METHYLBUTYLAMINE, **V, 736,** 759, 760
N-Methyl-*t*-butylamine, **V,** 760
METHYL *n*-BUTYL KETONE, **I, 352**
2-Methylbutyric acid, **V,** 742
O-METHYLCAPROLACTIM, **IV, 588**
4-METHYLCARBOSTYRIL, **III,** 194, **580**
Methyl Cellosolve, **IV,** 78, 240, 506
Methyl chloride, **II,** 248, **251; III,** 696; **IV,** 126, 647
Methyl chloroacetate, **IV,** 649
N-Methyl-*p*-chloroaniline, **IV, 422**
2-Methyl-5-chlorobenzaldehyde, **V,** 142
Methyl chlorocarbonate, **II,** 596
Methyl chloroformate, **II,** 596; **IV,** 413
Methyl chlorothiolformate, **V,** 166
METHYL COUMALATE, **IV, 532**
3-METHYLCOUMARILIC ACID, **IV, 591**
4-METHYLCOUMARIN, **III, 581**
3-METHYLCOUMARONE, **IV, 590**
Methyl crotonate, **IV,** 631; **V,** 762
METHYL ω-CYANOPELARGONATE, **III, 584**

Methylcyclohexane, **III,** 182; **V,** 36
1-METHYLCYCLOHEXANECAR-BOXYLIC ACID, **V, 22, 739**
2-METHYL-1,3-CYCLOHEXANE-DIONE, **V, 486, 743**
2-Methylcyclohexanol, **IV,** 19, 164; **V, 739**
2-METHYLCYCLOHEXANONE, **IV,** 19, **162; V,** 187
1-Methylcyclohexene, **IV,** 614
2-METHYL-2-CYCLOHEXENONE, **IV, 162**
Methylcyclopentadiene, **IV,** 475
METHYL CYCLOPENTANECAR-BOXYLATE, **IV, 594**
1-Methylcyclopentanecarboxylic acid, **V,** 742
2-METHYLCYCLOPENTANE-1,3-DIONE, **V, 747**
2-METHYLCYCLOPENTANE-1,3,5-TRIONE, **V, 748;** HY-DRATE, **V, 748;** 5-SEMI-CARBAZONE, **V, 749**
METHYL CYCLOPROPYL KETONE, **IV, 597**
9-Methyl-1-decalone, **V,** 190
2-METHYL-2,5-DECANEDIOL, **IV,** 350, **601**
METHYL DESOXYCHOLATE, **III, 237**
Methyl diazoacetate, **IV,** 426
Methyl α,γ-di-(2-hexyl)-acetoacetate, **V,** 765
N-METHYL-3,4-DIHYDROXYPHEN-YLALANINE, **III, 586**
N-METHYL-2,3-DIMETHOXYBEN-ZYLAMINE, **IV, 603**
N-Methyl-3,4-dimethoxybenzylamine, **IV,** 603
METHYL 5,5-DIMETHOXY-3-METHYL-2,3-EPOXYPEN-TANOATE, **IV, 649**
Methyl 5,5-dimethoxy-3-phenyl-2,3-epoxy-pentanoate, **IV,** 651
Methyl 3,5-dinitrobenzoate, **V,** 798
N-Methyl-3,4-dioxymethylenebenzylamine, **IV,** 605
N-METHYL-1,2-DIPHENYLETHYL-

AMINE AND HYDRO-
CHLORIDE, **IV, 605**
2-Methyl-1,1-diphenylpropane, **V,** 525
Methyl disulfide, **V,** 709
2-Methyldodecanoic acid, **IV,** 608, 616,
618
2-Methyl-2-dodecenoic acid, **IV,** 618
trans-2-METHYL-2-DODECENOIC
ACID, **IV, 608**
2-Methyl-2-eicosenoic acid, **IV,** 611
METHYLENEAMINOACETONI-
TRILE, **I, 298, 355; II,** 310
METHYLENE BROMIDE, **I, 357**
Methylene chloride, **IV,** 155, 254, 425, 816,
950
3-METHYLENECYCLOBUTANE-1,2-
DICARBOXYLIC ACID,
DIMETHYL ESTER, **V, 459**
3-METHYLENECYCLOBUTANE-1,2-
DICARBOXYLIC ANHY-
DRIDE, **V, 459**
Methylenecyclobutanes, **V,** 462
METHYLENECYCLOHEXANE, **IV,**
614; V, 751
METHYLENECYCLOHEXANE
OXIDE, **V, 755**
2-Methylenecyclohexanone dimer, **IV,** 165
2-METHYLENEDODECANOIC
ACID, **IV, 616**
5-METHYLENE-2-HEXANONE, **V. 767**
METHYLENE IODIDE, **I, 358; V,** 855
Methylene-α-naphthylmethylamine, **IV,**
691
Methylenetriphenylphosphine, **V,** 751, 752
4-METHYLESCULETIN, **I, 360**
Methyl ethers, **V,** 248
2-METHYL-4-ETHOXALYLCYCLO-
PENTANE-1,3,5-TRIONE, **V,**
747
dl-METHYLETHYLACETIC ACID, **I,**
361
N-METHYLETHYLAMINE, **V,** 758
Methyl *p*-ethylbenzoate, **IV,** 579, **580**
5-Methyl-5-ethylhydantoin, **III,** 324
Methyl ethyl ketone, **II,** 205; **IV,** 84, 86,
94, 441, 794

1-METHYL-3-ETHYLOXINDOLE, **IV,**
620
9-METHYLFLUORENE, **IV, 623**
N-Methylformamide, **V,** 772
N-METHYLFORMANILIDE, **III,** 98.
99, **590; IV,** 916
Methyl formate, **III,** 33; **V,** 365
Methyl 2'-formyl-2-diphenylcarboxylate,
V, 495
2-Methylfuran, **IV,** 626
3-METHYLFURAN, **IV, 628**
5-METHYLFURFURAL, **II, 393**
5-METHYLFURFURYLDIMETHYL-
AMINE, **IV, 626**
3-METHYL-2-FUROIC ACID, **IV, 628**
α-METHYL-*d*-GLUCOSIDE, **I, 364**
β-Methylglutaraldehyde, **IV, 661**
β-METHYLGLUTARIC ACID, **III, 591**
β-METHYLGLUTARIC ANHY-
DRIDE, **IV, 630**
METHYLGLYOXAL-ω-PHENYLHY-
DRAZONE, **IV, 633**
3-METHYLHEPTANOIC ACID, **V, 762**
N-Methylheptylamine, **V,** 738
2-Methyl-2-hexacosenoic acid, **IV,** 611
4-METHYLHEXANONE-2, **I, 352**
5-METHYL-5-HEXEN-2-ONE, **V, 767**
N-Methylhexylamine, **V,** 738
METHYL-*n*-HEXYLCARBINOL, **I,**
366, 418
METHYL HOMOVERATRATE, **II, 333**
METHYLHYDRAZINE SULFATE, **II,**
395
Methylhydrazinium hydrogen sulfate, **V,**
44
METHYL HYDRAZODICARBOXYL-
ATE, **IV, 413**
Methyl hydrogen adipate, **III,** 578; **IV,** 556
Methyl hydrogen glutarate, **III,** 171
METHYL HYDROGEN HENDECANE-
DIOATE, **IV, 635**
Methyl hydrogen β-methylglutarate, **IV,**
632
Methyl hydrogen sebacate, **III,** 613; **V,** 464
Methyl hydrogen succinate, **III, 169**
4-METHYL-7-HYDROXYCOUMARIN,

III, 282
2-METHYL-4-HYDROXYQUINOL-
    INE, III, 593
4-METHYL-6-HYDROXYPYRIMIDINE,
    IV, 638
METHYLIMINODIACETIC ACID, II,
    397
1-METHYL-2-IMINO-β-NAPHTHO-
    THIAZOLINE, III, 595
2-Methyl-1-indanone, V, 623, 624
1-METHYLINDOLE, V, 769
2-METHYLINDOLE, III, 597
α-Methylindole-3-acetic acid, V, 656
METHYL IODIDE, II, 399; III, 27, 398;
    IV, 325, 326, 585, 642, 668, 836;
    V, 315, 743, 758
5-METHYL ISATIN, I, 330
N-Methylisobutylamine, V, 760
METHYL ISOBUTYL KETONE, I,
    352; III, 291
METHYL ISOCYANIDE, V, 301, 772,
    774
N-Methylisopropylamine, V, 760
METHYL ISOPROPYL CARBINOL,
    II, 406
METHYL ISOPROPYL KETONE, II,
    408
2-METHYL-4-ISOPROPYLPHENYL
    ISOTHIOCYANATE, I, 449
1-METHYLISOQUINOLINE, IV, 641
METHYL ISOTHIOCYANATE, III,
    599, 618
METHYL ISOTHIOUREA SULFATE,
    II, 345, 411; III, 75
METHYLISOUREA HYDROCHLOR-
    IDE, IV, 645; V, 966
METHYL 4-KETO-7-METHYLOC-
    TANOATE, III, 601
Methyl ketones, V, 777
Methyl 4-ketopentyl sulfoxide, V, 792
Methyl lactate, III, 47
N-Methyllauramide, IV, 564–565
Methyl lithium, V, 775, 777, 860
Methylmagnesium bromide, IV, 351, 601,
    771; V, 729, 1016
Methyl malonate, II, 596

α-METHYL MANNOSIDE, I, 371
α-Methyl-d-mannoside, III, 541
Methyl mercaptan, II, 346; III, 75
2-METHYLMERCAPTO-N-METHYL-
    Δ²-PYRROLINE, V, 780
2-METHYLMERCAPTO-N-METHYL-
    Δ¹-PYRROLINIUM IODIDE,
    V, 781
Methylmercuric iodide, II, 403
N-Methyl-p-methoxybenzylamine, IV, 605
N-Methyl-2-methoxyethylamine, V, 760
N-METHYL-(3-METHOXY-4-
    HYDROXYPHENYL)-
    ALANINE, III, 588
Methyl methoxymagnesium carbonate, V,
    436
METHYL β-METHYLBUTYL
    KETONE, I, 352
Methyl cis-2-methyl-2-dodecenoate, IV,
    610
Methyl trans-2-methyl-2-dodecenoate, IV,
    610
Methyl 2-methylenedodecanoate, IV, 610
METHYL 3-METHYL-2-FUROATE,
    IV, 628, 649
METHYL 4-METHYL-4-NITROVAL-
    ERATE, IV, 357, 652; V, 445
N-Methylmyristamide, IV, 565
METHYL MYRISTATE, III, 605
Methylmyristylamine, IV, 565
α-Methylnaphthalene, III, 766
N-METHYL-1-NAPHTHYLCYAN-
    AMIDE, III, 608, 609
METHYL β-NAPHTHYL ETHER, I, 59
Methyl β-naphthyl ketone, II, 428
1-METHYL-1-(1-NAPHTHYL)-2-THIO-
    UREA, III, 595, 609
METHYL NITRATE, II, 412, 512
Methyl nitrite, II, 363
METHYL m-NITROBENZOATE, I, 372,
    391
3-Methyl-1-nitrobutane, IV, 726
2-Methyl-1-nitropropane, IV, 726
3-METHYL-4-NITROPYRIDINE-1-
    OXIDE, IV, 654
2-Methyl-2-nitroso-1-indanone dimer, V,
    625

N-Methyl-N-nitrosoterephthalamide, **V,**
  351
4-METHYL-4-NITROVALERIC ACID,
  V, 445
3-Methylnonanoicnitrile, **IV,** 122
Methylnonylamine, **IV, 565**
Methyl oleate, **III,** 785
Methyl orthoformate, **IV,** 422
METHYL OXALATE, **I,** 264; **II, 414**
3-METHYLOXINDOLE, **IV, 657**
2-Methyl-2-(3'-oxobutyl)-1,3-cyclohexane-
  dione, **V,** 487
Methyl 10-oxocyclodec-2-ene-1-carboxy-
  late, **V,** 278
METHYL PALMITATE, **III, 605**
N-Methylpelargonamide, **IV,** 565
5-Methyl-1,2,3,4,5-pentachlorocyclopenta-
  diene, **V,** 896
1-Methyl-2,3-pentamethylenediaziridine, **V,**
  899
3-METHYL-1,5-PENTANEDIOL, **IV,**
  660, 677
3-METHYLPENTANE-2,4-DIONE, **V,**
  785, 1022
3-METHYLPENTANOIC ACID, **II, 416**
4-METHYL-2-PENTANONE, **I,** 352; **V,**
  452
N-Methylpentylamine, **V,** 738
3-Methyl-1-pentyn-3-ol, **IV,** 794
p-Methylphenacylamine, hydrochloride, **V,**
  915
p-Methylphenylacetamide, **IV, 762**
2-(o-Methylphenyl)cycloheptanone, **IV,**
  783
2-(p-Methylphenyl)cycloheptanone, **IV,**
  783
β-METHYL-β-PHENYL-α,α'-DICYAN-
  OGLUTARIMIDE, **IV, 662,**
  664
4-Methyl-4-phenyl-m-dioxane, **IV,** 787
5-Methyl-4-phenyl-m-dioxane, **IV,** 787
Methyl β-phenylethyl ketone, **II,** 391
β-METHYL-β-PHENYLGLUTARIC
  ACID, **IV, 664**
α-METHYL-α-PHENYLHYDRAZINE,

II, 418
1-METHYL-3-PHENYLINDANE, **IV,**
  665
p-Methylphenylmagnesium bromide, **V,**
  1058
m-Methylphenylmethylcarbinol, **III, 201**
3-Methyl-3-phenylpentanenitrile, **IV,** 97
3-Methyl-3-phenylpentanoic acid, **IV,** 97
3-METHYL-1-PHENYLPHOSPHACY-
  CLOPENTENE 1-OXIDE, **V,**
  787
3-METHYL-1-PHENYLPHOSPHO-
  LENE 1,1-DICHLORIDE, **V,**
  787
3-METHYL-1-PHENYLPHOSPHO-
  LENE OXIDE, **V,** 501, **787**
2-Methyl-6-phenyl-4-pyrone, **V,** 722
Methyl phenyl sulfide, **V,** 791
Methyl phenyl sulfone, **IV,** 675
METHYL PHENYL SULFOXIDE, **V,**
  791
6-METHYLPICOLINONITRILE, **V,**
  269
Methyl propiolate, **V,** 278
2-Methyl-5,6-pyrazinedicarboxylic acid,
  IV, 827
3-Methylpyridine, **IV,** 655
3-Methylpyridine-1-oxide, **IV,** 654, **655**
1-METHYL-2-PYRIDONE, **II, 419**
4-METHYLPYRIMIDINE, **V, 794**
N-METHYL-2-PYRROLIDINE
  THIONE, **V, 780,** 781
N-Methyl-2-pyrrolidinone, **V,** 780
METHYL PYRUVATE, **III, 610**
2-Methylquinoxaline, **IV, 826**
METHYL RED, **I, 374**
METHYL SEBACAMATE, **III, 584, 613**
Methyl stearate, **IV, 855**
α-Methylstyrene, **IV,** 666, 787
trans-Methylstyrylcarbinol, **IV, 773**
METHYLSUCCINIC ACID, **III, 615**
Methylsuccinic acid, disodium salt, **IV,**
  671
Methyl sulfate, **II,** 209, 361, 387, 396, 404,
  411, 419, 619; **III,** 565, 754, 800
2-(METHYLSULFINYL)ACETO-

PHENONE, **V**, 937, **939**

2-Methyl-1,1,4,4-tetrafluorobutadiene, **V**, 239

4-Methyl-1-tetralone, **IV**, 899

METHYL 2-THIENYL SULFIDE, **IV**, **667**

METHYL β-THIODIPROPIONATE, **IV**, **669**

4-Methyl-2α-thiohomophthalimide, **V**, 1053

5-Methyl-2α-thiohomophthalimide, **V**, 1053

1-METHYLTHIOL-3-PHTHALAMIDO-PROPANE-3,3-DICARBOX-YLIC ACID, **II**, **385**

3-METHYLTHIOPHENE, **IV**, **671**, 921

1-(p-Methylthiophenyl)-3-phenyl-2-propanone, **IV**, 177

6-Methyl-1α-thiophthalimide, **V**, 1053

METHYLTHIOUREA, **III**, **617**

N-Methyl-p-toluenesulfonamide, **IV**, 943

METHYL p-TOLUENESULFONATE, **I**, **146**

N-Methyl-m-toluidine, **IV**, 422

METHYL p-TOLYL SULFONE, **IV**, **674**

1-METHYL-3-p-TOLYLTRIAZENE, **V**, **797**

2-Methyl-2,5-undecanediol, **IV**, 351

6-METHYLURACIL, **II**, **422**

β-METHYL-δ-VALEROLACTONE, **IV**, **677**

Methyl vinyl ether, **IV**, 311, 312

Methyl vinyl ketone, **IV**, 312; **V**, 486, 869

Methyl violet, **IV**, 382

Michael reaction, **V**, 572, 871

Milk, **II**, 120

Mineral oil, **IV**, 509, 671

MONOBENZALPENTAERYTHRI-TOL, **IV**, **679**

MONOBROMOPENTAERYTHRITOL, **IV**, **681**

MONOCHLOROMETHYL ETHER, **I**, **377**

Monochlorourea, **IV**, 157, 158

N-MONO- AND N,N-DISUBSTI-TUTED UREAS AND THI-OUREAS, **V**, **801**

Monoethyl malonate, **IV**, 418

MONOPERPHTHALIC ACID, **III**, **619**; **V**, **805**

Monosodium glutamate, **I**, 287

N-Monosubstituted thioureas, **V**, 804

N-Monosubstituted ureas, **V**, 804

MONOVINYLACETYLENE, **IV**, **683**

Morland salt, **II**, **555**

Morpholine, condensation with cyclo-, hexanone, **V**, 808

MORPHOLINE, 4-(1-CYCLOHEX-ENYL)-, **V**, **808**

MORPHOLINE, 4-NITRO-, **V**, **839**

1-MORPHOLINO-1-CYCLOHEXENE, **V**, **808**, 533, 869

Mucic acid, **I**, 473

MUCOBROMIC ACID, **III**, **621**; **IV**, **688**, 844

MUCONIC ACID, **III**, **623**

MYRISTIC ACID, **I**, 379; **III**, **607**

Myristyl alcohol, **II**, **374**

Naphtha, **IV**, 415

β-NAPHTHALDEHYDE, **III**, **626**; **IV**, **690**

Naphthalene, **I**, 121; **II**, 524; **III**, 195; **IV**, 691, 698

NAPHTHALENE, 2-BROMO-, **V**, **142**

β-NAPHTHALENEDIAZONIUM CHLORIDEMERCURIC CHLORIDE COMPOUND, **II**, 381, **433**

2,3-NAPHTHALENEDICARBOXYLIC ACID, **V**, **810**

2,6-NAPHTHALENEDICARBOXYLIC ACID, **V**, **813**

1,6-NAPHTHALENEDIONE, 1,2,3,4,6,-7,8,8a-OCTAHYDRO-8a-METHYL-, **V**, **486**

Naphthalene-1,5-disulfonic acid, disodium salt, **IV**, 693

NAPHTHALENE-1,5-DISULFONYL CHLORIDE, **IV**, **693**, 695

1,5-NAPHTHALENEDITHIOL, **IV**, **695**

NAPHTHALENE, 1,2,3,4-TETRA-PHENYL-, **V**, **1037**

2(3*H*)-NAPHTHALENONE,
    4,4*a*,5,6,7,8-HEXAHYDRO-,
    V, 869
1,2-NAPHTHALIC ANHYDRIDE, **II,**
    423; V, 813
α-NAPHTHOIC ACID, **II, 425; V,** 708
β-NAPHTHOIC ACID, **II, 428; V,** 812
β-Naphthoic acids, **V,** 708
α-Naphthol, **I,** 49; **II,** 40; **IV,** 887
β-Naphthol, **I,** 59, 381, 411; **II,** 36; **III,**
    132, 463; **IV,** 37, 136; **V,** 142
2-Naphthol-3,6-disulfonic acid, **II,** 148
β-NAPHTHOL PHENYLAMINO-
    METHANE, **I,** 381
1-Naphthol-4-sulfonic acid, **II,** 149
2-Naphthol-6-sulfonic acid, **II,** 148
α-NAPHTHONITRILE, **III, 631**
β-Naphthonitrile, **III,** 626
1,2-NAPHTHOQUINONE, **II, 430**
1,4-NAPHTHOQUINONE, **I,** 383; **III,**
    310; **IV,** 698
1,2-NAPHTHOQUINONE-4-SUL-
    FONIC ACID SALTS, **III,**
    633
NAPHTHORESORCINOL, **III, 637**
β-Naphthoyl chloride, **III,** 629
Naphthylacetamide, **IV, 762**
β-Naphthylamine, **I,** 500; **II,** 432
2-(α-Naphthyl)aziridine, **V,** 86
4-(α-Naphthyl)-2-butanone, **V,** 769
Naphthylcyanamide, **IV,** 174
α-(β-Naphthyl)-ethylamine, **II, 505**
β-Naphthyl ethyl ether, **III,** 99; **IV,** 903
α-Naphthyl isocyanate, **IV,** 215, 557
2-Naphthyl isocyanide, **V,** 1063
α-NAPHTHYL ISOTHIOCYANATE,
    IV, 700
β-NAPHTHYL ISOTHIOCYANATE, **I,**
    449; IV, 701
β-Naphthylmercuric bromide, **II,** 434
β-NAPHTHYLMERCURIC CHLOR-
    IDE, **II, 432**
α-Naphthylphenylacetylene, **IV,** 378
N-α-Naphthylpiperidine, **V,** 817
N-β-NAPHTHYLPIPERIDINE, **V, 816**
α-Naphthylthiourea, **IV,** 700, 701

NEOPENTYL ALCOHOL, **V, 818**
Neopentyl iodide, **IV,** 726
NEOPHYL CHLORIDE, **IV, 702**
Niacinamide, **IV,** 705
Nickel catalyst, **V,** 102, 743, 1130
Nickel chloride, hexahydrate, **I,** 305
Nicotinamide, **IV,** 45, 704, 706
NICOTINAMIDE-1-OXIDE, **IV,** 166,
    704
Nicotine, **I,** 385
NICOTINIC ACID, **I, 385; IV,** 88; **V,**
    822
NICOTINIC ACID HYDROCHLOR-
    IDE, **I,** 387
NICOTINIC ANHYDRIDE, **V, 822**
NICOTINONITRILE, **IV, 706**
Nicotinyl chloride, **IV,** 88
Nitramines, **V,** 842
Nitration, See Index No. 3, Type of
    Reaction
Nitric acid, anhydrous, **III,** 804
Nitric acid, fuming, **I,** 390, 408; **II,** 254,
    440, 449, 459, 466; **III,** 472, 837;
    IV, 412, 414, 654
Nitriles, **V,** 657
m-NITROACETOPHENONE, **II, 434**
o-NITROACETOPHENONE, **IV, 708**
p-Nitroacetophenone, **IV,** 709
p-NITRO-*p'*-ACETYLAMINODI-
    PHENYLSULFONE, **III, 240**
1-NITRO-2-ACETYLAMINONAPH-
    THALENE, **II, 438,** 451
3-Nitro-4-aminoanisole, **III,** 568
3-Nitro-4-aminotoluene, **I,** 415
5-Nitro-2-aminotoluene, **III,** 334
m-Nitroaniline, **I,** 163, 404; **IV,** 718
o-NITROANILINE, **I, 388; II,** 226, 501;
    IV, 74, 75
p-Nitroaniline, **II,** 196, 225, 453; **III,** 666;
    IV, 727
p-Nitroaniline hydrochloride, **IV,** 727
o-Nitroaniline-*p*-sulfonic acid, **I,** 388
9-NITROANTHRACENE, **IV,** 711
NITROANTHRONE, **I,** 390
o-Nitroazobenzenes, **V,** 943
NITROBARBITURIC ACID, **II, 440,** 617

o-NITROBENZALANILINE, V, 941
m-NITROBENZALDEHYDE, I, 398; II, 130; III, 454, 644; IV, 732
o-NITROBENZALDEHYDE, III, 57, 641; IV, 730; V, 825, 828, 941
p-NITROBENZALDEHYDE, II, 441; V, 364
m-NITROBENZALDEHYDE DIMETHYLACETAL, III, 59, 644
o-NITROBENZALDIACETATE, III, 641; IV, 713
p-NITROBENZALDIACETATE, II, 441; IV, 713
m-NITROBENZAZIDE, IV, 715
p-Nitrobenzazide, IV, 717
Nitrobenzene, I, 123, 445, 478; II, 57; III, 6, 103, 581, 668; IV, 720
p-Nitrobenzenediazonium borofluoride, III, 665
o-NITROBENZENEDIAZONIUM FLUOBORATE, II, 226
p-NITROBENZENEDIAZONIUM FLUOBORATE, II, 225
o-Nitrobenzenesulfinic acid, V, 60
m-Nitrobenzenesulfonyl chloride, V, 843
o-NITROBENZENESULFONYL CHLORIDE, II, 471
m-NITROBENZOIC ACID, I, 391; IV, 715
o-Nitrobenzoic acid, IV, 709
p-NITROBENZOIC ACID, I, 392, 394; III, 646; IV, 714
p-NITROBENZONITRILE, III, 646
m-NITROBENZOYL CHLORIDE, IV, 715, 716
o-Nitrobenzoyl chloride, IV, 708
p-NITROBENZOYL CHLORIDE, I, 394; III, 649
p-NITROBENZOYL PEROXIDE, III, 649
p-NITROBENZYL ACETATE, III, 650, 652
p-NITROBENZYL ALCOHOL, III, 652
p-NITROBENZYL BROMIDE, II, 443

p-Nitrobenzyl chloride, III, 650
p-NITROBENZYL CYANIDE, I, 396, 406
o-NITROBENZYLPYRIDINIUM BROMIDE, V, 825
m-NITROBIPHENYL, IV, 718
o-Nitrobiphenyl, V, 829
o-Nitrobiphenyls, V, 943
p-Nitrobromobenzene, V, 145
Nitrobutane, IV, 726
t-Nitrobutane, V, 355
2-NITROCARBAZOLE, V, 829
o-Nitrochlorobenzene, II, 446
p-Nitrochlorobenzene, II, 445
o-Nitro-p-chlorophenylsulfur chloride, II, 456
o-NITROCINNAMALDEHYDE, IV, 722
m-NITROCINNAMIC ACID, I, 398; IV, 731
o-Nitrocinnamic acid, IV, 732
p-Nitrocinnamic acid, IV, 732
2-NITRO-p-CYMEME, III, 63, 653
5-NITRO-2,3-DIHYDRO-1,4-PHTHALAZINEDIONE, III, 69, 656
m-NITRODIMETHYLANILINE, III, 658
2-Nitro-2,3-dimethylbutane, V, 847
1-Nitro-1,4-dimethylcyclohexene, V, 847
2-Nitro-2,4-dimethylpentane, V, 847
o-Nitrodiphenyl ether, II, 446
p-NITRODIPHENYL ETHER, II, 445
Nitroethane, IV, 573
2-NITROETHANOL, V, 833, 838
Nitroethylene, V, 838
2-NITROFLUORENE, II, 447; V, 30
Nitrogen, V, 1002
NITROGUANIDINE, I, 399; III, 73
Nitroheptane, IV, 726
Nitrohexane, IV, 726
3-Nitro-4-hydroxystyrene, IV, 733
5-NITROINDAZOLE, III, 660
NITROMESITYLENE, II, 449
NITROMETHANE, I, 401, 413; IV, 221, 223; V, 833
2-NITRO-4-METHOXYACETANIL-

IDE, **III, 661**
2-NITRO-4-METHOXYANILINE, **III, 661**
2-NITRO-6-METHOXYBENZONI-TRILE, **III, 293**
1-Nitro-1-methylcyclohexane, **V,** 847
1-(Nitromethyl)cyclohexanol, **IV,** 224
1-(Nitromethyl)cyclohexanol, sodio derivative, **IV,** 222
1-Nitro-1-methylcyclopentane, **V,** 847
2-Nitro-2-methylpropane, **V,** 847
N-NITROMORPHOLINE, **V, 839**
α-Nitronaphthalene, **III,** 664
1-NITRO-2-NAPHTHOL, **II, 451**
4-NITRO-1-NAPHTHYLAMINE, **III,** 341, **664**
Nitrones, **V,** 1127
NITRONIUM TETRAFLUOROBOR-ATE, **V, 480,** 484
1-NITROÖCTANE, **IV, 724**
p-Nitroperoxybenzoic acid, **V,** 908
p-Nitrophenacylamine, hydrochloride of, **V,** 915
m-NITROPHENOL, **I, 404**
o-Nitrophenol, **III,** 140
p-Nitrophenol, **II,** 173; **III,** 468
p-NITROPHENYLACETIC ACID, **I,** 53, **406**
p-NITROPHENYLARSONIC ACID, **III, 665**
o-NITROPHENYLAZIDE, **IV, 75,** 76
1-(*p*-NITROPHENYL)-1,3-BUTA-DIENE, **IV, 727**
4-(*m*-Nitrophenyl)-2-butanone, **V,** 769
1-(*p*-Nitrophenyl)-4-chloro-2-butene, **IV, 726**
cis-o-Nitro-α-phenylcinnamic acid, **IV,** 730
trans-o-NITRO-α-PHENYLCINNAMIC ACID, **IV,** 730
1-(*m*-NITROPHENYL)-3,3-DIMETH-YLTRIAZENE, **IV, 718**
m-NITROPHENYLDISULFIDE, **V, 843**
p-NITROPHENYL ISOCYANATE, **II, 453**
p-Nitrophenyl isocyanide, **V,** 1063
o-Nitrophenyl ketones, **V,** 943

p-Nitrophenyl salicylate, **IV, 179**
p-NITROPHENYL SULFIDE, **III, 667**
o-NITROPHENYLSULFUR CHLOR-IDE, **II, 455**
Nitrophenylsulfur trifluoride, **V,** 961
3-NITROPHTHALIC ACID, **I,** 56, **408,** 410; **III,** 656
4-NITROPHTHALIC ACID, **II, 457**
3-NITROPHTHALIC ANHYDRIDE, **I, 410**
4-NITROPHTHALIMIDE, **II,** 457, **459**
1-Nitropropane, **V,** 656
2-Nitropropane, **IV,** 652, 932
N-Nitrosamides, **V,** 338
Nitrosation, See Index No. 3, Type of Reaction
N-Nitroso-*o*-acetotoluidide, **V,** 651
NITROSOBENZENE, **III, 668,** 711
p-Nitrosodiethylaniline, **II, 224**
p-Nitrosodiethylaniline hydrochloride, **II, 224**
NITROSODIMETHYLAMINE, **II, 211**
p-Nitrosodimethylaniline, **V,** 826
p-NITROSODIMETHYLANILINE HYDROCHLORIDE, **II, 223**
N-NITROSO-β-METHYLAMINOISO-BUTYL METHYL KETONE, **III, 244**
N-NITROSOMETHYLANILINE, **II,** 418, **460**
N-Nitroso-*N*-methyl-*N'*-nitroguanidine, **V,** 246
NITROSOMETHYLUREA, **II,** 165, **461;** **IV,** 250
N-Nitrosomethylurethan, **III,** 119
NITROSOMETHYLURETHANE, **II, 464**
    warning, **V, 842**
NITROSO-β-NAPHTHOL, **I, 411; II,** 33, 42; **IV,** 981
NITROSO-5-NITROTOLUENE, **III, 334**
N-NITROSO-*N*-(2-PHENYLETHYL)-BENZAMIDE, **V,** 337
N-NITROSO-*N*-PHENYLGLYCINE, **V, 962**
NITROSOTHYMOL, **I, 511**

6-Nitrosoveratric acid, **IV**, 736
o-Nitrostilbenes, **V**, 943
   triethyl phosphite, **V**, 943
β-NITROSTYRENE, **I**, 413
m-NITROSTYRENE, **IV**, 731
o-Nitrostyrene, **IV**, 733
p-Nitrostyrene, **IV**, 733
Nitrosyl chloride, **V**, 266, 864
2-NITROTHIOPHENE, **II**, 466
m-NITROTOLUENE, **I**, 415; **IV**, 151
o-Nitrotoluene, **III**, 641; **IV**, 151, 714; **V**, 567
p-Nitrotoluene, **I**, 392; **II**, 441, 443; **IV**, 31, 114, 367, 713
4-NITRO-2,2,4-TRIMETHYLPENT-ANE, **V**, 845
NITROUREA, **I**, 417, 485; **IV**, 361
NITROUS ANHYDRIDE, **I**, 267
6-NITROVERATRALDEHYDE, **IV**, 735
NITROXIDE, DI-*t*-BUTYL, **V**, 355
3-NITRO-4,6-XYLENOL, **I**, 405
Nonamethylene bromide, **III**, 228
NONAMETHYLENE GLYCOL, **II**, 155
2,4-NONANEDIONE, **V**, 848
NONANE, 1,1,3-TRICHLORO-, **V**, 1076
γ-Nonanoic lactone, **IV**, 601
n-Nonyl fluoride, **IV**, 527
Norbornadiene, **V**, 151
2,5-NORBORNADIENE, 7-*t*-BUTOXY-, **V**, 151
2-NORBORNANONE, **V**, 852
2-NORBORNANONE, 7,7-DI-METHYL-1-VINYL-, D-, **V**, 877
2-NORBORNEN-7-ONE, **V**, 91
2-NORBORNEN-7-ONE DIMETHYL ACETAL, **V**, 424
2-*exo*-Norborneol, **V**, 854
NORBORNYLENE, **IV**, 738
2-*exo*-NORBORNYL FORMATE, **V**, 852
NORCAMPHOR, **V**, 852
NORCARANE, **V**, 855
exo/*endo*-7-NORCARANOL, **V**, 859
NORTRICYCLANOL, **V**, 863, 866
NORTRICYCLANONE, **V**, 866

NORTRICYCLYL ACETATE, **V**, 863
Nutmegs, **I**, 538
NW acid, **II**, 149

OCTADECAMETHYLENE GLYCOL, **II**, 155
OCTADECANEDIOIC ACID, DI-METHYL ESTER, **V**, 463
Octadecyl alcohol, **IV**, 624
Octadecyl bromide, **II**, 247
9-*n*-Octadecylfluorene, **IV**, 624
$\Delta^{1(9)}$-OCTALONE-2, **V**, 869
$\Delta^{9(10)}$-OCTALONE-2, **V**, 869
OCTANAL, **V**, 872
OCTANE, 2,7-DIMETHYL-2,7-DINI-TRO, **V**, 445
OCTANE, 1-OXASPIRO(2.4)-, **V**, 755
1-Octanol, **IV**, 138, 724
dl-OCTANOL-2, **I**, 366, 418; **IV**, 856
d- and *l*-OCTANOL-2, **I**, 418
Octanoyl fluoride, **V**, 69
n-Octyl alcohol, **I**, 30
s-OCTYL ALCOHOL, **I**, 366, 418
n-OCTYL BROMIDE, **I**, 30
n-Octyl fluoride, **IV**, 527
s-OCTYL HYDROGEN PHTHALATE, **I**, 418
n-Octyl iodide, **V**, 872
1-Octyl nitrite, **IV**, 724
Olefins, **V**, 879, 1127
cis-Olefins, **V**, 883
Oleic acid, **IV**, 317, 739, 851
Oleic acid-urea complex, **IV**, 742
OLEOYL CHLORIDE, **IV**, 739
OLEYL ALCOHOL, **II**, 468; **III**, 671
Olive oil, **II**, 469; **IV**, 852
OPTICALLY ACTIVE *s*-OCTYL AL-COHOLS, **I**, 418
Orange I, **II**, 40
Orange II, **II**, 36
ORTHANILIC ACID, **II**, 471
2-OXA-7,7-DICHLORONORCARANE, **V**, 874
Oxalic acid, **I**, 44, 197, 261, 263, 264, 421, 424; **II**, 414; **IV**, 198
OXALIC ACID (ANHYDROUS), **I**, 421

Oxalyl chloride, **V**, 111, 113, 204, 706
1,4-Oxathian 1-oxide, **V**, 792
OXAZIRIDINE, 2-*t*-BUTYL-3-
    PHENYL, **V, 191**
OXEPIN, 2,7-DIMETHYL, **V, 467**
Oxidation Reactions, See Index No. 3,
    Type of Reaction
Oximes, **V**, 86
OXIMINOMALONONITRILE, **V, 32**
OXINDOLE, 3-ACETYL-, **V, 12**
γ-Oxocapric acid, **IV, 432**
2-OXOCYCLOOCTANECARBOX-
    YLIC ACID, ETHYL ES-
    TER, **V, 198**
D-2-OXO-7,7-DIMETHYL-1-VINYLBI-
    CYCLO(2.2.1)HEPTANE, **V,
    877**
OXONIUM COMPOUNDS, **V, 1080,
    1096, 1099**
2-OXO-1,2*H*-PYRANE, **V, 982**
Oxygen, **III**, 673; **IV**, 367, 494, 579, 895
OZONE, **III, 673; IV, 484; V, 489, 493**

Palladium, **II**, 566
PALLADIUM CATALYSTS, **III, 685;
    V**, 30, 97, 278, 347, 376, 552,
    587, **880, 989**
Palladium chloride, **III**, 520, 685, 686, 687
Palladium oxide, **V**, 61
Palladous chloride, **V**, 880
PALLADOUS OXIDE CATALYST, **II,
    566**
Palmitic acid, **III, 607; IV**, 742, 862
Palmitonitrile, **IV**, 437
Palmitoyl chloride, **IV, 742**
Pancreatin, **II**, 612
PARABANIC ACID, **IV, 744**
(2.2)PARACYCLOPHANE, **V, 883**
Paraformaldehyde, **I**, 188, 425, 528, 531;
    II, 387; **III**, 195, 198, 305; **IV**,
    102, 281, 691, 907; **V**, 139, 624,
    833
Paraldehyde, **II**, 87; **III**, 438; **IV**, 293, 451,
    748, 980
PELARGONIC ACID, **II, 474**
Pelargonyl fluoride, **V**, 69

PENTAACETYL *d*-GLUCONONI-
    TRILE, **III, 101, 690**
2,3,4,5,6-PENTA-*O*-ACETYL-D-GLY-
    CONIC ACID, **V, 887**
2,3,4,5,6-PENTA-*O*-ACETYL-D-GLY-
    CONYL CHLORIDE, **V, 887**
PENTACHLOROBENZOIC ACID, **V,
    890**
1,2,3,4,5-PENTACHLORO-5-ETHYL-
    CYCLOPENTADIENE, **V,
    893**
Pentachlorophenylmagnesium chloride, **V,**
    890, 892
Pentacyanopropene, **V**, 1015
1,4-PENTADIENE, **IV, 746**
PENTAERYTHRITOL, **I, 425; II**, 476;
    IV, 679, 681, 753
Pentaerythrityl benzenesulfonate, **IV**, 753
PENTAERYTHRITYL BROMIDE, **II,
    476**
PENTAERYTHRITYL IODIDE, **II, 477**
PENTAERYTHRITYL TETRABROM-
    IDE, **IV, 753**
Pentalene, **V**, 1091
1-PENTALENECARBOXYLIC ACID,
    OCTAHYDRO-, **V, 93**
PENTAMETHYLBENZENE, **II, 248**
PENTAMETHYLENE BROMIDE, **I,
    428; III, 692**
3,3-PENTAMETHYLENEDIAZIRINE,
    V, 897
n-PENTANE, **II, 478**
1,5-Pentanediol, **III, 693; IV**, 748
1,5-Pentanediol diacetate, **IV, 747, 748**
2,4-Pentanedione, **IV**, 352, 869; **V**, 767,
    848
2,4-PENTANEDIONE, 3-METHYL-, **V,
    785**
PENTANE, 1,1-DIPHENYL-, **V, 523**
PENTANE, 2,2,4-TRIMETHYL-4-NI-
    TRO, **V, 845**
1,3,5-PENTANETRIONE, 1-(*p*-METH-
    OXYPHENYL)-5-PHENYL,
    V, 718
1,3,5-Pentanetriones, **V**, 720, 721
2-Pentanol, **I**, 430

Pentatricontanedioic acid, **V,** 538
2-PENTENE, **I, 430**
3-PENTENOIC ACID, 3-HYDROXY-
2,2,4-TRIMETHYL-, β-
LACTONE, **V, 456**
4-PENTEN-1-OL, **III, 698**
3-PENTEN-2-OL, **III, 696**
4-Penten-1-ol acetate, **IV,** 748
4-Penten-1-yn-3-ol, **IV,** 794
3-Pentyl α-bromopropionate, **IV,** 123
9-*n*-Pentylfluorene, **IV, 625**
4-PENTYN-1-OL, **IV, 755**
α-naphthylurethan, **IV,** 756
silver derivative, **IV,** 756
Peracetic acid, **IV,** 136, 828, 860. See also
Peroxyacetic acid, **V,** 417, 660,
661, 665
PERBENZOIC ACID, **I, 431,** 494 esti-
mation of active oxygen, **I,** 434.
See also Peroxybenzoic acid,
V, 191, 647, 900, 904, 906
Perchloric acid, **IV,** 382; **V,** 1106, 1108
PERCHLOROFULVALENE, **V, 901**
Perfluoroolefins, **V,** 951
Performic acid, **V,** 647
Performic acid, *in situ*, **IV,** 317
Periodic acid, **IV,** 14
Permutit, **I,** 24
Peroxides, in ether, **V,** 664, 925
Peroxyacetic acid, **V,** 414, 417, 660, 661,
665
Peroxy acids, **V,** 907
PEROXYBENZOIC ACID, **V,** 191, 647,
900, **904,** 906
Peroxystearic acid, **V,** 908
Peroxytrifluoroacetic acid, **V,** 369, 598,
601
Petroleum ether, **III,** 351, 503
PHENACYLAMINE HYDROCHLOR-
IDE, **V, 909**
PHENACYL BROMIDE, **II, 480**
Phenacyl chloride, **III,** 191, 192
Phenanthrene, **II,** 482; **III,** 134; **IV,** 313,
314, 757; **V,** 489, 493, 956
PHENANTHRENE-9-ALDEHYDE, **III,
701**

PHENANTHRENE, 9-PHENYL-, **V,
952**
PHENANTHRENEQUINONE, **IV,** 757
sodium bisulfite adduct, **IV,** 758
2- AND 3-PHENANTHRENESULF-
ONIC ACIDS, **II, 482**
o-Phenanthroline, **V,** 212
9-Phenanthryl isothiocyanate, **IV,** 701
PHENETHYLAMINE, α,α-DI-
METHYL, **V, 471**
α-Phenethyl chloride, **IV,** 963
p-Phenetidine hydrochloride, **IV,** 52
Phenol, **I,** 58, 128, 161, 296, 435, 490; **II,**
97, 445; **III,** 54, 714; **IV,** 390,
520, 592; **V,** 408, 918, 927
Phenolsulfonic acid, **I,** 247
Phenothiazines, **V,** 943
PHENOXTHIN, **II, 485**
β-PHENOXYETHYL BROMIDE, **I, 436**
Phenoxyethyltriphenylphosphonium brom-
ide, **V, 1146**
γ-PHENOXYPROPYL BROMIDE, **I,
435**
PHENOXY, 2,4,6-TRIPHENYL-, **V,
1130**
Phenylacetaldehyde oxime, **V,** 86
PHENYLACETAMIDE, **IV, 760**
PHENYLACETIC ACID, **I, 436; II,** 61,
156, 229, 390; **IV,** 730, 760, **761,**
777; **V,** 527
N-Phenylacetimide chloride, **IV,** 385
α-PHENYLACETOACETONITRILE,
II, 284, 391, **487**
Phenylacetone, **III,** 344
Phenylacetonitrile, **I, 107,** 270, 396, 436;
II, 287, 487, 512; **III,** 220, 347,
715, 720; **IV,** 175, 380, 387, 392,
461, 760
Phenylacetyl chloride, **III,** 119, 638; **V,**
1051
PHENYLACETYLENE, **I, 438; IV, 763,**
801, 802; **V,** 515, 880, 921
Phenylacetyl fluoride, **V,** 69
PHENYLACETYL ISOTHIOCYAN-
ATE, **V, 1051**
dl-β-PHENYLALANINE, **II, 489; III, 705**

L-Phenylalanine, **V**, 973
γ-PHENYLALLYLSUCCINIC ACID, **IV, 766**
γ-Phenylallylsuccinic anhydride, **IV**, 767
1-PHENYL-3-AMINO-5-PYRAZOL-ONE, **III, 708**
4-PHENYL-5-ANILINO-1,2,3-TRIA-ZOLE, **IV, 380**
N-PHENYLANTHRANILIC ACID, **II, 15; III, 53**
PHENYLARSONIC ACID, **II, 494**
PHENYL AZIDE, **III, 710; IV, 77**, 380
2-Phenylaziridine, **V**, 86
p-PHENYLAZOBENZOIC ACID, **III, 711**, 712
p-PHENYLAZOBENZOYL CHLOR-IDE, **III, 712**
N-PHENYLBENZAMIDINE, **IV, 769**
PHENYLBENZOYLDIAZOMETH-ANE, **II, 496**
α-PHENYL-β-BENZOYLPROPIONI-TRILE, **II, 498; III**, 358
1-Phenylbiguanide, **IV**, 29
1-Phenylbiguanide hydrochloride, **IV**, 29
4-PHENYL-2-BROMOBUTENE, **I, 187**, 192
PHENYLBROMOETHYNE, **V, 921**
α-Phenyl-γ-(4-bromo-phenyl)acetoacetonitrile, **IV, 175**
trans-1-PHENYL-1,3-BUTADIENE, **IV, 771**
4-Phenyl-2-butanone, **V**, 769
3-Phenyl-2-butanone oxime, **V**, 86
4-PHENYLBUTINE, **I, 192**
PHENYL t-BUTYL ETHER, **V, 924, 926**
γ-PHENYLBUTYRIC ACID, **II, 196, 499, 569; III**, 798; **IV**, 900
N-PHENYLCARBAZOLE, **I, 547**
α-PHENYL-α-CARBETHOXYGLUTAR-ONITRILE, **IV, 776**, 790
Phenyl chloroformate, **V**, 169
PHENYL CINNAMATE, **III, 714**
α-PHENYLCINNAMIC ACID, **IV, 777**, 857

β-PHENYLCINNAMIC ACID, **V, 509, 510**
α-PHENYLCINNAMONITRILE, **III, 715; IV**, 393
2-PHENYLCYCLOHEPTANONE, **IV, 780**
PHENYLCYCLOPROPANE, see Cyclopropylbenzene, **V, 328**, 513, **929**
2-Phenylcyclopropanol, **V**, 861
1-PHENYLDIALIN, **III, 729**
PHENYLDICHLOROPHOSPHINE, **IV, 784**
4-PHENYL-m-DIOXANE, **IV, 786**, 798
Phenyl disulfide, **V**, 959
o-PHENYLENE CARBONATE, **IV, 788**
o-PHENYLENEDIAMINE, **II**, 65, **501; III**, 106; **IV**, 569, 824
p-Phenylenediamine, **V**, 1018
o-PHENYLENEDIAMINE DIHY-DROCHLORIDE, **II, 502**
p-PHENYLENEDIAMINE, N,N,N',N'-TETRAMETHYL, **V, 1018**
p-Phenylene diisothiocyanate, **V**, 226
Phenyl ether, **II**, 485
d- AND l-α-PHENYLETHYLAMINE, **II, 506**
α-PHENYLETHYLAMINE, **II, 503, 506; III**, 717; **V**, 909
R(+) AND S(−)-α-PHENYLETHYL-AMINE, **V, 932**
β-PHENYLETHYLAMINE, **III, 720, 723; V**, 336
β-Phenylethylamines, **III**, 721
d-α-Phenylethylamine-l-malate, **II, 506**
l-α-Phenylethylamine-d-tartrate, **II**, 507
N-(2-PHENYLETHYL)BENZAMIDE, **V, 336**
2-Phenylethyl benzoate, **V**, 337
α-Phenylethyl chloride, **V**, 527
β-PHENYLETHYLDIMETHYL-AMINE, **III, 723**
PHENYLETHYLENE, **I, 440**
β-Phenylethylphthalimide, **II**, 84
Phenylethynyl n-butyl dimethyl ketal, **IV, 802**

Phenylethynyl methyl diethyl ketal, **IV**, 802

α-Phenylglutaric acid, **IV**, 790

α-PHENYLGLUTARIC ANHYDRIDE, **IV, 790**

Phenyl glycidyl ether, **IV**, 12

N-Phenylglycine, **V**, 962

PHENYLGLYOXAL, **II, 509; V, 937**

PHENYLGLYOXAL HEMIMERCAPTAL, **V, 937**

PHENYLGLYOXYLIC ACID, **I, 244**

PHENYLHYDRAZINE, **I, 442; III**, 360, 708, 710; **IV**, 657, 884, 885

2-Phenyl-2-hydroxyethane-1-sulfonic acid, sodium salt, **IV, 850**

N-Phenylhydroxylamine, **V**, 957, 1124

β-PHENYLHYDROXYLAMINE, **I**, 177, **445**

PHENYL α-HYDROXYSTYRYL KETONE, **I, 205**

2-PHENYLINDAZOLE, **V, 941**

2-PHENYLINDOLE, **III, 725**

Phenyl isocyanate, **IV**, 213, 561; **V**, 501

Phenyl isocyanide, **V**, 1063

PHENYL ISOTHIOCYANATE, **I, 447; IV**, 701; **V**, 226

Phenyllithium, **II, 518; IV**, 642; **V**, 453, 454

PHENYLMAGNESIUM BROMIDE, **I, 226, 550; IV**, 68, 96, 767, 912; **V**, 521, 924

N-PHENYLMALEIMIDE, **V, 944**, 957

Phenylmercuric chloride, **V**, 969

α-Phenyl-γ-(4-methoxyphenyl)acetoacetonitrile, **IV, 175**

PHENYLMETHYLGLYCIDIC ESTER, **III, 727**, 733

α-Phenyl-γ-(4-methylphenyl)acetoacetonitrile, **IV, 175**

α-Phenyl-γ-(4-methylthiophenyl)acetoacetonitrile, **IV, 175**

1-PHENYLNAPHTHALENE, **III, 729**

Phenyl-β-naphthylamine, **IV**, 772, 774

PHENYLNITROMETHANE, **II, 512**

2-PHENYL-2-OXAZOLIN-5-ONE, **V, 946**

2-PHENYL-5-OXAZOLONE, **V, 946**

1-Phenyl-1,4-pentanedione, **V**, 769

5-Phenyl-1,3,5-pentanetriones, 1-aryl, **V**, 720

1-PHENYL-1-PENTEN-4-YN-3-OL, **IV, 792**

2-Phenyl-2*H*-perfluoropropane, **V**, 950

2-PHENYLPERFLUOROPROPENE, **V, 949**

p-Phenylphenacylamine, hydrochloride, **V**, 915

9-PHENYLPHENANTHRENE, **V, 952**

α-Phenyl γ-phenylacetoacetonitrile, **IV, 175**

o-Phenylphenyl salicylate, **IV**, 179

p-Phenylphenyl salicylate, **IV**, 179

1-PHENYLPIPERIDINE, **IV, 795; V**, 817

3-PHENYL-1-PROPANOL, **IV, 798**

PHENYLPROPARGYL ALDEHYDE, **III, 731**

PHENYLPROPARGYL ALDEHYDE ACETAL, **III, 732**

PHENYLPROPARGYL ALDEHYDE DIETHYL ACETAL, **IV, 801**

PHENYLPROPIOLIC ACID, **II, 515; V**, 1045

α-PHENYLPROPIONALDEHYDE, **III, 733**

Phenyl propionate, **II**, 543, **544**

2-PHENYL-3-*n*-PROPYLISOXAZOLIDINE-4,5-*cis*-DICARBOXYLIC ACID *N*-PHENYLIMIDE, **V, 957**

5-Phenylpyrazoline, **V**, 929

2-PHENYLPYRIDINE, **II, 517**

Phenyl 4-pyridyl ketone, **IV**, 89

5-Phenyl-3-pyridyl-1,3,5-pentanetrione, **V**, 720

2-Phenyl-6-(3-pyridyl)-4-pyrone, **V**, 722

PHENYLPYRUVIC ACID, **II, 519**

Phenylpyruvic acids, **V**, 632

Phenyl salicylate, **I**, 552; **III**, 765

4-PHENYLSEMICARBAZIDE, **I, 450**

PHENYLSUCCINIC ACID, I, 451; IV,
    804
Phenylsuccinic anhydride, IV, 806
Phenylsulfinylacetic acid, V, 792
PHENYLSULFUR TRIFLUORIDE, V,
    396, 397, 859
3-PHENYLSYDNONE, V, 962
N-PHENYLSYDNONE, V, 962
PHENYL THIENYL KETONE, II, 520
1-PHENYL-2-THIOBIURET, V, 966
1-Phenyl-2-thio-4-ethylisobiuret, V, 968
[β-(Phenylthio)ethyl]triphenylphosphonium
    bromide, V, 1147
1α-Phenyl-2α-thiohomophthalimide, V,
    1053
1-PHENYL-2-THIO-4-METHYLISOBI-
    URET, V, 966
α-PHENYLTHIOUREA, III, 735; IV,
    181
Phenyl p-tolyl sulfone, IV, 36, 37
1-PHENYL-2,2,2-TRICHLORO-
    ETHANOL, V, 130
PHENYL(TRICHLOROMETHYL)-
    MERCURY, V, 969, 971
Phenyl(trihalomethyl)mercurials, V, 971
Phenyltrivinyltin, IV, 260
PHENYLUREA, I, 450, 453; V, 804
PHLOROACETOPHENONE, II, 522
PHLOROGLUCINOL, I, 455; II, 522;
    IV, 454
Phosgene, II, 453; III, 167, 488; IV, 522,
    788; V, 201, 822
PHOSPHINE, (p-BROMO-
    PHENYL)DIPHENYL-, V,
    496
PHOSPHINOUS CHLORIDE, DIISO-
    PROPYL-, V, 211
3-PHOSPHOLENE, 3-METHYL-1-
    PHENYL-, 1-OXIDE, V, 787
PHOSPHONIC DICHLORIDE,
    STYRYL-, V, 1005
PHOSPHONIUM BROMIDE, TRI-
    PHENYLVINYL-, V, 1145
Phosphonothioic dichlorides, V, 220
PHOSPHORANE, (DICHLOROME-
    THYLENE)TRIPHENYL-1,

V, 361
Phosphoric acid, II, 152, 604; III, 14, 195,
    301; IV, 321, 323, 350, 543, 691
Phosphorus, III, 476, 588
Phosphorus, red, I, 36, 37, 38, 224; II, 74,
    308, 322, 399, 490; IV, 348; V,
    255
    yellow, I, 36, 37, 38; II, 399
Phosphorus heptasulfide, IV, 671
Phosphorus oxychloride, I, 85, 217; II,
    109, 560; III, 53, 98, 99, 194,
    273, 476, 535; IV, 166, 178, 266,
    331, 384, 390, 446, 540, 784,
    831, 916; V, 172, 173, 215, 300,
    1060
Phosphorus pentachloride, I, 84, 394; II,
    133, 379, 528, 549, 573; III, 646,
    818; IV, 35, 166, 383, 554, 693,
    847, 900, 914; V, 171, 196, 267,
    365, 888, 1005, 1138
Phosphorus pentasulfide, III, 332; V, 780
Phosphorus pentoxide, III, 22, 493, 584;
    IV, 27, 70, 144, 214, 242, 321,
    323, 487, 544, 706, 938, 940
PHOSPHORUS TRIAMIDE, HEXA-
    METHYL-, V, 602
Phosphorus triamides, hexaalkyl, V, 603
Phosphorus tribromide, I, 428; II, 358,
    359, 476; III, 370, 793; IV, 107,
    616; V, 609
Phosphorus trichloride, I, 116; II, 156; III,
    523, 848; IV, 398, 784, 950, 955;
    V, 211, 602
Phosphorus trisulfide, II, 578, 579
Photocyclization, See Index No. 3, Type
    of Reaction
o-PHTHALALDEHYDE, IV, 807
PHTHALALDEHYDIC ACID, II, 523;
    III, 737
Phthalic acid, IV, 56
Phthalic acids, V, 1053
Phthalic anhydride, I, 57, 408, 418, 457,
    476, 517; II, 61, 528; III, 152,
    619, 796; IV, 107; V, 973
PHTHALIC MONOPEROXY ACID, V,
    805

PHTHALIDE, II, 526; III, 174, 737, 739;
    IV, 811
Phthalideacetic acid, IV, 138
PHTHALIMIDE, I, 119, 457; II, 83, 459,
    526
α-PHTHALIMIDO-*o*-TOLUIC ACID,
    IV, 810
Phthaloyl chloride, III, 422
Phthaloyl fluoride, V, 69
N-Phthalyl-β-alanine, V, 975
N-Phthalyl-*l*-alanine, V, 975
sym. and *unsym. o*-PHTHALYL CHLO-
    RIDE, II, 528
N-Phthalylglycine, V, 975
N-PHTHALYL-*l*-β-PHENYLALA-
    NINE, V, 973
Picene, V, 956
α-Picoline, III, 413, 740, 757
2-Picoline-1-oxide, V, 270
3-Picoline-1-oxide, IV, 655
PICOLINIC ACID HYDROCHLO-
    RIDE, III, 740
Picric acid, III, 205
PIMELIC ACID, II, 531
PIMELIC ACID, β-OXO-, DIETHYL
    ESTER, V, 384
Pinacol, III, 312, 313
PINACOL HYDRATE, I, 459, 462
Pinacol rearrangement, V, 647
PINACOLIN (see Pinacolone)
PINACOLONE, I, 81, 462, 527; III, 313;
    V, 601
PINACONE HYDRATE, I, 459
Pine oleoresin, V, 699
Piperazine, V, 88
PIPERAZINE, 1-BENZYL, V, 88
Piperidine, I, 101; III, 165, 317, 377; IV,
    210, 327, 409; V, 487, 575
Piperidine acetate, IV, 210, 408
PIPERIDINE, 1-ETHYL, V, 575
    1-(2-NAPHTHYL)-, V, 816
Piperonal, II, 538, 549; IV, 605
Piperonyl alcohol, II, 591
PIPERONYLIC ACID, II, 538
Pivalic acid, V, 1028
Pivaloin, II, 115

Pivalonitrile, V, 1028
PLATINIC OXIDE, I, 463
Platinum black, in peroxide decomposi-
    tion, IV, 612
PLATINUM CATALYST FOR RE-
    DUCTIONS, I, 463
Platinum oxide, V, 16, 568
Platinum oxide catalyst, II, 191, 491
Polyphosphoric acid, III, 798; V, 450, 550
Potassium, III, 71, 479, 510; IV, 132, 399,
    459, 617, 963
    powdered, II, 195
Potassium acetate, II, 228; III, 127, 426;
    V, 208
Potassium acid acetylenedicarboxylate, IV,
    329
Potassium acid sulfate, I, 475; III, 204
POTASSIUM AMIDE, IV, 963; V, 12,
    188
Potassium 4-amino-3,5-dinitrobenzenesul-
    fonate, IV, 365
POTASSIUM ANTHRAQUINONE-α-
    SULFONATE, II, 128, 539
POTASSIUM BENZOHYDROXAM-
    ATE, II, 67
Potassium bicarbonate, II, 557; IV, 412,
    804
Potassium bromate, I, 89; IV, 849
Potassium bromide, III, 814; IV, 526, 877
POTASSIUM *t*-BUTOXIDE, IV, 132,
    399, 459, 617; V, 319, 361, 362,
    468, 926, 1060, 1061
Potassium carbonate, I, 263, 545; II, 15,
    83, 341; III, 140, 418, 731; IV,
    211, 419, 558, 589, 781, 836
Potassium chlorate, III, 42
Potassium chloride, II, 483; III, 634
Potassium 4-chloro-3,5-dinitrobenzenesul-
    fonate, IV, 365
Potassium cyanate, II, 79; IV, 49
Potassium cyanide, I, 156; II, 292, 498;
    III, 174, 260, 293, 436; IV, 115,
    394, 439, 496, 642, 804; V, 572,
    578, 614
Potassium dichromate, I, 384; II, 542; III,
    325, 449

Potassium ethoxide, **V**, 990
Potassium ethyl malonate, **IV**, 417
Potassium ethyl xanthate, **III**, 809; **IV**, 569
Potassium ferricyanide, **II**, 419; **V**, 985
Potassium fluoride, **IV**, 525
Potassium hexacyanoferrate(III), **V**, 1132
Potassium hydroxide
    methanolic, **IV**, 367
Potassium hypochlorite, **III**, 303; **IV**, 485
Potassium iodide, **II**, 352, 356, 404, 604;
    III, 355; **IV**, 297, 319, 321, 323,
    543, 896, 935
Potassium methoxide, **III**, 71
·Potassium *p*-methoxybenzohydroxamate,
    II, **68**
Potassium *p*-methylbenzohydroxamate, **II**,
    68
Potassium methyl sulfate, **IV**, 589
Potassium nitrate, **IV**, 365, 498
Potassium 1-nitropropylnitronate, **IV**, 373
Potassium oxalate, **IV**, 808
Potassium permanganate, **I**, 159, 241; **II**,
    22, 53, 135, 307, 315, 524, 538;
    III, 740, 791; **IV**, 467, 825; **V**,
    394, 689, 845
Potassium persulfate, **III**, 334; **V**, 1037
POTASSIUM 3-PHENANTHRENE-
    SULFONATE, **II**, **482**
POTASSIUM PHTHALIMIDE, **I**, **119**,
    271; **II**, **25**; **IV**, 811
Potassium-sodium tartrate, **IV**, 193
Potassium sulfate, **I**, 16
Potassium thiobenzoate, **IV**, 924
Potassium thiocyanate, **IV**, 233; **V**, 562,
    1058
Potassium trithiocarbonate, **IV**, 967
Potassium xanthate, **III**, 667
PREGN-5-EN-18-OIC, 3β,20β-DIHY-
    DROXY, 18,20-LACTONE, 3-
    ACETATE, **V**, **692**
Pregnenolone acetate, **V**, 8, 692
PROPANE, 2,2-DIBUTOXY-, **V**, **5**
2-Propanol, **V**, 998
Propargyl alcohol, **IV**, 813
Proparglysuccinic anhydride, **V**, 460
1-PROPENE-1,1,2,3,3-PENTACAR-

BONITRILE, TETRAME-
    THYLAMMONIUM SALT,
    V, **1013**
Propenylbenzene, **IV**, 787
PROPIOLALDEHYDE, **IV**, **813**
PROPIONALDEHYDE, **II**, **541**
PROPIONIC ACID, 3-CHLORO-2,-
    2,3-TRIFLUORO-, **V**, **239**
    2,3-DIPHENYL-, **V**, **526**
    2-(2,4,5,7-TETRANITROFLU-
    OREN-9-YLIDENEAMI-
    NOOXY)-, (+)- AND (−)-, **V**,
    1031
Propionic anhydride, **IV**, 657
Propionitrile, **IV**, 518
Propionoin, **II**, 115
Propionyl fluoride, **V**, 69
β-PROPIONYLPHENYLHYDRA-
    ZINE, **IV**, **657**
o- AND *p*-PROPIOPHENOL, **II**, **543**
Propiophenone, **II**, 363; **V**, 624
n-Propylacetylene, **IV**, 118, **119**
n-Propyl alcohol, **I**, 37; **II**, 541; **III**, 119;
    IV, 284
N-Propylallenimine, **V**, **544**
N-Propylaniline, **IV**, 147, 284
n-PROPYLBENZENE, **I**, **471**
n-PROPYL BROMIDE, **I**, 37; **II**, **359**,
    548; **IV**, 119
N-Propylbutylamine, **IV**, 284
γ-*n*-PROPYLBUTYROLACTONE, **III**,
    742
n-PROPYL CHLORIDE, **I**, **143**
l-PROPYLENE GLYCOL, **II**, **545**
Propylene oxide, **IV**, 12
9-*n*-Propylfluorene, **IV**, **625**
n-Propyl iodide, **II**, **402**; **IV**, 324
n-Propylmagnesium bromide, **IV**, 96
C-(*n*-Propyl)-*N*-phenylnitrone, **V**, 957
n-Propyl phosphate, **II**, 110
n-Propyl sulfate, **II**, 112
n-PROPYL SULFIDE, **II**, **547**
n-Propyl sulfite, **II**, 113
N-Propyl-*m*-toluidine, **II**, 291
PROTOCATECHUALDEHYDE, **II**,
    549

PROTOCATECHUIC ACID, III, 745
PSEUDOIONONE, III, 747
PSEUDOPELLETIERINE, IV, 816
Pseudopelletierine hemihydrate, IV, 817
PSEUDOTHIOHYDANTOIN, III, 751
PUTRESCINE DIHYDROCHLORIDE, IV, 819
PYOCYANINE, III, 753
4H-PYRAN-4-ONE, 2,6-DIMETHYL-3,5-DIPHENYL-, V, 450
4H-PYRAN-4-ONE, 2-(p-METHOXY-PHENYL)-6-PHENYL-, V, 721
2,3-PYRAZINEDICARBOXYLIC ACID, IV, 824
PYRAZOLE, 3(OR 5)-AMINO-, V, 39
5-Pyrazolone of s-butyl α-n-caproylpropionate, IV, 122
Pyrene, IV, 484; V, 147
PYRENE, 1-BROMO-, V, 147
1-PYRENOL, V, 632
1-Pyrenyl isothiocyanate, IV, 701
Pyridine, I, 87, 99; II, 30, 109, 159, 375, 419, 518; III, 28, 29, 89, 366, 410, 425, 432, 476, 575, 631, 699, 783, 793; IV, 5, 56, 198, 244, 327, 381, 383, 446, 478, 479, 732, 753, 814, 828, 981; V, 43, 103, 105, 300, 585, 826, 977, 1013
3-Pyridinecarboxylic acid, IV, 88
PYRIDINE, 2,3-DIAMINO-, V, 346
2,5-Pyridinedicarboxylic acid, V, 815
Pyridine hydrochloride, IV, 982
PYRIDINE-N-OXIDE, IV, 828
Pyridine-1-oxide acetate, IV, 829
Pyridine-1-oxide hydrochloride, IV, 828
4-PYRIDINESULFONIC ACID, V, 977
PYRIDINIUM, 1-AMINO-, IODIDE, V, 43
Pyridinium bromide perbromide, V, 606
Pyridinium trifluoroacetate, V, 245
4-Pyridones, V, 722
1-(α-PYRIDYL)-2-PROPANOL, III, 757
N-(4-PYRIDYL)PYRIDINIUM CHLORIDE HYDROCHLORIDE, V, 977

PYRIMIDINE, 4-METHYL-, V, 794
2-PYRIMIDINETHIOL, V, 703
Pyrogallol, II, 304; IV, 27
Pyrogallol-1,3-dimethyl ether, IV, 866
PYROGALLOL 1-MONOMETHYL ETHER, III, 753, 759
Pyrolysis, See Index No. 3, Type of Reaction; Thermal Decomposition Reactions
PYROMELLITIC ACID, II, 551
Pyromellitic anhydride, II, 552
PYROMUCIC ACID, I, 276
α-PYRONE, V, 982
4-PYRONES, 2,6-disubstituted, V, 722
PYRROLE, I, 473; IV, 831
2-PYRROLEALDEHYDE, IV, 831
2,5-Pyrroledicarboxylic acid, V, 815
PYRROLE, 1-(2-METHOXYCARBONYLPHENYL)-, V, 716
Pyrroles, V, 717
PYRROLE, 2,3,4,5-TETRAMETHYL-, V, 1022
Pyrrolidine, IV, 466; V, 277, 486
2-(1-PYRROLIDYL)PROPANOL, IV, 834
2-PYRROLINE, 1-METHYL-2-METHYLTHIO-, V, 780
2,2'-Pyrrolylpyridine, V, 106
PYRUVIC ACID, I, 475; III, 610; IV, 468; V, 636
PYRUVIC ACID, p-HYDROXYPHENYL-, V, 627
Pyruvic aldehyde, sodium bisulfite addition product, IV, 826
PYRYLIUM TETRAFLUOROBORATE, 2,4,6-TRIPHENYL-, V, 1135

Quinacetophenone, IV, 836, 837
Quinacetophenone dimethyl ether, IV, 837
QUINACETOPHENONE MONOMETHYL ETHER, IV, 836
QUINIZARIN, I, 476
QUINOLINE, I, 478; IV, 609, 628, 732, 857; V, 772, 880

Quinoline-sulfur poison, **III, 629**
3-QUINOLINOL, **V, 635**
QUINONE, **I,** 317, **482; II,** 553 (see *p*-
    Benzoquinone)
QUINONE, TETRAHYDROXY-, **V,**
    1011
QUINOXALINE, **IV, 824**
p-QUINQUEPHENYL, **V, 985**
3-QUINUCLIDINONE, HYDRO-
    CHLORIDE, **V, 989**

Raney nickel alloy, **III,** 176, 181, 742; **IV.**
    137, 222, 283, 299, 314, 357,
    432, 603, 639, 660, 672
Rape-seed oil, **II,** 258
Rearrangements, See Index No. 3, Type
    of Reaction
Reduction Reactions, See Index No. 3,
    Type of Reaction
REINEKE SALT, **II, 555**
Reissert's compound, **IV, 641**
Replacement Reactions, See Index No. 3,
    Type of Reaction
RESACETOPHENONE, **III, 761**
R salt, **II,** 148
Resolution, **V,** 11, 932, 935, 936, 1033
Resorcinol, **II,** 557; **III,** 278, 282, 761; **V,**
    743
β-RESORCYLIC ACID, **II,** 100, **557**
REYCHLER'S ACID, **V, 194**
RHODANINE, **III, 763**
Rhodium-on-alumina catalyst, **V,** 591
Rhodium-palladium catalyst, **V,** 671
Ricinoleic acid, **II,** 53
Ricinoleoyl chloride, **IV, 742**
Ritter reaction, **V,** 73, 471
Rochelle salt, **IV,** 194
Rosin, wood, **IV,** 1
Ruthenium, **V,** 1002
RUTHENIUM, DICYCLOPENTA
    DIENYL-, **V, 1001**
Ruthenium trichloride, **V,** 1001
RUTHENOCENE, **V, 1001**

S acid, **II,** 149
Salicylaldehyde, **I,** 149; **III,** 165, 166

Salicylic acid, **I,** 54; **II,** 341, 343, 535; **IV,**
    178
Salicylaldehyde, **V,** 252
SALICYL-*o*-TOLUIDE, **III, 765**
Sand, **III,** 339; **IV,** 150
Sarcosine, **IV,** 5
Schaeffer's salt, **II,** 148
Schiemann reaction, **V,** 136
Schmidt reaction, **V,** 273
Sebacic acid, **II,** 276; **III,** 768; **V,** 538
SEBACIL, **IV,** 219, **838**
SEBACOIN, **IV,** 216, 218, 219, 838, **840**
SEBACONITRILE, **III,** 229, **768**
Sebacoyl chloride, **V, 536**
2,2'-SEBACOYLDICYCLOHEXAN-
    ONE, **V, 533**
3,4-SECO-Δ⁵-CHOLESTEN-3,4-DIOIC
    ACID, **IV, 191**
Seignette salt, **IV,** 194
Selenious acid, **III,** 438; **IV,** 229
Selenium, **II,** 238, **510; III,** 771; **IV,** 229,
    231
Selenium dioxide, **II,** 510; **IV,** 231
SELENOPHENOL, **III, 771**
Semicarbazide hydrochloride, **V,** 28, 749
SEMICARBAZIDE SULFATE, **I, 485**
dl-SERINE, **III, 774**
Silica gel, **IV,** 64
Silicon tetraisocyanate, **V,** 801
Silicon tetraisothiocyanate, **V,** 802
Silver, **IV,** 920, 973
Silver acetate, **III,** 102; **IV,** 548
Silver carbonate, **III,** 434
Silver chloride, **IV, 85**
Silver cyanide, **IV,** 438
Silver difluoride, **V,** 959
Silver monofluoride, **V,** 959
Silver nitrate, **III,** 578; **IV,** 41, 85, 548,
    614, 648, 725, 919, 972
Silver nitrate, complex, **V,** 317, 517
Silver nitrite, **IV,** 369, 425, **724**
Silver oxide, **I,** 22; **II,** 19, 28; **IV,** 493, 547,
    548, 919, **972; V,** 316, **885, 897**
N-Silver succinimide, **V,** 664
SILVER TRIFLUOROACETATE, **IV,**
    547

Sodamide, **I**, 191 (See Sodium amide)

Sodio-2-formyl-6-methylcyclohexanone, **V**, **188**

SODIUM ACETYLACETONATE, **IV**, **869**

Sodium acetylide, **III**, 417

SODIUM AMALGAM, **I**, 228, **554**; **III**, 587; **IV**, **508**
1%, **II**, 607, **609**; 3%, **II**, 609

SODIUM AMIDE, **III**, **44**, **291**, 320, 597, 727, **778**; **IV**, 117, 128, 296, 387, 404, 586, 755, 851, 970; **V**, **523**, **526**, 541, 559, **770**, 816, 848, 1043

SODIUM 2-AMINOBENZENESULFI-NATE, **V**, 61, 62

Sodium and ammonia, **V**, 355, 399, 419, 425, 428, 523, 526, 770

Sodium anthraquinone-α-sulfonate, **III**, 573

Sodium arsenite, **I**, 357, 358

SODIUM *p*-ARSONO-*N*-PHENYL-GLYCINAMIDE, **I**, **488**

Sodium azide, **III**, 846; **IV**, 76, 716; **V**, 179, 273, 586, 830

Sodium benzenesulfonate, **I**, 84, 85

Sodium benzyloxide, **I**, 104

Sodium bicarbonate, **I**, 103; **II**, 347; **III**, 73, 75, 232, 747; **IV**, 16, 469, 674

Sodium bisulfite, **I**, 243, 318, 336; **II**, 42, 205, 324; **III**, 275, 399, 438; **IV**, 32, 226, 469, 574, 688, 758, 826, 867

Sodium bromate, **I**, 89

Sodium bromide, **I**, 29, 31, 136; **II**, 132; **III**, 186; **IV**, 753

SODIUM 2-BROMOETHANESULF-ONATE, **II**, **558**, 563

Sodium chlorate, **II**, 128, 302, 553

Sodium chlorodifluoroacetate, **V**, **390**, **949**

Sodium cyanate, **IV**, 49, 515; **V**, 162, 239

Sodium cyanide, **II**, 7, 29, 150, 182, 376, 387; **III**, 66, 84, 88, 275, 372, 558, 615; **IV**, 59, 207, 274, 392, 506, 577

SODIUM CYANOACETATE, **I**, **181**

Sodium cyclopentadienide, **IV**, 474; **V**, **1001**

Sodium dichromate, **I**, 138, 211, 340, 392, 482, 543; **II**, 139; **III**, 1, 420, 669; **IV**, 150, 164, 189, 197; **V**, 810

Sodium dihydrogen phosphate, **IV**, 467

SODIUM DIMETHYLGLYOXIMATE, **II**, **206**

Sodium diphenylketyl, **IV**, 794

Sodium disulfide, **I**, 220

Sodium dodecylbenzenesulfonate, **V**, 40

Sodium ethoxide, **I**, 250; **III**, 215, 251, 300, 397, 715; **IV**, 141, 174, 184, 221, 245, 288, 291, 396, 427, 457, 461, 536, 566, 618, 631, 662, 932; **V**, **288**, 687

Sodium fluoride, **II**, 612

Sodium formate, **V**, 880

Sodium formylacetone, **IV**, 210

Sodium hydride, **IV**, 537; **V**, 198, 285, 379, 547, 718, 752, 755, 1064

Sodium hydrosulfide, **IV**, 32

Sodium hydrosulfite, **I**, 50; **II**, 33, 40; **III**, 69, 87, 639; **IV**, 16, 46, 248, 758

Sodium hydroxylamine monosulfonate, **II**, **205**

SODIUM *p*-HYDROXYPHENYLAR-SONATE, **I**, **490**

Sodium hypobromite, solution of, **IV**, 45

Sodium hypochlorite, **I**, 309; **III**, 486; **V**, 184
solution of, **IV**, 74, 346

Sodium hypophosphite, **I**, 74

Sodium iodide, **I**, 231; **II**, 477; **IV**, 84, 208, 577; **V**, 478, 489

Sodium isopropoxide, **V**, **40**, 41, 42, 285

Sodium metabisulfite, **II**, 131

Sodium metaperiodate, **V**, 791

Sodium methoxide, **I**, 205, 219, 431; **IV**, 29, 210, 278, 380, 382, 516, 594, 624, 638, 650, 744; **V**, 463, 874, 909, **912**

Sodium 2-methoxyethoxide, **V**, 684, 986

Sodium β-naphthalenesulfonate, **V**, 816

Sodium nitrate, **I**, 464; **II**, 566; **IV**, 498

SODIUM NITROMALONALDEHYDE
    MONOHYDRATE, **IV, 844**
    (Warning), **V, 1004**
Sodium perborate, **V,** 341
Sodium phenoxide, **IV,** 590, **592**
Sodium phenyl-*aci*-nitroacetonitrile, **II, 512**
Sodium 2-phenyl-2-hydroxyethane-1-sulfon-
    ate, **IV, 850**
Sodium polysulfide, **V,** 1067
SODIUM β-STYRENESULFONATE,
    IV, 846
Sodium sulfate, **III,** 130
Sodium sulfide, **I,** 220; **II,** 576, 580; **III,** 82,
    86
Sodium sulfide nonahydrate, **IV,** 31, 295,
    893
Sodium sulfite, **I,** 7, 443; **II,** 558, 564; **III,**
    186, 209, 341, 747; **IV,** 346, 529,
    674
Sodium thiocyanate, **II,** 366; **III,** 76
Sodium thiosulfate, **IV,** 105, 158, 319, 543,
    896, 935
SODIUM *p*-TOLUENESULFINATE, **I,**
    492, 519
Sodium *p*-toluenesulfonate, **I,** 175
Sommelet reaction, **V,** 668
SORBIC ACID, **III, 783**
Stannic chloride, **II,** 9; **III,** 326; **IV,** 881,
    900
Stannous chloride, **III,** 240, 453, 626, 818
Stannous chloride dihydrate, **II,** 36, 40,
    130, 255, 290, 393
    as antioxidant, **II,** 36, 40
Starch, **I,** 335; **IV,** 469, 896
Stearic acid, **IV,** 142, 854
STEAROLIC ACID, **III, 785; IV, 851**
STEARONE, **IV, 854**
cis-STILBENE, **IV, 857**
trans-STILBENE, **III,** 350, **786; IV,** 375,
    378, 858, 860; **V,** 606
meso-Stilbene dibromide, **V, 606**
Stilbene dibromide, **IV,** 858
trans-STILBENE, 4,4'-DIMETHOXY,
    V, 428
trans-STILBENE OXIDE, **IV,** 375, **860**
Stopper protection, **I,** 343, 423, 546

Stoppers, asbestos-sodium silicate, **I,** 20
STYRENE, **I, 440,** 494; **IV,** 665, 786, 846;
    V, 1005, 1124
Styrene dibromide, **IV,** 764, 858
STYRENE, β,β-DIFLUORO-α-(TRI-
    FLUOROMETHYL)-, **V, 949**
STYRENE OXIDE, **I, 494; IV,** 12
Styrenes, substituted, **III, 205**
β-Styrenesulfonic acid; benzylthiuronium,
    p-chlorobenzylthiuronium, ani-
    line, and *p*-toluidine salts, **IV,**
    850
β-STYRENESULFONYL CHLORIDE,
    IV, 846
STYRYLPHOSPHONIC DICHLOR-
    IDE, **V, 1005**
Suberic acid, **IV,** 225
SUCCINIC ACID, **I, 64; II,** 560, 562,
    578; **IV,** 390, 500
    disodium salt, **IV, 673**
SUCCINIC ACID, 2,2-DIFLUORO-, **V,**
    393
    α-ETHYL-α-METHYL-, **V, 572**
SUCCINIC ANHYDRIDE, **II,** 81, **560;**
    III, 6, 169; **IV,** 342
SUCCINIMIDE, **II,** 19, **562; IV,** 109, 254
SUCCINIMIDE, *N*-IODO-, **V, 663**
Sucrose, **I,** 335; **II,** 393, 545
Sugar, **II,** 545
    powdered, **II,** 393
Sulfanilamide, **III,** 262, 263
Sulfanilic acid, **V,** 47
Sulfanilic acid dihydrate, **II,** 35, 39
Sulfinic esters, **V,** 725
N-SULFINYLANILINE, **V, 504,** 506
Sulfinyl chlorides, **V, 714**
α-Sulfobehenic acid, **IV,** 864
o-SULFOBENZOIC ANHYDRIDE, **I,**
    495
o-Sulfobenzoic imide, **I,** 14
α-Sulfolauric acid, **IV,** 864
α-Sulfomyristic acid, **IV,** 864
Sulfonic acids, identification of, **II,** 483
Sulfonyl chlorides, **V,** 879
α-SULFOPALMITIC ACID, **IV, 862**
    monosodium salt, **IV,** 864

Sulfosalicylic acid, **I**, 247
α-Sulfostearic acid, **IV**, 864
Sulfoxides, **V**, 792
Sulfur, **I**, 220; **II**, 87, 424, 485, 580; **III**,
        560, 729; **IV**, 31, 244, 295, 308,
        667
Sulfur chloride, **II**, 242
Sulfur dioxide, **I**, 26, 318, 443; **II**, 76, 315,
        335; **IV**, 973, 974; **V**, 1005
Sulfuric acid, fuming, **I**, 10, 353; **II**, 540;
        III, 289, 796; **IV**, 201, 364, 610,
        848, 978
Sulfur tetrafluoride, **V**, 397, 1082, 1084
Sulfur trioxide, **IV**, 846, 863; **V**, 226
Sulfuryl chloride, **II**, 111; **III**, 76, 268; **IV**,
        162, 592; **V**, 475, 636
Sunflower-seed oil, **III**, 526
Sydnones, 3- and 3,4-substituted, **V**, 964
SYRINGIC ALDEHYDE, **IV**, 866

TAPA, See (+) and (−)-α-(2,4,5,7-Tetra-
        nitro-9-fluorenylideneamino-
        oxy)-propionic acid, **V**, 1034
d-Tartaric acid, **I**, 475, 497; **II**, 507; **III**,
        471; **IV**, 242, **V**, 932
dl-TARTARIC ACID, **I**, 497
meso-TARTARIC ACID, **I**, 497
TAURINE, **II**, 563
TEREPHTHALALDEHYDE, **III**, 788
TEREPHTHALIC ACID, **III**, 791; **V**,
        815
1,3,4,6-Tetra-O-acetyl-2-amino-2-deoxy-α-
        D-glucopyranose hydrochloride,
        V, 3
TETRAACETYLETHANE, **IV**, 869
2,3,4,6-TETRA-O-ACETYL-D-GLY-
        CONIC ACID MONOHY-
        DRATE, **V**, 887
Tetraallyltin, **V**, 454
TETRABROMOSTEARIC ACID, **III**,
        526
α,α,α′,a′-TETRABROMO-o-XYLENE,
        IV, 807
α-α-α′-α′-TETRABROMO-p-XYLENE,
        III, 788
Tetra-n-butyltin, **IV**, 882

1,1,1,2-Tetrachloro-2,2-difluoroethane, **IV**,
        269
dl-4,4′,6,6′-TETRACHLORODIPHENIC
        ACID, **IV**, 872
3,3′-4, 4′-Tetrachloro-α,α′-epoxybibenzyl,
        V, 360
Tetrachloroethane, **III**, 549, 584
Tetrachloroethylene, **II**, 312; **IV**, 863
2,3,5,6-Tetrachloro-4-methylacetophenone,
        V, 120
Tetracyanoethylene, **V**, 1007, **1008**, 1013,
        1014
TETRACYANOETHYLENE, **IV**, 244,
        276, **877**, 953
TETRACYANOETHYLENE OXIDE,
        V, **1007**, 1010
TETRADECAMETHYLENE GLY-
        COL, **II**, 155
Tetradecanedioic acid, **V**, 538
Tetradecyl bromide, **II**, **247**
n-Tetradecyl fluoride, **IV**, 527
1,1,1′,1′-TETRAETHOXYETHYL
        POLYSULFIDE, **IV**, **295**, 296
1,1,3,3-Tetraethoxypropane, **V**, 703
Tetraethylene glycol, **IV**, 614
Tetraethylthiuram disulfide, **IV**, 308
TETRAETHYLTIN, **IV**, **881**
2,2,3,3-Tetrafluoropropionic acid, **V**, 242
1,2,3,4-TETRAHYDROCARBAZOLE,
        IV, **884**
endo-TETRAHYDRODICYCLOPEN-
        TADIENE, **V**, 16
5,6,7,8-Tetrahydroflavone, **V**, 723
TETRAHYDROFURAN, **II**, **566**, 571;
        III, 187; **IV**, 258, 266, 321, 355,
        474, 792, 965 (Warning), **V**, **976**
Tetrahydrofurfuryl alcohol, **III**, 276, 693,
        698, 793, 833; **IV**, 500, 756
TETRAHYDROFURFURYL BROM-
        IDE, **III**, **793**
TETRAHYDROFURFURYL CHLOR-
        IDE, **III**, **698**; **IV**, 755
β-(TETRAHYDRO-
        FURYL)PROPIONIC ACID,
        III, **742**
4,5,6,7-TETRAHYDROINDAZOLE, **IV**,
        537

1,4,5,6-Tetrahydro-3-methoxybenzoic acid, **V, 402**

2,3,4,5-Tetrahydro-1-methyl-7-(methylthio)-1*H*-azepine, **V, 784**

1,2,3,4-Tetrahydro-1-methyl-6-(methylthio)pyridine, **V, 784**

1*a*,2,3,7*b*-Tetrahydro-1*H*-naphth(1,2-*b*)azirine, **V, 86**

ar-TETRAHYDRO-α-NAPHTHOL, **IV, 887**

5,6,7,8-Tetrahydro-2-naphthylacetamide, **IV, 762**

ac.-TETRAHYDRO-β-NAPHTHYL-AMINE, **I, 499**

cis-Δ⁴-TETRAHYDROPHTHALIC ANHYDRIDE, **IV, 890,** 304, 306

TETRAHYDROPYRAN, **III, 692, 794; IV, 796**

Tetrahydroquinoline, **III, 504**

TETRAHYDROTHIOPHENE, **IV, 892**

1,2,3,4-Tetrahydro-*p*-toluic acid, **V, 402**

TETRAHYDROXYMETHYL METHANE, **I, 425**

TETRAHYDROXYQUINONE, **V, 1011**

TETRAIODOPHTHALIC ANHYDRIDE, **III, 796**

Tetralin, **IV, 522, 895**

TETRALIN HYDROPEROXIDE, **IV, 895**

α-TETRALONE, **II, 569; III, 729, 798; IV, 885, 898**

β-TETRALONE, **IV, 903**
    bisulfite adduct, **IV, 904**

1-Tetralone oxime, **V, 86**

1,1,1′,1′-Tetramethoxyethyl polysulfide, **IV, 297**

α,α,α′,α′-Tetramethyladipic acid, **V, 1028**

α,α,α′,α′-Tetramethyladiponitrile, **V, 1028**

Tetramethylammonium chloride, **V, 1014**

TETRAMETHYLAMMONIUM 1,1,2,-3,3-PENTACYANOPROPENIDE, **V, 1013**

TETRAMETHYLBIPHOSPHINE, DISULFIDE, **V, 1016**

2,2,3,3-Tetramethylbutyric acid, **V, 742**

Tetramethyl-1,3-cyclobutanedione, **V, 456**

3,4,6,6-Tetramethyl-2,4-cyclohexadienone, **V, 600**

2,2,3,3-Tetramethylcyclopropanol, **V, 861**

TETRAMETHYLENE CHLOROHYDRIN, **II, 571**

Tetramethylethylene, **V, 299, 601**

2,3,4,6-TETRAMETHYL-*d*-GLUCOSE, **III, 800**

N,N,N′,N′-Tetramethylmethylenediamine, **V, 435**

2,2,6,6-TETRAMETHYLOLCYCLOHEXANOL, **IV, 907**

TETRAMETHYL-*p*-PHENYLENEDIAMINE, **V, 1018**

2,3,5,6-Tetramethyl-4*H*-pyran-4-one, **V, 452**

2,3,4,5-TETRAMETHYLPYRROLE, **V, 1022**

Tetramethylsuccinonitrile, **IV, 273**

2,2,5,5-TETRAMETHYLTETRAHYDRO-3-KETOFURAN, **V, 1024**

α,α,α′,α′-Tetramethyltetramethylenediamine, **V, 1028**

α,α,α′,α′-TETRAMETHYLTETRAMETHYLENE GLYCOL, **V, 1026**

2,4,5,7-TETRANITROFLUORENONE, **V, 1029, 1034**

(+)- AND (−)-α-(2,4,5,7-TETRANITRO-9-FLUORENYLIDENEAMINOOXY)-PROPIONIC ACID, **V, 1031**

TETRANITROMETHANE, **III, 803**

Tetraphenylarsonium bromide, **IV, 913**

TETRAPHENYLARSONIUM CHLORIDE HYDROCHLORIDE, **IV, 910**

TETRAPHENYLCYCLOPENTADIENONE, III, 806, 807; **V, 605, 1037**

1,1,2,2-Tetraphenylethane, **V, 525**

TETRAPHENYLETHYLENE, **IV, 914**

2,3,5,5-Tetraphenylisoxazolidine, **V, 1126**

1,2,3,4-TETRAPHENYLNAPHTHAL-

ENE, **V, 1037**
TETRAPHENYLPHTHALIC ANHY-
DRIDE, **III, 807**
2,2,3,3-Tetraphenylpropionitrile, **IV, 964**
Tetraphenyltin, **V,** 453
Tetraphosphorus decasulfide, **V,** 218
Tetra-*n*-propyltin, **IV,** 882
Tetravinyltin, **IV,** 260
TETROLIC ACID, **V, 1043**
2-THENALDEHYDE, **IV, 915**
3-THENALDEHYDE, **IV, 918,** 920
3-THENOIC ACID, **IV, 919**
3-THENYL BROMIDE, **IV,** 918, **921**
Thian 1-oxide, **V,** 792
Thianthrene, **II,** 243
Thiazolidine-2-thiones, **V,** 87
bis-(2-Thienyl)methane, **III,** 198
THIIRANE, 2-CHLORO-1,1-DIOXIDE,
V, 231
Thioanisole, **V,** 791
THIOBENZOIC ACID, **IV, 924**
THIOBENZOPPHENONE, **II, 573; IV,**
927
THIOBENZOYLTHIOGLYCOLIC
ACID, **V, 1046**
THIOCARBONYL PERCHLORIDE, **I,**
506
m-THIOCRESOL, **III, 809;** (warning), **V,**
1050
p-THIOCYANODIMETHYLANILINE,
II, 574
β-THIODIGLYCOL, **II, 576**
2*a*-Thio-1-homo-1,2-naphthalimide, **V,** 1053
2*a*-THIOHOMOPHTHALIMIDE, **V,**
612, 1051
THIOLACETIC ACID, **IV, 928**
2-THIO-6-METHYLURACIL, **IV, 638**
1*a*-Thio-1,2-naphthalimide, **V,** 1053
Thionyl chloride, **I,** 13, 147, 495; **II,** 113,
136, 159, 169, 569; **III,** 29, 170,
490, 547, 556, 613, 623, 699,
712, 714; **IV,** 88, 154, 169, 263,
333, 339, 436, 556, 561, 571,
621, 715, 739, 937; **V,** 197, 504,
977
purification of, **II,** 570

Thionyl fluoride, **V,** 1083
THIOPHENE, **II,** 9, 357, 466, 521, **578;**
III, 14, 198; **IV,** 545, 673, 916,
980; **V,** 149
recovery of, **II,** 357
THIOPHENE, 3-BROMO-, **V, 149**
THIOPHENE-2-OL, **V, 642**
Thiophene-2*a*-thio-2,3-dicarboximide, **V,**
1053
THIOPHENOL, **I, 504**
2(5*H*)-THIOPHENONE, **V, 642**
THIOPHOSGENE, **I,** 165, **506**
2-THIOPHOSPHENEALDEHYDE, **III,**
811
Thiophosphoryl chloride, **V,** 1016
THIOSALICYLIC ACID, **II, 580**
Thiosemicarbazide, **V,** 1070
Thiourea, **II,** 31, 411; **III,** 363, 440, 751;
IV, 401, 491, 566, 638; **V,** 703
Thorium oxide catalyst, **II,** 389, **390**
dl-THREONINE, **III, 813**
Thymol, **I,** 511
THYMOQUINONE, **I, 511**
Tiglyaldehyde diethyl acetal, **IV, 22**
Tin, **I,** 60, 456, 509; **II,** 132, 175, 617; **V,**
339
Titanium tetrachloride, **V,** 49
o-Tolidine, **II,** 145
o-TOLUALDEHYDE, **III, 818; IV, 932**
p-TOLUALDEHYDE, **II, 583,** 590
p-TOLUALDEHYDE, 2-BROMO-, **V,**
139
o-TOLUAMIDE, **II, 586,** 589; (warning),
V, 1054
o-Toluanilide, **III,** 818
Toluene, **I,** 229, 517; **II,** 264, 499, 583; **III,**
167, 220, 446, 544, 590, 649; **IV,**
38, 192, 204, 273, 461, 788
p-Toluenediazonium chloride, **V,** 797
TOLUENE, α,α-DIFLUORO-, **V, 396**
p-TOLUENESULFENYL CHLORIDE,
IV, 934
p-Toluenesulfinic acid, sodium salt, **IV,**
674
sodium salt, dihydrate, **IV, 937**
p-TOLUENESULFINYL CHLORIDE,

IV, 937
p-Toluenesulfonamide, III, 646; V, 181, 1064
p-Toluenesulfonic acid, III, 146, 203, 503, 610; V, 33, 277, 539, 808
p-Toluenesulfonic acid esters, III, 366
p-TOLUENESULFONIC ACID, HYDRAZIDE, V, 1055
p-Toluenesulfonic acid, monohydrate, IV, 304, 306, 534, 719, 940
p-TOLUENESULFONIC ANHYDRIDE, IV, 940
p-TOLUENESULFONYLANTHRANILIC ACID, IV, 34, 35
p-TOLUENESULFONYL AZIDE, V, 179, 180
p-Toluenesulfonyl chloride, I, 145, 146, 492; III, 366; IV, 35, 674, 943; V, 40, 179, 181, 555, 557, 772, 1055
p-TOLUENESULFONYLHYDRAZIDE, V, 258, 1055
p-Toluenethiol, IV, 934
o-TOLUIC ACID, II, 588; III, 820
p-Toluic acid, II, 589; III, 822
m-Toluidine, II, 290; III, 827
o-Toluidine, I, 135, 170, 514; III, 765, 824; IV, 42, 147; V, 650, 1062
p-Toluidine, I, 111, 136, 171, 330, 515; III, 76; V, 797
p-Toluidine salts of 2- and 3-phenanthrenesulfonic acids, II, 483
o-TOLUIDINESULFONIC ACID, III, 824
m-TOLUIDINE, α,α,α-TRIFLUORO-N,N-DIMETHYL-, V, 1085
o-TOLUNITRILE, I, 514; II, 586, 588; V, 481
p-TOLUNITRILE, I, 514; IV, 714
Toluquinone, IV, 151, 152
p-TOLUYL-o-BENZOIC ACID, I, 353, 517
m-TOLYLBENZYLAMINE, III, 827
m-Tolyl t-butyl ether, V, 925
o-Tolyl t-butyl ether, V, 925
p-TOLYL CARBINOL, II, 590

1-p-TOLYLCYCLOPROPANOL, V, 1058
p-Tolyl disulfide, IV, 936
α-p-Tolylethylamine, II, 505
N-o-Tolylformamide, V, 1060, 1062
p-Tolylglyoxal, V, 940
o-TOLYLISOCYANIDE, V, 1060
p-TOLYLMERCURIC CHLORIDE, I, 159, 231, 519
1-(p-Tolyl)-3-phenyl-2-propanone, IV, 177
Tolylphenylsulfur trifluoride, V, 961
1-m-Tolylpiperidine, IV, 797
1-o-Tolylpiperidine, IV, 797
1-p-Tolylpiperidine, IV, 797
2-(p-Tolylsulfonyl)benz(f)isoindoline, V, 1066
2-(p-TOLYLSULFONYL)DIHYDROISOINDOLE, V, 406, 1064
p-Tolylsulfonylmethylamide, IV, 253
p-TOLYLSULFONYLMETHYLNITROSAMIDE, IV, 225, 251, 943
1-(p-Tolylsulfonyl)pyrrolidine, V, 1066
p-Tolylthiourea, III, 77
m-Tolylurea, IV, 51
o-Tolylurea, IV, 51
p-Tolylurea, IV, 51
Trialkylnitromethanes, V, 847
Trialkyl phosphates, V, 1087
Triallylphenyltin, V, 454
2,4,5-TRIAMINONITROBENZENE, V, 1067
Tri-n-amyl carbinol, II, 603
TRIAZENE, 1-METHYL-3-p-TOLYL-, V, 797
1,2,4-TRIAZOLE, V, 1070
1H-1,2,4-TRIAZOLE, V, 1070
1,2,4-TRIAZOLE-3(5)-THIOL, V, 1071
1,3,5-TRIACETYLBENZENE, III, 829
TRIBIPHENYLCARBINOL, III, 831
p,α,α-TRIBROMOACETOPHENONE, IV, 110, 111
Tribromoaniline, II, 592
sym.-TRIBROMOBENZENE, II, 592
2,4,6-TRIBROMOBENZOIC ACID, IV, 947

1,2,3-TRIBROMOPROPANE, **I**, 209, **521**
2,3,5-Tribromothiophene, **V**, 149
Tri-*n*-butyl carbinol, **II**, **603**
Tributylphosphine, **V**, 388, 949
Tri-*n*-butylphosphine oxide, **V**, 389
Tri-*n*-butylvinyltin, **IV**, 260
TRICARBALLYLIC ACID, **I**, **523**
TRICARBETHOXYMETHANE, **II**, **594**
TRICARBOMETHOXYMETHANE, **II**, **596**
N-Tricarboxylic ester, **III**, 404
TRICHLORAMINE, **V**, **35**, 36
α,α,α-Trichloroacetamide, **V**, 1075
α,α,α-TRICHLOROACETANILIDE, **V**, **1074**
Trichloroacetyl chloride, **V**, 387
Trichloroacetyl fluoride, **V**, 69
1,2,4-Trichlorobenzene, **III**, 765
TRICHLOROETHYL ALCOHOL, **II**, **598**
Trichloroethylene, **IV**, 430
2-(TRICHLOROMETHYL)BICYCLO-(3.3.0)OCTANE, **V**, **93**
TRICHLOROMETHYLPHOSPHONYL DICHLORIDE, **IV**, **950**
1,1,3-TRICHLORO-*n*-NONANE, **V**, **1076**
1,1,2-TRICHLORO-2,3,3-TRIFLUORO-CYCLOBUTANE, **V**, **393**
1,1,2-Trichloro-1,2,2-trifluoroethane, **V**, 959
Tricontanoic acid, **V**, 538
p-TRICYANOVINYL-*N*,*N*-DIMETHYL-ANILINE, **IV**, **953**
TRICYCLO(3.3.1.1^{3,7})DECANE, **V**, **16**
TRICYCLO(2.2.1.0^{2,6})HEPTAN-3-OL, **V**, **863**
TRICYCLO(2.2.1.0^{2,6})HEPTAN-3-ONE, **V**, **866**
TRICYCLO(8.2.2.2^{4,7})HEXADECA-4,6,-10,12,13,15-HEXAENE, **V**, **883**
TRIDECAMETHYLENE GLYCOL, **II**, **155**
2,4-Tridecanedione, **V**, 851
Triethylamine, **IV**, 286, 557, 560, 561, 730,

777; **V**, 256, 297, 514, 822, 973
Triethylamine hydrochloride, **IV**, 557, 560, 561
Triethylbenzaldehyde, **III**, **551**
TRIETHYL CARBINOL, **II**, **602**
Triethylenetetramine, **V**, 36, 37
TRIETHYL 2-METHYL-1,1,3-PROPA-NETRICARBOXYLATE, **IV**, **630**
sodio derivative, **IV**, 631
Triethyl orthoacetate, **IV**, 802 (see Ethyl orthoformate)
Triethyl orthoformate, **V**, 115, 582
TRIETHYL OXALYSUCCINATE, **V**, **687**, 688
TRIETHYLOXONIUM FLUOBOR-ATE, **V**, **1080**, 1081, 1096
Triethyl phosphite, **V**, 893, 941, 943
TRIETHYL PHOSPHITE, **IV**, 326, **955**
Triethyl phosphonoacetate, **V**, 509, 547
TRIETHYL α-PHTHALIMIDO-ETHANE-α,α-β-TRICARBOX-YLATE, **IV**, **55**, 56
Triethylvinyltin, **IV**, 260
Trifluoroacetic acid, **IV**, 547
Trifluoroacetic anhydride, **V**, 367, 598
α,α,α-Trifluoroacetophenone, **V**, 949
(4,4,4-Trifluorobutyl)cyclohexane, **V**, 1084
1,1,1-Trifluorododecane, **V**, 1084
1,1,1-TRIFLUOROHEPTANE, **V**, 1082
1,1,1-Trifluorohexadecane, **V**, 1084
Trifluoromethanesulfonic acid, **V**, 1114
m-Trifluoromethylaniline, **V**, 1085
p-Trifluoromethylbenzaldehyde, **IV**, **933**
m-TRIFLUOROMETHYL-*N*,*N*-DI-METHYLANILINE, **V**, **1085**
p-Trifluoromethylnitrobenzene, **V**, 1084
m-(TRIFLUOROMETHYL)PHENYL ISOTHIOCYANATE, **I**, **449**
m-Trifluoromethylphenylmethylcarbinol, **III**, **201**
β,β,p-Trifluorostyrene, **V**, 392
1,1,1-Trifluoro-3,5,5-trimethylhexane, **V**, 1084
Tri-*n*-heptyl carbinol, **II**, **603**
1,2,4-Trihydroxybutane, **IV**, 534

1,2,5-TRIHYDROXYPENTANE, **III, 833**

1,2,3-TRIIODO-5-NITROBENZENE, **II, 604**

Triisopropylbenzaldehyde, **III,** 551

Triisopropyl phosphite, **IV,** 325, 326, 956; **V,** 901

1,3,5-Triketones, **V,** 720

1,3,5-Trimercaptobenzene, **V,** 420

2,4,6-Trimercaptomesitylene, **V,** 420

3,4,5-Trimethoxybenzonitrile, **V,** 657

TRIMETHYLACETIC ACID, **I, 524; V,** 21, 742

TRIMETHYLAMINE, **I, 528; II,** 85; **IV,** 85, 86, 99

TRIMETHYLAMINE HYDRO-CHLORIDE, **I, 531; IV,** 86

Trimethylamine oxide, **V,** 872

4,6,8-TRIMETHYLAZULENE, **V, 1088**

Trimethylbenzene, technical, **IV,** 144

Trimethyl borate, **V,** 818

2,2,3-Trimethylbutanoic acid, **V,** 22

2,6,6-TRIMETHYL-2,4-CYCLOHEX-ADIENONE, **V, 1092,** 1093

3,3,5-Trimethyl-1,2-cyclohexanedione, **IV,** 958

2,2,4-TRIMETHYLCYCLOPENTAN-ONE, **IV, 957,** 958

N,N,N-TRIMETHYLCYCLOOCTYL-AMMONIUM HY-DROXIDE, **V,** 316

N,N,N-TRIMETHYLCYCLOOCTYL-AMMONIUM IODIDE, **V,** 315

2,2,3-Trimethylcyclopropanol, **V,** 861

TRIMETHYLENE BROMIDE, **I, 30,** 435, 536

Trimethylene bromide, **III,** 214, **228**

TRIMETHYLENE CHLOROBROM-IDE, **I,** 156, **157; III,** 504; **IV,** 288

TRIMETHYLENE CHLOROHYDRIN, **I,** 168, **533; III,** 203

TRIMETHYLENE CYANIDE, **I,** 289, **536**

Trimethylene glycol, **I,** 31, 533

TRIMETHYLENE OXIDE, **III, 835**

TRIMETHYLETHYLENE DIBROM-IDE, **II, 409**

TRIMETHYLGALLIC ACID, **I, 537**

N-(3,5,5-Trimethylhexyl)allenimine, **V, 544**

1,1,1-Trimethylhydrazinium hydrogen oxa-late, **V,** 45

Trimethyl *n*-orthovalerate, **IV,** 802

TRIMETHYLOXONIUM FLUOBOR-ATE, **V, 1096,** 1097, 1098

TRIMETHYLOXONIUM 2,4,6-TRINI-TROBENZENESULFON-ATE, **V, 1099**

Trimethyloxosulfonium iodide, **V,** 755

2,2,4-Trimethyl-3-oxovaleric acid, **V,** 458

2,2,4-TRIMETHYL-3-OXOVALERYL CHLORIDE, **V, 1103**

2,4,6-Trimethylphenol, **V,** 408

1,1,3-Trimethyl-3-phenylindane, **IV, 666**

Trimethyl phosphate, **V,** 1085

2,4,6-TRIMETHYLPYRYLIUM PERCHLORATE, **V,** 1088, **1106, 1108,** 1110

2,4,6-TRIMETHYLPYRYLIUM TE-TRAFLUOBORATE, **V, 1112**

2,4,6-TRIMETHYLPYRYLIUM TRI-FLUOROMETHANESUL-FONATE, **V, 1114**

Trimethylvinyltin, **IV,** 260

TRIMYRISTIN, **I,** 379, **538**

1,3,5-TRINITROBENZENE, **I,** 220, **541**

2,4,6-Trinitrobenzenesulfonic acid, **V,** 1099

2,4,6-TRINITROBENZOIC ACID, **I,** 455, 541, **543; V,** 1101

2,4,7-TRINITROFLUORENONE, **III, 837**

2,4,6-Trinitrophenylhydrazine, **II, 229**

Trinitrotoluene, **I,** 543

Trioxane, **IV,** 518

Trioxymethylene, **III,** 196; **V,** 833

TRIPHENYLALUMINUM, **V, 1116**

TRIPHENYLAMINE, **I, 544**

2,4,6-TRIPHENYLANILINE, **V, 1130**

TRIPHENYLARSINE, **IV, 910**

TRIPHENYLARSINE OXIDE, **IV, 911**

2,3,3-Triphenylbutyronitrile, **IV, 964**

TRIPHENYLCARBINOL, **III, 839,** 841; **IV,** 47

TRIPHENYLCHLOROMETHANE, **II**,
    607; **III**, 432, 841
TRIPHENYLCINNAMYLPHOS-
    PHONIUM CHLORIDE, **V**,
    499
TRIPHENYLENE, **V**, **1120**
1,1,2-Triphenylethane, **V**, 525
TRIPHENYLETHYLENE, **II**, **606**; **V**,
    952
2,3,5-TRIPHENYLISOXAZOLIDINE,
    V, **1124**
TRIPHENYLMETHANE, **I**, **548**
**TRIPHENYLMETHYLPHOSPHON-
    IUM BROMIDE, V, 751**, 752
TRIPHENYLMETHYLSODIUM, **II**,
    268, **607**
2,4,6-TRIPHENYLNITROBENZENE,
    V, **1128**
2,4,6-TRIPHENYLPHENOL, **V**, **1131**
2,4,6-TRIPHENYLPHENOXYL, **V**,
    1130
Triphenylphosphine, **V**, 142, 249, 251, 362,
    390, 490, 751, 949, 985, 1145
Triphenylphosphine oxide, **V**, 362
1,1,2-Triphenylpropane, **V**, 525
1,1,3-Triphenylpropane, **V**, 525
α,β,β-TRIPHENYLPROPIONIC ACID,
    IV, **960**; **V**, 528
α,α,β-TRIPHENYLPROPIONITRILE,
    IV, **962**
2,4,6-TRIPHENYLPYRYLIUM TE-
    TRAFLUOROBORATE, **V**,
    1128, **1135**
TRIPHENYLSELENONIUM CHLOR-
    IDE, **II**, **240**
TRIPHENYLSTIBINE, **I**, **550**
Triphenyltin chloride, **V**, 453
Triphenylvinyltin, **IV**, 260
Tri-n-propyl carbinol, **II**, **603**
Tri-n-propylvinyltin, **IV**, 260
TRIPTYCENE, **IV**, **964**
2,2′,2″-Tripyridine, **V**, 166
sym.-TRITHIANE, **II**, **610**; **V**, 231
TRITHIOCARBODIGLYCOLIC
    ACID, **IV**, 7, **967**

TRI-p-TOLYLSTIBINE, **I**, **551**
Triton B, **IV**, 652
6-TRITYL-β-d-GLUCOSE-1,2,3,4-TE-
    TRAACETATE, **III**, **432**
Tropilidene, **V**, 1138
TROPYLIUM FLUOBORATE, **V**, **1138**
Trypan blue, **II**, 149
TRYPARSAMIDE, **I**, **488**
l-TRYPTOPHANE, **II**, **612**
l-TYROSINE, **II**, **612**

UNDECAMETHYLENE GLYCOL, **II**,
    155
γ-Undecanoic lactone, **IV**, 351
10-Undecenoic acid, **IV**, 969
Undecylenyl alcohol, **II**, **374**
n-Undecyl fluoride, **IV**, **527**
UNDECYL ISOCYANATE, **III**, **846**
10-UNDECYNOIC ACID, **IV**, **969**
URAMIL, **II**, **617**
Urea, **I**, 140, 279, 453; **II**, 60, 231, 422,
    461; **III**, 130, 152, 768; **IV**, 52,
    157, 247, 276, 513, 719, 744
UREA, CYCLOHEXYL-, **V**, **801**
UREA, 1-(2,6-DIMETHYLPHENYL)-2-
    THIO-, **V**, **802**
UREA, HYDROXY-, **V**, **645**
Urea nitrate, **I**, 417
Urethan, **III**, 415
Uric acid, **II**, 21; **III**, 42

Vacuum desiccator, **III**, 845
n-VALERIC ACID, **I**, **363**
γ-Valerolactone, **IV**, 280, 899
VALERYL CHLORIDE, 2,2,4-TRIME-
    THYL-3-OXO-, **V**, **1103**
Valeryl fluoride, **V**, 69
dl-VALINE, **III**, **848**
Vanadium pentoxide, **I**, 290; **II**, **302**, 553
VANILLIC ACID, **IV**, **972**
Vanillin, **II**, 619; **III**, 587, 745; **IV**, 203,
    972, 974
VERATRALDEHYDE, **II**, 55, **619**, 622;
    IV, 410, 605, 735
Veratraldoxime, **II**, **622**
VERATRIC AMIDE, **II**, **44**

Veratrole, **IV,** 548
VERATRONITRILE, **II,** 44, **622**
Vinyl acetate, **III,** 123; **IV,** 977; **V,** 235
VINYLACETIC ACID, **III, 851**
VINYLAMINE, 1,2,2-TRICHLORO-
N,N-DIETHYL-, **V, 387**
Vinyl bromide, **IV,** 258
Vinyl caprate, **IV, 978**
Vinyl caproate, **IV, 978**
Vinyl caprylate, **IV, 978**
VINYL CHLOROACETATE, **III, 853**
VINYL ESTERS, **IV, 977**
Vinyl 10-hendecenoate, **IV, 978**
Vinylidene chloride, **V,** 684
VINYL LAURATE, **IV, 977**
Vinyllithium, **V,** 455
Vinylmagnesium bromide, **IV, 260**
Vinyl myristate, **IV, 978**
Vinyl oleate, **IV, 978**
Vinyl palmitate, **IV, 978**
Vinyl pelargonate, **IV, 978**
Vinyl stearate, **IV, 978**
2-VINYLTHIOPHENE, **IV, 980**
VINYL TRIPHENYLPHOSPHONIUM
BROMIDE, **V, 1147**
Vinyl undecylenate, **IV, 978**
Viscolizer, **I,** 310

Water separator, **III,** 383, 503
Wittig reaction, **V,** 361, 499, 547, 751
Witt-Utermann solution, **IV,** 44
Wolff-Kishner reduction, **V,** 534
XANTHONE, **I, 552,** 554
XANTHYDROL, **I, 554**
α-p-Xenylethylamine, **II, 505**

Xylene, **I,** 231; **II,** 248; **III,** 17, 19, 367,
387, 510, 552, 628, 829; **IV,** 313,
474, 840
o-Xylene, **III,** 138, 820; **IV,** 807, 933, 984;
V, 467
p-Xylene, **III,** 326, 789
D-Xylose, **IV,** 506
o-Xylyl bromide, **IV,** 932
p-XYLYLENE-
BIS(TRIPHENYLPHOSPHO
NIUM) CHLORIDE, **V, 985**
o-XYLYLENE DIBROMIDE, **IV, 984;**
V, 1064

Yeast, **II,** 545

Zinc, **I,** 90, 445, 492, 504; **II,** 203, 212,
243, 320, 418, 446, 448, 501,
526, 581; **III,** 408, 410, 444, 514,
527, 532, 668, 786; **V,** 149, 374,
394, 1023
Zinc, amalgamated, **II, 499; IV,** 203, 204,
695
dust, **IV,** 196, 198, 218, 268, 642, 750;
III, 73, 103
foil, **IV,** 121, 445
turnings, **IV,** 348
Zinc chloride, **I,** 142, 143, 173; **II,** 304,
522; **III,** 78, 141, 187, 422, 725,
761, 833; **IV,** 268, 750, 802; **V,**
888, 1103
ZINC-COPPER COUPLE, **II,** 184, **185;**
V, 330, **855,** 857
Zinc cyanide, **III,** 549
Zinc iodide, **IV,** 801, 802
Zinc nitrate, **IV,** 802

APPENDIX A

CONTENTS OF ANNUAL VOLUMES, 50–54

This index shows the Title Names of the compounds for which procedures are given in the recent *Annual Volumes,* 50 through 54. These volumes published from 1970–1974 bring up-to-date the checked directions published since the appearance of *Collective Volume* V, which had the revised procedures for *Annual Volumes,* 40 through 49.

The syntheses of certain classes of compounds by selective unique reactions are shown in these volumes. For example, different methods for preparing molecules containing the aldehyde functional group are given. The consolidated index in the back of Volume 54 should be consulted to locate citations to these special reactions.

ACETIC FORMIC ANHYDRIDE, **50,** 1
ACETONE HYDRAZONE, **50,** 3
ACETOPHENONE HYDRAZONE, **50,** 102
3β-ACETOXY-5α-CYANOCHOLESTAN-7-ONE, **52,** 100
p-ACETYL-α-BROMOHYDROCINNAMIC ACID, **51,** 1
2-ACETYLCYCLOPENTANE-1,3-DIONE, **52,** 1
3-ACETYL-2,4-DIMETHYLFURAN, **53,** 1
2-ACETYL-6-METHOXYNAPHTHALENE, **53,** 5
ADAMANTANONE, **53,** 8
ANDROSTAN-17-β-OL, **52,** 122
o-ANISALDEHYDE, **54,** 42
[18]ANNULENE, **54,** 1
AZETIDINE, **53,** 13
AZOETHANE, **52,** 11

BENZYL CHLOROMETHYL ETHER, **52,** 16
1-BENZYLINDOLE, **54,** 58, 60
2-BENZYL-2-METHYLCYCLOHEXANONE, **52,** 39

2-BENZYL-6-METHYLCYCLOHEXANONE, **52,** 39
BICYCLO[1.1.0]BUTANE, **51,** 55
BICYCLO[3.2.1]OCTAN-3-ONE, **51,** 60
BIPHENYL, **51,** 82
1,1-BIS-(BROMOMETHYL)CYCLOPROPANE, **52,** 22
BIS(CHLOROMETHYL) ETHER (Hazard note), **51,** 148
BIS(TRIFLUOROMETHYL)DIAZOMETHANE, **50,** 6
2-BORNENE, **51,** 66
2-BROMOACETYL-6-METHOXYNAPHTHALENE, **53,** 111
1-BROMO-3-CHLOROCYCLOBUTANE, **51,** 106
t-BUTYL AZIDOFORMATE, **50,** 9
cis-4-*t*-BUTYLCYCLOHEXANOL, **50,** 13
2-*t*-BUTYL-1,3-DIAMINOPROPANE, **53,** 21
3-*n*-BUTYL-2-METHYLHEPT-1-EN-3-OL, **52,** 19
t-BUTYLOXYCARBONYL-L-PROLINE, **53,** 25
3-*n*-BUTYL-2,4-PENTANEDIONE, **51,** 90
t-BUTYL-*p*TOLUATE, **51,** 96

CARBONYL CYANIDE, **51**, 70
3-CHLOROCYCLOBUTANECARBOX-
YLIC ACID, **51**, 73
m-CHLOROPERBENZOIC ACID, **50**, 15
CHOLESTANE, **53**, 86
5.β-CHOLEST-3-ENE-5-ACETALDE-
HYDE, **54**, 71
CINNAMONITRILE, **50**, 18
1-CYANO-6-METHOXY-3,4-DIHY-
DRONAPHTHALENE, **52**, 96
CYCLOBUTADIENEIRON TRICAR-
BONYL, **50**, 21
CYCLOBUTANECARBOXALDEHYDE,
51, 11
CYCLOBUTANONE, **51**, 76; **54**, 84
CYCLOHEXYLIDENEACETALDE-
HYDE, **53**, 104
CYCLOPROPYLDIPHENYLSULF-
ONIUM FLUOROBORATE,
54, 27

1-DECALOL, **51**, 103
n-DECANE, **53**, 107
trans,trans-1,4-DIACETOXY-1,3-BUTA-
DIENE, **50**, 24
DIAMANTANE: PENTACY-
CLO[7.3.1⁴,¹²0²,⁷0⁶,¹¹]TETRA-
DECANE, **53**, 30
1,10-DIAZACYCLOOCTADECANE, **54**,
88
DIAZOACETOPHENONE, **53**, 35
2-DIAZOCYCLOHEXANONE, **51**, 86
2-DIAZOPROPANE, **50**, 27
DIBENZO-18-CROWN-6 POLYETHER,
52, 66
trans-1,2-DIBENZOYL-CYCLOPROP-
ANE, **52**, 33
2,2-DIBROMOACETYL-6-METHOXY-
NAPHTHALENE, **53**, 111
α,α'-DIBROMODIBENZYL SULF-
ONE, **50**, 31
cis-3,4-DICHLOROCYCLOBUTENE, **50**,
36
DI-(*p*-CHLOROPHENYL)ACETIC ACID
(Correction), **51**, 148

DICYCLOHEXYL-18-CROWN-6 POLY-
ETHER, **52**, 66
DIDEUTERIODIAZOMETHANE, **53**, 38
DIETHYLALUMINUM CYANIDE, **52**,
90
DIETHYL *t*-BUTYLMALONATE, **50**, 38
DIETHYL 2-(CYCLOHEXYLAM-
INO)VINYLPHOSPHON-
ATE, **53**, 44
DIETHYL *trans*-Δ⁴-TETRAHYDRO-
PHTHALATE, **50**, 43
DIHYDROCARVONE, **53**, 63
trans-1,2-DIHYDROPHTHALIC ACID,
50, 50
2,4-DIMETHOXYBENZONITRILE, **50**,
52
(2-DIMETHYLAMINO-5-METHYL-
PHENYL)DIPHENYL CAR-
BINOL, **53**, 56
4,5-DIMETHYL-1,2-BENZOQUIN-
ONE, **52**, 88
N,N-DIMETHYL-5.β-CHOLEST-3-ENE-
5-ACETAMIDE, **54**, 77
4,4-DIMETHYL-2-CYCLOHEXEN-1-
ONE, **53**, 48
N,N-DIMETHYLCYCLOHEXYLAM-
INE, **52**, 124
N,N-DIMETHYLDODECYLAMINE OX-
IDE, **50**, 56
2,2-DIMETHYL-4-PHENYLBUTYRIC
ACID, **50**, 58
2,2-DIMETHYL-3-PHENYLPROPION-
ALDEHYDE, **54**, 46
3,5-DINITROBENZALDEHYDE, **53**, 52
2,3-DIPHENYL-1,3-BUTADIENE, **50**, 62
DIPHENYLKETENE, **52**, 36
2,3-DIPHENYLVINYLENE SULF-
ONE, **50**, 65
1,3-DITHIANE, **50**, 72
2,2'-DITHIENYL SULFIDE, **50**, 75
n-DODECANE, **53**, 107

ETHYL 5.β-CHOLEST-3-ENE-5-ACET-
ATE, **54**, 74
2,2-(ETHYLENEDITHIO)CYCLO-
HEXANONE, **54**, 37

ETHYLENE DITHIOTOSYLATE, **54**, 33
ETHYL 3-ETHYL-5-METHYL-4-ISOXA-
    ZOLECARBOXYLATE, **53**, 59
ETHYL 1-HYDROXYCYCLOHEXYL-
    ACETATE, **53**, 66
ETHYL 4-METHYL-E-4,8-NONADI-
    ENOATE, **53**, 116
ETHYL 6-METHYLPYRIDINE-2-ACET-
    ATE, **52**, 75
ETHYL 1-NAPHTHYLACETATE, **50**, 77
ETHYL PYRROLE-2-CARBOXYL-
    ATE, **51**, 100

p-FORMYLBENZENE SULFON-
    AMIDE, **51**, 20

GERANYL CHLORIDE, **54**, 63, 68

1-HEPTANAL, **52**, 5
HEXAFLUOROACETONE IMINE, **50**,
    81
1-HEXANOL, **53**, 77
trans-4-HYDROXY-2-HEXENAL, **54**, 19
threo-4-HYDROXY-3-PHENYL-2-HEPT-
    ANONE, **54**, 49

IODOCYCLOHEXANE, **51**, 44
IODODURENE, **51**, 94
ISOCROTONIC ACID, **53**, 123
(+)-ISOPINOCAMPHEOL, **52**, 59

METHALLYLBENZENE, **52**, 115
1,6-METHANO[10]ANNULENE, **54**, 11
6-METHOXY-β-TETRALONE, **51**, 109
1-d-2-METHYLBUTANAL, **51**, 31
3-METHYL-2-BUTEN-2-YL TRIFLATE,
    54, 79
5-METHYLCOPROST-3-ENE, **52**, 109
1-METHYLCYCLOHEXANOL, **53**, 94
1-METHYL-4,4*a*,5,6,7,8-HEXAHY-
    DRONAPHTHALEN-2(3*H*)-
    ONE, **53**, 70
METHYL (*trans*-2-IODO-1-TE-
    TRALIN)CARBAMATE, **51**,
    112

2-METHYL-2-NITROSOPROPANE AND
    ITS DIMER, **52**, 77
2-METHYL-3-PHENYLPROPIONALDE-
    HYDE, **51**, 17
N-METHYL-2-PHENYL-Δ^2-TETRA-
    HYDROPYRIDINE, **54**, 93

NAPHTHALENE, **52**, 62
2-NAPHTHALENETHIOL, **51**, 139
NEOPENTYL IODIDE, **51**, 44
1-NITROCYCLOOCTENE, **50**, 84
3-NITROPHTHALIC ACID (Hazard
    Note), **53**, 129
D-NORANDROST-5-EN-3β-OL-16-CAR-
    BOXYLIC ACIDS, **52**, 53

$\Delta^{9,10}$-OCTALIN, **50**, 88
1-OCTANOL, **53**, 77
ORCINOL MONOMETHYL ETHER, **53**,
    90

n-PENTADECANAL, **51**, 39
trans-3-PENTEN-2-ONE, **51**, 115
β-PHENYLCINNAMALDEHYDE, **50**, 66
1-PHENYLCYCLOPENTANECARBOX-
    ALDEHYDE, **51**, 24
1-PHENYLCYCLOPENTYLAMINE, **51**,
    48
cis-2-PHENYLCYCLOPROPANECAR-
    BOXYLIC ACID, **50**, 94
2-PHENYLFURO[3;2-*b*]PYRIDINE, **52**,
    128
PHENYL HEPTYL KETONE, **53**, 77
1-PHENYL-1,4-PENTADIYNE AND 1-
    PHENYL-1,3-PENTADIYNE,
    50, 97
1-PHENYL-2,4-PENTANEDIONE, **51**,
    128
1-PHENYL-4-PHOSPHORINANONE, **53**,
    98
4-PHENYL-1,2,4-TRIAZOLINE-3,5-
    DIONE, **51**, 121
trans-PINOCARVEOL, **53**, 17

QUADRICYCLANE, **51,** 133

1,2,3,4-TETRAHYDRO-β-CARBOLINE, **51,** 136
1,2,3,4-TETRAHYDRONAPHTHAL-ENE(1,2)IMINE, **51,** 53
2,2,3,3-TETRAMETHYLIODOCYCLO-PROPANE, **52,** 132
4*H*-1,4-THIAZINE 1,1-DIOXIDE, **52,** 135
2-THIOPHENETHIOL, **50,** 104
TRI-*t*-BUTYLCYCLOPROPENYL FLUOROBORATE, **54,** 97

3,4,5-TRIMETHOXYBENZALDEHYDE, **51,** 8
TRIMETHYL-*p*-BENZOQUINONE, **52,** 83
2,2-(TRIMETHYLENEDITHIO)CYCLO-HEXANONE, **54,** 39
TRIMETHYLENE DITHIOTOSYL-ATE, **54,** 33
TRIMETHYLOXONIUM TETRAFLUO-ROBORATE, **51,** 142
2,4,4-TRIMETHYLPENTANAL, **51,** 4
TRIMETHYLSILYL AZIDE, **50,** 107

ORGANIC SYNTHESES

Origin, Development, Organization, Operations

Prior to 1914, the industrial production of organic chemicals in the United States was very limited both in the number of compounds and quantities. Petroleum refining was primarily by distillation; there were no cracking processes and no petrochemical plants. Replacement of beehive coke ovens by by-product coking ovens to recover aromatic chemicals had just started. Most organic compounds were imported from Europe; research chemicals for use in universities and industrial laboratories were imported from Germany (Kahlbaum's Chemicals), Great Britain (Boots Ltd.), and France. There were only a few small scientific supply houses that distributed small amounts of imported chemicals. Indeed, organic research in universities and industry was limited to a few schools and very few companies (1).

In 1914, the outbreak of the war in Europe led to embargoes, blockades, and destruction of shipping which meant that chemical supplies in the United States were quickly exhausted. The escalation of World War I (2), with United States involvement in 1917, demanded immediate production of tremendous amounts of food, grains, meat, oils, coke, iron, steel, nonferrous metals, ships, trucks, guns, tanks, airplanes, gasoline, kerosene, lubricating oils, war gases, phenol, toluene, glycerol and nitric acid, protective agents, dyes and drugs. Since all the industrial plants and laboratories were in use, the chemistry staff at the universities began to increase their "student preps" to make chemicals needed for research.

Clarence G. Derick of the Chemistry Department at the University of Illinois in Urbana, actually initiated "Summer Preps" with about five students in 1914 before the war started. In the summer of 1915, Ernest H. Volwiler, a graduate student, joined Derick's prep group and was placed in charge during 1916 and 1917. Oliver Kamm, a member of the teaching staff after 1915, also helped in the prep work.

423

Carl S. Marvel, starting graduate study in 1915, began making compounds in June of 1916 and worked full time until August 1919. He was a most skillful operator and "speedily" (3) built up a reputation for modifying poor procedures so that they would work. Roger Adams joined the chemistry staff in 1916, and enthusiastically took up the idea of synthesizing research chemicals in larger quantities: one-half to several kilos. The compounds made during 1917–1918 were those needed in the World War I effort.

Dr. William A. Noyes, Head of the Chemistry Department of the University of Illinois, persuaded the Illinois Administration to provide a revolving "Organic Chemical Manufactures" fund which was used to purchase chemicals and to pay the summer preps chemists. These graduate students, numbering from 10 to 12, worked full time, 8–10 hours per day for the eight-week summer session. Their pay started at 25¢ per hour in 1915 and gradually rose over the years but the students received one unit of graduate credit for their work. Adams and Marvel put the operation on a sound cost basis by requiring all students making preps to keep careful notebook records of the cost of chemicals, apparatus, and the time needed for each preparation. The compounds made were then sold to anyone who needed them and the money returned to the fund. In 1917 Roger Adams (4) published a list of 43 organic chemicals available for purchase, and in 1918 a note listing 59 compounds as available at once, 37 to be made and 29 more which would probably be available by the end of the summer.

When the importation of dyes for sensitizing photographic film stopped in 1914, Hans T. Clarke, who had just joined the research division of the Eastman Kodak Co., was called on to synthesize the dyes. The lack of organic raw materials for this project and others led Hans T. Clarke and C. E. K.Mees to recommend to Mr. George Eastman the formation of an Eastman Organic Chemicals Division. It would assist research chemists by repackaging commercial chemicals in small lots, purifying industrial chemicals, and synthesizing any needed but nonavailable chemicals. Clarke visited Adams and Marvel at the University of Illinois and spent several weeks observing how "Summer Preps" was operated. The Eastman Organic Chemicals Division began operations at the end of 1918 and contributed greatly to the advancement of organic chemical research. Its synthesis group worked out many good procedures and designed unique laboratory apparatus and techniques. After Clarke left in 1928, William W. Hartman took charge.

The production and distribution of Pyrex laboratory glassware by the Corning Glass Works, Corning, New York in 1915 was a very important factor in the above preps. Pyrex® labware was far superior to the old lime-soda glass against breakage by mechanical or thermal shock, and resistance to reagents. It surpassed even the pre-war Jena glass which had been imported from Germany prior to 1914. Pyrex® round-bottomed reaction flasks became available in large 5 l., 12 l., and 22 l. sizes, in addition to smaller sizes. Glass blowing with Pyrex was easily mastered; hence special distilling flasks, fractionating columns, the now familiar three-necked flasks for use with mechanical stirrer, and the reflux condenser and dropping funnel were made and used as standard items in the prep labs. Also in 1914, when shipments of laboratory porcelain ware from Germany ceased, the Coors Porcelain Company of Golden, Colorado, converted their ovenware and pottery plant to chemical porcelainware. High-quality Coors U.S.A.® glazed laboratory porcelain evaporating dishes, Büchner funnels, casseroles, mortars, and pestles became available.

The armistice of November 11, 1918 ended the war but did not end the shortage of research chemicals. Hence, the synthesis of special research compounds, not available commercially, was continued during the summers under the direction of Carl S. Marvel who became a member of the Organic staff after completing his graduate study. The expanding organic and biochemical research divisions of universities and commercial concerns requested the compounds to be made in the "preps" lab.

About 1940, Harold R. Snyder took over operations from Carl S. Marvel and carried the synthetic work through the difficult World War II years (1941–1946). The prep group made unclassified starting compounds and intermediates needed by any of the various war-time agencies. Leonard E. Miller of the Organic Chemistry Department at Illinois directed the Summer Prep work during 1948–1950. After 1950 the program was discontinued because by that time many organic and biochemical supply companies had been established for the synthesis of specialty chemicals. The Summer Prep operation had provided a superior education for over 500 graduate students (and some seniors) for 36 years. Other universities also incorporated advanced organic preparations in their graduate programs. These well-trained chemists contributed to the pool of expert synthetic organic chemists for the organic chemical industries which had established real research laboratories from about 1922 onward.

The foregoing account is incomplete, however. What were the sources of the procedures, operating directions, techniques for carrying out reactions, isolating and purifying the products? Most of the compounds made were not new; they had been described in the various journals, both American and European; some were described in patents. Beilstein's "Handbuch" gave only a sentence or two summarizing the method. Houben-Weyl's "Methoden" were likewise limited. The previous literature procedures were so incomplete that frequently a synthesis, using what seemed to be a simple reaction, became a research problem of weeks or months. Four laboratory manuals available in 1915–16 which proved helpful were: Ludwig Gatterman's *Die Praxis der organischen Chemie,* 1st Ed., 1894, later revised by H. Wieland (21st to 24th Eds.). L. Vanino's *Handbuch der preparativen Chemie,* Part II, summarized the literature preparations of several hundred organic compounds. E. Fischer's *Anleitung zur Darstellung Organische Preparative* (1908) was useful as was J. B. Cohen's *Practical Organic Chemistry,* 2nd Ed., (1908), London. These manuals, designed for the first course in organic chemistry were very useful but limited in scope. It was common experience that many procedures in the chemical literature could not be duplicated; indeed, certain ones were hazardous.

Hence, from the very beginning of Summer Preps in 1914, and continuing through all the years, each student had to write out in detail the procedures he used, add precautionary notes, and references to the literature. The procedures were carefully filed and used in succeeding years; each prep man added his observations plus data on yield and purity. The first batch of directions culminated in the publication of four pamphlets; "Organic Chemical Reagents", by Roger Adams, O. Kamm, and C. S. Marvel. These were bulletins published by the University of Illinois Press, Urbana, Illinois, from 1919 to 1922, containing directions for preparing a total of 111 compounds. Although not advertized, these bulletins were quickly sold out, as their availability became known at meetings of the Organic Division of the ACS and citations in articles published in the journals.

The success of these little booklets, and the accumulation of several hundred additional good directions for the syntheses of organic compounds, led Roger Adams (6) to consider the publication of an annual volume of satisfactory methods. He discussed this project with James B. Conant of Harvard, Hans T. Clarke of Eastman Kodak, and

Oliver Kamm of Parke Davis. The unique feature was the preparation of sets of directions which, if carefully followed, could be duplicated by an advanced student (senior or graduate). Moreover, before publication, each preparation must be checked in the laboratory of an editor and always in a laboratory other than that of the submitter. In addition, the fact that this original group represented both industrial and university laboratories constituted excellent support for the project.

The first annual volume of *Organic Syntheses* was published in 1921. The procedures were collected, checked, edited by the first Board of Editors; Roger Adams (University of Illinois), James B. Conant (Harvard), Hans T. Clarke (Eastman Kodak Co.), and Oliver Kamm (Parke Davis). Publication was made possible through the friendship of Mr. Edward P. Hamilton of John Wiley & Sons, Inc. This was a most unusual publication venture for those times; there was no assurance that the publisher could recover the costs of the printing, binding, and distribution of this slender little "pamphlet" of 84 pages. Each of the first four members of the Editorial Board acted as Editor-in-Chief of one or two volumes. Then the Editorial Board was expanded during the next ten years to include Carl S. Marvel (University of Illinois), Frank C. Whitmore (Northwestern University, Pennsylvania State University), Henry Gilman, (Iowa State University), and Carl R. Noller (Stanford). In 1929, C. F. H. Allen was appointed Secretary to the Board when the number of chemists contributing preparations rose from 8 to 24, thereby causing a great increase in correspondence and record keeping. Each of the new editors took turns in preparing volumes. The policy of changing membership on the editorial board by selecting additional organic chemists to serve on the active board and moving those who had already served a term and edited one or more volumes to an Advisory Board of Editors was adopted. A new secretary to the Board of Editors was appointed every ten years. Beginning with *Collective Volume II* the retiring secretary became the Editor-in-Chief of the collective volume for the years in which he served. Thus, this project involved many different university and industrial research chemists so as to make it representative of as many institutions as possible. These policies continue today (6).

In addition to the first eight editors mentioned above, there are 51 other organic chemists who have served on the Boards of Editors. Their names are listed on the title page of this volume. They are an

ROGER ADAMS
"The Chief"

HANS T. CLARKE
"Hans"

JAMES B. CONANT
"Jim"

CARL S. MARVEL
"Speed"

EDWARD P. HAMILTON
John Wiley & Sons, Inc.

enthusiastic group of chemists working with their students in universities and co-workers in industry, dovetailing their regularly assigned work with writing up procedures, and editing and checking them in their "spare time."

None of the contributors of procedures, editors, or checkers received any pay or any of the royalties from the sale of the volumes. The starting chemicals needed for checking procedures were contributed by the chemistry departments of the universities or the research departments of industrial companies, and the products of the syntheses then were added to the research stocks of the contributors or editors. The products were always more valuable than the crude commercial starting materials so this was an economical way of getting valuable intermediates for research.

From 1921 to 1939 the *Organic Syntheses* Editorial Boards operated in a very informal fashion. However, changes in the income tax laws led to the formal incorporation of *Organic Syntheses* as a "Membership Corporation" under the laws of the State of New York on December 11, 1939. The certificate specified:

The purposes for which the corporation is to be formed are the following:

To collaborate in the writing, editing and causing to be published from time to time of books and articles dealing with the methods of preparation of organic chemicals and other subject matter connected with organic chemistry; the royalties or other proceeds received from them to be placed in a fund, the principal and income thereof to be used exclusively (apart from bona fide expenses of operation of the corporation) for the establishment of fellowships, scholarships and other benefits for students in organic chemistry in various colleges and universities;

To acquire property, both real and personal, for the conduct of its corporate purposes.

The corporation is to be organized and operated exclusively for strictly scientific, educational and charitable purposes, and not for pecuniary profit, and no part of its net earnings will inure to the benefit of any member, director or officer other than as reasonable compensation for services in effecting one or more of such purposes, or to any other individual except as a proper beneficiary of its strictly charitable purposes, and no part of its activities will be the carrying on of propaganda or otherwise attempting to influence legislation.

The First Board of Directors consisted of Roger Adams (University of Illinois), President; William W. Hartman (Eastman Kodak Co.), Treasurer; A. Harold Blatt (Queens College), Secretary to the Editorial Board; Louis Fieser (Harvard), and John R. Johnson (Cornell). Royalties from the sale of the *Annual Volumes* and *Collective Volumes* were paid to the *Organic Syntheses* treasurer and used to pay postage and typing expenses in collecting preparations and editing the volumes. Periodically any balance in the fund was invested in stocks in the growing chemical industries. A set of By-Laws of the Corporation was adopted and filed with the State of New York. They were amended from time to time as conditions changed, but always conformed to the abve cited nonprofit purposes. The Board of Directors for 1974–75 consists of the following officers and members:

Richard T. Arnold, President
(Southern Illinois University)

Ralph L. Shriner, Vice-President
(Southern Methodist University)

Wayland E. Noland, Secretary
(University of Minnesota)

William E. Parham, Treasurer
(Duke University)

William G. Dauben
(University of California, Berkeley)

William D. Emmons
(Rohm & Haas Co.)

Nelson J. Leonard
(University of Illinois, Urbana)

Blaine C. McKusick
(E.I.duPont de Nemours & Co.)

Norman Rabjohn
(University of Missouri)

Richard S. Schreiber
(Kalamazoo, Michigan)

The corporation membership is composed of all the editors.

The Board of Directors has responsibility for:

1. Supervising all operations of the corporation so that they are in conformity with the Certificate of Incorporation and the By-Laws;

2. Authorizing those expenditures from the Treasurer's funds which are essential to its scientific, educational and charitable purposes; and

3. Conforming to Acts of Congress concerning nonprofit corporations.

For efficient operation the Board has delegated certain duties to its officers, committees, Current Active Editorial Board, Editors of Annual Volumes, Editors of Collective Volumes and Editors of Cumulative Indices. Expenditures from its funds (royalties plus investment income) have been used for:

1. Expenses for the current Annual Volumes of *Organic Syntheses*.

2. Secretarial help for the Editors-in-Chief of the *Collective Volumes* and *Cumulative Index*.

3. A biennial award of $7,500 to an outstanding organic chemist who presents a scientific educational lecture at the biennial Symposium of the Division of Organic Chemistry of the American Chemical Society (a nonprofit corporation chartered by an Act of Congress.) This award, known as the Roger Adams Award, amounts to $10,000, the additional $2,500 being contributed by Organic Reactions, Inc., a nonprofit corporation, also founded by Roger Adams.

4. Subsidies to enable undergraduate or graduate students majoring in any field of chemistry and to postdoctoral fellows and research associates in chemistry at schools in the United States and Canada to purchase volumes of *Organic Syntheses* at one-half the list price (7).

5. Special awards in recognition of outstanding contributions for the advancement of the purposes of Organic Syntheses, Inc.

6. Corporation expenses for accounting and legal assistance to the Treasurer of the Corporation.

To complete the story of 54 years work of *Organic Syntheses,* the prefaces to the *Annual Volumes,* the *Collective Volumes,* and this volume and its dedication page should be read.

R.L.S.
R.H.S.

Dallas, Texas
May 1975

References

1. Fisher, Harry L., "Organic Chemistry, 1876–1951" in "Chemistry: Key to Better Living," *Diamond Jubilee Volume,* pp. 52–57 (1951) American Chemical Society, Washington, D.C.

2. Browne, Charles Albert, and Weeks, Mary Elvira, "The American Chemical Society and the First World War," in *A History of the American Chemical Society, Seventy-five Eventful Years,* Chap IX, pp. 108–126 (1951).

3. Carl S. Marvel is known to all organic chemists as "Speed" Marvel. There are many legends as to origin of this nickname; lecturing, eating, driving a Marmon car, hunting, trapshooting, birding, and the present text.

4. Adams, Roger, *J. Ind. Eng. Chem.* **9,** 685(1917).

5. Adams, Roger, *J. Amer. Chem. Soc.,* **40,** 869(1918).

6. Adams, Roger, "Fifty years of Organic Syntheses," *Org. Syn.,* **50,** (1970).

7. Certification forms and the latest special student discount price lists for all volumes of *Organic Syntheses* in print are mailed out in October of each year by John Wiley & Sons, Inc. to chairmen of chemistry departments and to professors of organic chemistry at colleges and universities in the United States and Canada. If the certification forms and price lists are not available from these sources, they may be obtained by writing to Dr. Wayland E. Noland, Secretary of *Organic Syntheses, Inc.*, School of Chemistry, University of Minnesota, Minneapolis, Minn., 55455.